An Integrated Approach to Materials Science and Engineering

An Integrated Approach to Materials Science and Engineering

Editor: Heather Dale

NY RESEARCH
PRESS

New York

Published by NY Research Press
118-35 Queens Blvd., Suite 400,
Forest Hills, NY 11375, USA
www.nyresearchpress.com

An Integrated Approach to Materials Science and Engineering
Edited by Heather Dale

© 2020 NY Research Press

International Standard Book Number: 978-1-63238-729-5 (Hardback)

Cataloging-in-Publication Data

An integrated approach to materials science and engineering / edited by Heather Dale.
    p. cm.
Includes bibliographical references and index.
ISBN 978-1-63238-729-5
1. Materials science. 2. Materials. 3. Engineering. I. Dale, Heather.
TA403 .I58 2020
620.11--dc23

Contents

Preface..VII

Chapter 1 **Displacement Investigation of KNN-Bitumen-Based Piezoceramics in
Asphalt Concrete**..1
Ning Tang, Kaikai Yang, Wenhao Pan, Limei Wu, Qing Wang and Yanwen Chen

Chapter 2 **Fracture Toughness Improvement of Poly(lactic acid) Reinforced with
Poly(ε-caprolactone) and Surface-Modified Silicon Carbide**........................ 8
Jie Chen, Tian-Yi Zhang, Fan-Long Jin and Soo-Jin Park

Chapter 3 **Behavior of Concrete-Filled Steel Tube Columns Subjected to Axial Compression**............18
Pengfei Li, Tao Zhang and Chengzhi Wang

Chapter 4 **Effect of Recycling Agents on Rheological and Micromechanical Properties of
SBS-Modified Asphalt Binders**...33
Meng Guo, Yiqiu Tan, Daisong Luo, Yafei Li, Asim Farooq,
Liantong Mo and Yubo Jiao

Chapter 5 **Structural and Magnetic Properties of $Ba_3[Cu_{0.8-x}Zn_xMn_{0.2}]_2Fe_{24}O_{41}$
Z-Type Hexaferrites**...45
Eman S. Al-Hwaitat, Sami H. Mahmood, Mahmoud Al–Hussein and Ibrahim Bsoul

Chapter 6 **The Impact of Magnetic Materials in Renewable Energy-Related Technologies in
the 21st Century Industrial Revolution**...56
Wallace Matizamhuka

Chapter 7 **Effect of High-Temperature Curing Methods on the Compressive Strength
Development of Concrete Containing High Volumes of Ground Granulated
Blast-Furnace Slag**..65
Wonsuk Jung and Se-Jin Choi

Chapter 8 **Electrical and Vibrational Studies of $Na_2K_2Cu(MoO_4)_3$**.......................... 71
Wassim Dridi, Mohamed Faouzi Zid and Miroslaw Maczka

Chapter 9 **Carbon Nanofoam by Pulsed Electric Arc Discharges** 79
David Saucedo-Jimenez, Isaac Medina-Sanchez and Carlos Couder Castañeda

Chapter 10 **Effect of Alternative Wood Species and First Thinning Wood on Oriented
Strand Board Performance**..89
Fabiane Salles Ferro, Amós Magalhães Souza, Isabella Imakawa de Araujo,
Milena Maria Van Der Neut de Almeida, André Luis Christoforo and
Francisco Antonio Rocco Lahr

Chapter 11 **Formation and Physical Properties of h-BN Atomic Layers: A First-Principles
Density-Functional Study**...96
Yoshitaka Fujimoto

Chapter 12   **Physical and Compaction Properties of Granular Materials with Artificial Grading behind the Particle Size Distributions**..102
Ming-liang Chen, Gao-jian Wu, Bin-rui Gan, Wan-hong Jiang and Jia-wen Zhou

Chapter 13   **Thermodynamic Study of Tl$_6$SBr$_4$ Compound and Some Regularities in Thermodynamic Properties of Thallium Chalcohalides**..122
Dunya Mahammad Babanly, Qorkhmaz Mansur Huseynov, Ziya Saxaveddin Aliev, Dilgam Babir Tagiyev and Mahammad Baba Babanly

Chapter 14   **Experimental Investigation on Embedding Strength Perpendicular to Grain of Parallel Strand Bamboo**..131
Junwen Zhou, Dongsheng Huang, Yang Song and Chun Ni

Chapter 15   **Superposed Incremental Deformations of an Elastic Solid Reinforced with Fibers Resistant to Extension and Flexure**..139
Chun IL Kim

Chapter 16   **Shape Stability of Polyethylene Glycol/Acetylene Black Phase Change Composites for Latent Heat Storage**..150
Jingjing Zhang, Hairong Li, Junyang Tu, Ruonan Shi, Zhiping Luo, Chuanxi Xiong and Ming Jiang

Chapter 17   **Representative Stress-Strain Curve by Spherical Indentation on Elastic-Plastic Materials**..159
Chao Chang, M. A. Garrido, J. Ruiz-Hervias, Zhu Zhang and Le-le Zhang

Chapter 18   **Study on the Thermal Properties of Hollow Shale Blocks as Self-Insulating Wall Materials**..168
Guo-liang Bai, Ning-jun Du, Ya-zhou Xu and Chao-gang Qin

Chapter 19   **Additive Manufacturing Enabled by Electrospinning for Tougher Bio-Inspired Materials**..180
Komal Agarwal, Yinning Zhou, Hashina Parveen Anwar Ali, Ihor Radchenko, Avinash Baji and Arief S. Budiman

Chapter 20   **Thermomechanical Properties of Jute/Bamboo Cellulose Composite and its Hybrid Composites: The Effects of Treatment and Fiber Loading**..189
Fui Kiew Liew, Sinin Hamdan, Md. Rezaur Rahman and Mohamad Rusop

Chapter 21   **Experiment on Behavior of a New Connector used in Bamboo (Timber) Frame Structure under Cyclic Loading**..199
Junwen Zhou, Dongsheng Huang, Chun Ni, Yurong Shen and Longlong Zhao

Permissions

List of Contributors

Index

Preface

This book has been an outcome of determined endeavour from a group of educationists in the field. The primary objective was to involve a broad spectrum of professionals from diverse cultural background involved in the field for developing new researches. The book not only targets students but also scholars pursuing higher research for further enhancement of the theoretical and practical applications of the subject.

Materials science is a multi-disciplinary field that focuses on designing and discovering new materials. It uses concepts from physics, chemistry and engineering. Materials science is an interdisciplinary field which combines areas such as metallurgy, solid-state physics, ceramics and chemistry. It is concerned with the processing of any material and how it influences the structure, properties and performance of the material. This understanding of processing, structure and properties of the material is known as materials paradigm. This paradigm is helpful in getting a better understanding of various research areas such as metallurgy, nanotechnology and biomaterials. Materials science is an important part of forensic engineering and failure analysis which includes investing products, materials, components or structures that do not function as expected. This book is a valuable compilation of topics, ranging from the basic to the most complex advancements in the field of materials science and engineering. Different approaches, evaluations, methodologies and advanced studies in this discipline have been included in it. The book will serve as a valuable source of reference for graduate and postgraduate students.

It was an honour to edit such a profound book and also a challenging task to compile and examine all the relevant data for accuracy and originality. I wish to acknowledge the efforts of the contributors for submitting such brilliant and diverse chapters in the field and for endlessly working for the completion of the book. Last, but not the least; I thank my family for being a constant source of support in all my research endeavours.

Editor

Displacement Investigation of KNN-Bitumen-Based Piezoceramics in Asphalt Concrete

Ning Tang[ID],[1] Kaikai Yang[ID],[1] Wenhao Pan[ID],[1,2] Limei Wu,[1] Qing Wang[ID],[1] and Yanwen Chen[ID][1]

[1]School of Materials Science and Engineering, Shenyang Jianzhu University, Shenyang, China
[2]School of Materials Science and Engineering, Northeast University, Shenyang, China

Correspondence should be addressed to Qing Wang; wangqingmxy@126.com

Academic Editor: Fabrizio Pirri

Piezoelectric material has excellent characteristics of electromechanical coupling so that it could be widely applied in structural health monitoring field. Nondestructive testing of piezoelectric technique becomes a research focus on piezoelectric field. Asphalt concrete produces cumulative damage under the multiple repeated vehicle load and natural situation, so it is suited material and structure for nondestructive application. In this study, a test system was established by driving power of piezoceramic, laser displacement sensor, computer, and piezo-embedded asphalt concrete. Displacement, hysteresis, creeps, and dynamic behavior of KNN piezoceramic element embedded in asphalt concrete were tested. The results indicate that displacement output attained 0.4 μm to 0.7 μm when the loads were from 0 N to 150 N. The hysteresis was not obvious when the load was from 0 N to 100 N, aside from higher loads. The creep phenomenon can be divided into two parts: uptrend and balance. The more serious the asphalt binder ageing is, the larger the displacement is, when piezo-asphalt concrete has already been in serious ageing.

1. Introduction

Piezoelectric material is a kind of intelligent material which can translate between mechanical energy and electrical energy. At present, the yearly sales amount of piezoceramics is over 10 billion, and the application filed is from consumer electronics to aerospace, naval sonar, and high-speed train [1]. Hundreds of years, lead zirconate titanate piezoceramic (PZT) still reign supreme due to excellent performance, but PZT has lead harm for body or environment, so that its application is gradually limited in the Europe, Japan, United States, and China. Dealing with this problem, researchers intend to discover a new piezoceramic to replace PZT which uses "lead-free" components [2]. Hence, many researchers have invested heavily in research and development of high performance lead-free piezoelectric materials. The study focused on the piezoceramics of potassium-sodium niobate (KNN), sodium bismuth titanate (BNT), and barium titanate (BT) which can improve piezoceramic property and temperature stability like PZT [3–8]. However, compared with PZT, the performance of almost all of lead-free piezoceramics is poor so that cannot replace it.

At present, major type of pavement in road engineering is asphalt concrete. As a result of heavy traffic, climate, and environment in a long period, the strain and kinetic energy were accumulated in the asphalt concrete [9–11]. Furthermore, those strain and kinetic energy are absorbed by asphalt and changed to thermal energy of pavement so that it increases the risk of crack and track in pavement [12, 13]. In fact, structural health monitoring and damage detection using piezoceramics are important research topics due to the advantages of piezoceramics, such as quick response, broadband frequency, and low price [14–16]. Although researchers have obtained some achievements of piezoceramic application in engineering, some problems need to be further studied, like the displacement properties of piezoceramics embedded in the structure.

In the present study, the properties of KNN-bitumen-based lead-free piezoceramics (KNN) were focused after embedding in asphalt mixtures. The KNN was a stacked-type piezoelectric material that was prepared by a KN_bO_3–$N_aN_bO_3$

ceramic doped with Li and bitumen as a binder. Encapsulated piezoelectric elements were embedded into asphalt mixture. The experimental results and analyses include the following aspects: displacement, hysteresis, creep properties, and displacement changes of piezoceramics after ageing.

2. Materials and Methods

2.1. Materials. Potassium carbonate (K_2CO_3), sodium carbonate (Na_2CO_3), lithium carbonate (Li_2CO_3), and niobium oxide (Nb_2O_5) were produced by China Sinopharm Chemical Reagent Co., Ltd, which are raw materials used in the preparation of KNN. They are all powders and analytically pure.

In order to stack pieces of KNN each other, electrode layer was designed. The Young modulus, Poisson ratio, and electric conductivity of electrode materials have to be considered. To cost effectiveness, brass is a better one compared to others. The thickness of brass is 0.1 mm as thin as possible because the brass only serves as an electrode.

The metal panel improves the mechanical properties of the piezoelectric materials. From the choice of metal panel materials, the duralumin was chosen. It has a few good features such as corrosion resistance, good electrical and thermal conductivity, positive alloy strength, heat-resistant properties, and so on. The thickness of duralumin is 1.6 mm.

A base bitumen was procured from China with a penetration of 78 (0.1 mm at 25°C, 100 g, and 5 s), ductility of 49.4 cm (at 15°C), and softening point of 45.7°C. Basalt with particle size less than 19 mm was supplied from China. Limestone powder was used as the mineral filler from China, too.

2.2. Methods

2.2.1. Preparation of KNN. $KNbO_3$–$NaNbO_3$–$0.035Li$ (KNN) obtained by the traditional solid-phase reaction method. In order to remove the adsorbed water from the powder, the raw materials of K_2CO_3, Na_2CO_3, Li_2CO_3, and Nb_2O_5 were applied by air-drying in the oven at 105°C for 24 hours. Following this, the dried materials were weighed according to the stoichiometric ratio of the formula and milled by the planetary ball mill for 12 hours. According to the content of Li, the dry materials were mixed with Li_2CO_3, and the presintering temperature is 920°C in 4 hours. Afterwards, the calcined powder was milled by the planetary ball mill for 12 hours once again and compressed into the ring. The diameter of the KNN piece was 18 mm, and thickness was 2 mm. Finally, the KNN piece was sintered at 1135°C for 2 hours. The performance of KNN was as follows: piezoelectric constant d_{33} was 163 pc/N and d_{33}^* was 212 pm/V, electromechanical coupling factor k_p was 0.384, and dielectric loss $\tan\delta$ was 0.12%.

2.2.2. Preparation of KNN Stacked Element. The stacked piezoelectric specimen was composed of KNN pieces, bitumen and electrodes. The KNN piezoceramics were connected tandem. The brass sheets were pasted on two sides of each piece as electrodes by epoxy and bitumen. Duralumin

FIGURE 1: Sample of the KNN-bitumen stacked piezoceramic.

sheets were pasted on two sides of the stacked piezoelectric specimen. At last, bitumen was heated to encapsulate the piezoceramics and wires in a mold as shown in Figure 1. Encapsulation size of the stacked piezoceramic is 20 mm × 20 mm × 30 mm like an aggregate of asphalt mixture, which contains 12 pieces of the KNN piezoceramic.

2.2.3. Preparation of Asphalt Concrete. According to the experimental method of T0703-2011 (JTG E20-2011, China), a slab of asphalt mixture was prepared. Afterwards, the slab was drilled in the center by the drill of 150 mm. Finally, new hot asphalt mixture and stacked piezoceramic filled the space in the center of the slab. The upper surface of the stacked piezoceramic kept same flat with the surface of asphalt slab. After compaction and cooling, the specimen can be applied to experiments as follows. The mix proportion of AC-13 and the best bitumen-aggregate ratio of asphalt mixture were taken in this research as shown in Table 1. The void content was controlled at 4% [14].

2.2.4. Experimental Method.

(i) Test System. Figure 2 shows the diagram of the test system of piezoceramic asphalt. Driving power of piezoceramic was employed using the equipment HVA-200 produced by the Xinmingtian Company of China. The displacement of the stacked piezoceramic was tested by a noncontact laser displacement sensor named M72L/0.5 produced by Melsensor in Germany. The testing precision is 0.1 μm. The displacement of the stacked piezoceramic was generated after driving voltage was applied; afterwards, the data of displacement output were transferred from the displacement sensor to the destination computer.

(ii) Displacement and Hysteresis under Different Loads. The input voltage increased from 0 V to 150 V by 10 V each step and then decreased to 0 V. The input voltage and displacement data were collected and recorded by the

TABLE 1: Mixture gradations.

Mesh (mm)	16	13.2	9.5	4.75	2.36	1.18	0.6	0.3	0.15	0.075	Bitumen-aggregate ratio
AC-13	100.0	86.1	75.3	56.7	37.3	24.3	15.9	111	8.7	6.8	4.8%

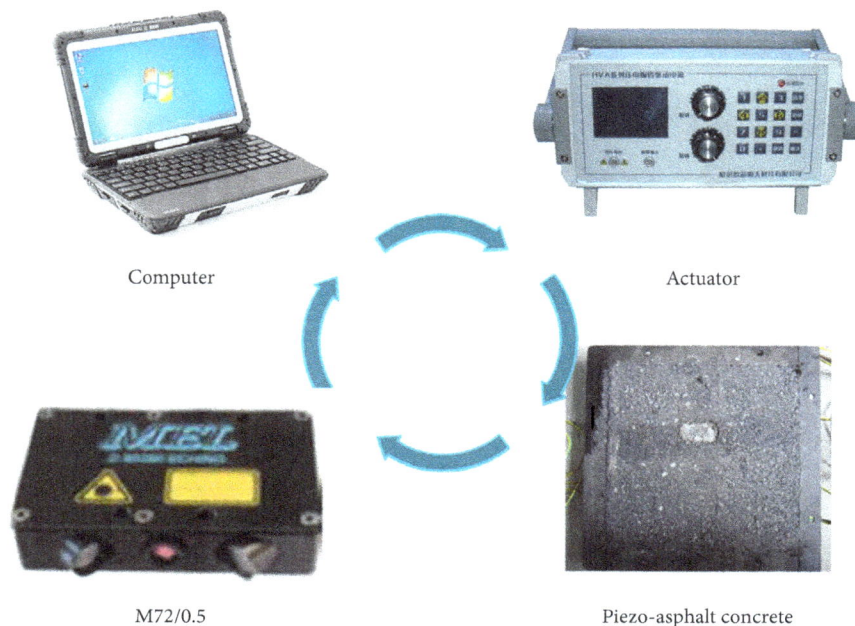

FIGURE 2: Diagram of the test system.

abovementioned test system. The stress of vehicles and the surface area of KNN element have to be considered. Therefore, the load was 0 N, 50 N, 100 N, and 150 N.

(iii) Creep. The input voltage was 50 V, 100 V, and 150 V. The displacement data were collected per 30 seconds for 20 times.

(iv) Dynamic Behavior. The input voltage was 150 V, and the bias voltage was 75 V. According to the vibrational frequency of pavement, a sinusoidal voltage signal was input with different frequencies of 5 Hz, 10 Hz, and 20 Hz.

(v) Ageing of Piezoelectric-Asphalt Concrete. The SHRP plan provides several methods for ageing, such as oven ageing, delayed mixing, microwave heating, pressurized oxidation, and ultraviolet ageing. Oven ageing is considered to be the most effective method for ageing of asphalt mixes in the laboratory. Hence, short-term oven ageing (STOA) and long-term oven ageing (LTOA) were conducted. For STOA, the piezoelectric-asphalt concrete was placed in a forced ventilated oven at 135°C for 4 hours. For LTOA, the piezoelectric-asphalt concrete was aged through STOA first, afterwards was placed in a forced ventilated oven, and underwent a long-term ageing of 5 days at a temperature of 85°C. After ageing, the asphalt concrete which filled in the space center of slab was picked up, and ageing bitumen was reclaimed. The extraction procedure of bitumen followed the Abson method (T0726-2011, JTG E20-2011, China). The

solution used for extraction was trichloroethylene. Finally, the softening point, penetration, and ductility should be tested.

3. Results and Discussion

3.1. Displacement under Different Loads. One of important parameters is the displacement output of KNN element under different loads. In order to analyze displacement properties close to the real pavement, the loading test was carried out under slow loading speed at ambient temperature. As shown in Figure 3, compared with no load, the KNN element had a greater displacement output under different loads. Displacement output attained 0.4 μm to 0.7 μm when the load was from 0 N to 150 N. In the experimental range, sensitivity of displacement has nonlinear relationship with the excitation voltage. With the excitation voltage increasing and decreasing, the phenomenon of hysteresis is becoming more and more serious. Ideally, according to converse piezoelectric effect, the relationship between displacement output of KNN element and excitation voltage should be linear. In reality, there is no superposition between the rising curve and the decline curve like the curve shown in the Figure 3. This nonlinear relationship between displacement and excitation voltage was named hysteretic loops.

3.2. Hysteresis. Figure 4 shows the hysteresis curve under different loads. Changes of hysteresis from 0 N to 100 N were

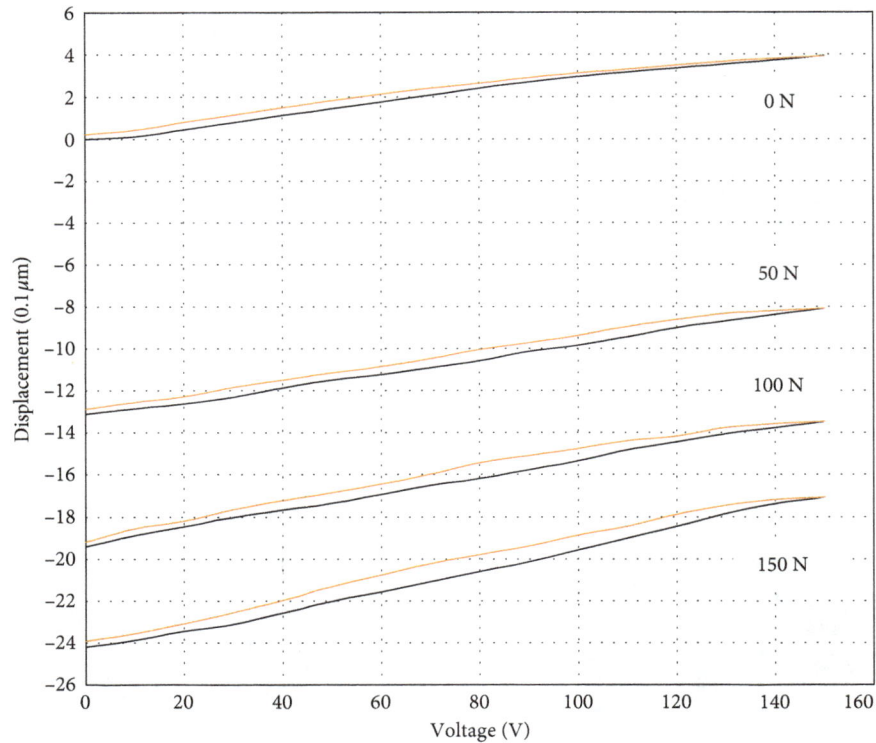

FIGURE 3: Displacement of KNN element in the asphalt concrete under different loads.

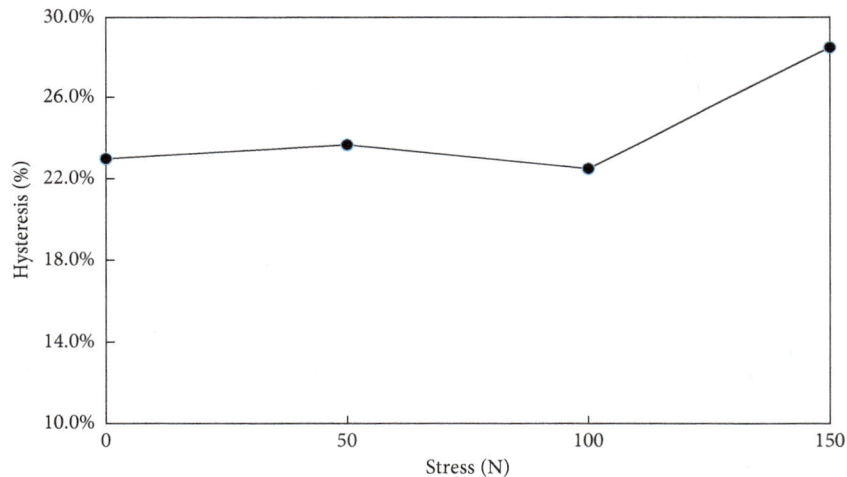

FIGURE 4: Hysteresis of KNN element in the asphalt concrete under different loads.

not obvious, but hysteresis was 28.3% when the load was 150 N. On the micro level, according to the polarization of the piezoceramic, the cause of the hysteresis phenomenon of KNN element is the viscosity among the electric dipoles in the dielectric medium based on the analysis of electric domain formation of piezoceramics. The shape of the hysteresis curve is influenced by the value and frequency of excitation voltage and own properties of piezoceramics. On the macro level, the cause of the hysteresis phenomenon of KNN element is electrostriction effect, ferroelectric effect, and damping effect of bitumen. Bitumen has plasticity and

viscoelasticity and blocks the motion of KNN element in the asphalt concrete.

3.3. Creep. Creep phenomenon was occurred when excitation voltage is applied to the KNN element. This is because displacement output is more volatile immediately. The displacement always increases to a certain value at the beginning and then grows slowly to the final value in a short time. The final displacement cannot be completed immediately when KNN element exposed to an electric field,

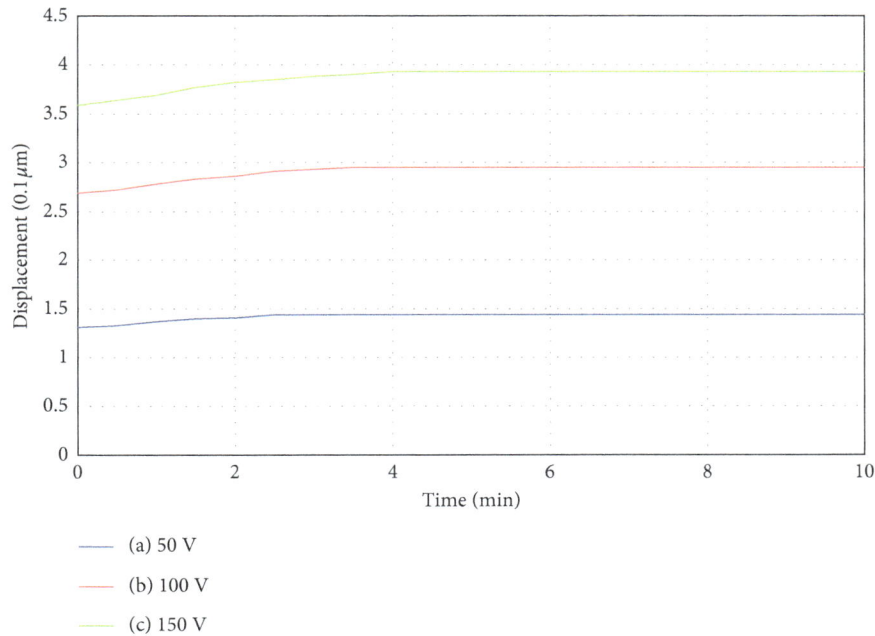

FIGURE 5: Creep of KNN element in the asphalt concrete under different voltages.

because of friction among internal lattices of the piezoceramic. Hence, the piezoceramic has to fully meet the final deformation in a short time and shows the hysteresis of displacement output from a macroscopic view.

The creep test of KNN element embedded in the asphalt concrete was carried out under voltages of 50 V, 100 V, and 150 V respectively. Figure 5 reveals the creep of KNN element. In Figure 5, the creep phenomenon can be divided into two parts. First, the displacement output was close to the maximum value within a few milliseconds. Afterwards, the displacement output attains the maximum value in a matter of minutes. In addition, the creep rate related to the input voltage, and the creep time increases with the input voltage.

3.4. Dynamic Behavior.

The vibration frequency of the pavement structural layer is from 1 Hz to 20 Hz when a vehicle travels forwards with different speeds. In common, the frequency of 10 Hz always be used in the test of asphalt mixture, or approximately a vehicle moves on the pavement with a speed of 60–65 km/h. The dynamic behavior was characterized by amplitude. Figure 6 shows the displacement amplitude of the piezoceramic which was drove by single-phase sinusoidal alternating voltage. In Figure 6, compared with the peak and trough of the sinusoid, the waveform of displacement amplitude was distorted a little. The displacement amplitude of the piezoceramic was the same and around 0.4 μm under voltage of 150 V without any load. This indicates that the displacement was not related to the frequency.

3.5. Displacement after Ageing.

The ageing of bitumen is a complex physical change and chemical change. The rheological properties also have a large change after ageing, eventually leading to cracking and damage of pavement, seriously affecting the durability of the road. Table 2 shows the properties of bitumen and asphalt samples with different ageing. The results indicated that the penetration became smaller, the softening point increased, and the ductility was greatly reduced. Furthermore, the conventional performance of the asphalt concrete sample was tested, and the Marshall stability decreased.

In order to study the relationship between bitumen ageing and displacement of KNN element, the displacement was measured without any load. Then, the voltage was applied from 0 to 150 V by 10 V each step. The input voltage and displacement data were collected and recorded by the test system, and the voltage-displacement curve was drawn, as shown in Figure 7. It reveals that the displacement increases with the aggravated ageing process. In addition, the displacement curve of the sample was not smooth after STOA and LTOA. Three curves were close to each other at the beginning stage; afterwards, the curves were separated due to different ageing. The more serious the asphalt binder ageing, the larger the displacement output.

This phenomenon can be explained in two ways. One is the effect of bitumen ageing. Bitumen is composed of saturates, aromatics, resins, and asphaltenes. For ageing, acceleration movement of bitumen molecules causes the evaporation of internal light-weight components so that a series of physical and chemical changes occurred and the content of asphaltenes increases. Hence, the bitumen becomes hard and brittle, and the viscosity decreases. In a word, the capability of impeding relative movement decreases. The other one is adhesion between bitumen and KNN element. Saturates and aromatics are Low-molecular weight and nonpolar compounds and play the role of providing lubrication and flexibility. These two compositions mainly adhere

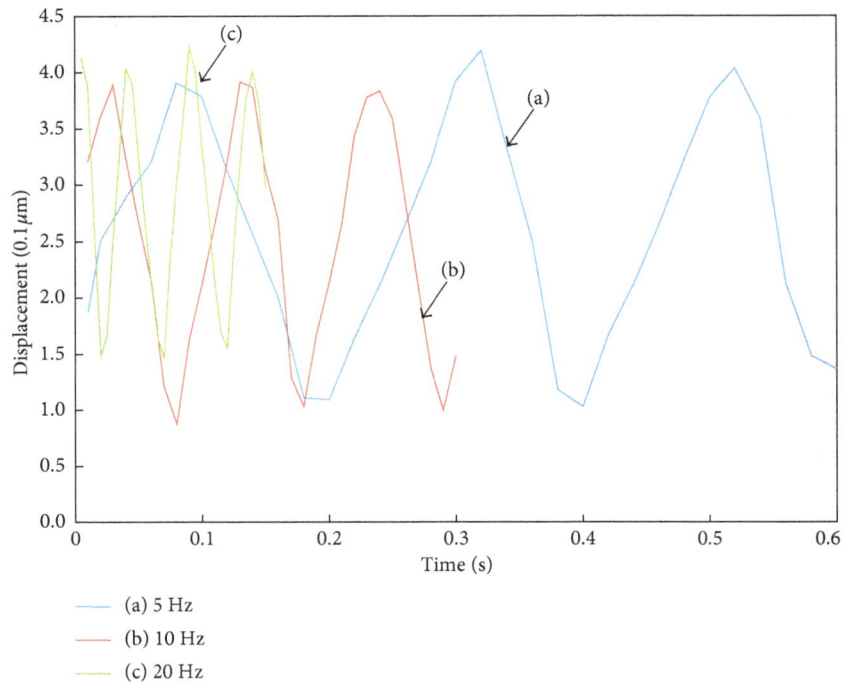

FIGURE 6: Dynamic behavior of KNN element in the asphalt concrete under different frequencies.

TABLE 2: Properties of bitumen and asphalt samples with different ageing.

Index	Softening point (°C)	Penetration (dmm, 25°C)	Ductility (cm, 15°C)	Marshall stability (kN)
No-ageing	45.7	78	59.4	9.34
STOA	48.5	72	41.7	8.17
LTOA	51.4	63	33.8	6.45

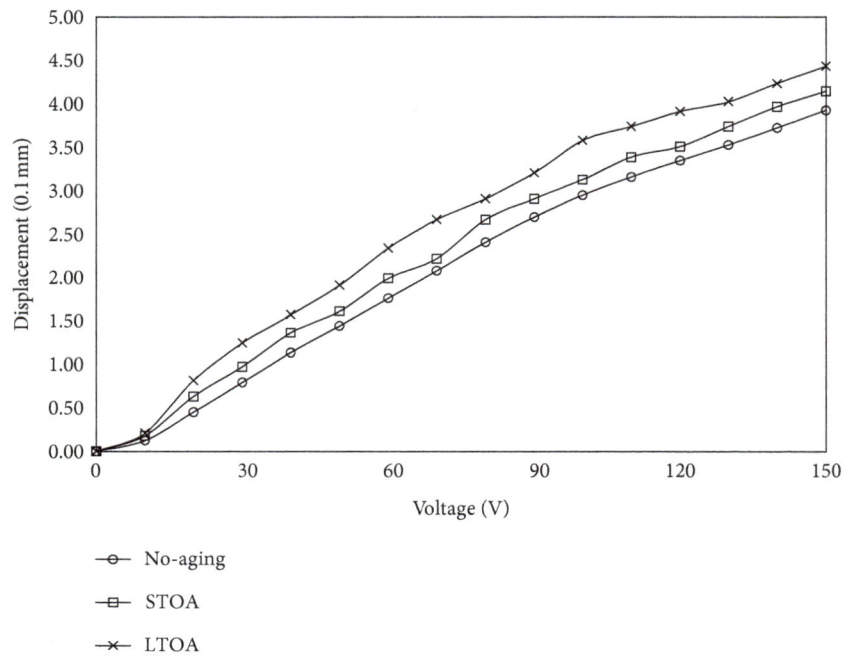

FIGURE 7: Displacement-voltage curve of KNN element in asphalt concrete after ageing.

to KNN element by the Van der Waals force. Resins and asphaltenes are polar and surfactivity compounds and contain asphaltic acid and anhydrides. These two compositions mainly adhere to KNN element through chemisorption. On the whole, the content of saturates and aromatics decreased due to bitumen ageing so that adhesion became worse and ultimately led to displacement changes.

4. Conclusions

The theoretical analysis and experimental verification are combined to conduct the research, and conclusions are as follows:

(a) Displacement output of KNN element embedded in the asphalt concrete attained $0.4 \mu m$ to $0.7 \mu m$ when the loads were from 0 N to 150 N. Changes of hysteresis from 0 N to 100 N were not obvious, but hysteresis attained 28.3% when the load was 150 N.

(b) The creep phenomenon of displacement of KNN element embedded in the asphalt concrete can be divided into two parts. First, the displacement output was close to the maximum value within a few milliseconds. Second, displacement output attains the maximum value in a matter of minutes.

(c) Displacement output of KNN embedded in the asphalt concrete increases with the aggravated ageing process. The more serious the asphalt binder ageing, the larger the displacement.

Conflicts of Interest

The authors declare that there are no conflicts of interest regarding the publication of this paper.

Acknowledgments

This work was financially supported by National Natural Science Foundation of China (no. 51508344), Educational Commission of Liaoning Province of China (no. L2015449), China Postdoctoral Science Foundation-Funded Project (no. 2016M591458), Doctoral Scientific Research Foundation of Liaoning Province (no. 201601212), State Key Laboratory of Silicate Materials for Architectures (Wuhan University of Technology, no. SYSJJ2015-14), and Hong Kong Polytechnic University (no. G-YZ1A) through the Hong Kong Scholar Program. Thanks are due to Science Program and Discipline Content Education Project of Shenyang Jianzhu University and visiting scholar invitation of Politecnico di Milano and China Scholarship Council.

References

[1] W. Jo, R. Dittmer, and M. Acosta, "Giant electric-field-induced strains in lead-free ceramics for actuator applications–status and perspective," *Journal of Electroceramics*, vol. 29, no. 1, pp. 71–93, 2012.

[2] Y. Saito, H. Takao, and T. Tani, "Lead-free piezoceramics," *Nature*, vol. 432, no. 7013, pp. 84–87, 2004.

[3] E. Hollenstein, M. Davis, and D. Damjanovic, "Piezoelectric properties of Li-and Ta-modified $(K_{0.5}Na_{0.5})NbO_3$ ceramics," *Applied Physics Letters*, vol. 87, no. 18, p. 182905, 2005.

[4] S. Zhang, R. Xia, and T. R. Shrout, "Piezoelectric properties in perovskite $0.948(K_{0.5}Na_{0.5})NbO_3$–$0.052LiSbO_3$ lead-free ceramics," *Journal of Applied Physics*, vol. 100, no. 10, pp. 104–108, 2006.

[5] J. Rödel, W. Jo, and K. T. Seifert, "Perspective on the development of lead-free piezoceramics," *Journal of the American Ceramic Society*, vol. 92, no. 6, pp. 1153–1177, 2009.

[6] B. J. Chu, D. R. Chen, and G. R. Li, "Electrical properties of $Na_{1/2}Bi_{1/2}TiO_3$–$BaTiO_3$ ceramics," *Journal of the European Ceramic Society*, vol. 22, no. 13, pp. 2115–2121, 2002.

[7] Y. Kawakami, M. Watanabe, and K. I. Arai, "Effects of substrate materials on piezoelectric properties of $BaTiO_3$ thick films deposited by aerosol deposition," *Japanese Journal of Applied Physics*, vol. 55, no. 10S, pp. 1–11, 2016.

[8] S. Wada, K. Yako, and K. Yokoo, "Domain wall engineering in barium titanate single crystals for enhanced piezoelectric properties," *Ferroelectrics*, vol. 334, no. 1, pp. 17–27, 2006.

[9] F. Xiao, S. N. Amirkhanian, and B. Wu, "Fatigue and stiffness evaluations of reclaimed asphalt pavement in hot mix asphalt mixtures," *Journal of Testing and Evaluation*, vol. 39, no. 1, pp. 50–58, 2010.

[10] N. Li, A. Pronk, and A. Molenaar, "Comparison of uniaxial and four-point bending fatigue tests for asphalt mixtures," *Transportation Research Record: Journal of the Transportation Research Board*, vol. 2373, pp. 44–53, 2013.

[11] Q. Ye, S. Wu, and N. Li, "Investigation of the dynamic and fatigue properties of fiber-modified asphalt mixtures," *International Journal of Fatigue*, vol. 31, no. 10, pp. 1598–1602, 2009.

[12] Y. C. Tsai, C. Jiang, and Y. Huang, "Multiscale crack fundamental element model for real-world pavement crack classification," *Journal of Computing in Civil Engineering*, vol. 28, no. 4, p. 04014012, 2012.

[13] M. Ameri, A. Mansourian, and M. H. Khavas, "Cracked asphalt pavement under traffic loading–a 3D finite element analysis," *Engineering Fracture Mechanics*, vol. 78, no. 8, pp. 1817–1826, 2011.

[14] N. Tang, S. Wu, and L. Pang, "Preparation and electrical properties of piezoelectric-embedded asphalt mixture," *Journal of Testing and Evaluation*, vol. 42, no. 5, pp. 1119–1126, 2014.

[15] N. Kaur, L. Li, and S. Bhalla, "A low-cost version of electromechanical impedance technique for damage detection in reinforced concrete structures using multiple piezo configurations," *Advances in Structural Engineering*, vol. 20, no. 8, pp. 1247–1254, 2017.

[16] C. Zhang, X. Yu, and L. Alexander, "Piezoelectric active sensing system for crack detection in concrete structure," *Journal of Civil Structural Health Monitoring*, vol. 6, no. 1, pp. 129–139, 2016.

Fracture Toughness Improvement of Poly(lactic acid) Reinforced with Poly(ε-caprolactone) and Surface-Modified Silicon Carbide

Jie Chen,[1] Tian-Yi Zhang,[2] Fan-Long Jin ⓘ,[1] and Soo-Jin Park ⓘ[3]

[1]*Department of Polymer Materials, Jilin Institute of Chemical Technology, Jilin City 132022, China*
[2]*Key Laboratory of Ministry of Education for Enhancing Oil and Gas Recovery Ratio, Northeast Petroleum University, Daqing 163318, China*
[3]*Department of Chemistry, Inha University, Nam-gu, Incheon 402-751, Republic of Korea*

Correspondence should be addressed to Fan-Long Jin; jinfanlong@163.com and Soo-Jin Park; sjpark@inha.ac.kr

Academic Editor: Maria Laura Di Lorenzo

In this study, bio-based poly(lactic acid) (PLA)/polycaprolactone (PCL) blends and PLA/PCL/silicon carbide (SiC) composites were prepared using a solution blending method. The surface of the SiC whiskers was modified using a silane coupling agent. The effects of the PCL and SiC contents on the flexural properties, fracture toughness, morphology of PLA/PCL blends, and PLA/PCL/SiC composites were investigated using several techniques. Both the fracture toughness and flexural strength of PLA increased by the introduction of PCL and were further improved by the formation of SiC whiskers. Fracture surfaces were observed by scanning electron microscopy, which showed that the use of PCL as a reinforcing agent induces plastic deformation in the PLA/PCL blends. The SiC whiskers absorbed external energy because of their good interfacial adhesion with the PLA matrix and through SiC-PLA debonding in the PLA/PCL/SiC composites.

1. Introduction

Poly(lactic acid) (PLA) is a biodegradable, bioabsorbable, and renewable thermoplastic polyester that can be obtained by the ring-opening polymerization of lactide [1–4]. PLA is perhaps the most useful and promising biopolymer because of its abundance, outstanding mechanical performance, and high chemical resistance. PLA has been used in a variety of applications in the automotive, medical, and food industries [5–7]. Despite its high tensile modulus and strength, PLA has poor toughness due to its stiff backbone chain, which limits its use in many fields. Thus, blending with a ductile biodegradable polymer and addition of inorganic fillers has previously been reported as methods to improve the toughness of PLA [8–10].

Polycaprolactone (PCL) is a petroleum-derived, semi-crystalline, and linear aliphatic polyester that is biodegradable and biocompatible. PCL has high flexibility, with a low glass transition temperature and melting point. The toughness of PLA is improved when blended with PCL, which acts as a plasticizer, and the resulting blend is likely to retain its biodegradability [11–14].

Several researchers have reported the preparation of biodegradable PLA/PCL blends with outstanding mechanical and physical properties. The morphological and thermal tests reported by Patrício and Bártolo indicated that PLA and PCL are immiscible polymers, and rheological tests showed that PLA/PCL blends prepared by a physical blending process have good thermal stability [15]. Tsuji et al. reported that the biodegradability of PLA/PCL blends prepared by melt blending could be manipulated by tuning the conditions of melt blending or the sizes and morphologies of the PLA- and PCL-rich domains [16]. Chen et al. demonstrated improved tensile extensibility of PLA/PCL blends [17]. Their results suggested that the combination of solution-coagulation and crosslinking resulted in a good, stable dispersion of PCL in the PLA matrix, improving its tensile toughness. Harada et al. studied the reactive compatibilization of PLA/PCL blends with the addition of reactive processing agents [18]. The impact strength of the blends increased considerably at 20 wt.% PCL. Morphological characterization showed that the PLA/PCL blends form sea-island structures, in which PLA adopts a continuous phase and PCL a dispersed phase induced by the reactive processes. Urquijo et al. studied the effect of melt processing conditions on phase structure, morphology, and mechanical properties of PLA/PCL blends [19]. Injection-molded specimens were ductile and broke at elongation values close to 140%. The elongation at break of the hot-pressed specimens was lower, likely due to the large size of the PCL particles. However, PCL, being softer, decreased the tensile strength of the PLA/PCL blends in the high impact strength region (>30 wt.% PCL).

Generally, the addition of inorganic fillers to a polymer matrix can improve its mechanical properties, such as toughness, stiffness, and heat distortion temperature. Silicon carbide (SiC) is a ceramic material with good erosion and oxidation resistance, high strength, excellent thermal stability, and a high melting point. SiC is widely used in many industrial applications, such as the production of structural and functional materials [20, 21]. Dorigato et al. [22] studied the effect of various kinds of fumed silica nanoparticles on the mechanical performance of PLA. Their results showed that the fracture toughness of PLA was significantly increased with the addition of functionalized silica nanoparticles. Rashmi et al. [23] investigated the toughening of PLA with polyamide 11 and halloysite nanotubes. The impact strength of PLA was increased by the introduction of the polyamide 11 and further improved significantly with the addition of halloysite nanotubes.

In this study, bio-based PLA/PCL blends and PLA/PCL/SiC composites with low PCL and SiC contents were prepared by a solution blending method. The effects of the PCL and SiC contents on the mechanical properties, fracture toughness, and morphology of the prepared PLA/PCL blends and PLA/PCL/SiC composites were investigated by mechanical testing and scanning electron microscopy (SEM).

2. Experimental

2.1. Materials. PLA pellets (Lehua Plastic Material Firm, Dongguan, China) were obtained with a weight-average molecular weight of 200,000 g/mol. PCL with a glass transition temperature of −60°C and melting point range of 59–64°C was obtained from Mingyuanxinchong Co., Ltd. (Dongguan, China). SiC whiskers were supplied by Xuzhou Hongwu Nano Materials Inc., (China). The average diameter of the SiC whiskers was 0.05–2.50 μm, and the length-to-diameter ratio was ≥20. The silane coupling agent KH570 was supplied by Xingfeilong Chem. Co., Jinan (China). Dichloromethane (CH_2Cl_2, YongDa Chemical Reagent Co., Tianjin, China) was used as the solvent.

2.2. Surface Modification of the SiC Whiskers. The silane coupling agent (1 g), ethanol (100 mL), and deionized water (10 g) were mixed in a 250 mL glass flask. The pH of the coupling agent solution was adjusted to 8-9 using a sodium hydroxide solution. The solution was sonicated at 30°C for 5 min. The SiC whiskers (1 g) were added to the solution, and the mixture was stirred at 70°C for 3 h. The surface-modified SiC whiskers were obtained after filtering, drying, and grinding the aforementioned mixture.

2.3. Sample Preparation

2.3.1. Preparation of the PLA/PCL Blends. PLA and PCL were blended at various ratios according to weight fraction, i.e., PLA/PCL = 100/0, 98.75/1.25, 97.5/2.5, 95/5, and 92.5/7.5. The desired amounts of PLA and PCL were mixed in CH_2Cl_2 and stirred with a magnetic stirring bar at room temperature for 5 h. The CH_2Cl_2 was removed from the solution by heating it to 190°C under reduced pressure. The mixture was then injected into a preheated mold, sprayed with a mold release agent, and compression molded at temperatures ranging from 180°C to 200°C and at a pressure of 5 MPa.

2.3.2. Preparation of the PLA/PCL/SiC Composites. The weight content of PCL was set at 2.5 wt.%, and the SiC content was varied from 0.25 to 1 wt.%. The desired amounts of PLA, PCL, and SiC were mixed in CH_2Cl_2 and stirred at room temperature for 5 h. The CH_2Cl_2 in the mixture was removed by heating to 190°C and under reduced pressure. Then, the mixture was injected into a preheated mold, sprayed with a mold release agent, and compression molded at temperatures ranging from 180°C to 200°C at a pressure of 5 MPa.

2.4. Characterization and Measurements. The surface properties of SiC whiskers before and after surface modification were investigated using Fourier-transform infrared spectroscopy (FT-IR, Bio-Rad Co., Digilab FTS-165) (wavenumber range: 400–4000 cm^{-1}, resolution: 4 cm^{-1}, scans in triplicate, KBr pellet) and X-ray photoelectron spectroscopy (XPS, Thermo K-Alpha).

Flexural tests were performed according to ASTM D790-86 using a three-point bend configuration. The sample size was 5 mm × 10 mm × 100 mm. The preload, span-to-span ratio, and cross-head speed were 0.1 N, 16 : 1, and 2.1 mm/min, respectively. The range of maximum deflection was sample fracture. The flexural strength (σ_f) and elastic modulus (E_b) were determined using a three-point bending test and calculated as follows [24–26]:

$$\sigma_f = \frac{3PL}{2bd^2}, \tag{1}$$

$$E_b = \frac{L^3}{4bd^3}\frac{\Delta P}{\Delta m}, \tag{2}$$

where P is the applied load (in N), L is the span length (in mm), b is the width of the specimen (in mm), and d is the thickness of the specimen (in mm). ΔP is the change in force in the linear portion of the load-deflection curve (in N), and Δm is the corresponding change in deflection (in mm). The flexural strength and elastic modulus were obtained from the average of seven experimental values.

The critical stress intensity factor (K_{IC}) of the samples was characterized via single-edge notched (SEN) testing in a three-point bending flexure. The three-point bending test was performed on a universal testing machine (Instron Model WDW3010) according to ASTM D-5045. The sample size used for these tests was $5\,mm \times 10\,mm \times 50\,mm$. The span-to-span ratio and cross-head speed were $4:1$ and $1\,mm/min$, respectively. For the three-point flexural test, the K_{IC} was calculated as follows [27–29]:

$$K_{IC} = PBW^{1/2}Y, \qquad (3)$$

where P is the rupture force (in kN), B is the specimen thickness (in cm), W is the specimen width (in cm), and Y is the geometrical factor. The fracture toughness was obtained from the average of seven experimental values.

After the fracture tests, the morphologies of the blends were examined using field-emission scanning electron microscopy (FE-SEM, SU 8010/HITACHI).

3. Results and Discussion

3.1. Surface Modification of the SiC Whiskers. Figure 1 shows the FT-IR spectra of the SiC whiskers before and after surface modification. The peaks attributed to the CH_2- and $C=O$ groups were observed at 884 and $1095\,cm^{-1}$, respectively. The areas of the CH_2- and $C=O$ peaks before and after modification were calculated from the FT-IR spectra [30, 31], and the results are shown in Table 1. These peaks increased significantly upon the introduction of organic functional groups onto the surface of the SiC whiskers [32, 33].

Figure 2 shows the X-ray photoelectron spectra of the SiC whiskers before and after surface modification. The O_{1s} peak was observed at 533.8 eV, and it decreased significantly in intensity after surface modification due to the reaction of the silane coupling agent with the hydroxyl group on the SiC surfaces [34, 35]. These results demonstrated that the surface modification introduces organic functional groups to the surface of the SiC whiskers.

3.2. Flexural Properties. The mechanical performance of the PLA/PCL blends was evaluated based on flexural strength and elastic modulus measurements. Figure 3(a) shows the flexural strength of the PLA/PCL blends as a function of the PCL content. The flexural strength of the blends increased upon increasing the PCL content up to 2.5 wt.% PCL. The flexural strength of neat PLA was 80.8 MPa, whereas that of the blend containing 2.5 wt.% PCL was 36% higher (110 MPa) because of the good interfacial adhesion between the PCL and PLA matrix in the PLA/PCL blends [36, 37]. The flexural strength of the blends decreased above 2.5 wt.% PCL because of the dispersion state of PCL in the PLA matrix deteriorated at the high PCL content.

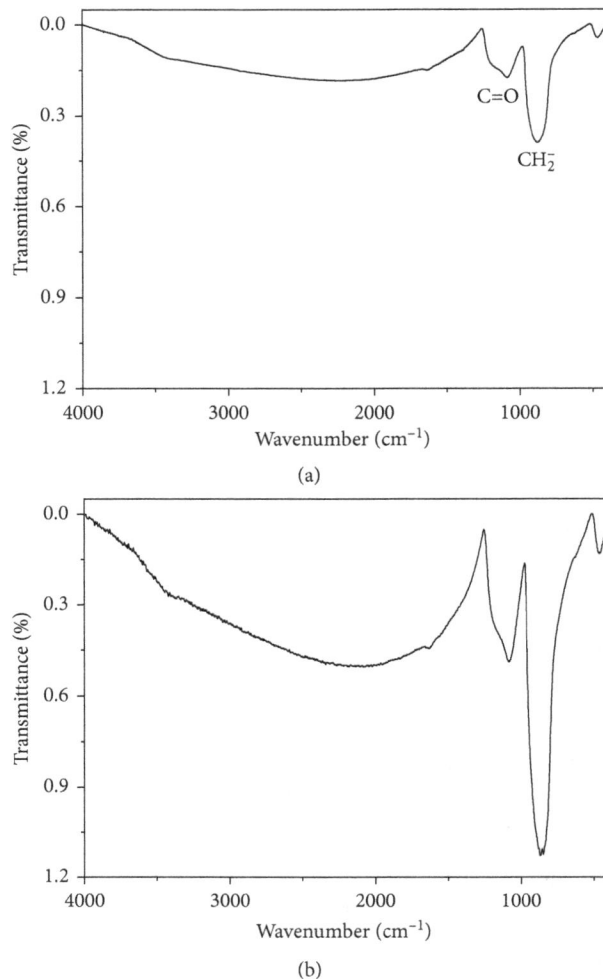

(a)

(b)

FIGURE 1: FT-IR spectra of SiC whiskers (a) before and (b) after surface modification.

TABLE 1: Areas of CH_2- and $C=O$ peaks before and after modification from calculated TF-IR.

Peak	CH_2-	$C=O$
Before modification	34.6	72.0
After modification	96.0	222.4

Figure 3(b) shows the elastic modulus of the PLA/PCL blends. The elastic modulus of the blends decreased gradually from 3.87 to 3.75 GPa for the 2.5 wt.% PCL blend (a 2.6% decrease) and to 3.23 GPa for the 10 wt.% PCL blend (a 16.5% decrease). These results could be attributed to the low glass transition temperature of PCL (61°C) [13, 38].

Figure 4(a) shows the flexural strength of PLA/PCL/SiC composites with 2.5 wt.% PCL as a function of SiC content. The flexural strength of the composites increased from 110 to 126.1 MPa with increasing SiC content. This could be attributed to the good interfacial adhesion between the SiC whiskers and PLA matrix [39, 40].

Figure 4(b) shows the elastic modulus of the PLA/PCL/SiC composites as a function of SiC content. The elastic

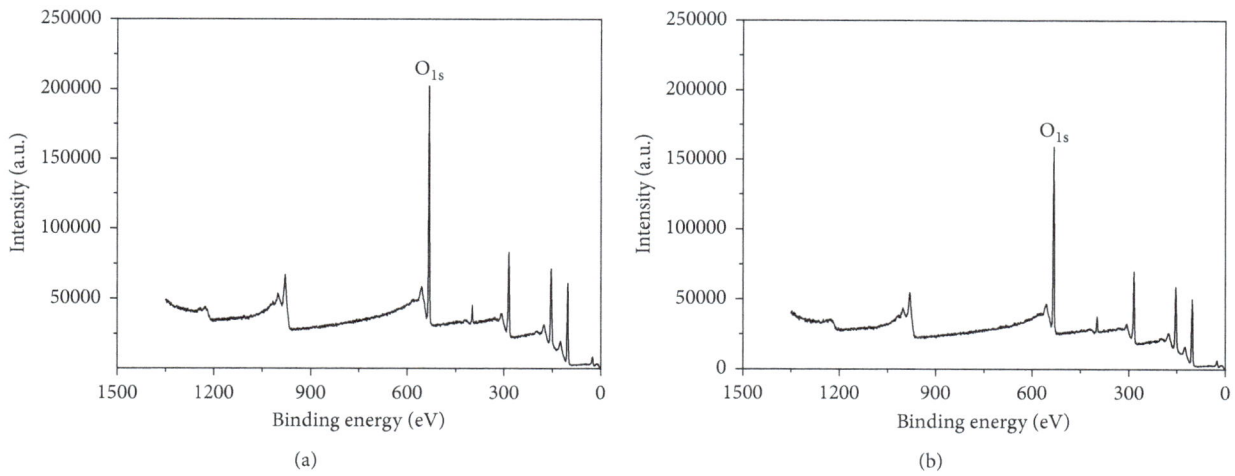

FIGURE 2: XPS spectra of SiC whiskers (a) before and (b) after surface modification.

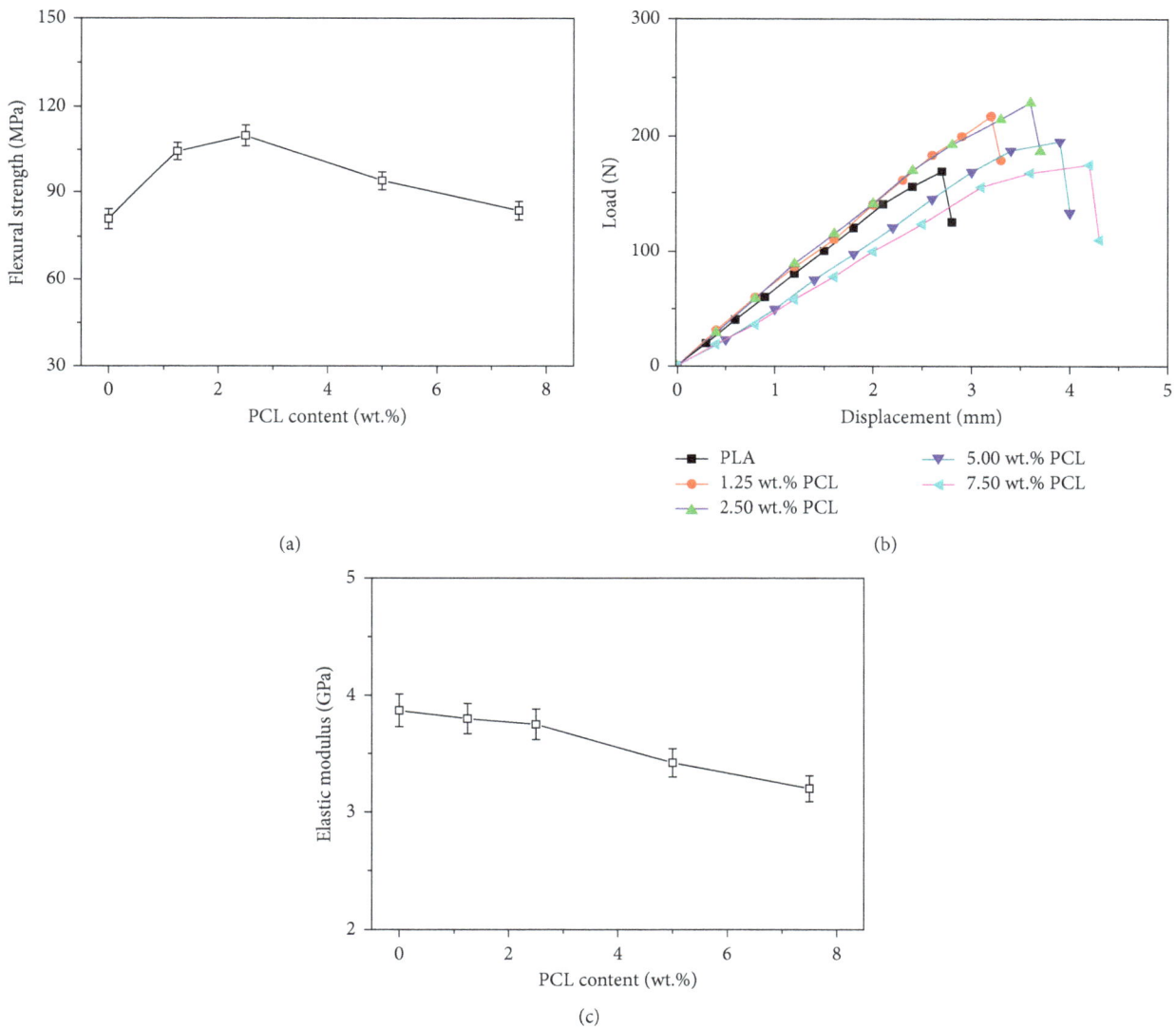

FIGURE 3: Flexural strength (a), load (b), and elastic modulus (c) of PLA/PCL blends.

(a)

(b)

(c)

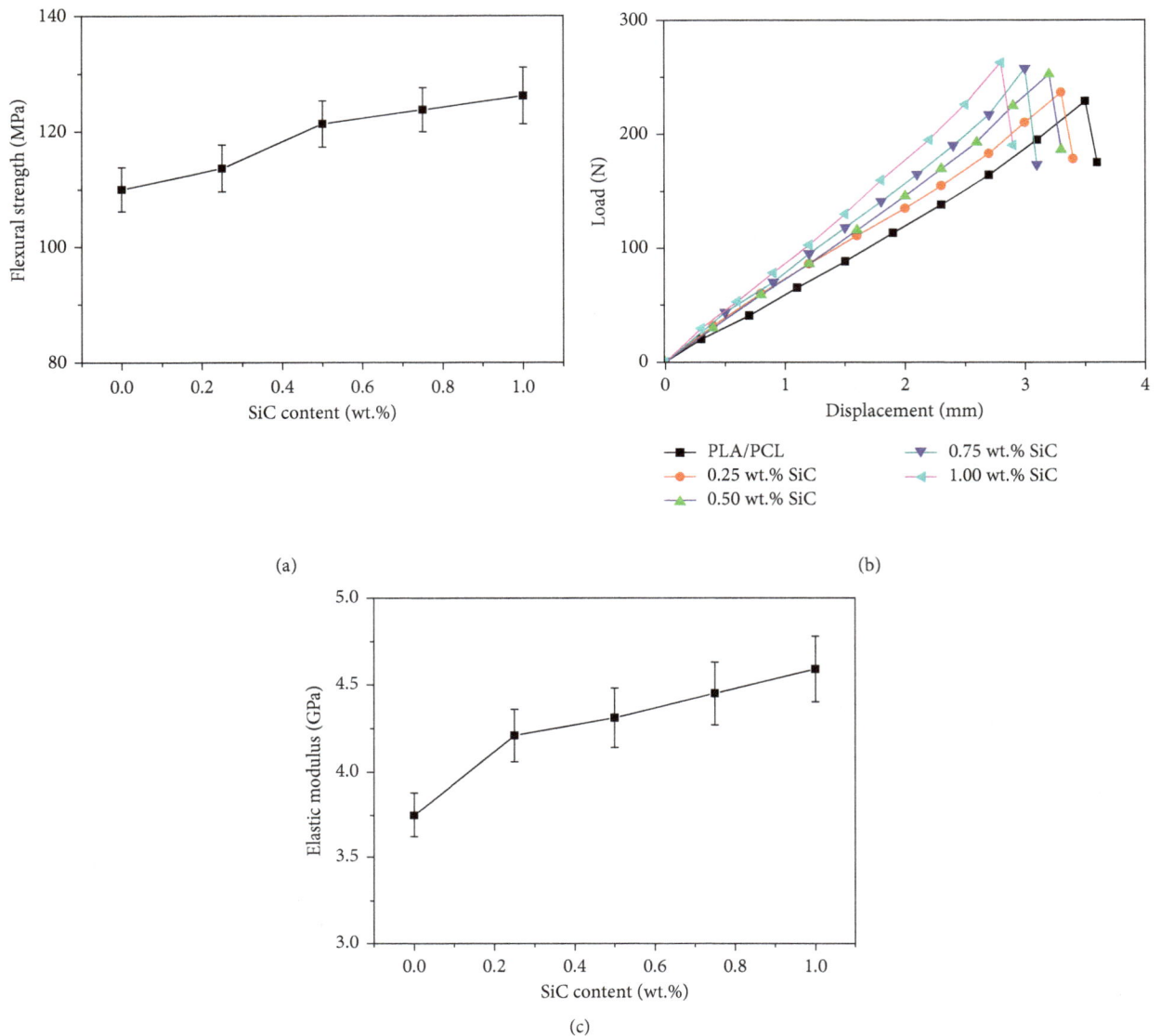

FIGURE 4: Flexural strength (a), load (b), and elastic modulus (c) of PLA/PCL/SiC composites.

modulus of the composites increased from 3.75 to 4.59 GPa with increasing SiC content. This was due to the restriction of the mobility of polymer chains in PLA under load by the dispersed rigid SiC whiskers [21, 41].

3.3. Fracture Toughness.

The fracture toughness of the PLA/PCL blends was investigated by K_{IC} measurements. Figure 5 shows the K_{IC} values of the PLA/PCL blends as a function of PCL content. It was clear that K_{IC} was significantly improved upon the addition of PCL. Neat PLA was very brittle with a K_{IC} of 1.22 MPa·m$^{1/2}$, whereas the blend containing 2.5 wt.% PCL had a K_{IC} of 1.79 MPa·m$^{1/2}$, which was 47% higher than that of neat PLA. This increase in K_{IC} may be due to the elasticity and flexibility of PCL, which also shows good energy-absorbing capacity and stretches to accommodate the cracks produced by external forces [36, 42]. The K_{IC} value of the blends decreased slightly above 2.5 wt.% PCL as

the dispersion state of PCL in the PLA matrix deteriorated at high PCL content.

Figure 6 shows the K_{IC} values of the PLA/PCL/SiC composites as a function of SiC content. The K_{IC} value significantly improved with the addition of SiC. K_{IC} of PLA/PCL blend containing 2.5 wt.% PCL was 1.79 MPa·m$^{1/2}$, whereas K_{IC} of the composite containing 2.5 wt.% PCL and 0.5 wt.% SiC was 2.6 MPa·m$^{1/2}$, which was 45% higher than that of the blend without SiC. This trend was attributed to the good energy-absorbing capacity of the SiC whiskers dispersed in the PLA matrix and the SiC–PLA debonding. The debonding effectively transferred stress and increased the resistance to deformation and crack propagation, resulting in the increased fracture toughness of the PLA/PCL/SiC composites [43, 44]. The K_{IC} value of the composites decreased slightly above 0.5 wt.% SiC because of the slight agglomeration of SiC whiskers in the PLA matrix at a high SiC content.

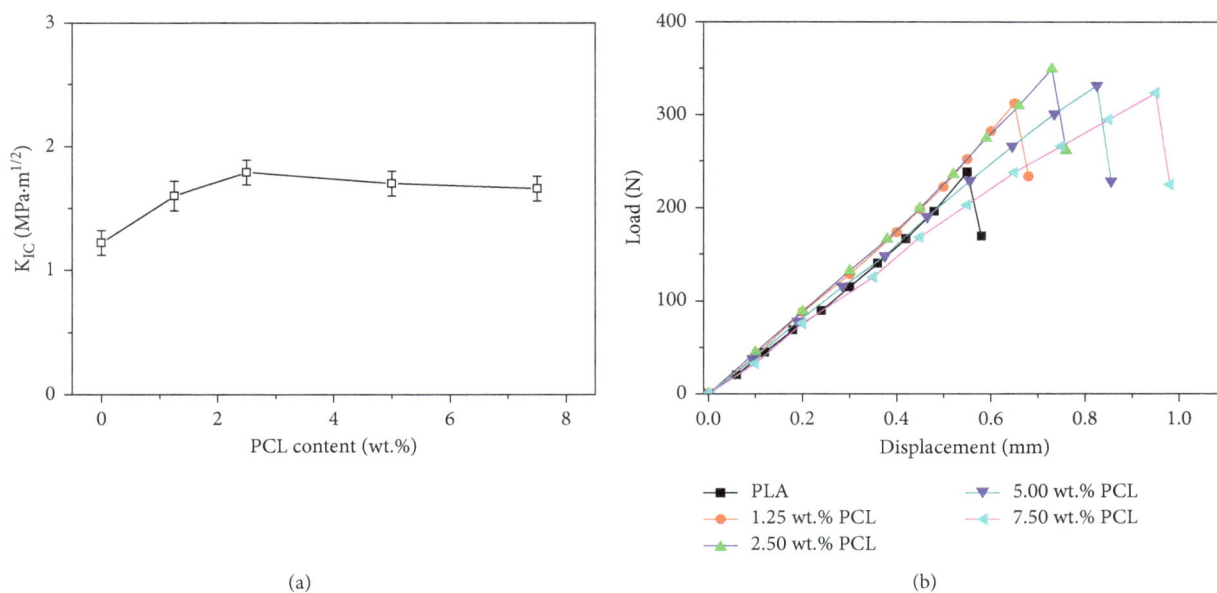

(a)

(b)

FIGURE 5: Fracture toughness of PLA/PCL blends as a function of PCL content.

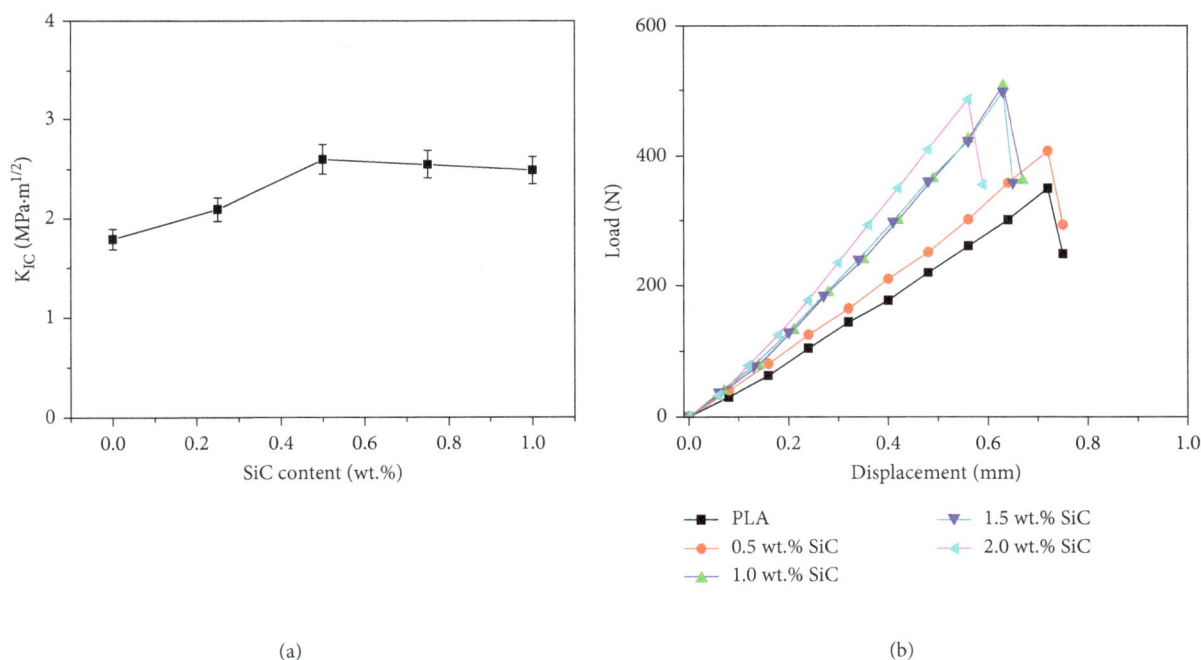

(a)

(b)

FIGURE 6: Fracture toughness of PLA/PCL/SiC composites as a function of SiC content.

3.4. *Morphology*. The toughness behavior of the PLA/PCL blends and PLA/PCL/SiC composites was further characterized by SEM observations. Figure 7 shows the SEM images of the PLA/PCL blends after the K_{IC} tests. Neat PLA exhibited a smooth and flat morphology (Figure 7(a)), indicating a brittle fracture surface [45]. Conversely, the SEM images of the blends show cracks and many ridges, indicating plastic deformation prior to fracturing. This accounts for the high fracture toughness of the blends, as shown in Figures 7(b)–7(e) [36, 46]. PCL is thermodynamically incompatible with PLA [13, 15]. As shown in Figures 7(b)–7(e), PCL was relatively well dispersed in the PLA matrix at low PCL content.

Figure 8 shows the SEM images of the PLA/PCL/SiC composites after the K_{IC} tests. The SiC whiskers were bonded in a continuous PLA matrix and showed good interfacial adhesion. The introduction of SiC whiskers led to increased resistance to deformation and crack propagation through energy absorption by the SiC whiskers and SiC-PLA debonding in the PLA/PCL/SiC composites. This is in

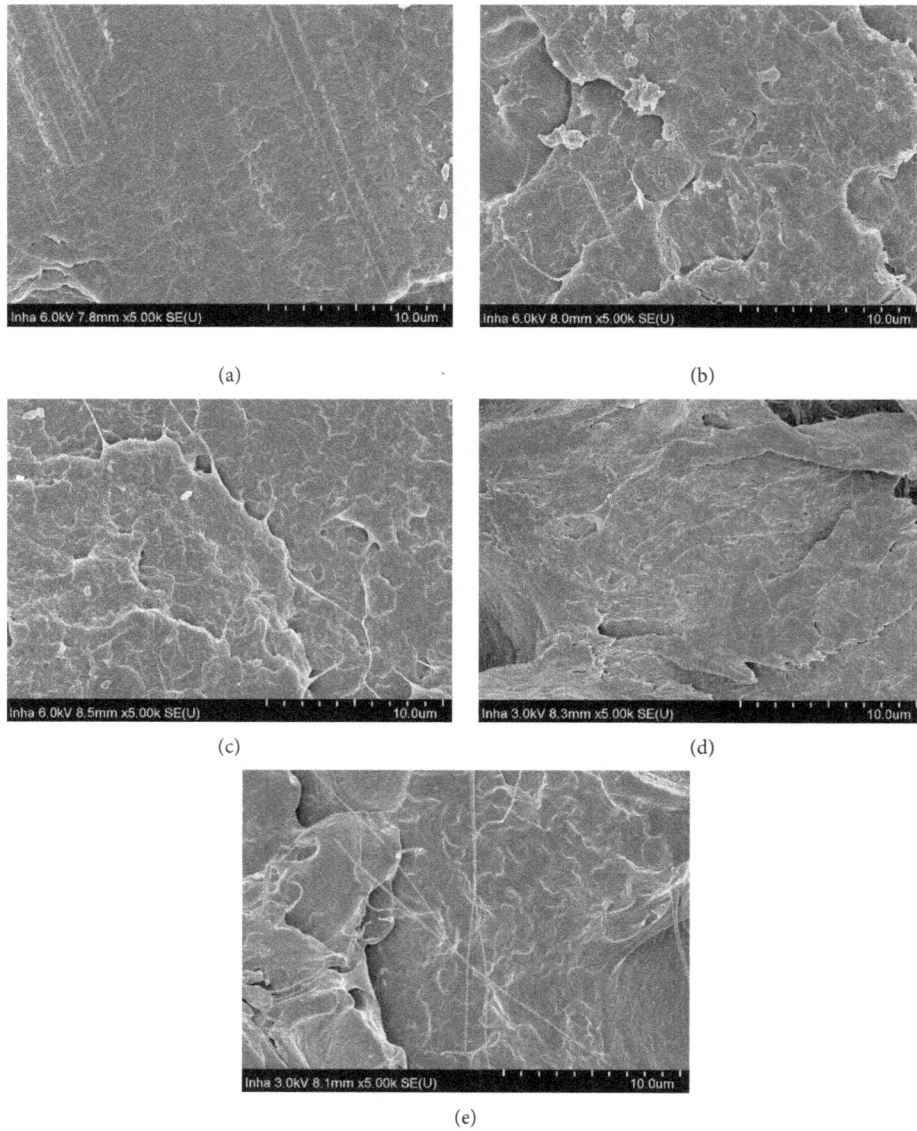

(a)

(b)

(c)

(d)

(e)

Figure 7: SEM micrographs of PLA/PCL blends after K_{IC} tests: (a) neat PLA; (b) 1.25 wt.% PCL; (c) 2.5 wt.% PCL; (d) 5 wt.% PCL; (e) 7.5 wt.% PCL (magnification of 5000).

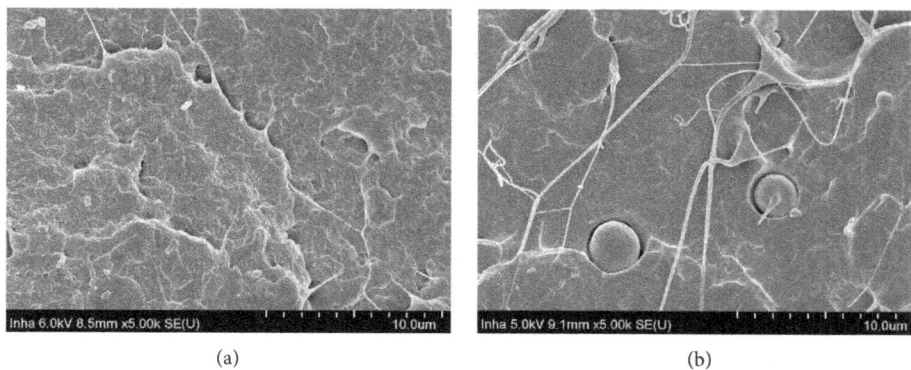

(a)

(b)

Figure 8: Continued.

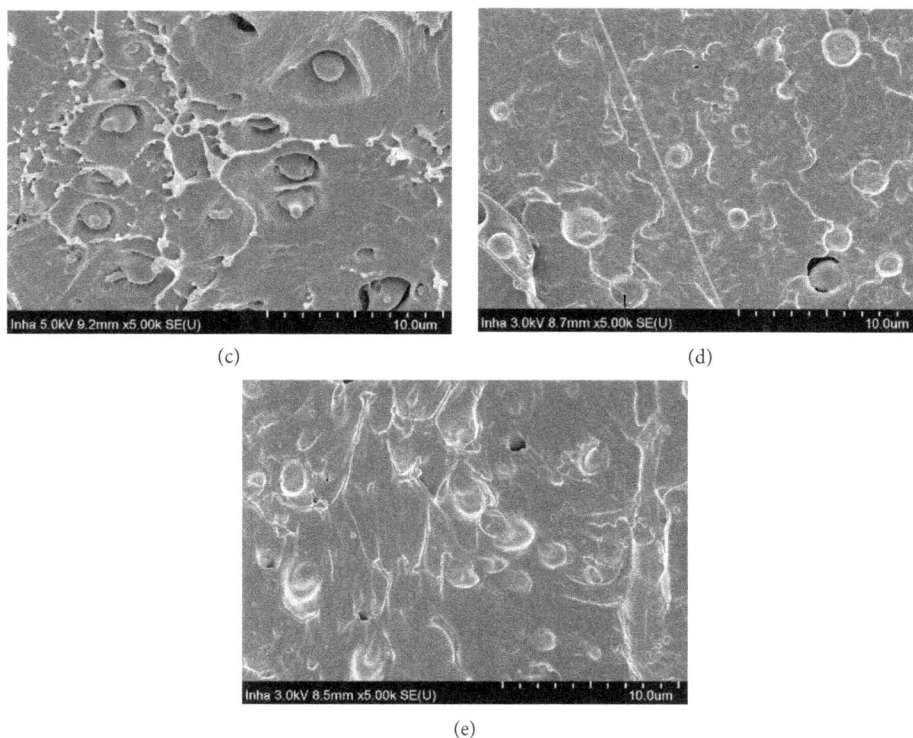

FIGURE 8: SEM micrographs of PLA/PCL/SiC composites after K_{IC} tests: (a) PLA/PCL (2.5 wt.% PCL); (b) 0.25 wt.% SiC; (c) 0.5 wt.% SiC; (d) 0.75 wt.% SiC; (e) 1.0 wt.% SiC (magnification of 5000).

agreement with the results showing that the SiC whiskers-reinforced PLA composites exhibited higher fracture toughness than that of the neat PLA or PLA/PCL blends [47, 48]. Generally, SiC whiskers were almost completely dispersed in the PLA matrix [13]. SiC was well dispersed in the PLA matrix at low SiC content (0.5 wt.%), and some SiC agglomeration was seen above 0.5 wt.% SiC.

4. Conclusions

The mechanical properties, fracture toughness, and morphology of the prepared bio-based PLA/PCL blends and PLA/PCL/SiC composites were examined using multiple techniques. The surface of the SiC whiskers was modified using a silane coupling agent. The fracture toughness and flexural strength of the PLA were increased by the incorporation of PCL and further improved by the addition of SiC whiskers. The elastic modulus of PLA decreased after the introduction of PCL but increased after the addition of SiC whiskers. SEM images revealed that reinforcement of PLA with PCL induced plastic deformation. Simultaneously, the SiC whiskers absorbed external energy through good interfacial adhesion between the SiC whiskers and PLA matrix and debonding between the two materials in the PLA/PCL/SiC composites.

Conflicts of Interest

The authors declare that they have no conflicts of interest.

Acknowledgments

This research was supported by Traditional Culture Convergence Research Program through the National Research Foundation of Korea (NRF) funded by the Ministry of Science, ICT & Future Planning (2016M3C1B5952897).

References

[1] R. Scaffaro, F. Lopresti, and L. Botta, "PLA based biocomposites reinforced with Posidonia Oceanica leaves," *Composites Part B: Engineering*, vol. 139, pp. 1–11, 2018.

[2] Y. Zhou, L. Lei, B. Yang, J. Li, and J. Ren, "Preparation and characterization of polylactic acid (PLA) carbon nanotube nanocomposites," *Polymer Testing*, vol. 68, pp. 34–38, 2018.

[3] J. Y. Zhu, C. H. Tang, S. W. Yin, and X. Q. Yang, "Development and characterization of novel antimicrobial bilayer films based on polylactic acid (PLA)/pickering emulsions," *Carbohydrate Polymers*, vol. 181, pp. 727–735, 2018.

[4] W. Pattanasuttichonlakul, N. Sombatsompop, and B. Prapagdee, "Accelerating biodegradation of PLA using microbial consortium from dairy wastewater sludge combined with PLA-degrading bacterium," *International Biodeterioration and Biodegradation*, vol. 132, pp. 74–83, 2018.

[5] X. Meng, V. Bocharova, H. Tekinalp et al., "Toughening of nancelluose/PLA composites via bio-epoxy interaction: mechanistic study," *Materials and Design*, vol. 139, pp. 188–197, 2018.

[6] S. Qian, H. Zhang, W. Yao, and K. Sheng, "Effects of bamboo cellulose nanowhisker content on the morphology, crystallization, mechanical, and thermal properties of PLA matrix

biocomposites," *Composites Part B: Engineering*, vol. 133, pp. 203–209, 2018.

[7] M. Shabanian, M. Khoobi, F. Hemati et al., "New PLA/PEI-functionalized Fe3O4 nanocomposite: preparation and characterization," *Journal of Industrial and Engineering Chemistry*, vol. 24, pp. 211–218, 2015.

[8] Y. Zhou, Z. Huang, X. Diao, Y. Weng, and Y. Wang, "Characterization of the effect of REC on the compatibility of PHBH and PLA," *Polymer Testing*, vol. 42, pp. 17–25, 2015.

[9] R. Al-Itry, K. Lamnawar, A. Maazouz, N. Billon, and C. Combeaud, "Effect of the simultaneous biaxial stretching on the structural and mechanical properties of PLA, PBAT and their blends at rubbery state," *European Polymer Journal*, vol. 68, pp. 288–301, 2015.

[10] Y. I. Hsu, K. Masutani, T. Yamaoka, and Y. Kimura, "Strengthening of hydrogels made from enantiomeric block copolymers of polylactide (PLA) and poly(ethylene glycol) (PEG) by the chain extending diels–alder reaction at the hydrophilic PEG terminals," *Polymer*, vol. 67, pp. 157–166, 2015.

[11] M. Félix, A. Romero, J. E. Martín-Alfonso, and A. Guerrero, "Development of crayfish protein-PCL biocomposite material processed by injection moulding," *Composites Part B: Engineering*, vol. 78, pp. 291–297, 2015.

[12] J. Rodenas-Rochina, A. Vidaurre, I. C. Cortázar, and M. Lebourg, "Effects of hydroxyapatite filler on long-term hydrolytic degradation of PLLA/PCL porous scaffolds," *Polymer Degradation and Stability*, vol. 119, pp. 121–131, 2015.

[13] J. P. Mofokeng and A. S. Luyt, "Dynamic mechanical properties of PLA/PHBV, PLA/PCL, PHBV/PCL blends and their nanocomposites with TiO2 as nanofiller," *Thermochimica Acta*, vol. 613, pp. 41–53, 2015.

[14] S. Jain, M. M. Reddy, A. K. Mohanty, M. Misra, and A. K. Ghosh, "A new biodegradable flexible composite sheet from poly(lactic acid)/poly(ε-caprolactone) blends and micro-talc," *Macromolecular Materials and Engineering*, vol. 295, no. 8, pp. 750–762, 2010.

[15] T. Patrício and P. Bártolo, "Thermal stability of PCL/PLA blends produced by physical blending process," *Procedia Engineering*, vol. 59, pp. 292–297, 2013.

[16] H. Tsuji, G. Horikawa, and S. Itsuno, "Melt-processed biodegradable polyester blends of poly(L-lactic acid) and poly(ε-caprolactone): effects of processing conditions on biodegradation," *Journal of Applied Polymer Science*, vol. 104, no. 2, pp. 831–841, 2007.

[17] L. Chen, J. Yang, K. Wang, F. Chen, and Q. Fu, "Largely improved tensile extensibility of poly(L-latic acid) by adding poly(ε-caprolactone)," *Polymer International*, vol. 59, pp. 1154–1161, 2010.

[18] M. Harada, K. Iida, K. Okamoto, H. Hayashi, and K. Hirano, "Reactive compatibilization of biodegradable poly(lactic acid)/poly(ε-caprolactone) blends with reactive processing agents," *Polymer Engineering and Science*, vol. 48, no. 7, pp. 1359–1368, 2008.

[19] J. Urquijo, G. Guerrica-Echevarría, and J. I. Eguiazábal, "Melt processed PLA/PCL blends: effect of processing method on phase structure, morphology, and mechanical properties," *Journal of Applied Polymer Science*, vol. 132, no. 41, 2015.

[20] G. C. Pradhan, S. Dash, and S. K. Swain, "Barrier properties of nano silicon carbide designed chitosan nanocomposites," *Carbohydrate Polymers*, vol. 134, pp. 60–65, 2015.

[21] K. Kueseng and K. I. Jacob, "Natural rubber nanocomposites with SiC nanoparticles and carbon nanotubes," *European Polymer Journal*, vol. 42, pp. 220–227, 2016.

[22] A. Dorigato, M. Sebastiani, A. Pegoretti, and L. Fambri, "Effect of silica nanoparticles on the mechanical performances of poly(lactic acid)," *Journal of Polymers and the Environment*, vol. 20, no. 3, pp. 713–725, 2012.

[23] B. J. Rashmi, K. Prashantha, M. F. Lacrampe, and P. Krawczak, "Toughening of poly(lactic acid) without sacrificing stiffness and strength by melt-blending with polyamide 11 and selective localization of halloysite nanotubes," *Express Polymer Letters*, vol. 9, pp. 721–735, 2015.

[24] F. L. Jin and S. J. Park, "Improvement in fracture behaviors of epoxy resins toughened with sulfonated poly(ether sulfone)," *Polymer Degradation and Stability*, vol. 92, no. 3, pp. 509–514, 2007.

[25] F. L. Jin and S. J. Park, "Thermal properties and toughness performance of hyperbranched polyimide-modified epoxy resins," *Journal of Polymer Science Part B: Polymer Physics*, vol. 44, no. 23, pp. 3348–3356, 2006.

[26] F. L. Jin, C. J. Ma, and S. J. Park, "Thermal and mechanical interfacial properties of epoxy composites based on functionalized carbon nanotubes," *Materials Science and Engineering: A*, vol. 528, no. 29-30, pp. 8517–8522, 2011.

[27] S. J. Park, F. L. Jin, and J. S. Shin, "Physicochemical and mechanical interfacial properties of trifluorometryl groups containing epoxy resin cured with amine," *Materials Science and Engineering: A*, vol. 390, no. 1-2, pp. 240–245, 2005.

[28] S. J. Park, F. L. Jin, and J. R. Lee, "Thermal and mechanical properties of tetrafunctional epoxy resin toughened with epoxidized soybean oil," *Materials Science and Engineering: A*, vol. 374, no. 1-2, pp. 109–114, 2004.

[29] S. J. Park, F. L. Jin, J. H. Park, and K. S. Kim, "Synthesis of a novel siloxane-containing diamine for increasing flexibility of epoxy resins," *Materials Science and Engineering: A*, vol. 399, no. 1-2, pp. 377–381, 2005.

[30] I. Poljanšek and M. Krajnc, "Characterization of phenol-formaldehyde prepolymer resins by in line FT-IR spectroscopy," *Acta Chimica Slovenica*, vol. 52, pp. 238–244, 2005.

[31] B. Strzemiecka, A. Voelkel, M. Hinz, and M. Rogozik, "Application of inverse gas chromatography in physicochemical characterization of phenolic resin adhesives," *Journal of Chromatography A*, vol. 1368, pp. 199–203, 2014.

[32] Y. Qi, Q. Luo, J. Shen, L. Zheng, J. Zhou, and W. Chen, "Surface modification of BMN particles with silane coupling agent for composites with PTFE," *Applied Surface Science*, vol. 414, pp. 147–152, 2017.

[33] M. Dinari and A. Haghighi, "Surface modification of TiO2 nanoparticle by three dimensional silane coupling agent and preparation of polyamide/modified-TiO2 nanocomposites for removal of Cr (VI) from aqueous solutions," *Progress in Organic Coatings*, vol. 110, pp. 24–34, 2017.

[34] S. J. Park and B. J. Kim, "Roles of acidic functional groups of carbon fiber surfaces in enhancing interfacial adhesion behavior," *Materials Science and Engineering: A*, vol. 408, no. 1-2, pp. 269–273, 2005.

[35] S. J. Park and S. Y. Jin, "Effect of ozone treatment on ammonia removal of activated carbons," *Journal of Colloid and Interface Science*, vol. 286, no. 1, pp. 417–419, 2005.

[36] B. S. Bouakaz, A. Habi, Y. Grohens, and I. Pillin, "Organomontmorillonite/graphene-PLA/PCL nanofilled blends: new strategy to enhance the functional properties of PLA/PCL blend," *Applied Clay Science*, vol. 139, pp. 81–91, 2017.

[37] F. L. Jin, Q. Q. Pang, T. Y. Zhang, and S. J. Park, "Synergistic reinforcing of poly(lactic acid)-based systems by polybutylene

succinate and nano-calcium carbonate," *Journal of Industrial and Engineering Chemistry*, vol. 32, pp. 77–84, 2015.

[38] J. P. Mofokeng and A. S. Luyt, "Morphology and thermal degradation studies of melt-mixed poly(lactic acid) (PLA)/poly(ε-caprolactone) (PCL) biodegradable polymer blend nanocomposites with TiO_2 as filler," *Polymer Testing*, vol. 45, pp. 93–100, 2015.

[39] F. L. Jin and S. J. Park, "Interfacial toughness properties of trifunctional epoxy resins/calcium carbonate nanocomposites," *Materials Science and Engineering: A*, vol. 475, no. 1-2, pp. 190–193, 2008.

[40] M. Shabanian, M. Hajibeygi, K. Hedayati, M. Khaleghi, and H. A. Khonakdar, "New ternary PLA/organoclay-hydrogel nanocomposites: design, preparation and study on thermal, combustion and mechanical properties," *Materials and Design*, vol. 110, pp. 811–820, 2016.

[41] H. Alamri and I. M. Low, "Effect of water absorption on the mechanical properties of n-SiC filled recycled cellulose fibre reinforced epoxy eco-nanocomposites," *Polymer Testing*, vol. 31, no. 6, pp. 810–818, 2012.

[42] J. Zhou, Z. Yao, C. Zhou, D. Wei, and S. Li, "Mechanical properties of PLA/PBS foamed composites reinforced by organophilic montmorillonite," *Journal of Applied Polymer Science*, vol. 131, no. 18, 2014.

[43] N. Graupner, D. Labonte, and J. Müssig, "Rhubarb petioles inspire biodegradable cellulose fibre-reinforced PLA composites with increased impact strength," *Composites Part A: Applied Science and Manufacturing*, vol. 98, pp. 218–226, 2017.

[44] L. Wang, W. M. Gramlich, and D. J. Gardner, "Improving the impact strength of poly(lactic acid) (PLA) in fused layer modeling (FLM)," *Polymer*, vol. 114, pp. 242–248, 2017.

[45] B. K. Chen, C. H. Shen, S. C. Chen, and A. F. Chen, "Ductile PLA modified with methacryloyloxyalkyl isocyanate improves mechanical properties," *Polymer*, vol. 51, no. 21, pp. 4667–4672, 2010.

[46] M. R. Kamal and V. Khoshkava, "Effect of cellulose nanocrystals (CNC) on rheological and mechanical properties and crystallization behavior of PLA/CNC nanocomposites," *Carbohydrate Polymers*, vol. 123, pp. 105–114, 2015.

[47] M. Bulota and T. Budtova, "PLA/algae composites: morphology and mechanical properties," *Composites Part A: Applied Science and Manufacturing*, vol. 73, pp. 109–115, 2015.

[48] S. S. Yao, Q. Q. Pang, R. Song, F. L. Jin, and S. J. Park, "Fracture toughness improvement of poly(lactic acid) with silicon carbide whiskers," *Macromolecular Research*, vol. 24, no. 11, pp. 961–964, 2016.

3

Behavior of Concrete-Filled Steel Tube Columns Subjected to Axial Compression

Pengfei Li⊚, **Tao Zhang**⊚, **and Chengzhi Wang**⊚

Chongqing Jiaotong University, Chongqing 400074, China

Correspondence should be addressed to Pengfei Li; lipengfei@cqjtu.edu.cn

Academic Editor: José A. Correia

The behavior of concrete-filled steel tube (CFST) columns subjected to axial compression was experimentally investigated in this paper. Two kinds of columns, including CFST columns with foundation and columns without foundation, were tested. Columns of pure concrete and concrete with reinforcing bars as well as two steel tube thicknesses were considered. The experimental results showed that the CFST column with reinforcing bars has a higher bearing capacity, more effective plastic behavior, and greater toughness, and the elastoplastic boundary point occurs when the load is approximately 0.4–0.5 times of the ultimate bearing capacity. The change of rock-socketed depth and the presence of steel tube will affect the ultimate bearing capacity of rock-socketed pile. The bearing capacities of the rock-socketed CFST columns are lower than those of rock-socketed columns without a steel tube under a vertical load; besides, the greater the rock-socketed depth, the greater the bearing capacity of the rock-socketed piles. In addition, a numerical comparison between the ultimate load and the theoretical value calculated from the relevant specifications shows that the ultimate load is generally considerably greater than the theoretical calculation results.

1. Introduction

In the last few decades, high-rise, large-span, and large-scale building structures have become more common. Concrete-filled steel tubular (CFST) members are well recognized for their excellent performance owing to the combined merits of steel and concrete materials [1]. Therefore, concrete-filled steel tubes are being increasingly used in high-rise buildings and in large-span structures. CFST columns have been used in earthquake-resistant structures and bridge piers subject to impact from traffic and used to support storage tanks, decks of railways, and high-rise buildings as well as being used as piles. Concrete-filled steel tubes require additional fire-resistant insulation if the fire protection of the structure is necessary [2].

Studies of CFST have also been frequently performed, and tests of CFST with rectangular, square, and circular cross sections filled with high-strength concrete have been reported [3–6], including compression [7–12], bending [13–17], or torsion [18–20] tests. The influence of the behavior and strength capacity of CFST on the bonding and local buckling of the steel profile, confinement of the concrete, and strength of the materials were discussed. Pull-out tests on CFST columns have shown the contribution of shear connectors to the force transference mechanism that occurs in beam-column connections. Local buckling has been widely studied, and it is possible to consider its influence on the strength capacity of the CFST columns. The confinement effect is influenced by the shape of the cross section, the thickness of the steel profile, the type of loading, the slenderness of the composite column, and the strength of the materials used. The studies on CFST rock-socketed columns and predicting the ultimate capacity of CFST columns subjected to axial compression are limited, and there is no standard method for calculating the strength of concrete-filled steel tube with reinforcing bars.

There are several design codes for concrete-filled steel columns, such as the America Institute of Steel Construction (AISC) Load and Resistance Factor Design (LRFD) [21], Eurocode 4 (EC4) [22], Brazilian Code NBR 8800 [23, 24], and GB50396-2014 [25]. Since the AISC is focused on steel columns, use of the AISC [21] specifications is limited to composite columns with steel yield stress and concrete cylinder strength no greater than 415 and 55 MPa, respectively. Modified values for both the yielding strength (f_{my}) and elasticity modulus (E_m) are assumed. These modified terms take into account the presence of the concrete filling the steel profile. EC4 [22] and NBR 8800 [23, 24] determine the resistance capacity of a section by adding the contribution of the steel tube and the concrete core. For columns with circular cross sections, the confinement effect is considered by using coefficients that increase the uniaxial compressive strength of the concrete (f_{ck}) and reduce the yielding strength of the steel (f_y). The instability in these standard codes is considered by using a coefficient, which depends on the slenderness of the composite column. Limits for the steel tube slenderness are also considered to avoid local buckling.

In China, reinforced CFST piles are widely used in foundations of piers and bridges for its good mechanical properties, for instance, Orchard Container pier in Chongqing Port, Cuntan pier, Xintian Container pier in Wanzhou Port, and Yangshuo pier in Wuhan Port. The remarkable feature of this new structure is that the thickness of the steel pipe is thin and the ratio of diameter to thickness is very large [26]. It greatly exceeds the allowable range of the much existing specifications [27], for example, the ferrule coefficient of circular-section concrete-filled steel tubular members specified in Chinese specification [28] should be limited to between 0.3 and 3.0, and the ratio of diameter to thickness (D/t) should be limited to the following range: $20\sqrt{235/f_y} \leq D/t \leq 85\sqrt{235/f_y}$; Eurocode 4 (EC4) [22] explicit limits the allowed D/t values in the European context to $90 * 235/f_y$. The maximum allowable value of diameter-thickness ratio (D/t) specified in AISC [21] is $0.31 * E/f_y$. So, it is an attempt to estimate the ultimate load of reinforced CFST piles by above codes, and it may provide a reference for the design and research of the new structure with large diameter-to-thickness ratio.

The present paper is thus an attempt to study the behavior of concrete-filled steel tube columns with foundation and columns without foundation subjected to axial compression. The main objectives of this paper were twofold: first, to discuss the bearing capacity of CFST columns under vertical load with regard to the thickness of steel tube, reinforcement, and the depth of rock-socketing; second, to discuss the ultimate load experimental results and theoretical calculations.

2. Experimental Study

2.1. Columns without Foundation

2.1.1. Test Program. The experimental study of concrete-filled steel tube columns subjected to axial compression was accomplished with several objectives, namely, the evaluation of the accuracy of several design codes to predict the resistance capacity, to determine the maximum load capacity of the concentric loading specimens, and to investigate the failure pattern before and after the ultimate load is reached.

The specimens were tested; a summary of their properties is presented in Table 1, and the schematic of pile is shown in Figure 1. The cross sections of the tested columns are circular, the length of each column specimen is 1200 mm, and the diameters of the specimens are either 150 mm (symbolized with "1") or 165 mm (symbolized with "2"), thickness of the steel tube are either 1.2 mm (symbolized with "A") or 1.6 mm (symbolized with "B").

2.1.2. Properties of the Materials. The concrete was produced in the laboratory of the Structural Engineering Department, and the mechanical properties of the concrete core were determined by cylinder tests (150 mm × 150 mm × 150 mm). In addition, the concrete mix ratio is shown in Table 2. The concrete grade is C30, the mixing ratio of cement: sand: stone: water for ordinary Portland cement with a cement strength grade of M32.5 is 1 : 2.28 : 4.07 : 0.65 (mass ratio); the gravel underwent natural drying after washing, and it ranged from 5 to 10 mm in diameter. During the casting of the specimen, three standard concrete cube specimens of 150 mm × 150 mm × 150 mm are reserved for each batch of concrete, and the standard cube test block is sprinkled for 28 days for concrete axial compressive strength test. The compressive strength of concrete cube specimens is shown in Table 3. According to the yield strength test of steel bars, the stress-strain curve of steel bars is obtained. The yield strength of steel bars is 375 MPa, and the ultimate yield tensile strength is 410 MPa, as is shown in Table 3.

In Table 4, E is equal to the modulus of elasticity. A HRB335 steel plate is used for the steel-retaining material, and it is welded to a steel sleeve with a thickness of either 1.2 mm or 1.6 mm, and the steel bar is a two-stage rib bar with a diameter of 6 mm.

2.1.3. Test Setup. All of the tests were performed in a vertical column-testing machine with a static load capacity of 5000 kN, as is shown in Figure 2. The upper and lower surfaces of the column are, respectively, connected with the loading device and the rigid base plate by using a rigid pad. The load is divided into 12 stages according to the estimated ultimate carrying capacity, but the load grading is not consistent: during the initial stage, the loading is larger, but the loading grading is gradually reduced after the plastic stage of deformation is reached. This loading scheme allows the study of the post-peak behavior of the composite columns. Therefore, it is possible to study the elastic behavior, the axial load capacity, and the ductility of the columns under axial loading.

For this test, three layers of strain gauges were arranged vertically along the column, and the data were collected by a signal acquisition system to explore the variation in the strain of the steel casing with an applied load. Besides, the distance between the first strain gauge and the column top and the distance between the bottom strain gauge and the pile bottom were both 30 cm, and the distance between adjacent strain gauge layers was 30 cm. The 4 strain gauges in each layer were arranged symmetrically along the column surface. And an additional strain gauge was embedded in the

TABLE 1: List of columns for CFST columns without foundation.

Specimens	$D \times \delta \times L$ (mm)	Ac (mm^2)	Ar (mm^2)	As (mm^2)	α_{sc} (%)
PL-2	$165 \times 0 \times 1200$	21382	0	0	0
NS-A-2	$165 \times 1.2 \times 1200$	21382	0	622	2.9
NS-B-2	$165 \times 1.6 \times 1200$	21382	0	829	3.9
RN-2	$165 \times 0 \times 1200$	21382	170	0	0.8
RS-A-2	$165 \times 1.2 \times 1200$	21382	170	622	3.7
RS-B-2	$165 \times 1.6 \times 1200$	21382	170	829	4.7
PL-1	$150 \times 0 \times 1200$	17671	0	0	0
NS-A-1	$150 \times 1.2 \times 1200$	17671	0	570	3.2
RS-A-1	$150 \times 1.2 \times 1200$	17671	170	570	4.1
RS-B-1	$150 \times 1.6 \times 1200$	17671	170	762	5.2

Notation: D: outside diameter of the circular steel tube; δ: wall thickness of the steel tube; L: length of the specimens; Ac: concrete cross-sectional area; Ar: reinforced cross-sectional area; As: steel cross-sectional area; α_{sc}: steel ratio; PL: plain concrete; RS: concrete-filled steel tube column with reinforcing bars; NS: concrete-filled steel tube column without reinforcing bars; and RN: concrete column with reinforcing bars. A: thickness of steel tube is 1.2 mm; B: thickness of steel tube is 1.6 mm; 1: diameter is 150 mm; 2: diameter is 165 mm.

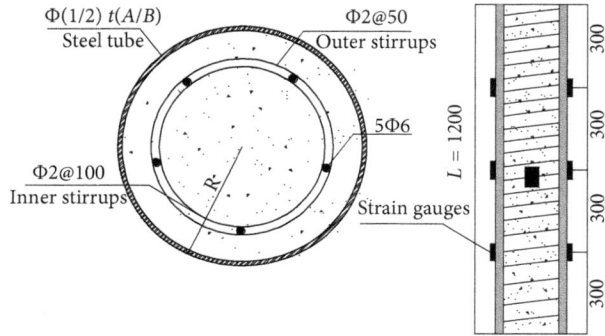

FIGURE 1: Schematic of the specimens and arrangement of the strain gauges around the columns.

TABLE 2: Concrete mix ratio.

Grade	Concrete mix ratio	Cement (kg/m^3)	Water (kg/m^3)	Stone (kg/m^3)	Sand (kg/m)
C30	0.65	300	195	1220	685

TABLE 3: Mechanical properties of the concrete and steel tube.

Specimens	f_{cu} (MPa)	f_y (MPa)
PL-2	32.1	—
NS-A-2	34.6	375
NS-B-2	34.9	375
RN-2	31.2	375
RS-A-2	33.7	375
RS-B-2	33.5	375
PL-1	34.8	—
NS-A-1	32.9	375
RS-A-1	33.4	375
RS-B-1	34.1	375

TABLE 4: Material properties of the concrete and steel tube.

Types	Grade	E (MPa)	N
Concrete	C30	$2.75E+04$	0.2
Reinforcement	HRB335	$2.10E+05$	0.3
Steel	HRB335	$2.10E+05$	0.3
Stirrups	HRB335	$2.10E+05$	0.3

concrete-filled steel tube columns without foundation; the cross sections of these columns are also circular, and all the columns were filled with C30 concrete. The specimens were tested, and a summary of their properties is presented in Table 5. The length of each column specimen is 800 mm, the diameter of each is 100 mm, and the size of foundation is 2.1 m × 1.3 m × 1 m.

2.2.2. Properties of the Materials. The concrete for the columns was produced with high-strength M30 cement mortar, with a cement : quartz : sand : water mix ratio of 1 : 1.76 : 0.32 (mass ratio). The diameter of the steel tube is 100 mm, its thickness is 1 mm, and its elastic modulus is approximately $E = 200$ GPa. Additionally, the main bar of secondary reinforcement has a diameter of 6 mm, and the stirrups have a diameter of 1 mm. The stirrups spacing is 3 cm in the column.

Due to the large size of the test foundations, they cannot be excavated to the laboratory, so concrete was used to simulate the rock foundation. Mudstone is susceptible to weathering, which leads to a reduction in the bearing capacity of the pile; therefore, the concrete strength is low to simulate strong weathering mudstones. The concrete mix ratio is 12.5% cementitious material, 12.5% stone, and 50% sand. After 28 days of curing, testing results showed that the compressive strength of the standard concrete was 300 KPa, and the elastic modulus was 16 MPa.

The pile was formed as shown in Figure 3. In addition, the material properties of the concrete core and steel tube are shown in Table 6.

2.2.3. Test Setup. The loading system uses a split hydraulic jack with a limit load of 200 kN. When the loading device is installed, the load sensor can be placed on a piece of steel plate on top of the pile, and the loading apparatus is a separate hydraulic jack that is placed on top. The top of the

concrete, and the strain of the concrete was measured with the change in the load. In these tests, each strain gauge was divided into vertical and horizontal measuring points; therefore, 12 strain gauges, a total of 24 points, were placed on the outside of each specimen. Additionally, four displacement gauges are arranged symmetrically on the pile top and the pile bottom to measure the change in the compression of the pile with the load during the loading process; these results were used as the basis of the loading control method.

2.2. Columns with Foundation

2.2.1. Test Program. The test program for the columns with foundation is approximately consistent with that of the

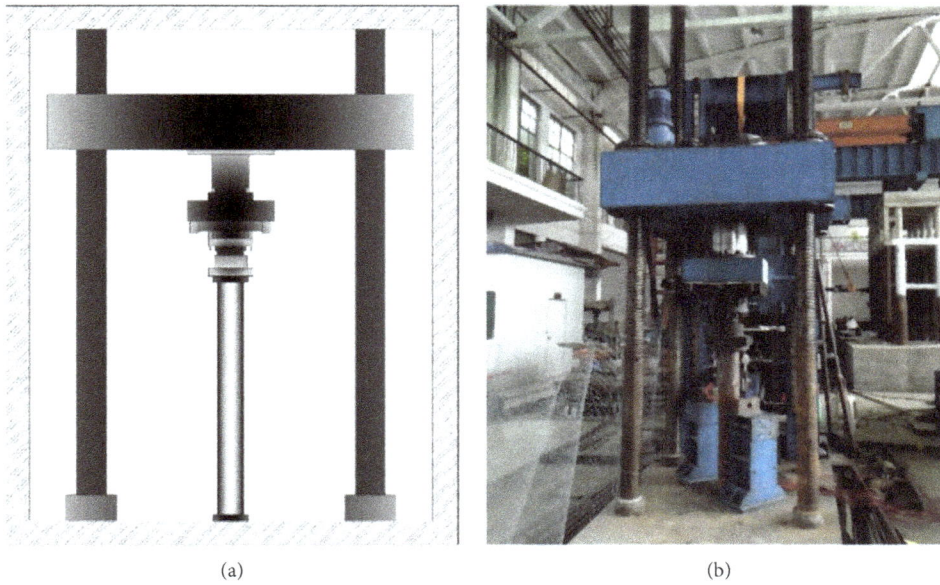

FIGURE 2: Test setup. (a) General view. (b) CFST detail.

TABLE 5: List of columns for CFST columns with foundation.

Specimens	D (mm)	h (mm)	n ($n = h/D$)	Steel tube (mm)
Z1P1	100	300	3	1
Z1P2	100	300	3	0
Z2P1	100	400	4	1
Z2P2	100	400	4	0
Z3P1	100	500	5	1
Z3P2	100	500	5	0

Notation: Z1: rock-socketed depth is 300 mm; Z2: rock-socketed depth is 400 mm; Z3: rock-socketed depth is 500 mm; P1: steel tube is 1 mm; P2: steel tube is 0 mm; h: rock-socketed depth; n: ratio of rock-socketed depth, $n = h/D$.

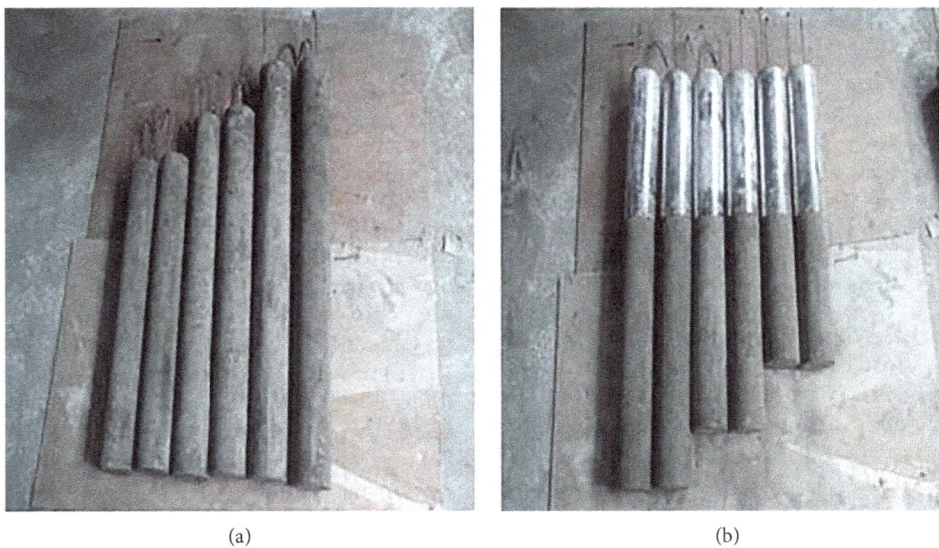

FIGURE 3: Properties of the materials. (a) General prefabricated piles. (b) Steel tube prefabricated piles.

TABLE 6: Material properties of the concrete core and steel tube.

Steel tube	Concrete	Reinforced bar	Steel tube (mm)
$D = 100$ mm	M30	$D = 6$ mm	$E_c = 16$ MPa
$t = 1$ mm	—	—	$f_y = 300$ KPa
$E_s = 200$ GPa	$C : S : W = 1 : 1.76 : 0.32$	—	—

jack is fixed to the reaction beam, and the device diagram is shown in Figure 4. During step-by-step loading, to keep each load the same, the load is calculated according to 10% of the estimated carrying capacity. The first stage load is 2 times the grading load; the unloading is reduced by a consistent 20% of the carrying capacity. The requirements are that the transmission load is uniform, continuous, and stable and that the time of loading is at least 1 minute.

3. Test Results and Discussion

3.1. Columns without Foundation

3.1.1. Failure Modes. The typical failure modes of the specimens in this test are as follows. The failure modes of the plain concrete specimen (PL-2) and the reinforced concrete specimen (RN-2) are shown in Figures 5(a) and 5(b). The main failure mode is the fracture of the top concrete, and the crushing range is about one quarter of the length of the pile.

The typical failure mode of the unreinforced CFST specimens (NS-A-2 and NS-B-2) is shown in Figure 5(c), which is mainly represented by the local buckling of the specimen after concrete crushing.

The reinforced CFST piles exhibit two different failure modes. The specimens with the thickness of 1.6 mm of steel casing (RS-B-1 and RS-B-2) locally exhibit buckling deformation during loading, the local deformation is gradually increased with the loading, and the specimen is bent and deformed until the bearing capacity is finally lost, as shown in Figure 6(a). The specimens with the thickness of 1.2 mm (RS-A-1 and RS-A-2) showed the same deformation characteristics as the specimen RS-B-1 in the early stage of loading, but when the load reaches 850 kN, the weld of the steel casing suddenly opened. The specimen is destroyed instantaneously, as shown in Figure 6(b). In summary, two different types of damage may occur in reinforced CFST when axially compressed:

(1) When the weld strength of the steel casing is greater than the axial load limit of the composite member, the failure mode is the bending failure caused by the local buckling deformation, which is a ductile failure feature

(2) When the weld strength of the steel casing is less than the axial load limit of the composite member, the failure mode is the open splitting of the steel casing weld under the expansion stress of the concrete, which is a brittle failure feature

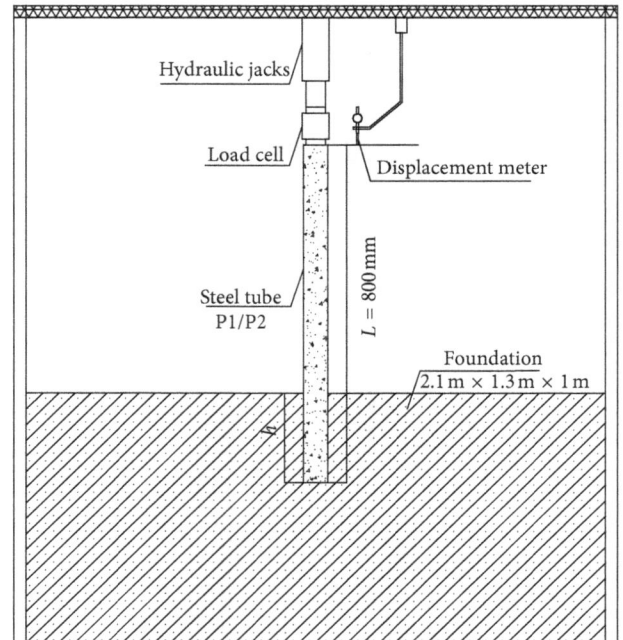

FIGURE 4: Test setup (*h* means rock-socketed depth; P1: steel tube is 1 mm; P2: steel tube is 0 mm).

3.1.2. Load-Displacement Curve

(1) Reinforced and Unreinforced CFST Members. The load-vertical displacement curves of reinforced CFST specimens with steel casing thickness of 1.2 mm (RS-A-2) and 1.6 mm (RS-B-2) are shown in Figure 7. It can be seen from the test results that when the thickness of the steel casing increases to 1.6 mm, the displacement of the top of the specimen with the load is the same as that of the thickness of 1.2 mm, and both exhibit obvious plastic failure properties, the proportional limit of two thicknesses is relatively close, which is approximately 600 kN. Before the load reaches 600 kN, the load-displacement curves of the two test pieces are basically coincident. When the load exceeds 600 kN, that is, after entering the plastic zone, the absolute value of the displacement increment under the same load increment is smaller than the case where the thickness of the steel casing is 1.2 mm under the same condition. It indicates that the restraining effect of the steel casing on the core concrete is mainly reflected in the plastic zone of the component, while the effect on the elastic zone is small. The maximum displacement that can be achieved with a thickness of 1.6 mm is 13.8 mm, which is greatly improved compared to the thickness of the steel casing of 1.2 mm. Therefore, the thickness of the steel casing has a great influence on the ductility of the reinforcing member and the improvement of its bearing capacity.

The load-vertical displacement curves of CFST specimens with steel tube thickness is 1.2 mm (RS-A-2 and NS-A-2) and *t* steel tube thickness is 1.6 mm (RS-B-2 and NS-B-2) composite members are shown in Figures 8 and 9.

(a)   (b)

(c)

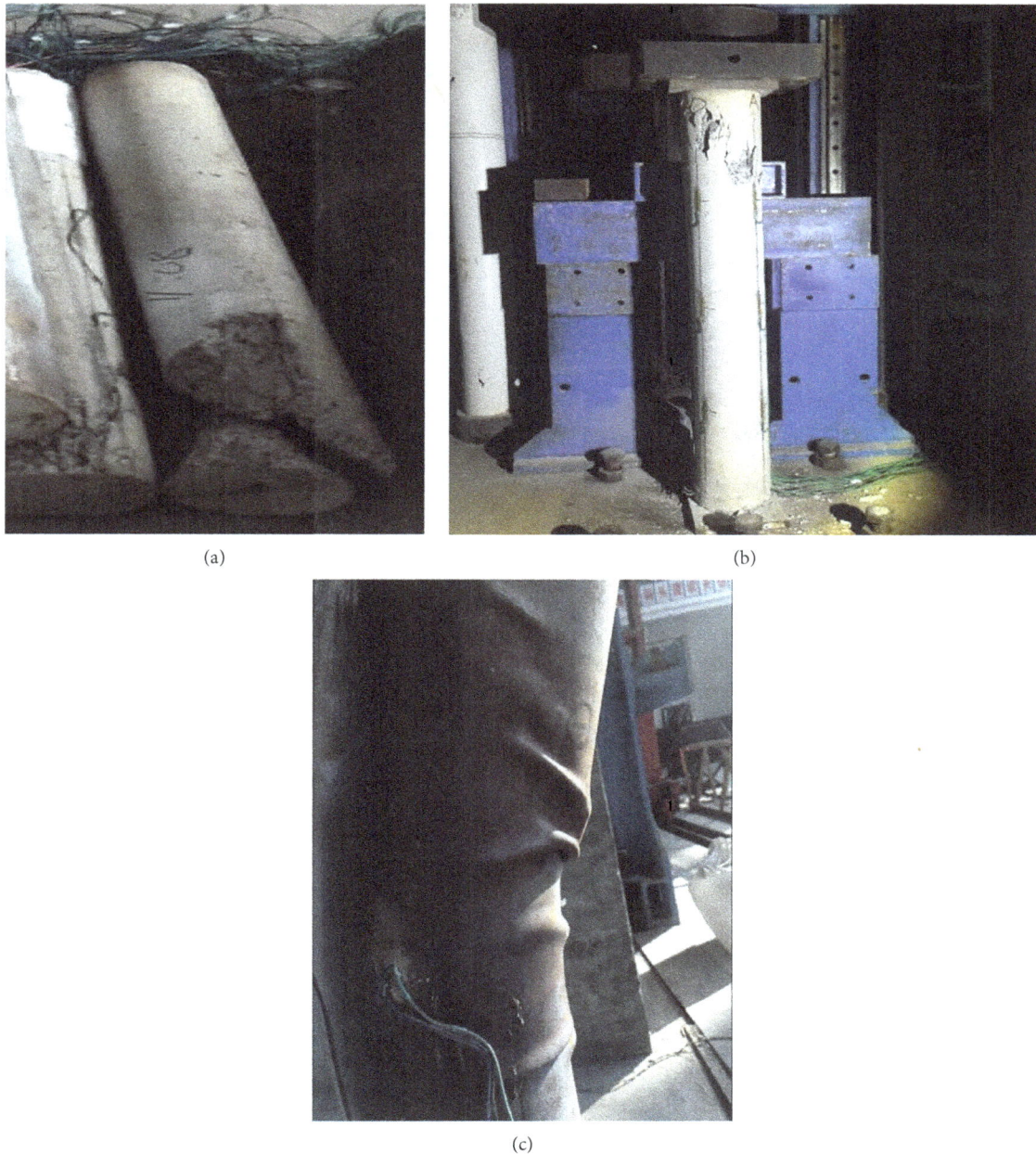

FIGURE 5: Failure modes. (a) Plain concrete pile. (b) Reinforced concrete pile. (c) Unreinforced CFST pile.

It can be obtained from the test results that the proportional limit of the load-vertical displacement curve of reinforced CFST piles is slightly higher than unreinforced ones and the proportional limit of the specimen RS-A-2 and RS-B-2 is about 600 kN, greater than 500 kN of test NS-A-2 and NS-B-2. At the same time, the ultimate deformation capacity of reinforced CFST composite members is better than that of unreinforced CFST specimens. It can be seen that in the elastic stress stage, the presence of the stressed steel bars delays the generation of concrete cracks, thereby increasing the proportional limit of the test. In the plastic stage, the stressed steel bars play a further restraining role on the core concrete, which inhibits the formation and expansion of the crack and improves the ductility and ultimate deformation ability of the test.

(2) Encased and Nonencased Members. The load-vertical displacement curves of plain concrete specimen (PL-2) and reinforced concrete specimen (RN-2) and reinforced CFST composite members (RS-A-2 and RS-B-2) are shown in Figure 10. It can be seen from the test results that the steel tube has a great influence on the bearing capacity and deformation capacity of the test. For the plain concrete specimen (PL-2), the ultimate bearing capacity and ultimate deformation are small, and the maximum displacement is 2.87 mm, showing the characteristics of brittle failure. For the

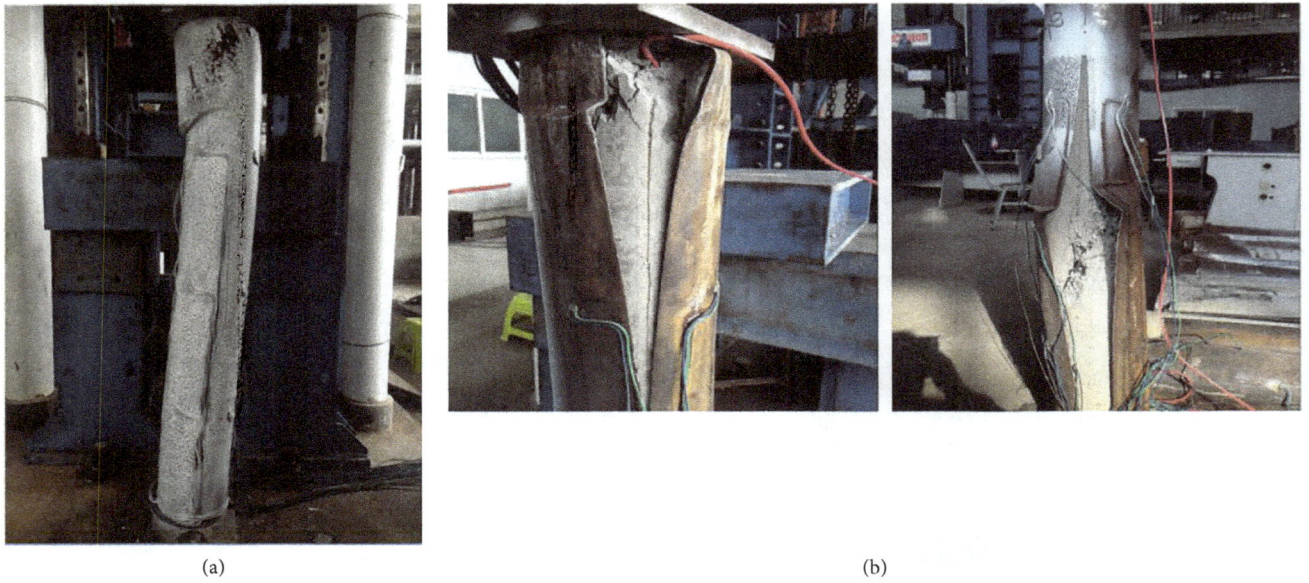

(a)   (b)

FIGURE 6: Failure modes of reinforced CFST pile. (a) Buckling deformation. (b) Splitting of the steel casing weld.

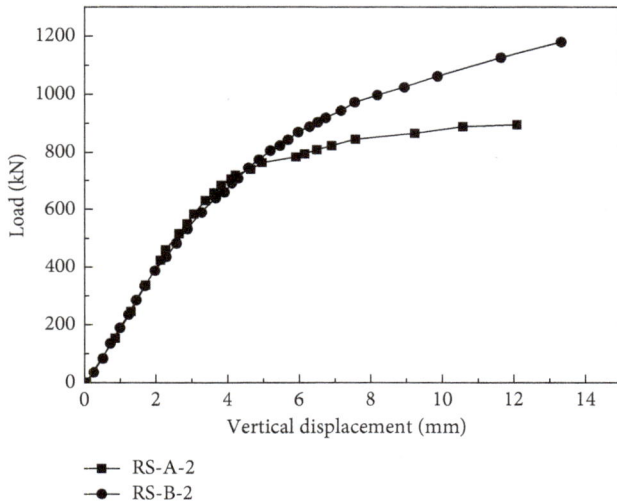

FIGURE 7: Vertical displacement-load curves of two thickness steel tube-reinforced CFST piles.

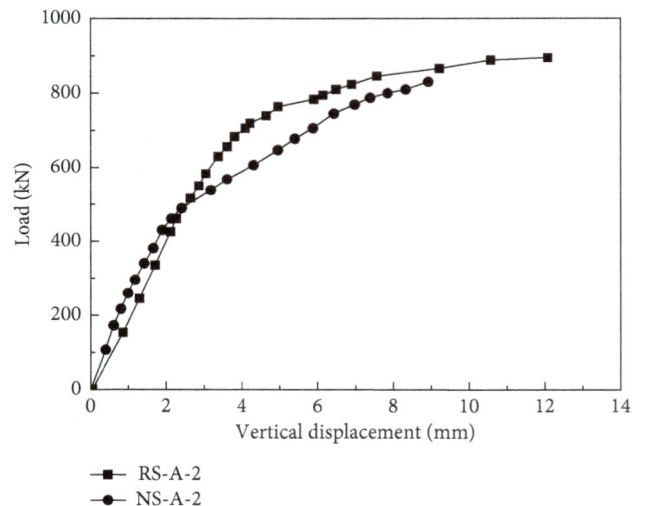

FIGURE 8: Comparison of reinforced and unreinforced CFST piles with thickness is 1.2 mm.

reinforced concrete specimen (RN-2), the load-displacement curve shows a certain ductility characteristic, but the ultimate bearing capacity and ultimate deformation of the specimen are significantly smaller than the reinforced CFST composite member.

3.1.3. Strain Development. In this paper, the strain law of the steel casing and concrete core under the condition of an axial compression load is described for the specimens with a steel tube thickness of 1.2 mm (specimen RS-A-2) and 1.6 mm (specimen RS-B-2). During the test, the vertical strain and hoop strain on the side of the steel casing were monitored and the development rule of the compressive strain inside

the core concrete was monitored. Among them, four strain gauges are arranged on the side of the steel casing at the top and middle and top of the test specimen in the circumferential direction, and the vertical strain and lateral strain of the steel casing at this position are measured, respectively. In the data processing, the strain values of four measuring points in the same plane are averaged to obtain the average strain of the steel casing at the plane. The load-strain curve and average stress-strain curve of the test piece are given below. Among them, the average stress is obtained by dividing the axial load on the specimen by the cross-sectional area.

The developments of the longitudinal strain, circumferential strain, and compression behavior of the steel tube are shown in Figures 11–16. Figures 11–13 are laws of strain

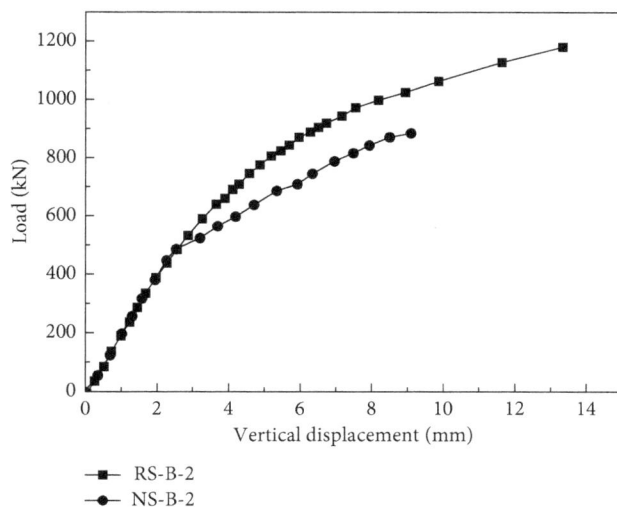

FIGURE 9: Comparison of reinforced and unreinforced CFST piles with thickness is 1.6 mm.

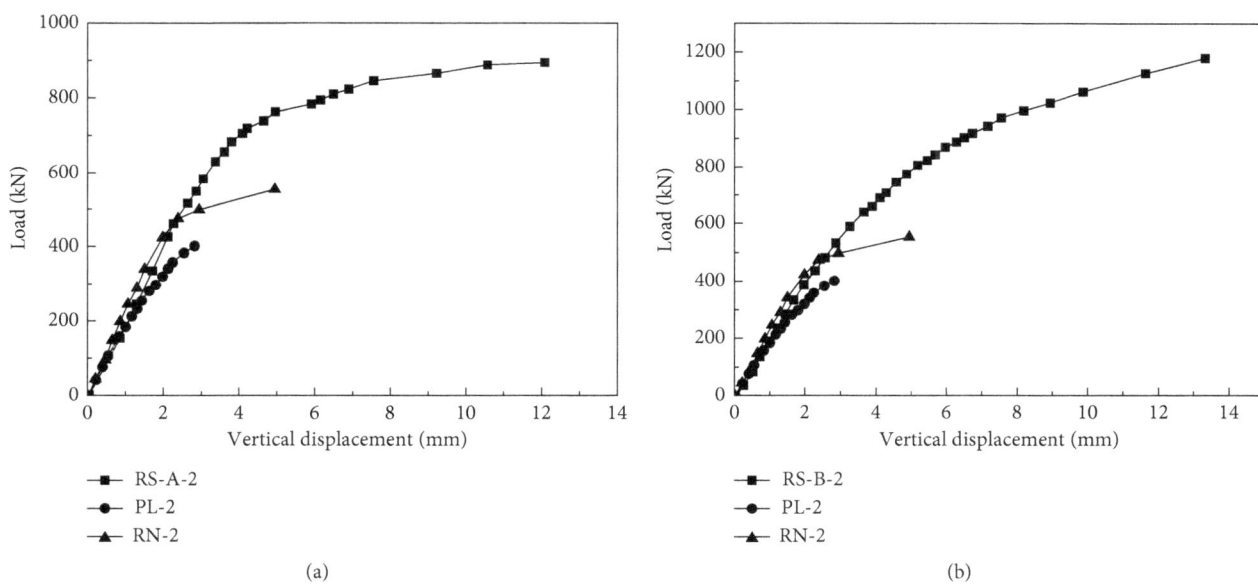

(a)

(b)

FIGURE 10: Comparison of plain concrete, reinforced concrete, and reinforced CFST piles.

development of specimen RS-A-2, and Figures 14–16 are laws of strain development of specimen RS-B-2.

From the above test results, it can be seen that for the steel casing-reinforced concrete composite member, the strain value in the middle of the specimen is slightly larger than the top and bottom strain values. The strain development of steel casings at each measuring point shows elastic-plastic characteristics with the increase of load. Before the axial load value is less than 0.45 times, the ultimate load of the model pile foundation, the load, and strain show a linear relationship, while the load reaches 0.45 times the ultimate load carrying capacity. After the capacity, under the same load increment, the strain increment gradually increases, showing a nonlinear characteristic. From the whole load-strain curves of the reinforced CFST specimens, it exhibited the characteristics of ductile failure.

In summary, the use of a steel tube plays a very important role in improving the compressive deformation and mechanical properties of the components. For a joint steel tube-reinforced concrete structure, the bearing capacity and the ductility of the components are greatly improved compared to the specimens with no steel, which is due to the existence of the internal structure of the composite structure, effectively enhancing the constraints between the concrete particles to limit the development of microcracks and the effective stress.

3.2. Columns with Foundation. It can be seen from Figure 17 that the variation in the results of the different model columns under the load at the pile top is consistent, and it can be considered that the influence of a rockless cylinder on the

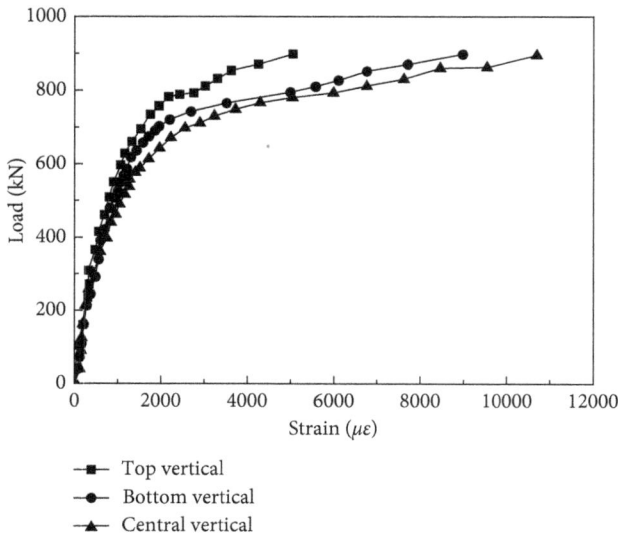

(a)

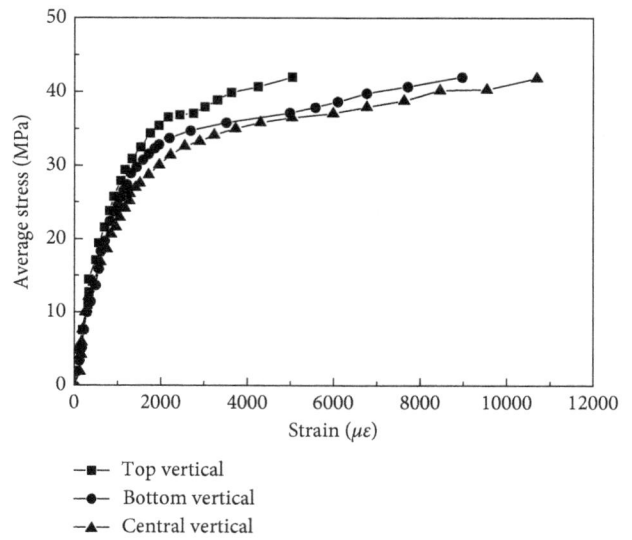

(b)

FIGURE 11: Vertical strain development laws of specimen RS-A-2 with steel casing. (a) Relationship of vertical strain development with load. (b) Relationship of vertical strain development with average stress.

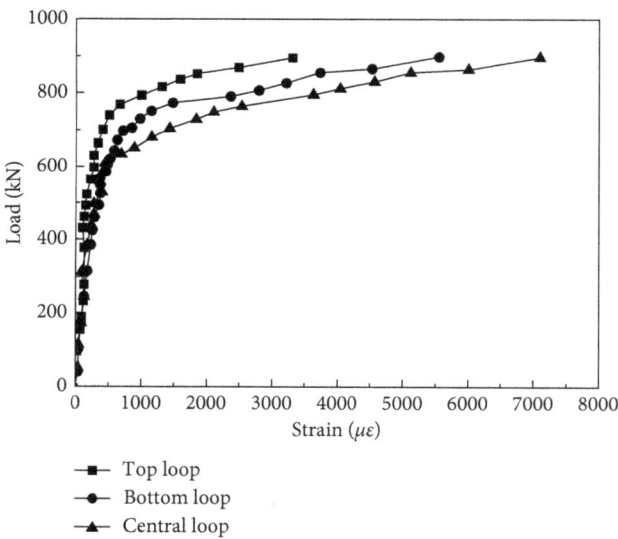

(a)

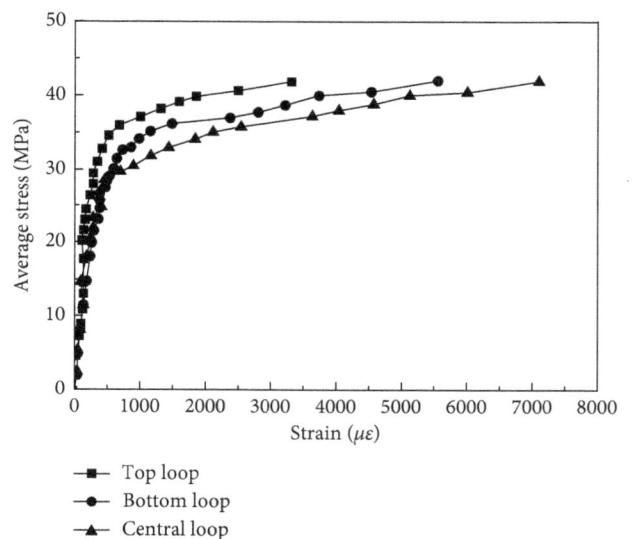

(b)

FIGURE 12: Loop strain development of specimen RS-A-2 with steel casing. (a) Relationship of loop strain development with load. (b) Relationship of loop strain development with average stress.

settlement of a rock-socketed column is also consistent. It can be seen from Figure 17 that the displacements of Z1P1 and Z1P2 are basically the same from the load values of zero to 9 kN, the load-vertical displacement curve is approximately a straight line, and the two piles are in the elastic stage. After the load exceeds 9 kN, the displacement rate of the Z1P1 pile exceeds that of the Z1P2 pile and the settlement of the pile top increases. The load-vertical displacement curves of the two piles change slowly, and the process of steep drop is not obvious. Figure 17 shows several settlings that may be due to the inhomogeneity of the model

foundation. According to the above analysis, it can be seen that the vertical bearing capacities of CFST rock-socketed piles are weaker than those rock-socketed piles without a steel tube; the difference is mainly because the rock-socketed pile without a steel tube has a concrete-rock interface with a coefficient of friction greater than that of the CFST rock-socketed pile with a steel-rock interface.

As seen from Figure 18, the greater the rock-socketed depth, the greater the load-bearing capacity of the rock-socketed piles. In addition, when the depth of the pile in the rock-socketed increased from 3D to 5D, the vertical ultimate

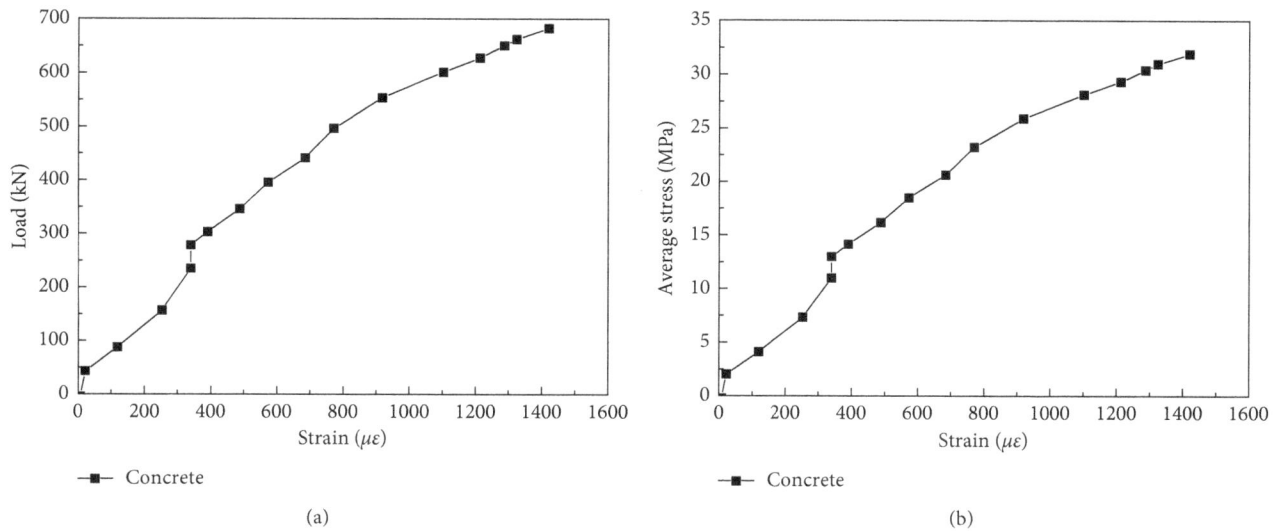

FIGURE 13: Axial compressive strain development of core concrete of specimen RS-A-2. (a) Relationship of vertical strain development with load. (b) Relationship of vertical strain development with average stress.

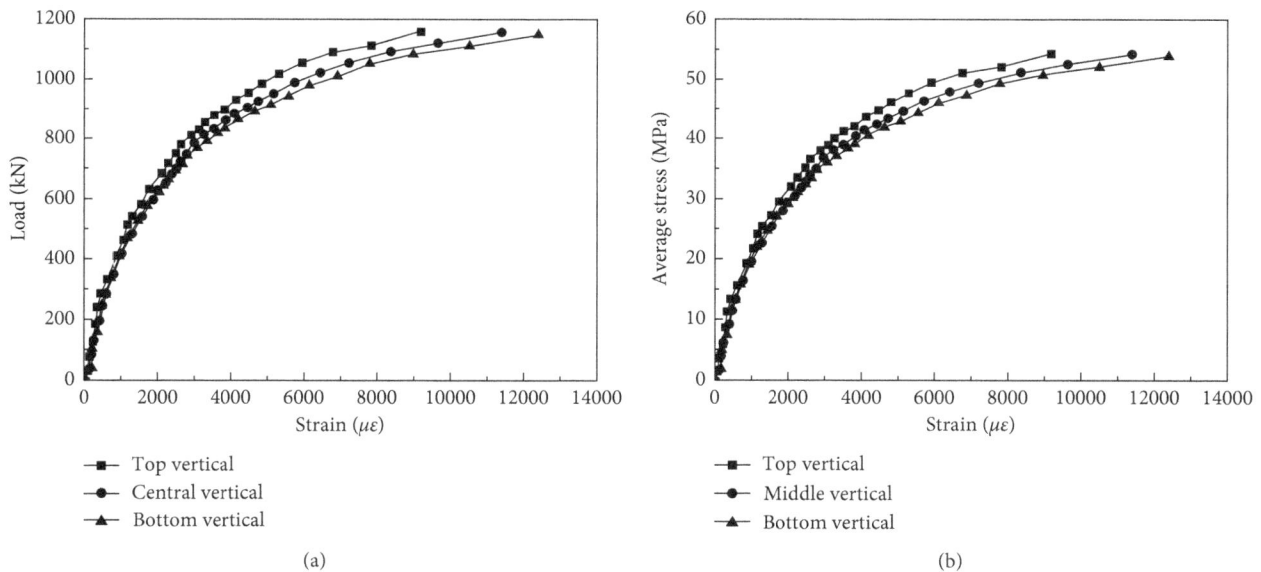

FIGURE 14: Vertical strain development laws of specimen RS-B-2 with steel casing. (a) Relationship of loop strain development with load. (b) Relationship of loop strain development with average stress.

bearing capacity increased by approximately 40%. The main reason for this effect is that with the increase in the rock-socketed depth, the contact area will increase, and although the side friction coefficient does not change, the total friction of the pile will significantly increase; therefore, the vertical bearing capacity of the pile is significantly improved. It is found that the vertical ultimate bearing capacity of rock-socketed piles without a steel tube (steel-free cylinder) is approximately 10% higher than that of the steel-retaining tube when the depth of rock-embedment is the same. The main reason for this phenomenon is that the outer wall of the pile is very smooth; since its roughness is much smaller than the concrete, the corresponding side friction is lower,

and the ultimate bearing capacity of the pile is equal to the pile side friction and pile end resistance, decreasing the ultimate bearing capacity of the pile.

4. Analysis of the Ultimate Load

According to EC4 [22], the axial load capacity ($N_{\mathrm{pl,Rd}}$) of a CFST column can be determined by summing the yield load of both the steel section and concrete core. For concrete-fined tubes of circular cross section, account may be taken of increase in strength of concrete caused by confinement provided that the relative slenderness $\bar{\lambda}$ does not exceed 0, 5, and $e/d < 0$, 1, where e is the eccentricity of

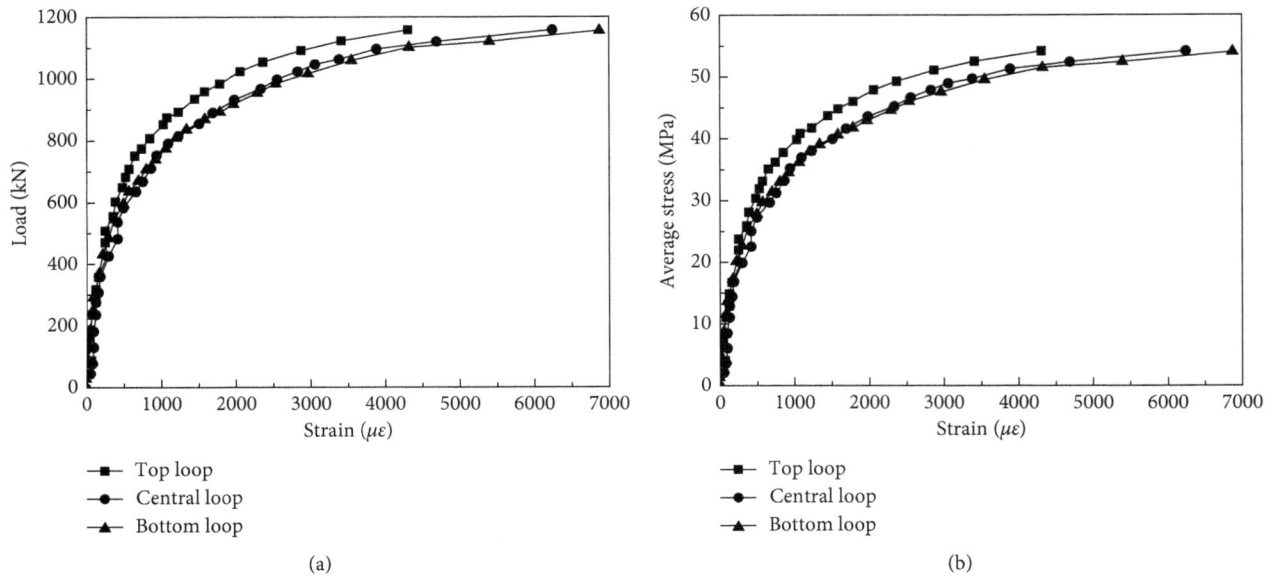

FIGURE 15: Loop strain development of specimen RS-B-2 with steel casing. (a) Relationship of strain development with load. (b) Relationship of strain development with average stress.

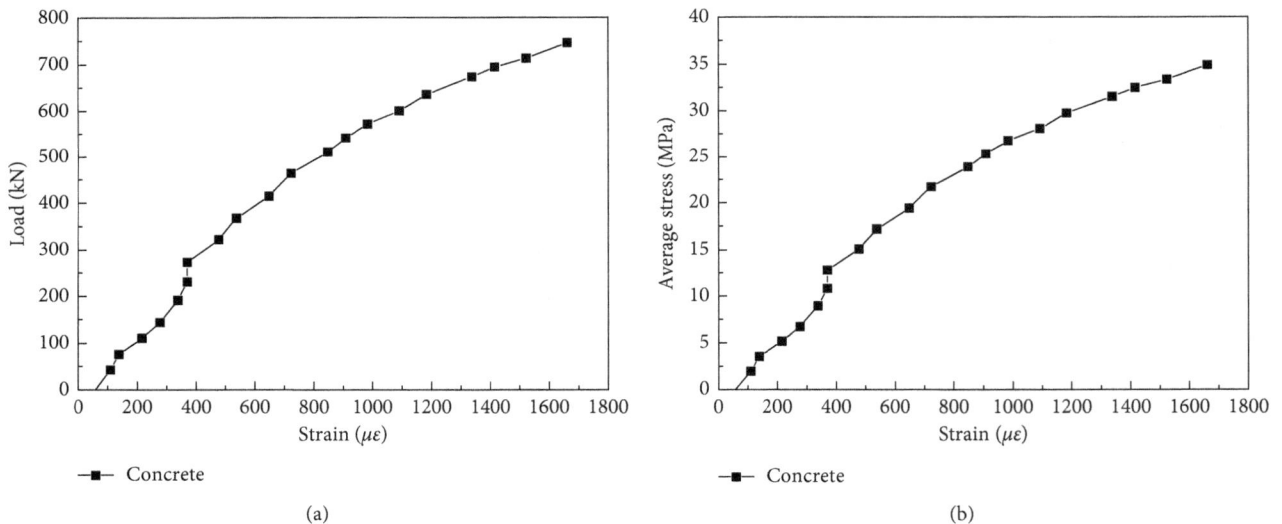

FIGURE 16: Axial compressive strain development of core concrete of specimen RS-B-2.

loading given by M_{Ed}/N_{Ed} and d is the external diameter of the column. The plastic resistance to compression may then be calculated from the following expression in equations (2)–(5). The application of EC4 [22] is restricted to composite columns with steel yield stress and concrete cylinder strength of less than 355 and 50 MPa, respectively. The following presents the formulations of the AISC [21], EC4 [22], NBR 8800 [23, 24], and GB50396-2014 [25] standard codes to determine the resistance capacity of axially loaded CFST columns. And r_s is the radius of gyration of the steel hollow section, l_e is the length of the column, E_c is the modulus of elasticity of the concrete, and E_s is the modulus of elasticity of the steel. The formulations presented were used to calculate the compressive strength of the axially loaded CFST columns investigated in the present study. In addition, the boundary conditions of the testing approach are upper if the pile is loading and lower if the pile is fixed.

In GB50396-2014 [25], θ means the hoop coefficient of concrete-filled steel tubular members $\theta = \alpha_{sc} f/f_c$, α_{sc} represents the steel content of concrete-filled steel tubular members, in the design of compressive strength of concrete-filled steel tube piles, B and C are the influence factors of the cross-sectional shape on the hoop effect, and about circular cross section, $B = 0.176 * f/213 + 0.974$, $C = -0.104f_c/14.4 + 0.031$. In AISC [21], the axial compression member is considered to be the overall stability of the member, the

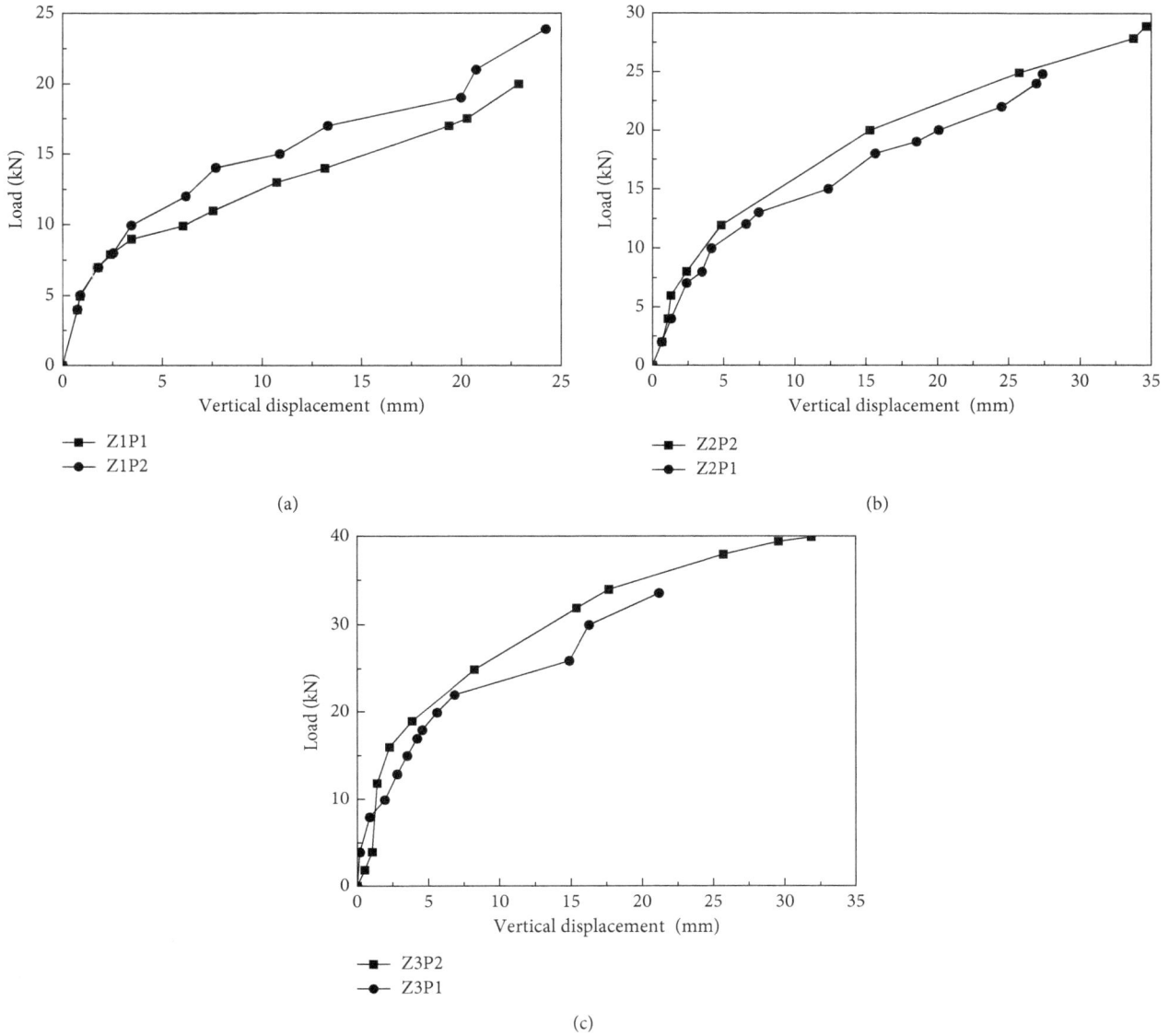

FIGURE 17: Relationship of vertical displacement and load of three depths of rock-socketed columns. (a) Rock-socketed depth is 300 mm (P1: steel tube is 1 mm; P2: steel tube is 0 mm). (b) Rock-socketed depth is 400 mm. (c) Rock-socketed depth is 500 mm.

strength of the concrete is converted into the steel, the nominal compressive strength f_{cr} of the steel is obtained, and the bearing capacity of the concrete-filled steel tubular member is calculated from f_{cr}, f_{ys} is the yield limit of steel, f_{ck} means the compressive strength of concrete cylinder, A_s is the cross-sectional area of steel pipe, A_c is the cross-sectional area of concrete, and λ_c is the ratio of slenderness to slenderness. In Eurocode 4 (EC4) [22], the relative slenderness $\bar{\lambda}$ for the plane of bending being considered is given by $\bar{\lambda} = \sqrt{N_{PL,Rk}/N_{cr}}$, where $N_{pl,Rk}$ is the characteristic value of the plastic resistance to compression if, instead of the design strengths, the characteristic values are used and N_{cr} is the elastic critical normal force for the relevant buckling mode, calculated with the effective flexural stiffness $(EI)_{cff}$.

However, it should be stated that the maximum allowable value of diameter-thickness ratio (D/t) specified in AISC [21] is $0.31 * E/f_y$; the ratio of diameter-thickness of circular-section concrete-filled steel tubular members specified in Chinese specification [28] should be limited to the following range: $20\sqrt{235/f_y} \leq D/t \leq 85\sqrt{235/f_y}$; and Eurocode 4 (EC4) [22] explicit limits the allowed D/t values in the European context to $90 * 235/f_y$. The diameter-thickness ratio of the specimens studied in this paper is very large, exceeding the allowable range of the specifications. The paper attempts to estimate these specimens with large diameter-thickness ratio by using the above codes. And the ultimate strength calculation results of relevant specifications are shown in Table 7. Apparently, it can be found that value of the ultimate load is generally much greater than the theoretical calculation results.

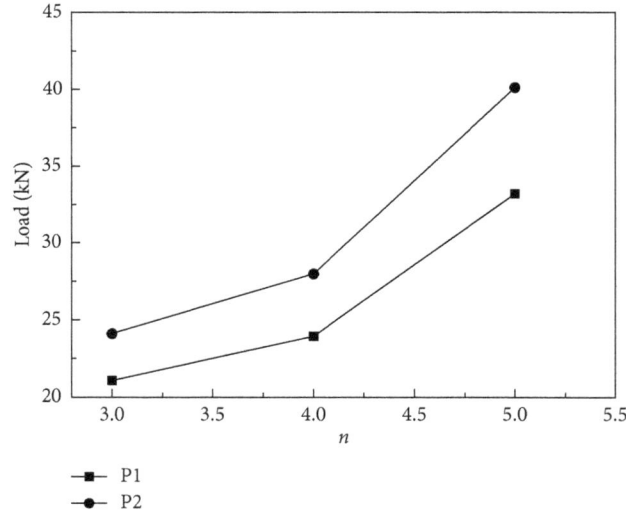

FIGURE 18: Relationship between n and load (n: ratio of rock-socketed depth, $n = h/D$).

TABLE 7: The ultimate strength calculation results of relevant specifications.

Specimens	Ultimate load (kN)	NBR 8800 and EC4 (kN)	AISC (kN)	GB50396-2014 (kN)
PL-2	406	412	419	429
NS-A-2	830	493	479	560
NS-B-2	885	566	535	603
RN-2	560	430	416	465
RS-A-2	895	493	479	596
RS-B-2	1190	566	535	639
PL-1	480	513	519	530
NS-A-1	800	428	409	478
RS-A-1	895	428	409	478
RS-B-1	1125	452	474	498

AISC [21] method:

$$
\begin{aligned}
E_m &= E_s + 0.4E_c\frac{A_c}{A_s}, \\
f_{my} &= f_{ys} + 0.85f_{ck}\frac{A_c}{A_s}, \\
f_{cr} &= \left(0.658^{\lambda_c^2}\right)f_{my} \quad \text{if } \lambda_c \leq 1.5, \\
f_{cr} &= \frac{0.877}{\lambda_c^2}f_{my} \quad \text{if } \lambda_c \leq 1.5, \\
\lambda_c &= \frac{l_c}{r_c\pi}\left(\frac{f_{my}}{E_m}\right)^{0.5}, \\
N_{pl.Rd} &= 0.85A_s f_{cr}.
\end{aligned}
\tag{1}
$$

EC4 and NBR 8800 [23, 24] methods:

$$
N_{pl.Rd} = \eta_2 A_s f_y + A_c f_{ck}\left(1 + \eta_1\frac{t f_y}{D f_{ck}}\right),
\tag{2}
$$

$$
\begin{aligned}
\eta_1 &= \eta_{10}, \\
\eta_2 &= \eta_{20},
\end{aligned}
\tag{3}
$$

$$
\eta_{10} = 4.9 - 18.5 \times \bar{\lambda} + 17\left(\bar{\lambda}\right)^2 \geq 0,
\tag{4}
$$

$$
\eta_{20} = 0.25\left(3 + 2\bar{\lambda}\right) \leq 1.0.
\tag{5}
$$

GB50396-2014 [25] method:

$$
\begin{aligned}
N_0 &= A_{sc} f_{sc}, \\
f_{sc} &= \left(1.212 + B\theta + C\theta^2\right)f_c, \\
\alpha_{sc} &= \frac{A_s}{A_c}, \\
\theta &= \alpha_{sc}\frac{f}{f_c}.
\end{aligned}
\tag{6}
$$

5. Conclusion

The present study is an attempt to study the behavior of concrete-filled steel tube columns subjected to axial compression. Based on the results of this study, the following conclusions can be drawn within the scope of these tests:

(1) The reinforced CFST piles exhibit two different failure modes. When the weld strength of the steel casing is greater than the axial load limit of the composite member, the failure mode is the bending failure caused by the local buckling deformation, which is a ductile failure feature. When the weld strength of the steel casing is less than the axial load limit of the composite member, the failure mode is the open splitting of the steel casing weld under the expansion stress of the concrete, which is a brittle failure feature.

(2) The bearing capacity of rock-socketed CFST columns is lower than that of rock-socketed columns without a steel tube under the same conditions of a vertical load. The depth of the rock inlay is clearly affecting by the bearing capacity of the two rock-socketed columns, and the greater the rock-socketed depth, the greater the bearing capacity of the rock-socketed piles. In addition, the overall sliding of the steel tube and the core concrete of pile is small, and the steel tube-concrete interface is basically in a bonding state under the vertical load.

(3) A comparison of failure loads between the tests and the design codes was presented. The results show that the provisions in EC4 [22], NBR 8800 [23, 24], AISC [21], and GB50396-2014 [25]conservatively estimate the ultimate capacities of the specimens, and it can be found that value of the ultimate load is generally much greater than the theoretical calculation results. Since the diameter-thickness ratio of the specimens studied in this paper is very large, exceeding the allowable range of the specifications, the test results provide a reference for design and performance of the new structure with large diameter-thickness ratio.

Conflicts of Interest

The authors declare that there are no conflicts of interest regarding the publication of this paper.

Acknowledgments

This study was financially supported by the National Natural Science Foundation of China (nos. 2011550031 and 51709026) and the Western Project of China (nos. 2011660041 and 2011560076).

References

[1] S. Jegadesh and S. Jayalekshmi, "Load-bearing capacity of axially loaded circular concrete-filled steel tubular columns," *Proceedings of the Institution of Civil Engineers: Structures and Buildings*, vol. 169, no. 7, pp. 508–523, 2016.

[2] N. E. Shanmugam and B. Lakshmi, "State of the art report on steel–concrete composite columns," *Journal of Constructional Research*, vol. 57, no. 10, pp. 1041–1080, 2001.

[3] S. De Nardin, "Theoretical-experimental study high strength concrete-filled steel tubes," M.S. thesis, EESC-USP, Sao Carlos-SP, 1999, in Portuguese.

[4] K. Cederwall, B. Engstrom, and M. Grauers, "High-strength concrete used in composite columns," in *Proceedings of High-Strength Concrete–2nd International Symposium*, pp. 195–214, Detroit, MI, USA, 1990.

[5] L. Han, W. Li, and R. Bjorhovde, "Developments and advanced applications of concrete-filled steel tubular (CFST) structures: members," *Journal of Constructional Steel Research*, vol. 100, pp. 211–228, 2014.

[6] L. Han, H. Shan-Hu, and F. Liao, "Performance and calculations of concrete filled steel tubes (CFST) under axial tension," *Journal of Constructional Steel Research*, vol. 67, no. 11, pp. 1699–1709, 2011.

[7] M. A. Bradford, "Design strength of slender concrete filled rectangular steel tubes," *ACI Structural Journal*, vol. 93, no. 2, pp. 229–235, 1996.

[8] M. D. O'Shea and R. Q. Bridge, "Design of thin-walled concrete filled steel tubes," *Journal of Structural Engineering*, vol. 126, no. 11, pp. 1295–1303, 2000.

[9] B. Uy, "Local and postlocal buckling of fabricated steel and composite cross sections," *Journal of Structural Engineering*, vol. 127, no. 6, pp. 666–677, 2001.

[10] K. K. Choi and Y. Xiao, "Analytical studies of concrete-filled circular steel tubes under axial compression," *Journal of Structural Engineering*, vol. 136, no. 5, pp. 565–573, 2010.

[11] Z. Ou, B. Chen, K. H. Hsieh, M. W. Halling, and P. J. Barr, "Experimental and analytical investigation of concrete filled steel tubular columns," *Journal of Structural Engineering*, vol. 137, no. 6, pp. 635–645, 2011.

[12] T. Perea, R. T. Leon, J. F. Hajjar, and M. D. Denavit, "Full-scale tests of slender concrete-filled tubes: axial behavior," *Journal of Structural Engineering*, vol. 139, no. 7, pp. 1249–1262, 2013.

[13] L. H. Han, "Flexural behavior of concrete-filled steel tubes," *Journal of Constructional Steel Research*, vol. 60, no. 2, pp. 313–337, 2004.

[14] L. H. Han, "Further study on the flexural behavior of concrete-filled steel tubes," *Journal of Constructional Steel Research*, vol. 62, no. 6, pp. 554–565, 2006.

[15] H. Lu, L. H. Han, and X. L. Zhao, "Analytical behavior of circular concrete-filled thin-walled steel tubes subjected to bending," *Thin-Walled Structures*, vol. 47, no. 3, pp. 346–358, 2009.

[16] C. W. Roeder, D. E. Lehman, and E. Bishop, "Strength and stiffness of circular concrete-filled tubes," *Journal of Structural Engineering*, vol. 136, no. 12, pp. 1545–1553, 2010.

[17] J. Moon, C. W. Roeder, D. E. Lehman, and H. Lee, "Analytical modeling of bending of circular concrete-filled steel tubes," *Engineering Structures*, vol. 42, pp. 349–361, 2012.

[18] L. H. Han, G. H. Yao, and Z. Tao, "Performance of concrete-filled thin-walled steel tubes under pure torsion," *Thin-Walled Structures*, vol. 45, no. 1, pp. 24–36, 2007.

[19] J. Nie, Y. Wang, and J. Fan, "Experimental study on seismic behavior of concrete filled steel tube columns under pure torsion and compression–torsion cyclic load," *Journal of Constructional Steel Research*, vol. 79, no. 12, pp. 115–126, 2013.

[20] Y. Wang, J. Nie, and J. Fan, "Theoretical model and investigation of concrete filled steel tube columns under axial force–torsion combined action," *Thin-Walled Structures*, vol. 69, no. 1, pp. 1–9, 2013.

[21] American Institute of Steel Construction, *AISC–LRFD: Metric Load and Resistance Factor Design Specification for Structural Steel Buildings*, AISC, Chicago, IL, USA, 1994.

[22] Europe en Comite de Normalisation, *ENV 1994-1-1: Euro-code 4—Design of Composite Steel and Concrete Structures, Part 1.1: General Rules and Rules for Buildings*, CEN, Brussels, Belgium, 1994.

[23] Brazilian Society of Standard Codes, *NBR 8800: Design and Constructional Details of Steel Structures of Buildings*, Brazilian Society of Standard Codes, Rio de Janeiro, Brazil, 1986.

[24] Brazilian Society of Standard Codes, *NBR 8800 Brazilian Standard Code Review: Design and Constructional Details of Steel and Composite Structures of Buildings*, BSSC, Belo Horizonte, Brazil, 2003.

[25] National Standards of the People's Republic of China, *GB 50936, Technical Code for Concrete Filled Steel Tubular Structures*, National Standards of the People's Republic of China, Beijing, China, 2014, in Chinese.

[26] Q. Liu, D. Wang, R. Huang et al., "Study on failure mode of inland river large water level dock under ship impact," *Port Engineering Technology*, vol. 47, no. 4, 2010, in Chinese.

[27] J. Xu, "Structural design of overhead vertical wharf for inland river large water level," *Water Transport Engineering*, vol. 2006, pp. 62–67, 2006, in Chinese.

[28] China Association for Engineering Construction Standardization, *CECS 28:90: Specification for Design and Construction of Concrete-Filled Steel Tubular Structures*, China Association for Engineering Construction Standardization, Beijing, China, 2012, in Chinese.

4

Effect of Recycling Agents on Rheological and Micromechanical Properties of SBS-Modified Asphalt Binders

Meng Guo [ID],[1] Yiqiu Tan,[2] Daisong Luo,[2,3] Yafei Li,[3] Asim Farooq [ID],[4] Liantong Mo,[5] and Yubo Jiao[1]

[1]The Key Laboratory of Urban Security and Disaster Engineering of Ministry of Education, Beijing University of Technology, Beijing 100124, China
[2]School of Transportation Science and Engineering, Harbin Institute of Technology, Harbin 150090, China
[3]Research & Consulting Department of Road Structure & Materials Research Center, China Academy of Transportation Sciences, Beijing 100029, China
[4]University of Science and Technology Beijing, Beijing 100083, China
[5]State Key Lab of Silicate Materials for Architectures, Wuhan University of Technology, Wuhan 430070, China

Correspondence should be addressed to Meng Guo; mguo@ustb.edu.cn

Guest Editor: Ghazi G. Al-Khateeb

Individual effect of aging and rejuvenator recycling on basic properties of asphalt is readily recognized, but there is only limited understanding about whether the recycling of SBS- (styrene-butadiene-styrene-) modified asphalt is an inverse process of aging or not. To compare the effects of aging and rejuvenator on microproperties and molecular composition of SBS-modified asphalt, comprehensive performance tests and physical-chemistry experiments were conducted. The results of infrared spectroscopy tests demonstrate that the reticular crosslinking structure of asphalt was destroyed and SBS's modification effect was gradually lost after aging. This can cause the strengthening of high-temperature performance and reduction of the low-temperature anticrack property of SBS-modified asphalt. Scanning electron microscope shows that the island structure of SBS-modified asphalt disappeared after aging. Energy spectrum analysis shows that the C (carbon) content of aged SBS-modified asphalt has decreased, while the O (oxygen) content and S (sulfur) content have increased obviously. Results of the fluorescence microscope, SEM, and rheological tests show that the epoxy functional group compounds of aliphatic glycidyl ether resin had high reactivity; the triblock molecular structure of SBS and the mechanical performance of SBS-modified asphalt were recovered.

1. Introduction

Traffic flow has been growing with the rapid development of China's highway construction, and many of the asphalt pavements built early in China have been damaged under heavy traffic pressure. Many high-grade road highways have approached or entered their maintenance period according to their design life and actual use conditions. It is estimated that, from now on, about 12% of asphalt pavement will need overhauling every year. At the same time, faced with the resourceful provinces' rising demand for transport capacity of the high-grade highway, China has been carrying out the reconstruction and widening project of the high-grade

highways in recent years. Both the overhaul and medium maintenance and reconstruction and widening of asphalt pavement produce a lot of reclaimed asphalt pavement (RAP). According to statistics, more than 35 million cubic meters of RAP are produced in overhaul and medium maintenance and widening of highways in China, and the figure grows at a rate of 15% per year and will reach 100 million cubic meters in ten years. The stacking, discarding, and degradation of such a huge amount of waste materials will aggravate the increasing serious environment pressure and resource problems. Therefore, using the RAP more efficiently is one big challenge in the field of highway construction.

Researchers have done a lot of research works, respectively, in the aspects of the physical properties decay law, component change law, and molecular structure changes. However, these were mainly done through indoor aging tests, and many factors affecting asphalt aging remain to be studied further, such as influence mechanism of water on asphalt aging, the relationship, and the difference between simulated and natural aging. Therefore, we also need to further study the asphalt aging performance and mechanism using advanced modern analytical test methods and various aging test methods to provide a theoretical basis for asphalt pavement recycling [1–7]. Zhang et al. studied the three types of aging methods on the performance of SBS- (styrene-butadiene-styrene-) modified asphalt, including thin film oven test (TFOT), pressure aging vessel (PAV), and ultraviolet (UV) radiation. They found that SBS-modified asphalt with penetration 90 had a higher retained penetration and ductility as well as the lower viscosity aging index compared to SBS-modified asphalt with penetration 70 [8].

At present, no definitive conclusion has been drawn from the mechanism of interaction between the recycling agent, aged asphalt, and new asphalt. For example, the measured data of some recycling projects show that the stiffness of recycled asphalt pavement is lower than the designed value. One explanation for this phenomenon is that the recycling agent fails to diffuse into the reclaimed asphalt. Therefore, it is of great significance for the asphalt recycling technology to study the law of recycling agent's diffusion into asphalt and recycling agent's peptization effect on asphalt and reveal the microproperties of recycling asphalt [9–15]. Xiao et al. studied the low-temperature performance characteristics of RAP mortars containing sieved RAP and soft binders at three aged states; they found that RAP mortar with a higher old binder content had a higher minimum low temperature regardless of the RAP source [16]. RAP mortars with virgin soft binder had the best low-temperature resistance followed by the RAP mortars with RTFO and PAV binders. However, it is still in doubt whether SBS-modified asphalt is recyclable. Supporters believe that SBS-modified asphalt is produced only by physical modification of base asphalt, while asphalt recycling involves more chemical reactions than a physical reaction. Therefore, SBS-modified asphalt recycling is only the recycling of aged asphalt and does not have effect on the SBS copolymer [17, 18]. Opponents believe that the aging process of SBS-modified asphalt includes the aging of both asphalt and SBS copolymer. Therefore, the recycling of SBS-modified asphalt inevitably includes the recycling of the SBS copolymer. To this end, recycling agents developed by researchers have different application scopes, and it is unknown whether they can be used for SBS-modified asphalt recycling [19–23]. In view that SBS-modified asphalt is adopted for surface and middle courses of most high-grade highways in China, it is of great practical significance to find a recycling agent applicable to SBS-modified asphalt.

The objective of this study was to investigate the mechanisms of aging and recycling of SBS-modified asphalt by conducting comprehensive performance tests and physical-chemistry experiments and further to demonstrate whether the two processes were an inverse or not.

2. Materials and Methods

2.1. Materials. SBS-modified asphalt was used in this research. Two typical rejuvenators were selected to recover the aged asphalt binder in this research. The molecular structural formulas of rejuvenator A and rejuvenator B and the curing agents used in this research are shown in Figure 1.

2.2. Laboratory Tests. In this research, both conventional and advanced tests have been conducted. The conventional tests are used to evaluate the effect of the aging-recycling cycle on rheology of SBS-modified asphalt at the macrolevel. The advanced tests are used to investigate the mechanism of effect of the aging-recycling cycle on rheology of SBS-modified asphalt at the microlevel.

2.2.1. Penetration. Penetration is a widely used method in the world to measure asphalt stiffness. The smaller the stiffness, the greater the penetration would be. The greater the stiffness, the smaller the penetration would be. It is sure that stiffness change can reflect not only the asphalt aging but also the effect of SBS modification. Three replicates were tested in this research, and the data shown in this paper were the average of the three replicates.

2.2.2. Softening Point. The softening point is the temperature at which a material softens beyond some arbitrary softness. A ring and ball apparatus can also be used for the determination of the softening point of bituminous materials. Two replicates were tested in this research, and the data shown in this paper were the average of the two replicates.

2.2.3. Ductility. Ductility is a solid material's ability to deform under tensile stress. The ductility of the bituminous material is defined as the distance in centimeters, to which it will elongate before breaking when two ends of a briquet specimen of the material are pulled apart at a specified speed and a specified temperature. Three replicates were tested in this research, and the data shown in this paper were the average of the three replicates.

2.2.4. Viscosity. The rotational viscometer was used to measure binder viscosity at 135°C. First, a sample was placed in a chamber that was heated to the test temperature. Second, the sample viscosity was measured by rotating the spindle immersed in the binder. The rotational speed was set at 20 revolutions per minute (RPM), and the viscosity reading was recorded in units of centipoise. Two replicates were tested in this research, and the data shown in this paper were the average of the two replicates.

2.2.5. Gel Permeation Chromatography (GPC). Gel permeation chromatography (GPC) has low requirements for flow

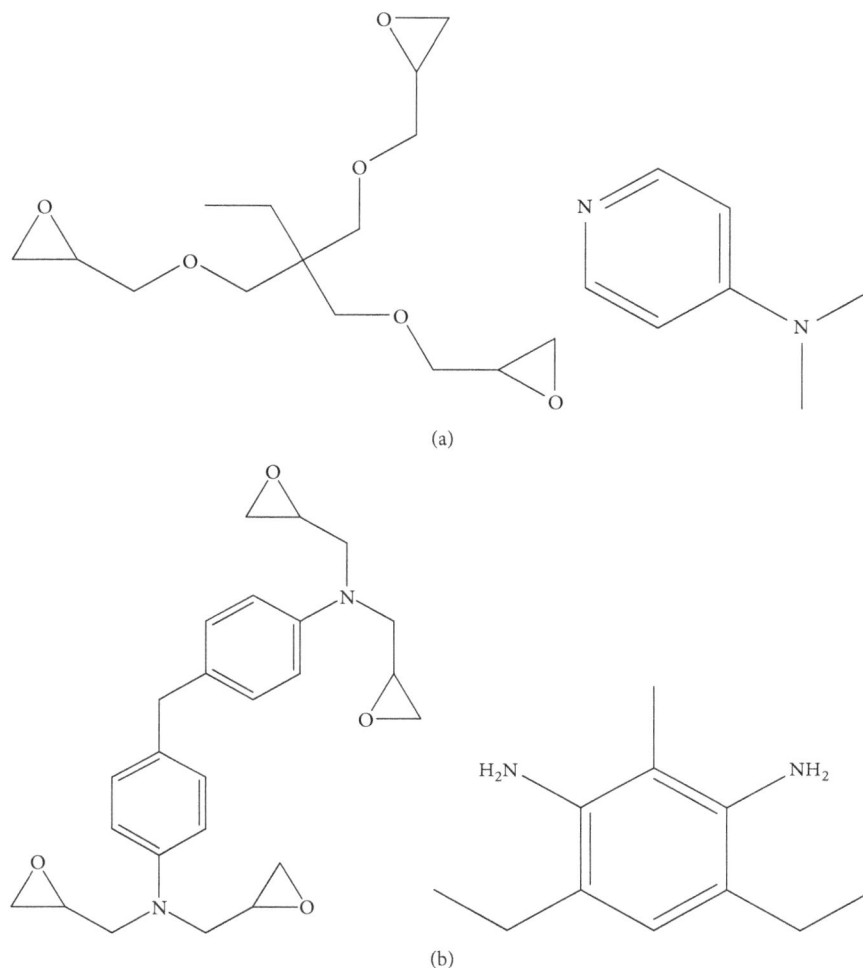

FIGURE 1: Molecular structural formulas of rejuvenators and the curing agents. (a) Rejuvenator A: fat glycidyl ether epoxy resin (left) and its curing agent (right). (b) Rejuvenator B: glycidyl amine epoxy resin (left) and its curing agent (right).

phase and features moderate test conditions, good repeatability, and fast analysis speed. It is currently the most widely used method for determining the molecular weight distribution of high polymer materials. Its separation basis is different hydrodynamic volumes of solute molecules in solution. The molecular elution volume of the solute depends on the physical parameters such as molecular dimension, filler aperture, porosity, and column volume. Two replicates were tested in this research, and the data shown in this paper were the average of the two replicates.

2.2.6. Thermogravimetric Analysis (TGA). To further reveal the changes before and after SBS-modified asphalt aging, the thermogravimetric (TG) test was conducted on SBS-modified asphalt and aged SBS-modified asphalt to analyze the thermal gravimetric property. Q600SDT TGA produced by TA Instruments was adopted for the TG test, with a temperature range of room temperature to 1500°C, a sensitivity of 0.1 μg, and a temperature accuracy higher than 2%. The test was conducted in a nitrogen atmosphere at a temperature rise rate of 10°C/min and at a temperature range from room

temperature to 1,000°C. Samples were made with SBS-modified asphalt before and after aging, respectively, and put in TGA for comprehensive thermal gravimetric analysis. Two replicates were tested in this research, and the data shown in this paper were the average of the two replicates.

2.2.7. Hydrogen Nuclear Magnetic Resonance (1HNMR) Spectrometry. Hydrogen nuclear magnetic resonance (1HNMR) spectrometry is conducted with Bruker AV-400 MHz NMR, with CDCl3 as the solute, TMS as the internal reference, and a test temperature of 25°C. Two replicates were tested in this research, and the data shown in this paper were the average of the two replicates.

2.2.8. Scanning Electron Microscope (SEM). Scanning electron microscope (SEM) is a method observing micro properties. It can be directly used for microimaging of the matter properties of the sample surface material. In principle, SEM scans the sample with a very thinly focused high-energy electron beam to stimulate various physical information, which is received, amplified, and imaged to get the observation results of the sample surface.

2.2.9. Energy Dispersive Spectrometer (EDS). Energy dispersive spectrometer is used for analyzing the types and content of microarea elements of the material, which is used together with an electronic microscope and transmission electron microscope. Different elements have their own characteristic wavelengths of X-ray, which depend on the characteristic energy $\triangle E$ released during energy transition. Energy spectrometer conducts composition analysis according to different characteristic energies of X-ray photons of different elements.

When X-ray photons enter the detector, they will stimulate a certain number of electron-hole pairs in the Si (Li) crystal. The minimum average energy ε for producing an electron-hole pair is constant (3.8 eV at low temperature), and the number of electron-hole pairs produced by one X-ray photon is $N = \triangle E/\varepsilon$. Therefore, the greater energy the incident X-ray photons have, the bigger the N will be. Biases installed on both ends of the crystal are used to collect electron-hole pairs, which are converted to current pulses by the preamplifier. The height of current pulses depends on the size of N. Current pulses are then converted into voltage pulses by the main amplifier and enter the multichannel pulse height analyzer, which classified and counted the pulses by height. Thus, a diagram of X-ray distribution according to energy is obtained. Two replicates were tested in this research, and the data shown in this paper were the average of the two replicates.

3. Results and Discussion

3.1. Aging of SBS-Modified Asphalt. In order to find the role of SBS-modified asphalt aging, the laboratory adopted I-D modified asphalt meeting the requirements of Technical Specification for Construction of Highway Asphalt Pavements (JTG E20-2011) as a sample and put it in an RTFOT at 163°C for aging for 16 h and then put the asphalt aged by RTFOT in a PAV at 100°C for aging for 20 h. Finally, the changes in performance indicators, functional groups, components, and molecular weight of the aged SBS-modified asphalt were analyzed.

3.1.1. Effects of Aging on the Thermodynamic Performance of SBS-Modified Asphalt. Figure 2 presents TG curves before and after SBS-modified asphalt aging. This figure shows that the thermal weight loss of SBS-modified asphalt in a nitrogen atmosphere had one phase. Little residue carbon is left at 540°C, and combustion is completed. The thermogravimetric processes are consistent before and after SBS-modified asphalt aging, indicating that the aging effect is weak.

3.1.2. Effects of Aging on the Chemical Properties of SBS-Modified Asphalt

(1) FTIR Tests. The aging process of SBS-modified asphalt is essentially a slow process of chemical reaction. A shift of functional groups inevitably accompanies the chemical reaction of organic matters. In the infrared spectrogram, the absorption band spectra shown are also different. Figure 3 contains the infrared spectrograms of SBS-modified asphalt before and after aging in room temperature conditions.

In general, the stretching vibration absorption peak of hydroxyl is at 3450 cm^{-1}, mainly characterized by a wide peak and that the higher degree of association, the closer the peak to low wavenumber; the stretching vibration absorption peak of alkyl hydrocarbon bonds is at 2800 cm^{-1}~3000 cm^{-1}, including –CH$_3$, –CH$_2$, and –CH; carbonyl absorption peaks in ketone and carboxylic acid are near 1600 cm^{-1}. When the carbonyl group and a benzene vibrate, the absorption peak moves towards the low wavenumber. The bending vibration absorption peak of alkyl hydrocarbon bonds is at 1350 cm^{-1}~1480 cm^{-1}, including –CH$_3$ (1370 cm^{-1} and 1450 cm^{-1}), –CH$_2$ (1465 cm^{-1}), and –CH(CH$_3$)$_2$ (1365 cm^{-1}~1385 cm^{-1}); the stretching vibration absorption peak of organic sulfoxide and sulphone is at 1120 cm^{-1}~1160 cm^{-1} and 1030 cm^{-1}~1070 cm^{-1}; the C–H bond out-of-plane bending vibration absorption peak of olefins (polybutadiene) is at 900 cm^{-1}~950 cm^{-1}; the C–H bond out-of-plane bending vibration absorption peak of a series of heteroaromatic compounds is at 650 cm^{-1}~810 cm^{-1}, whereas the C–H bond out-of-plane bending vibration absorption peak of benzene rings of styrene and fatty aldehyde is at 700 cm^{-1}. Asphalt mainly consists of elemental carbon, hydrogen, oxygen, nitrogen, and sulfur. The infrared spectrogram shows that, with the increase of the aging period, the absorption intensity of different groups has different degrees of increase and decrease. These long-chain polymers often contain carbon-carbon double bonds and carbon-carbon triple bonds, which are very unstable and easily have addition reaction or oxidation reaction to produce hydroxyl groups and carbonyl groups. Some hydroxyl groups, due to the chain end effect, will be further oxidized to produce ketone or carboxylic acid.

After aging of SBS-modified asphalt, a new absorption peak appears at 1120 cm^{-1}~1160 cm^{-1} and 1030 cm^{-1}~1070 cm^{-1}. This is because elemental sulfur in the modified asphalt is oxidized to sulfoxide (S=O) and sulphone (O=S=O). The absorption intensity of the two absorption peaks increases with the aging time of asphalt. Sulfur mainly exists in the form of sulfur ether and mercaptan in asphalt molecules. Peroxide easily oxidizes sulfur ether and mercaptan into sulfoxide base, a sulfur atom which still has a pair of lone electrons. In the case of sufficient oxygen, sulfoxide groups will be further oxidized into sulfonates. In addition to such polar functional groups as ketones, aldehydes, and carboxyl, a series of heteroaromatic compounds are also produced in the aging of SBS-modified asphalt. These heteroaromatic compounds are from the cleavage oxidation and local cleavage polymerization of SBS polystyrene chain segments. The deeper the asphalt is aged, the higher the heteroaromatic compound's absorption intensity is, indicating a higher decomposition degree of polystyrene.

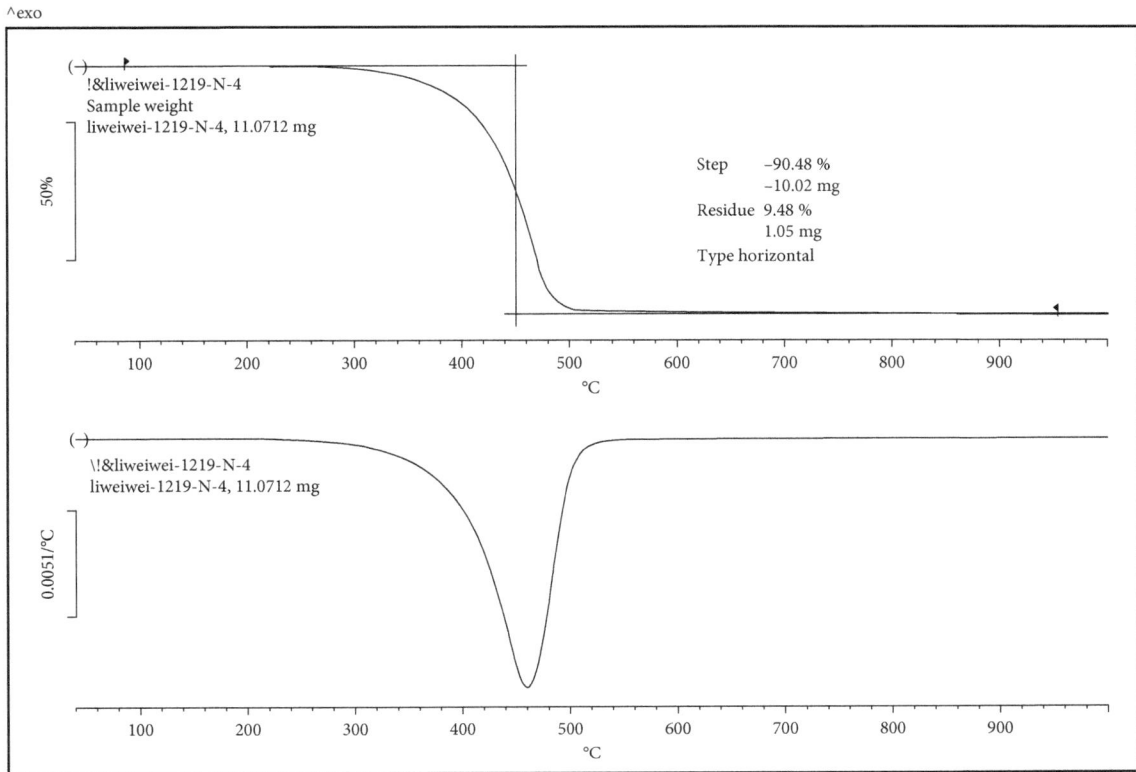

(a)

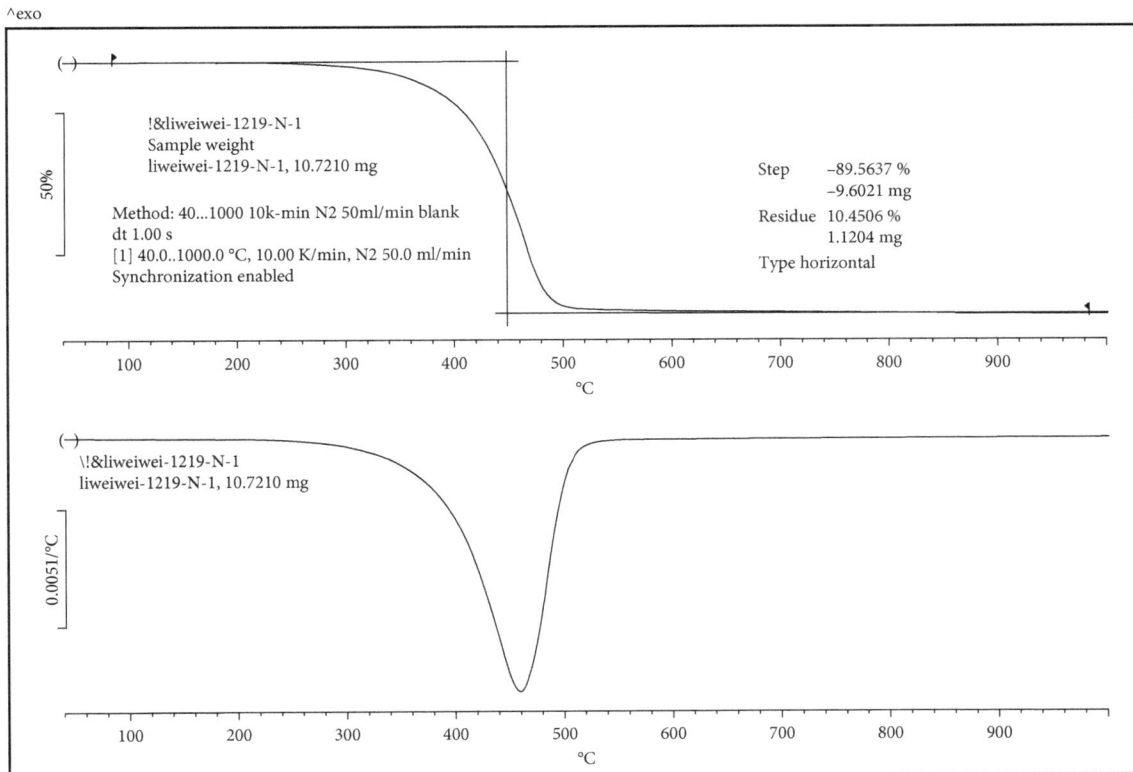

(b)

FIGURE 2: TG curve of aged SBS-modified asphalt. (a) Before aging. (b) After aging.

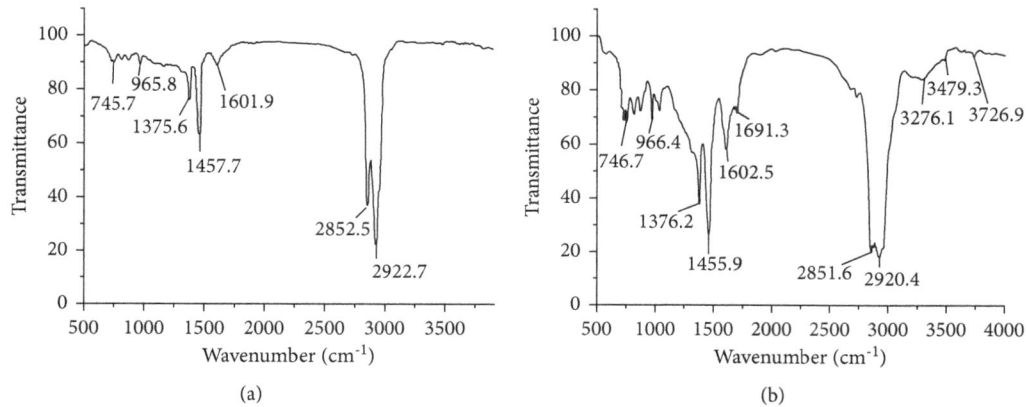

FIGURE 3: Infrared spectrogram of SBS-modified asphalt. (a) Before aging. (b) After aging.

Infrared spectroscopy shows that the reason for the embrittlement and hardening of SBS-modified asphalt after aging is SBS failure and the generation of a large number of polar groups. Polar groups have strong intermolecular interaction, and some form an association body, which reduces asphalt ductility and increases stiffness modulus.

(2) NMR Tests. Figure 4 shows the nuclear magnetic resonance spectrograms of SBS-modified asphalt before and after aging. The peak at 1 ppm-2 ppm belongs to hydrogen on methyl in the sample, and the one at 3.7 ppm belongs to hydrogen on methene (alkene) and methine (alkyne). Comparison between the two figures shows that the characteristic peaks of methene and methine at 3.7 ppm have disappeared in the nuclear magnetic resonance spectrogram of SBS-modified asphalt after aging. The reason is that, with the progress of aging, the double bonds of SBS polybutadiene segments are opened when oxidation occurs. Some double bonds are oxidized into alcoholic hydroxyl groups, carbonyl groups, or carboxyl groups, and some are added to produce saturated carbon bonds; some small molecules may also have mutual addition to producing macromolecules with unsaturated carbon-carbon bonds, namely, alkyl produced by alkene or alkyne through hydrogenation (generally oxygen addition).

In the nuclear magnetic resonance spectrogram of SBS-modified asphalt after aging, the characteristic peaks of methene and methine at 3.7 ppm have disappeared, indicating that SBS decomposition and degradation have completed.

3.1.3. Effects of Aging on the Surface Microtopography of SBS-Modified Asphalt. The SEM images of SBS-modified asphalt before and after aging are shown in Figures 5(a) and 5(b).

Figure 5(a) shows that the SBS asphalt is uniform. It indicates that modifier SBS can be evenly distributed in asphalt to form a subhomogeneous island structure; namely, styrene-butadiene-styrene triblock forms a network structure in the base asphalt. Figure 5(b) shows the SEM image of SBS-modified asphalt after aging. Observed under SEM, it is

very homogeneous, with few impurities in local parts, and the subhomogeneous island structure has disappeared. It is inferred that SBS has had severe cracking and decomposition and produced small molecules, or has had oxygen absorption reaction and produced highly polar substances. This coincides with the results of the GPC test above.

3.1.4. Effects of Aging on the Elementary Composition of SBS-Modified Asphalt. The analysis results of the SEM-EDS of SBS-modified asphalt before and after aging are shown in Table 1.

As shown in Table 1, the C content of aged SBS-modified asphalt has decreased compared with that of original SBS-modified asphalt, while the O content and S content have increased obviously. According to analysis, the reason is that, with the deepening of aging, sulfur elements in modified asphalt were oxidized into sulfoxide (S=O) and sulphone (O=S=O). In addition to such polar functional groups as ketones, aldehydes, and carboxyl, a series of heteroaromatic compounds are also produced in the aging of SBS-modified asphalt. These heteroaromatic compounds are from the cleavage oxidation and local cleavage polymerization of SBS polystyrene chain segments. The deeper the asphalt is aged, the higher the heteroaromatic compound's absorption intensity is, indicating a higher decomposition degree of polystyrene. These long-chain polymers often contain carbon-carbon double bonds and carbon-carbon triple bonds, which are very unstable and easily have addition reaction or oxidation reaction to produce hydroxyl groups and carbonyl groups. Some hydroxyl groups, due to the chain end effect, will be further oxidized to produce ketone or carboxylic acid.

3.2. Recycling of Aged SBS-Modified Asphalt

3.2.1. Effects of Rejuvenators on the Basic Properties of Aged SBS-Modified Asphalt. In this section, the effect of the two rejuvenators on the macroproperties of aged SBS-modified asphalt was explored. The contents of rejuvenators were 6%, 8%, 10%, 12%, and 14% of base asphalt binder by mass.

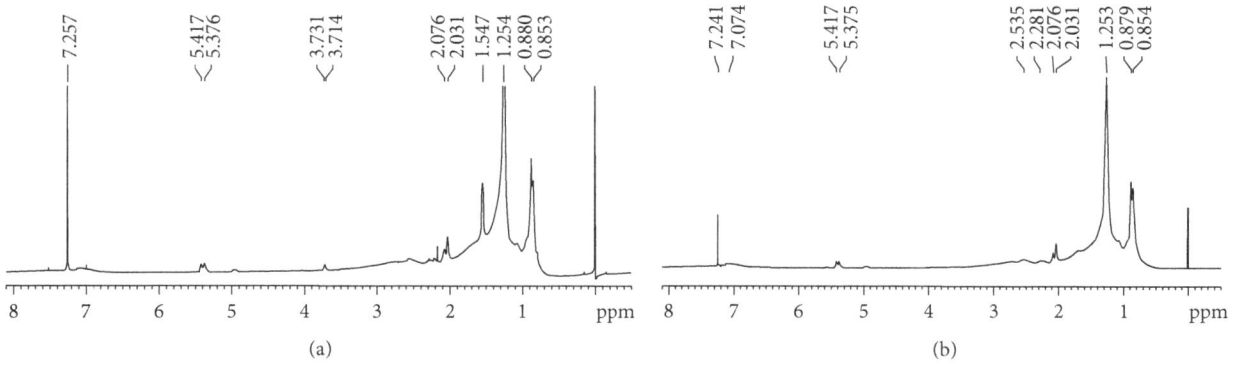

FIGURE 4: Nuclear magnetic resonance spectrograms of SBS-modified asphalt. (a) Before aging. (b) After aging.

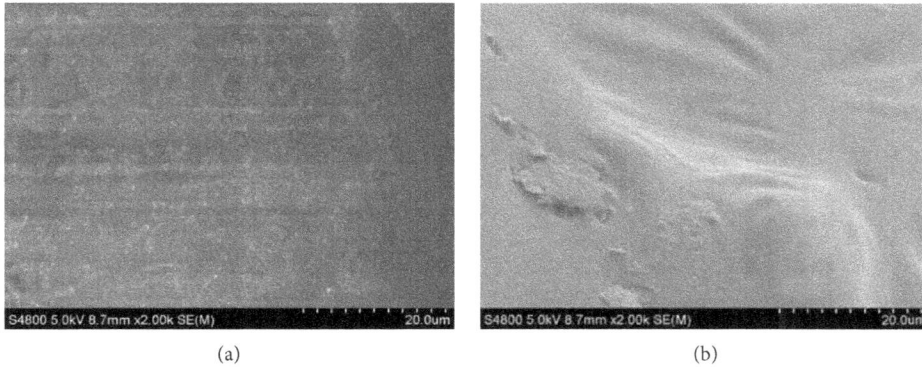

FIGURE 5: SEM images of SBS-modified asphalt. (a) Before aging. (b) After aging.

TABLE 1: Effects of aging on the elementary composition of SBS-modified asphalt.

Element type	Element content of SBS-modified asphalt (%)	
	Before aging	After aging
C	90.0	62.2
O	7.4	22.5
S	0.8	1.1

(1) Viscosity. Figure 6 shows the changes in the viscosity of aged SBS-modified asphalt after epoxy resin rejuvenators A and B are added. According to the figure, the viscosity of aged SBS-modified asphalt decreases with the increase of the additive amount of the epoxy resin rejuvenators A and B. This is because these two rejuvenators contain both low-viscosity catalytic cracking oil slurry and epoxy functional group compounds, whose viscosity is lower than that of SBS-modified asphalt, and it makes the viscosity of recycled SBS-modified asphalt decreases with the increasing additive amount of rejuvenator. The rejuvenators A and B have different effects on the viscosity of aged SBS-modified asphalt. The viscosity of asphalt modified by rejuvenator A is lower than that modified by rejuvenator B. According to the molecular structures of the two rejuvenators, the molecular backbone of rejuvenator A is a flexible aliphatic, while that of rejuvenator B is a rigid benzene ring structure. It leads to that viscosity of asphalt modified by rejuvenator A is lower than that modified by rejuvenator B.

(2) Penetration. Figure 7 shows the penetration improvement of the aged SBS-modified asphalt after the epoxy resin recycling agents A and B are added. According to the figure, the penetration of recycled SBS-modified asphalt increased with the increase of the content of the recycling agent. We speculate that the reason is that these two recycling agents contain both low-viscosity catalytic cracking oil slurry and epoxy functional group compounds, whose viscosity is lower than that of SBS-modified asphalt, and it makes the penetration of recycled SBS-modified asphalt decreases with the increasing additive amount of recycling agent.

Regarding the molecular structure, the molecular backbone of recycling agent A is a flexible aliphatic, while that of recycling agent B is a rigid benzene ring structure. However, recycling agent B does not recover to the triblock molecular chain structure of SBS, so the effect of recycling agent B on the penetration of aged SBS-modified asphalt is not as evident as that of recycling agent A.

(3) Ductility. Figure 8 shows the effects of recycling agents A and B on the ductility of aged SBS-modified asphalt.

It can be seen from Figure 8 that the ductility of aged SBS-modified asphalt increased with the increasing additive amount of recycling agent A. The possible reason was that

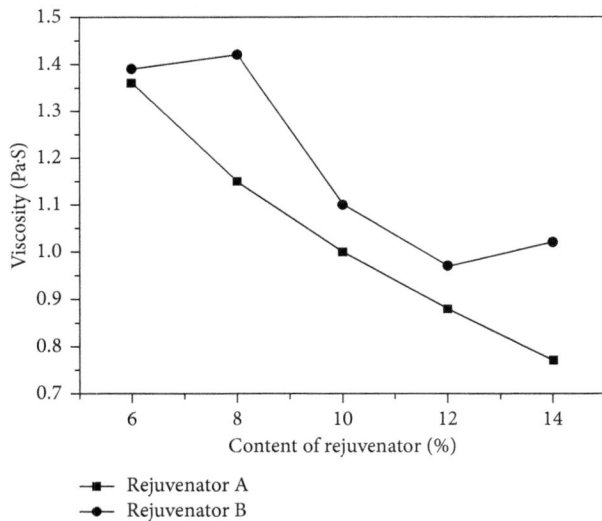

FIGURE 6: Effect of rejuvenators A and B on a viscosity of aged SBS-modified asphalt binder.

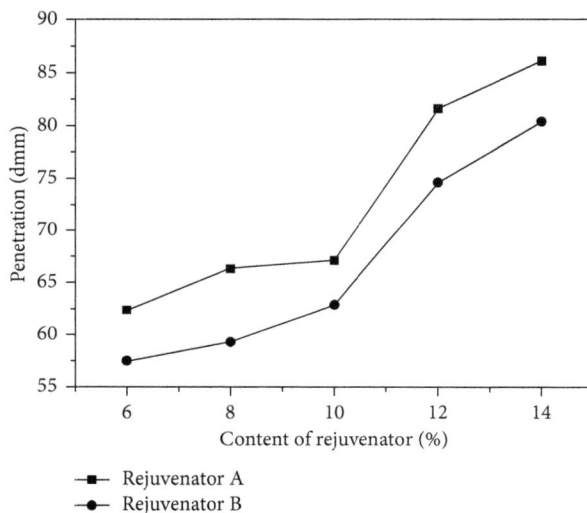

FIGURE 8: Effect of rejuvenators A and B on the ductility of aged SBS-modified asphalt binder.

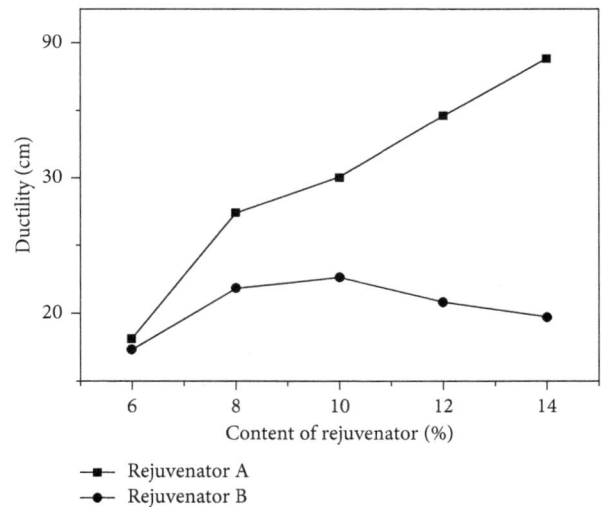

FIGURE 7: Effect of rejuvenators A and B on the penetration of aged SBS-modified asphalt binder.

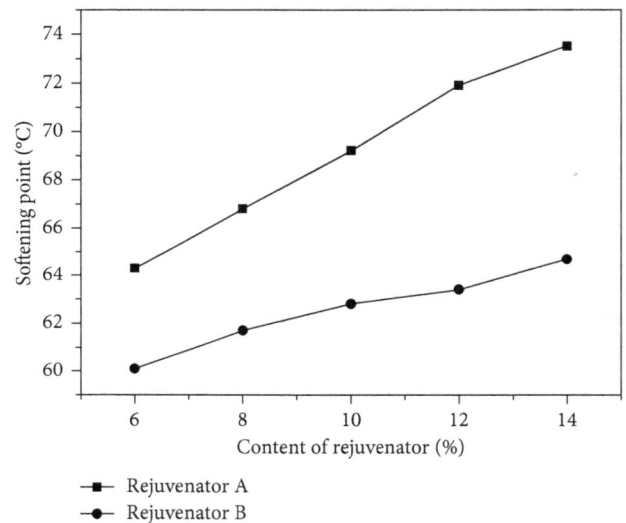

FIGURE 9: Effect of rejuvenators A and B on the softening point of aged SBS-modified asphalt binder.

the epoxy functional groups of recycling agent A reacted with the cracked carboxyl at the SB molecular chain end. It made biblock copolymer SB re-crosslinked into triblock copolymer SBS by epoxy resin molecular chains. That is, the triblock molecular structure of SBS was recovered, with the macro performance that the ductility of aged SBS-modified asphalt was obviously recovered. Recycling agent B did not have the recovered triblock molecular chain structure of SBS, so the addition of recycling agent B had an insufficient effect on the ductility of aged SBS-modified asphalt.

(4) Softening Point. The softening point improvement of the aged SBS-modified asphalt after the epoxy resin recycling agents A and B are added is shown in Figure 9.

According to Figure 9, the softening point of recycled SBS-modified asphalt rises with the increasing additive amount of recycling agent. Compared with recycling agent B, the softening point rises more sharply by adding recycling agent A because recycling agent A recovers the triblock molecular structure of SBS and accelerates the rising trend of the softening point. Although recycling agent B has a low-viscosity component, it does not recover the molecular chain structure of SBS, resulting in a very limited increase of softening point.

3.2.2. Effects of Rejuvenators on the Molecular Weight Distribution of Aged SBS-Modified Asphalt. The GPC graphs of SBS-modified asphalt after rejuvenators A and B are added are shown in Figure 10.

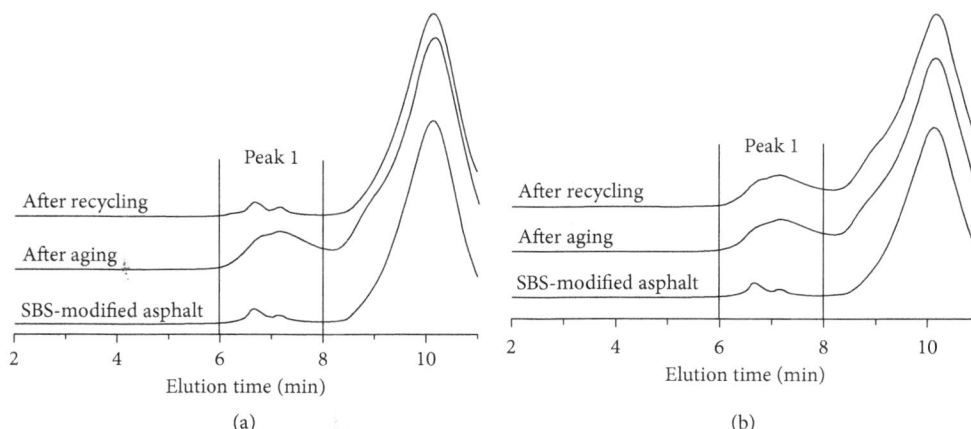

FIGURE 10: Effect of aging and recycling on molecular weight distribution of SBS-modified asphalt binder. (a) Rejuvenator A. (b) Rejuvenator B.

According to Figure 10, the SBS molecular chain cracks after the aging of SBS-modified asphalt, which is characterized by the trailing in the macromolecular part at Peak 1 of the GPC curve. After recycling agent A is added, the GPC curve at Peak 1 has migrated overall; that is, the previous macromolecular chain is recovered. It is inferred that the epoxy functional groups of recycling agent A react with the cracked carboxyl at the SB molecular chain end, which makes biblock copolymer SB re-crosslinked into triblock copolymer SBS by epoxy resin molecular chains. That is, the triblock molecular structure of SBS is recovered. After recycling agent B is added, the GPC curve at Peak 1 has no noticeable change. It indicates that epoxy functional groups of recycling agent B did not have coupling reaction with the SB molecular chain end, resulting in the very limited effect on the recycling of SBS-modified asphalt.

3.2.3. Effects of Rejuvenators on the Microstructure of Aged SBS-Modified Asphalt

(1) Fluorescence Microscope. Figures 11(a) and 11(b) show the microscopic morphology of SBS-modified asphalt recycled by adding the epoxy resin recycling agents A and B, respectively. Under a fluorescence microscope, SBS is the bright area in the figure. In Figure 11(b), there are very few particles in the bright area, and it is inferred that serious chain fracture occurred in the SBS triblock copolymer after asphalt aging, indicating serious aging of SBS. Figures 11(a)–(C) show the aged SBS-modified asphalt after recycling agent A is added. According to the figure, the distribution of bright area has been obviously improved and is close to that before aging. It indicates that the recycling agent A has obviously restored the SBS distribution and thus improved the aging degree of SBS-modified asphalt. Figures 11(b)–(C) show that SBS distribution has no obvious changes after recycling agent B is added, indicating very limited recycling effect. The microproperties are consistent with the molecular weight result tested by GPC.

Therefore, the recycling agent A has an obviously better recycling effect than the recycling agent B.

(2) SEM. The microscopic morphologies of SBS-modified asphalt recycled by adding the epoxy resin recycling agents A and B are shown in Figures 12(a) and 12(b), respectively.

It is can be seen from Figures 12(a)–(A) and 12(b)–(A) that SBS is distributed in the base asphalt, which can be called the island structure in the SEM image. With the progress of aging, serious chain fracture occurs in the SBS triblock copolymer, which is manifested by the missing of islands in Figure 12(a)–(B). Figure 12(a)–(C) shows the aged SBS-modified asphalt after recycling agent A is added. According to the figure, the islands are recovered and similar to those before aging. It indicates that the rejuvenator A has obviously restored the SBS triblock structure and thus improved the aging degree of SBS-modified asphalt. Figure 12(a)–(A) shows that the SEM image has no obvious changes after rejuvenator B is added, indicating a very limited recycling effect. The microproperties are consistent with the molecular weight result tested by GPC and fluorescence microscopy results. The rejuvenator A has higher reactivity and better recycling effect than rejuvenator B.

4. Conclusions

Based on the testing and analysis presented in this paper, the conclusions of the study are summarized as follows:

(1) Infrared spectrogram analysis results show that, with the aging progress of asphalt, the area of the carbonyl absorption peak and carbonyl content increase continuously; sulfur compounds in asphalt react with oxygen to produce sulfoxide base functional groups during asphalt aging. The butadiene absorption peak gradually disappears, indicating that base asphalt components and SBS copolymer age simultaneously during asphalt aging; with the degradation of the SBS copolymer, the reticular

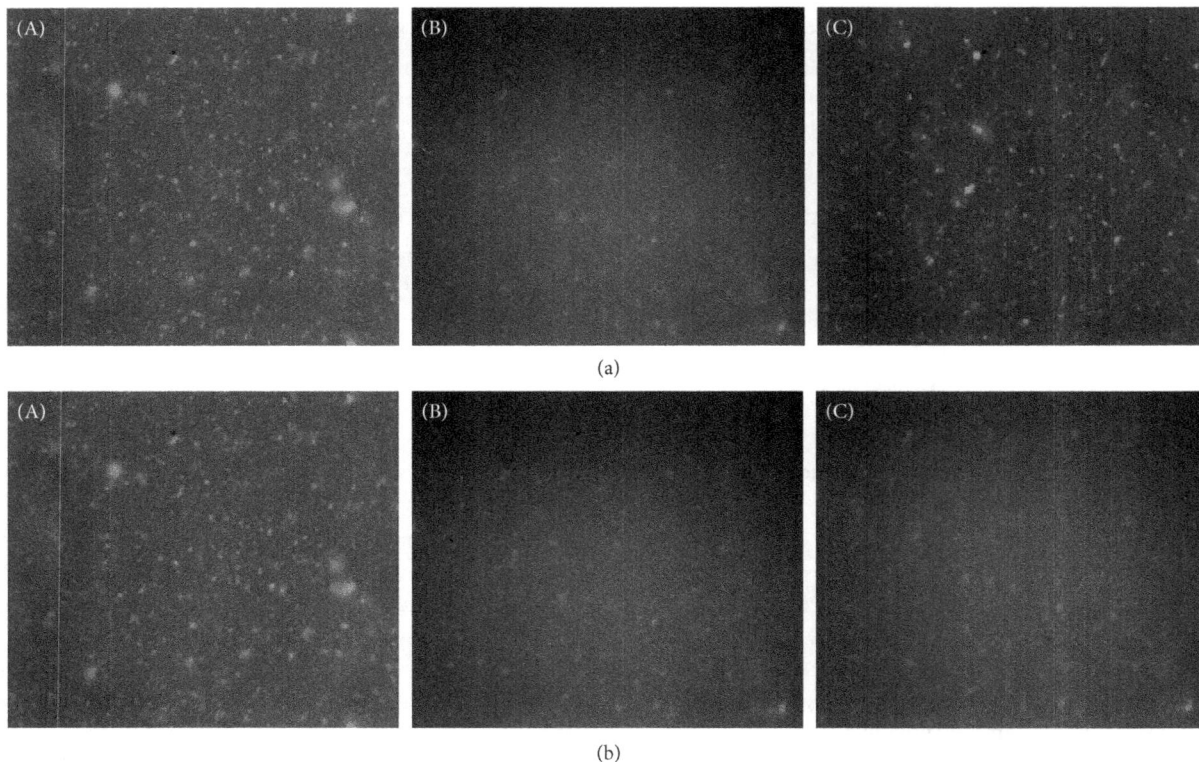

FIGURE 11: Effect of rejuvenators on microstructure of aged SBS-modified asphalt binder by fluorescence microscope (measuring scale is 1 : 10000): (A) SBS-modified asphalt, (B) aged SBS-modified asphalt, and (C) recycled SBS-modified asphalt of (a) rejuvenator A and (b) rejuvenator B.

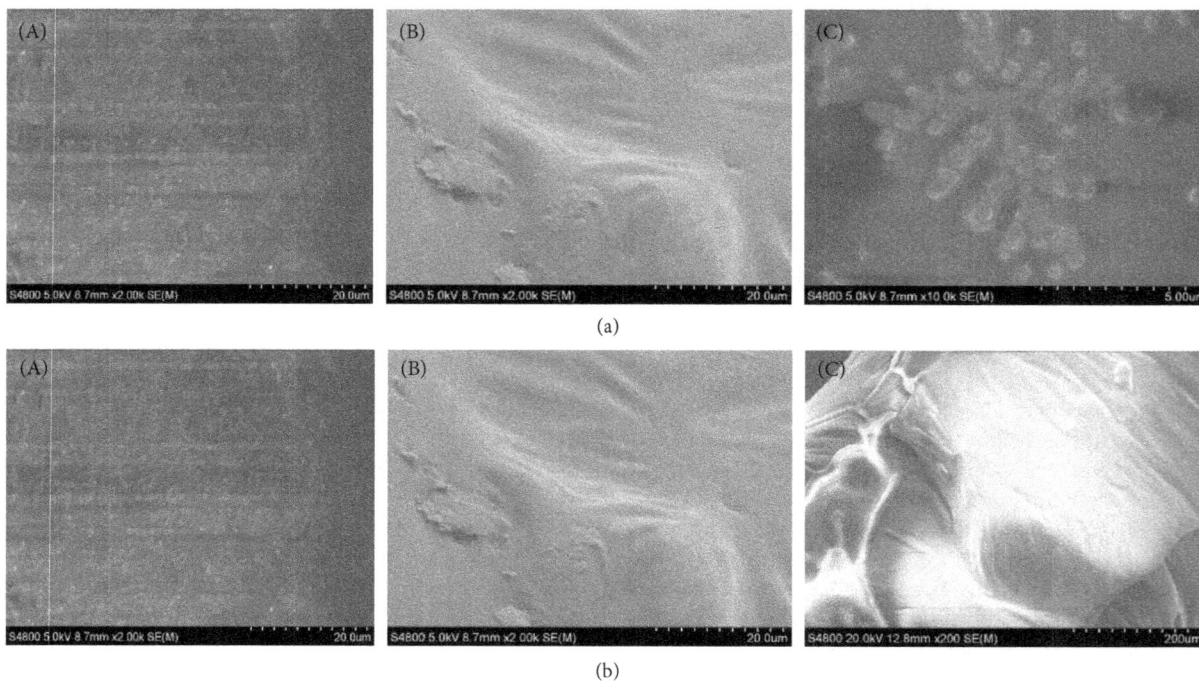

FIGURE 12: Effect of rejuvenators on microstructure of aged SBS-modified asphalt binder by SEM: (A) SBS-modified asphalt, (B): aged SBS-modified asphalt, and (C): recycled SBS-modified asphalt of (a) rejuvenator A and (b) rejuvenator B.

crosslinking structure of asphalt is destroyed and SBS's modification effect is gradually lost. The increase of polar functional groups of SBS-modified asphalt and the decomposition and failure of the SBS modifier are the immanent causes for continuous change in viscosity, strengthening of high-temperature performance, and reduction of low-temperature anticrack property.

(2) According to the SEM analysis results, modifier SBS can be evenly distributed in asphalt to form a sub-homogeneous island structure; namely, styrene-butadiene-styrene triblock forms a network structure in the base asphalt. According to the SEM image of SBS-modified asphalt after aging, observed under SEM, it is very homogeneous, with few impurities in local parts, and the subhomogeneous island structure has disappeared. It is inferred that SBS has had severe cracking and decomposition and produced small molecules, or has had oxygen absorption reaction and produced highly polar substances. Energy spectrum analysis shows that, with the progress of aging, the C content of aged SBS-modified asphalt has decreased compared with that of original SBS-modified asphalt, while the O content and S content have increased obviously.

(3) Fluorescence microscope and SEM were used to observe the changes in molecular structure and distribution of the SBS modifier before and after recycling. It shows that the epoxy functional group compounds of aliphatic glycidyl ether resin have high reactivity, and the triblock molecular structure of SBS is recovered. The pavement performance of recycled SBS-modified asphalt also shows that it has a good recycling effect. GPC demonstrated that epoxy functional groups of recycling agent B did not have coupling reaction with the SB molecular chain end, resulting in the very limited effect on the recycling of SBS-modified asphalt.

Based on the findings of this research, chemical-related tests are recommended conducting to select one proper rejuvenator for the aged SBS-modified asphalt. This can increase recycling efficiency.

Conflicts of Interest

The authors declare no conflicts of interest.

Authors' Contributions

Meng Guo and Daisong Luo conceived and designed the experiments. Meng Guo performed the experiments. Liantong Mo and Meng Guo analyzed the data. Yiqiu Tan and Yafei Li contributed reagents/materials/analysis tools. Yubo Jiao and Asim Farooq wrote the paper.

Acknowledgments

This study was supported by the Beijing Natural Science Foundation (8174071) and National Natural Science Foundation of China (51808016).

References

[1] S. Kim, S. H. Lee, O. Kwon, J. Y. Han, Y. S. Kim, and K. W. Kim, "Estimation of service-life reduction of asphalt pavement due to short-term ageing measured by GPC from asphalt mixture," *Road Materials and Pavement Design*, vol. 17, no. 1, pp. 153–167, 2015.

[2] J.-Y. Yu, P.-C. Feng, H.-L. Zhang, and S.-P. Wu, "Effect of organo-montmorillonite on aging properties of asphalt," *Construction and Building Materials*, vol. 23, no. 7, pp. 2636–2640, 2009.

[3] Y. F. Li, J. Chen, J. Yan, and M. Guo, "Influence of buton rock asphalt on the physical and mechanical properties of asphalt binder and asphalt mixture," *Advances in Materials Science and Engineering*, vol. 2018, Article ID 2107512, 7 pages, 2018.

[4] S.-p. Wu, L. Pang, L.-t. Mo, Y.-c. Chen, and G.-j. Zhu, "Influence of aging on the evolution of structure, morphology and rheology of base and SBS modified bitumen," *Construction and Building Materials*, vol. 23, no. 2, pp. 1005–1010, 2009.

[5] Y. Tan and M. Guo, "Interfacial thickness and interaction between asphalt and mineral fillers," *Materials and Structures*, vol. 47, no. 4, pp. 605–614, 2013.

[6] M. Guo, A. Motamed, Y. Tan, and A. Bhasin, "Investigating the interaction between asphalt binder and fresh and simulated RAP aggregate," *Materials & Design*, vol. 105, pp. 25–33, 2016.

[7] D. S. Luo, M. Guo, Y. Q. Tan, and Y. F. Li, "Study on effects of aging on SBS modified asphalt based on GPC and rheological methods," in *RILEM 252-CMB Symposium. RILEM 252-CMB 2018. RILEM Bookseries*, L. Poulikakos, A. Cannone Falchetto, M. Wistuba, B. Hofko, L. Porot, and H. Di Benedetto, Eds., Springer, Cham, Switzerland, 2019.

[8] D. Zhang, H. Zhang, and C. Shi, "Investigation of aging performance of SBS modified asphalt with various aging methods," *Construction and Building Materials*, vol. 145, pp. 445–451, 2017.

[9] M. Mohajeri, A. A. A. Molenaar, and M. F. C. Van de Ven, "Experimental study into the fundamental understanding of blending between reclaimed asphalt binder and virgin bitumen using nanoindentation and nano-computed tomography," *Road Materials and Pavement Design*, vol. 15, no. 2, pp. 372–384, 2014.

[10] C. Castorena, S. Pape, and C. Mooney, "Blending measurements in mixtures with reclaimed asphalt: use of scanning electron microscopy with X-ray analysis," *Transportation Research Record: Journal of the Transportation Research Board*, vol. 2574, pp. 57–63, 2018.

[11] S. Zhao, B. Huang, X. Shu, and M. E. Woods, "Quantitative evaluation of blending and diffusion in high RAP and RAS mixtures," *Materials & Design*, vol. 89, pp. 1161–1170, 2016.

[12] H. Qiu and M. Bousmina, "New technique allowing the quantification of diffusion at polymer/polymer interfaces using rheological analysis: theoretical and experimental results," *Journal of Rheology*, vol. 43, no. 3, pp. 551–568, 1999.

[13] H. Qiu and M. Bousmina, "Determination of mutual diffusion coefficients at nonsymmetric polymer/polymer interfaces from rheometry," *Macromolecules*, vol. 33, no. 17, pp. 6588–6594, 2000.

[14] Y. He, Z. Alavi, J. Harvey, and D. Jones, "Evaluating diffusion and aging mechanisms in blending of new and age-hardened binders during mixing and paving," *Transportation Research Record: Journal of the Transportation Research Board*, vol. 2574, pp. 64–73, 2018.

[15] F. Y. Rad, N. R. Sefidmazgi, and H. Bahia, "Application of diffusion mechanism," *Transportation Research Record: Journal of the Transportation Research Board*, vol. 2444, pp. 71–77, 2018.

[16] F. Xiao, R. Li, H. Zhang, and S. Amirkhanian, "Low temperature performance characteristics of reclaimed asphalt pavement (RAP) mortars with virgin and aged soft binders," *Applied Sciences*, vol. 7, no. 3, p. 304, 2017.

[17] C. Zhu, H. Zhang, D. Zhang, and Z. Chen, "Influence of base asphalt and SBS modifier on the weathering aging behaviors of SBS modified asphalt," *Journal of Materials in Civil Engineering*, vol. 30, no. 3, article 04017306, 2018.

[18] H. Zhang, Z. Chen, G. Xu, and C. Shi, "Evaluation of aging behaviors of asphalt binders through different rheological indices," *Fuel*, vol. 221, pp. 78–88, 2018.

[19] X. Liu, F. Cao, F. Xiao, and S. Amirkhanian, "BBR and DSR testing of aging properties of polymer and polyphosphoric acid-modified asphalt binders," *Journal of Materials in Civil Engineering*, vol. 30, no. 10, article 04018249, 2018.

[20] P. Cong, W. Luo, P. Xu, and H. Zhao, "Investigation on recycling of SBS modified asphalt binders containing fresh asphalt and rejuvenating agents," *Construction and Building Materials*, vol. 91, pp. 225–231, 2015.

[21] Y. H. Nie, S. H. Sun, Y. J. Ou, C. Y. Zhou, and K. L. Mao, "Experimental investigation on asphalt binders ageing behavior and rejuvenating feasibility in multicycle repeated ageing and recycling," *Advances in Materials Science and Engineering*, vol. 2018, Article ID 5129260, 11 pages, 2018.

[22] D. Sun, T. Lu, F. Xiao, X. Zhu, and G. Sun, "Formulation and aging resistance of modified bio-asphalt containing high percentage of waste cooking oil residues," *Journal of Cleaner Production*, vol. 161, pp. 1203–1214, 2017.

[23] L. Sun, Y. Wang, and Y. Zhang, "Aging mechanism and effective recycling ratio of SBS modified asphalt," *Construction and Building Materials*, vol. 70, pp. 26–35, 2014.

5

Structural and Magnetic Properties of Ba$_3$[Cu$_{0.8-x}$Zn$_x$Mn$_{0.2}$]$_2$Fe$_{24}$O$_{41}$ Z-Type Hexaferrites

Eman S. Al-Hwaitat,[1] **Sami H. Mahmood** [iD]**,**[1] **Mahmoud Al–Hussein,**[1] **and Ibrahim Bsoul**[2]

[1]*Department of Physics, The University of Jordan, Amman 11942, Jordan*
[2]*Department of Physics, Al al-Bayt University, Mafraq 13040, Jordan*

Correspondence should be addressed to Sami H. Mahmood; s.mahmood@ju.edu.jo

Academic Editor: Ling B. Kong

We report on the synthesis and characterization of Ba$_3$[Cu$_{0.8-x}$Zn$_x$Mn$_{0.2}$]$_2$Fe$_{24}$O$_{41}$ ($x = 0.0$, 0.2, 0.4, 0.6, and 0.8) barium hexaferrites. The samples were prepared by high-energy ball-milling technique and double-sintering approach. The effects of Zn substitution for Cu on the structural and magnetic properties of the prepared samples were investigated using X-ray diffraction (XRD), scanning electron microscopy (SEM), and vibrating sample magnetometer (VSM). XRD patterns of the samples revealed the presence of a major Z-type hexaferrite phase, together with secondary M-type and Y-type phases. The magnetic results indicated that the saturation magnetization increased slightly with increasing the Zn content, while the coercivity and magnetocrystalline anisotropy field exhibited a decreasing tendency with the increase of Zn content. The thermomagnetic curves revealed the complex magnetic structure of the prepared samples and confirmed that the Curie temperature of the magnetic phases decreased with increasing x as a result of the reduction of the strength of the superexchange interactions.

1. Introduction

The demands of rapidly developing modern technologies have driven and directed the efforts of a large sector of scientists and engineers toward the search for new materials with improved performance for miniaturized high-frequency devices [1]. Hexaferrites belonging to an important class of magnetic oxides have demonstrated potential for a plethora of permanent magnets as well as soft magnetic materials applications due to their favorable properties including chemical stability, low eddy current losses, and cost-effectiveness [2–10]. Z-type hexaferrites had attracted a great attention due to their high permeability in frequency regions higher than 300 MHz, and cutoff frequency up to 2 GHz, rendering these materials promising for inductive core at microwave frequencies, and ultrahigh-frequency communications devices [1, 11–13].

Synthesis of a pure phase of Z-type barium hexaferrite is a difficult process which involves a progressive transformation through intermediate ferrites before achieving the final complex crystalline structure [14–16]. This structure can be viewed as the sum of two simpler hexagonal ferrites, namely, M-type (BaFe$_{12}$O$_{19}$) and Y-type (Ba$_2$Me$_2$Fe$_{12}$O$_{22}$), which crystallize prior to the formation of the Z phase through solid-state reaction [11]. Accordingly, realization of the Z-type ferrite may occur through topotactic reaction involving alternate stacking of platelets of simpler hexagonal ferrite phases [2, 17].

The magnetic and dielectric properties of hexaferrites can be tuned by controlled substitution scenarios for the divalent (Me^{2+}) or trivalent (Fe^{3+}) ions [12, 16]. It was reported that specific choices of divalent metal ions in Z-type hexaferrite may improve their properties [12, 18–23]. Specifically, the substitution of Co^{2+} by Cu^{2+}, Zn^{2+}, and Ni^{2+} ions was adopted for enhancing the electromagnetic properties of Co$_2$Z for miniaturized antenna applications [24], where the substitution of Zn^{2+} improves the saturation magnetization, while the substitution of Cu^{2+} lowers the sintering temperature. Also, Cu and Zn-substituted Co$_2$Z hexaferrite demonstrated improved properties for multilayer chip inductor applications [25, 26]. In addition, various scenarios of Cu and/or Mn substitutions were adopted in the synthesis of M-type and Y-type hexaferrites (which are intermediate components of Z-type hexaferrites) and were reported to modify the magnetic properties of these hexaferrites for potential practical applications [27–37]. Also, the substitution of Co by Zn in Co_2Z hexaferrite demonstrated an increase in saturation magnetization, initial permeability, and resistivity, which makes Zn-substituted Co_2Z hexaferrites potentially important materials for high-frequency microchip inductor components [38, 39]. Further, the substitution of Mn for Fe in Co_2Z was reported to induce a shift of the relaxation peak in the dielectric behavior of Co_2Z hexaferrites toward higher frequencies [22].

Previous experimental results revealed that the optimal sintering temperature for the synthesis of Z-type hexaferrites using conventional solid-state method was in the range of 1200°C–1350°C [40]. Lower sintering temperatures of 1150°C–1200°C, however, were sufficient for the production of Z-type hexaferrites using wet chemical methods [13, 24, 40, 41]. The difficulty of obtaining a pure Z-type phase, and its tendency to coexist with other (M-, Y-, and W-type) hexaferrites and spinel phases, has led to adopting different modified synthesis routes, where synthesis of pure Co_2Z hexaferrite was reported to be achieved at 1250°C sintering temperature [42, 43]. Also, it was reported that a Z-type phase can be obtained using two-step solid-state reaction where precursor materials are initially sintered at temperatures between 980 and 1180°C, then ground, and again sintered at about 1230–1300°C [23, 44].

In this study, divalent ion dopants Zn^{2+} and Mn^{2+} were incorporated into the structure of Cu_2Z hexaferrite to enhance the magnetic properties. The samples were prepared using two-step sintering solid-state reaction of raw materials. This method is suitable to produce high-quality samples, since intermediate M and Y phases are formed at ~ 1000°C, and topotactical reaction of these intermediate phases at higher sintering temperature is expected to lead to crystallization of the Z-type phase.

2. Experimental Techniques

Precursor powders of $Ba_3[Cu_{0.8-x}Zn_xMn_{0.2}]_2Fe_{24}O_{41}$ ($x = 0$, 0.2, 0.4, 0.6, and 0.8) were prepared by ball-milling stoichiometric ratios of the starting Fe_2O_3, CuO, ZnO, MnO, and $BaCO_3$ powders. The starting powder mixtures were milled for 16 h at a rotational speed of 250 rpm in a planetary ball-mill, using a powder to ball ratio of 1 : 12. The resulting powder was preheated to 1000°C, and then ground for 1 h and pressed into disks of ~1 cm diameter and ~2 mm thickness under a force of 50 kN. The disks were then sintered at 1200°C for 2 h in air using a heating rate of 10°C/min.

The structure of the samples was investigated by X-ray diffraction (XRD), using a 7000 X-ray diffractometer, with Cu-K$_\alpha$ radiation ($\lambda = 0.154$ nm). XRD patterns for the samples were recorded over the angular range $20° < 2\theta < 70°$ with scanning step of 0.01° and scan speed 0.5 deg./min. A powder diffraction software package was used to identify the structural phases in the prepared samples. The morphology of the prepared samples was examined using scanning electron microscope (SEM), and the sample composition was examined by energy-dispersive X-ray spectroscopy (EDX). The magnetic measurements were carried out at room temperature in an applied field up to 10 kOe using a vibrating sample magnetometer (VSM). The sample for VSM measurements was needle-shaped to reduce sample shape anisotropy.

3. Results and Discussion

3.1. XRD Results. Figure 1 shows the diffraction patterns of $Ba_3[Cu_{0.8-x}Zn_xMn_{0.2}]_2Fe_{24}O_{41}$. Rietveld analysis of the patterns using FullProf software indicated that the samples generally consisted of mixtures of the BaM hexaferrite phase, Y-type hexaferrite phase, and Z-type hexaferrite phase. The diffraction peaks corresponding to M- and Y-type intermediate phases were relatively strong in the sample with $x = 0$, and their intensities seemed to be weakened by Zn substitution for Cu, especially at $x = 0.2$–0.6, indicating the production of Z-type hexaferrites with higher purity at these compositions. Although noticeable changes of the relative intensities of the diffraction peaks of Z-type hexaferrites were observed as a consequence of Zn and Mn substitution, the structure remained consistent with the standard pattern. These results indicated that while the high-purity Z-type phase was not successfully achieved in the two-end binary compounds $[Cu_{0.8}Mn_{0.2}]_2Z$ and $[Zn_{0.8}Mn_{0.2}]_2Z$, the ternary $[Cu,Zn,Mn]_2Z$ ferrites crystallized with high purity at sintering temperature of 1200°C, which is lower than the sintering temperature of 1250°C, 1300°C used for the production of the Z-type hexagonal phase by the sol-gel autocombustion method and ball-milling method, respectively [45]. The enhancement of the relative intensity of the (0018) reflection at 30.8° in comparison with the relative intensity of the corresponding reflection in the standard powder diffraction pattern is an indication that the possibility of crystallographic texture along the c-axis in these samples cannot be excluded. This result is consistent with the results reported by others [1, 46].

The refined lattice parameters and cell volume of the Z-type hexaferrite phase in the samples, as well as the bulk density and X-ray density, are listed in Table 1. The bulk densities of the samples were measured using Archimedes' principle, while the X-ray density of each sample was determined using the relation:

$$\rho_x = \frac{Z(M_w)}{N_A V},\qquad(1)$$

where Z is the number of molecules per unit cell, M_w is the molecular weight, V is the cell volume, and N_A is Avogadro's number.

The lattice parameter a of the Z-type phase remains almost constant (5.88 ± 0.02 Å) with substitution, while c revealed a maximum value at $x = 0.4$. The cell volume (V) fluctuated (within ~1.5%) with a tendency to increase with the increase of x. The increasing trend of cell parameters and cell volume can be associated with the substitution of Cu^{2+}

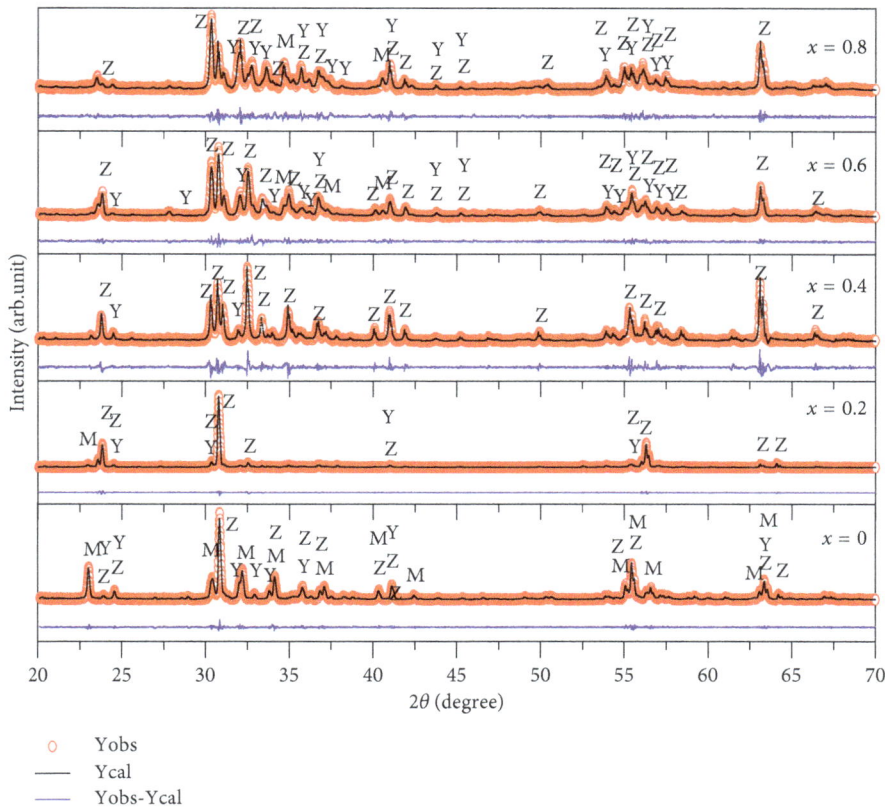

FIGURE 1: XRD patterns of $Ba_3[Cu_{0.8-x}Zn_xMn_{0.2}]_2Fe_{24}O_{41}$ samples: the peaks corresponding to Z-type, M-type, and Y-type hexaferrite are labeled by (Z) M and (Y), respectively.

TABLE 1: The lattice parameters (a and c), cell volume (V), bulk density (ρ_b), and X-ray density (ρ_x) for the $Ba_3[Cu_{0.8-x}Zn_x Mn_{0.2}]_2Fe_{24}O_{41}$ samples.

x	$a = b$ (Å)	c (Å)	V (Å^3)	ρ_b (g/cm^3)	ρ_x (g/cm^3)
0.0	5.87	52.16	1554	4.89	5.41
0.2	5.89	52.27	1569	5.22	5.36
0.4	5.90	52.37	1577	4.86	5.33
0.6	5.88	52.29	1568	4.97	5.37
0.8	5.89	52.34	1571	4.85	5.36

by Zn^{2+} with larger ionic radius [47], whereas the unsystematic fluctuations of these parameters can be associated with uncontrolled lattice distortions in the respective samples [32].

The bulk density of sintered disk samples was found to be in the range of (4.91 ± 0.06) g/cm^3 for all samples except that with $x = 0.2$, which demonstrated a relatively higher density of 5.22 g/cm^3. The X-ray density remains almost constant in the range (5.37 ± 0.04) g/cm^3 for the Z phase in all samples. These results indicated that our synthesis route yielded highly dense magnets ($\geq 90\%$ of theoretical density) for all compositions. The small decrease in X-ray density for the sample with $x = 0.4$ can be attributed to the small increase in the cell volume. The X-ray density of the Z-type phase in the samples is consistent with the reported value of 5.37 g/cm^3 for Zn_2Z hexaferrite [48].

3.2. SEM and EDX Results. Typical SEM images were collected for each sample to investigate the particle shape and particle size distribution. Furthermore, EDX measurements at randomly selected spots of the samples were carried out to identify the particular hexaferrite phase to which the measured particle belongs. In some cases, the results of EDX measurements on different spots of the same sample were discussed for comparison, limiting our discussion to special cases for brevity purposes, and clarification of the main findings.

Figure 2(a) shows SEM image for the $Ba_3[Cu_{0.8-x}Zn_xMn_{0.2}]_2Fe_{24}O_{41}$ ($x = 0$) sample. The SEM image shows platelets with varying sizes in the range of 4 to 11 μm, including a large hexagonal platelet with sharp edges. Discontinuous grain growth (DGG) could be responsible for the occurrence of anomalously large hexagonal crystals in this sample [2]. Also, the SEM image shows small irregular particles which could be due to poor crystallization of starting components or as a result of the fractured surface of the sample.

To check the chemical composition of the particles, EDX spectra were collected at different spots of the sample. A typical spectrum collected from the encircled spot is shown in Figure 2(b), and the atomic ratios are presented in Table 2. The observed Ba : Fe atomic ratio of 1 : 8.04 is in excellent agreement with the stoichiometric ratio of 1 : 8.00 for $Ba_3[Cu_{0.8}Mn_{0.2}]_2Fe_{24}O_{41}$ hexaferrite. Also, the observed

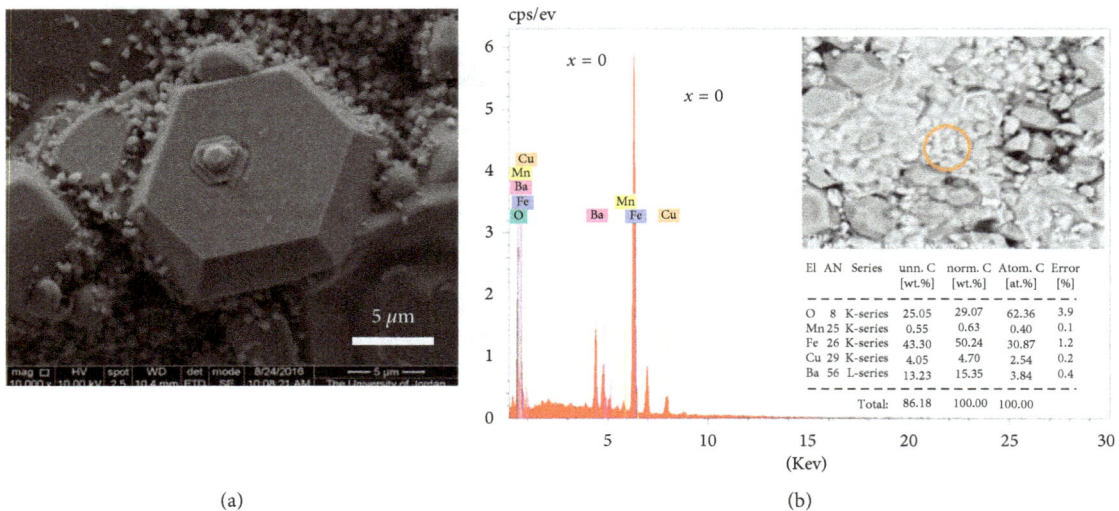

FIGURE 2: (a) SEM image and (b) EDX measurement for the $Ba_3[Cu_{0.8-x}Zn_xMn_{0.2}]_2Fe_{24}O_{41}$ ($x = 0$) sample.

TABLE 2: Atomic concentrations and atomic ratios of the different elements in the sample $Ba_3[Cu_{0.8-x}Zn_xMn_{0.2}]_2Fe_{24}O_{41}$ ($x = 0$).

Element light particles	Concentration (C)%	#Atom/mol.	Stoichiometric ratio Metal : Ba
Ba	3.84	1.00	1.00
Fe	30.87	8.04	8.00
Cu	2.54	0.66	0.53
Mn	0.40	0.10	0.13
O	62.36	16.2	13.7

Ba : Cu and Ba : Mn atomic ratios of 1 : 0.66 and 1 : 0.10 are close to the stoichiometric ratios of 1 : 0.53 and 1 : 0.13, respectively, for $Ba_3[Cu_{0.8}Mn_{0.2}]_2Fe_{24}O_{41}$ hexaferrite.

For reproducibility purposes, EDX measurements were carried out at two different spots of the same sample. The measured Ba : Fe atomic ratios were 1 : 9.00 and 1 : 7.00, the Ba : Cu ratios were 1 : 0.64 and 1 : 0.80, and the Ba : Mn atomic ratios were 1 : 0.17 and 1 : 0.12. The average ratio of Ba : (Fe + Cu + Mn) in the two regions is 1 : 8.87, which agrees well with the theoretical value of 1 : 8.67 for $Ba_3[Cu_{0.8}Mn_{0.2}]_2Fe_{24}O_{41}$ hexaferrite.

Figure 3 shows two SEM images (a, b) for $Ba_3[Cu_{0.8-x}Zn_xMn_{0.2}]_2Fe_{24}O_{41}$ ($x = 0.2$) sample. Sharp-edged hexagonal platelets with in-plane dimensions of several micrometers are observed in Figure 3(a), while much larger hexagonal platelets evolving through the DGG process are observed in Figure 3(b). The majority of the sample, however, is composed of large, nonparticulate masses, which may explain the appreciable increase in sample density. EDX spectra (not shown) were collected at different measuring spots to examine the chemical composition of the sample. The observed Ba : Fe atomic ratios at two different measuring spots were (1 : 7.00) and (1 : 8.00), while the atomic ratios of Ba : Zn were 1 : 0.22 and 1 : 0.14, the atomic ratios of Ba : Cu were 1 : 0.56 and 1 : 0.52, and the atomic ratios of Ba : Mn were 1 : 0.15 and 1 : 0.16. These atomic ratios are in agreement with the stoichiometric Ba : Fe ratio of 1 : 8.00, Ba : Zn ratio of 1 : 0.13, Ba : Cu ratio of 1 : 0.40, and Ba : Mn ratio of 1 : 0.13 for $Ba_3[Cu_{0.6}Zn_{0.2}Mn_{0.2}]_2Fe_{24}O_{41}$ hexaferrite. Also,

the average Ba: metal, Ba : (Fe + Zn + Cu + Mn) ratio was 1 : 8.4, which is close to the theoretical value of 1 : 8.7 for $Ba_3[Cu_{0.6}Zn_{0.2}Mn_{0.2}]_2Fe_{24}O_{41}$ hexaferrite.

The SEM image for $Ba_3[Cu_{0.8-x}Zn_xMn_{0.2}]_2Fe_{24}O_{41}$ ($x = 0.4$) sample in Figure 3(c) shows that a large fraction of the sample consists of sharp-edged hexagonal platelets with typical in-plane size of few micrometers, in addition to some larger particles and nonparticulate masses which may have evolved through DGG process. EDX measurements on a selected spot of the sample revealed that the Ba:Fe atomic ratio is 1 : 5.20, the Ba : Zn atomic ratio is 1 : 0.38, the Ba : Mn atomic ratio is 1 : 0.10, and the Ba : Cu atomic ratio is 1 : 0.47. These ratios are lower than the stoichiometric metal ratios in the Z-type phase and are closer to the stoichiometric Ba: metal ratios of 1 : 6.0, 1 : 0.4, 1 : 0.2, and 1 : 0.4, respectively, in $Ba_2Cu_{0.8}Zn_{0.8}Mn_{0.4}Fe_{12}O_{22}$ Y-type hexaferrite. This result confirmed the presence of Y-type hexaferrite crystallizing in the form of separate particles.

Figure 3(d) shows a typical SEM image for the samples with $x = 0.6$, which also revealed the presence of a large fraction of sharp-edged hexagonal platelets with a typical in-plane size in the range of 1.5 to 4 μm, in addition to large layered formations which may have evolved through the DGG process and topotactical reactions. EDX measurement on a selected spot of the sample revealed that the atomic ratio of Ba : Fe was 1 : 8.0, while the atomic ratios of (Ba : Cu, Zn, Mn) were 1 : 0.16, 1 : 0.28, and 1 : 0.07, respectively. These ratios are in agreement with the Ba : Fe, Cu, Zn, and Mn atomic ratios 0f 1 : 8.0, 1 : 0.13, 1 : 0.40, and 1 : 0.13 for

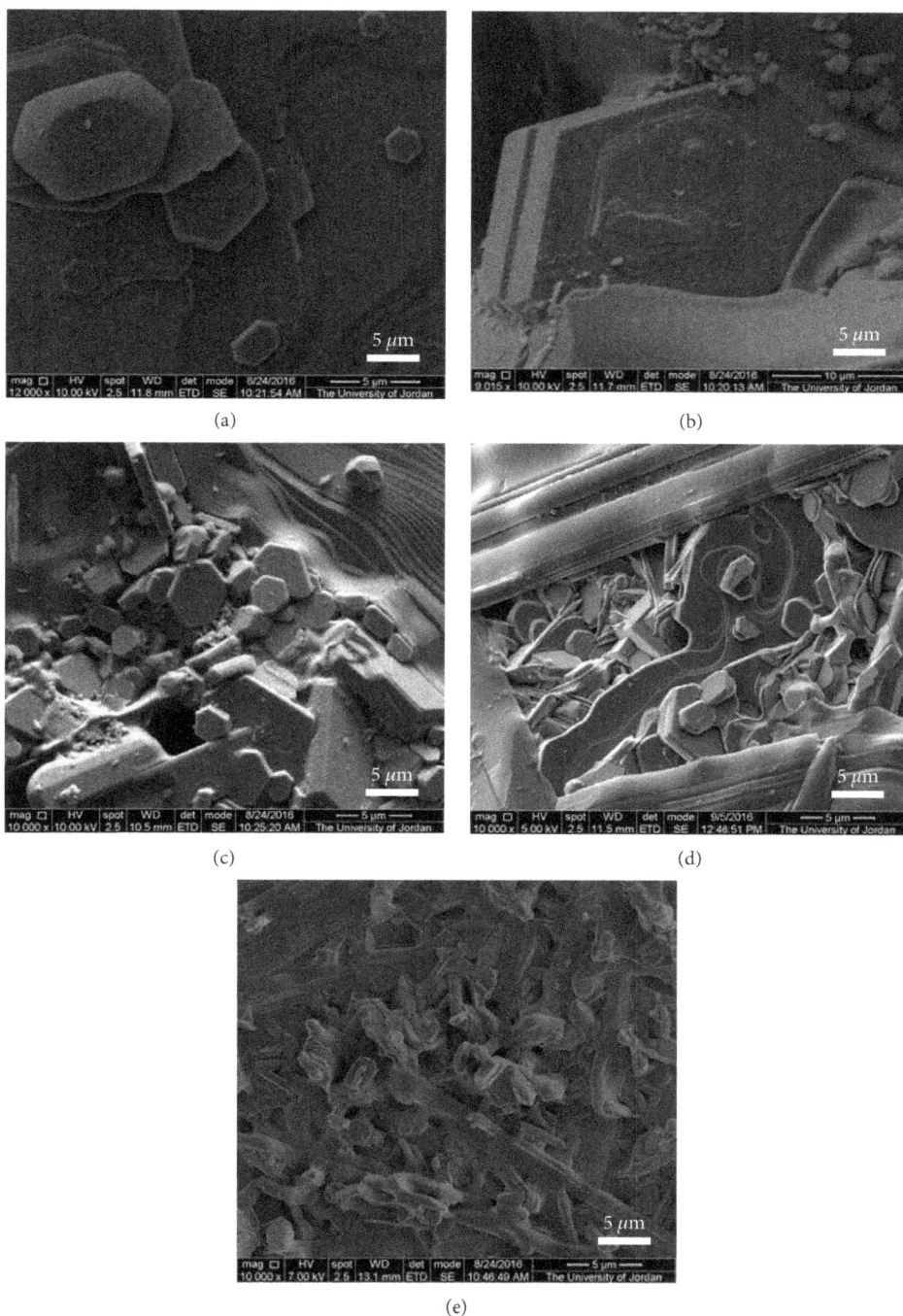

FIGURE 3: SEM images for the $Ba_3[Cu_{0.8-x}Zn_xMn_{0.2}]_2Fe_{24}O_{41}$ ($x = 0.2$. 0.4, 0.6, and 0.8) samples.

$Ba_3[Cu_{0.2}Zn_{0.6}Mn_{0.2}]_2Fe_{24}O_{41}$ hexaferrite. EDX measurement on a different spot of this sample, however, revealed that the Ba : Fe ratio was 1 : 6.59, which is close to the Ba:Fe ratio of 1 : 6.0 in Y-type hexaferrite. This could be an evidence of the presence of traces of the Y-type hexaferrite phase in the form of separate particles.

Figure 3(e) shows a typical SEM image for the sample with $x = 0.8$, which revealed crystallization of mainly hexagonal platelets with almost similar in-plane particle sizes in

the range from 1.67 to 2.3 μm. EDX measurements on a selected spot of the sample demonstrated that the atomic ratios of Ba : Fe, Zn, and Mn were 1 : 7.0, 1 : 1.0, and 1 : 0.10. The Ba : Zn ratio is consistent with Zn_2Y ($Ba_2Zn_2Fe_{12}O_{22}$) stoichiometry, while the observed Ba : Fe and Ba : Mn ratios are between their corresponding stoichiometric ratios in Zn_2Y and $Ba_3Zn_{1.6}Mn_{0.4}Fe_{24}O_{41}$ Z-type hexaferrite, which may indicate that the measured region of the sample contained a mixture of Y-type and Z-type hexaferrite phases.

3.3. VSM Results

3.3.1. Isothermal Magnetization.
The magnetic properties of $Ba_3[Cu_{0.8-x}Zn_xMn_{0.2}]_2Fe_{24}O_{41}$, $x = 0$, 0.2, 0.4, 0.6, and 0.8 samples were studied by VSM at room temperature in an applied field up to 10 kOe. The hysteresis loops of all samples (Figure 4) were typical smooth S-shaped curves revealing characteristics of relatively soft magnetic materials. An expanded view of the loops is shown in Figure 5, from which the remnant magnetization and coercive field for each sample were obtained directly. The saturation magnetization, however, was determined by applying the law of approach to saturation in the high-field range. In this range, the magnetization of the sample is dominated by magnetization rotation processes of the magnetic domains, where the magnetization is given by [49–51]:

$$M = M_s\left(1 - \frac{A}{H} - \frac{B}{H^2}\right) + \chi H, \quad (2)$$

where M_s is the spontaneous saturation magnetization of the domains per unit volume, A is a constant representing the contributions of inclusions and microstress, B is a constant representing the contributions of magnetocrystalline anisotropy, and χH is the forced magnetization term. For hexagonal crystals, the constant B is given by

$$B = \frac{H_a^2}{15}. \quad (3)$$

A plot of M versus $1/H^2$ in the field region 8 kOe < H < 10 kOe for each sample gave a straight line, indicating that the magnetization in this field range is dominated by the magnetocrystalline term. According to Equation (2), the intercept of the straight line with the M-axis determines the saturation magnetization, while the slope determines the constant B, from which the anisotropy field (H_a) can be determined using Equation (3). The magnetic parameters derived from the hysteresis loops are presented in Table 3. The metallic ions (Fe^{3+}, Mn^{2+}, Cu^{2+}, and Zn^{2+}) in the Z-type hexaferrite phase occupy 10 non-equivalent crystallographic sites, listed in Table 4 together with the number of metal ions per formula unit and magnetic sublattice spin orientation. The partial substitution of Cu^{2+} ions in Cu_2Z by Zn^{2+} or Mn^{2+} ions may lead to complex magnetic structure determined by the cationic preferential site occupation. The preferred site for Cu^{2+} ion is the tetrahedral site, while that for the Mn^{2+} ion is the octahedral site [52]. The Zn^{2+} ion generally prefers to occupy $4f_{IV}$ tetrahedral site, with some tendency to enter the $4f_{VI}$ octahedral site due to the large ionic diameter (0.74 Å) [53].

Figure 6 shows the effect of zinc content on the saturation magnetization and the coercivity of $Ba_3[Cu_{0.8-x}Zn_xMn_{0.2}]_2Fe_{24}O_{41}$. The saturation magnetization M_s for the sample with $x = 0.0$ was found to be 45 emu/g, which is close to the value of 46 emu/g reported for Cu_2Z hexaferrite [48]. As x increases, the saturation magnetization demonstrated increasing tendency, reaching the value of 53 emu/g for the sample with $x = 0.8$. While this value is in good agreement with recently reported values of 52.9 emu/g for Zn_2Y hexaferrite [45], it is lower than previously reported

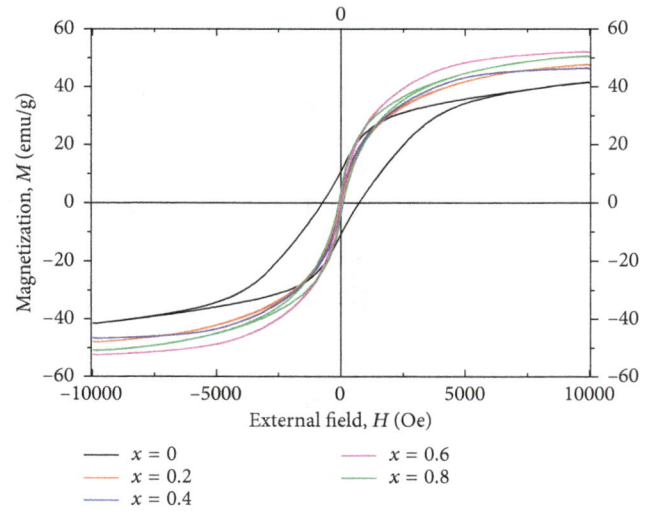

FIGURE 4: Hysteresis loops of $Ba_3[Cu_{0.8-x}Zn_xMn_{0.2}]_2Fe_{24}O_{41}$ ($x = 0.0$–0.8) samples.

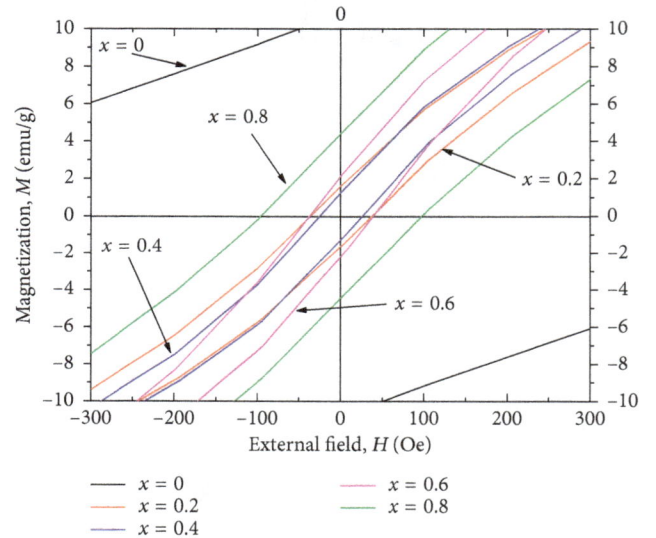

FIGURE 5: Expanded view of the hysteresis loops of $Ba_3[Cu_{0.8-x}Zn_xMn_{0.2}]_2Fe_{24}O_{41}$ samples.

TABLE 3: Saturation magnetization (M_s), remanence (M_r), squareness ratio M_{rs} (=M_r/M_s), coercive field (H_c), anisotropy field (H_a), initial susceptibility (χ_i), and initial permeability (μ_i) for the $Ba_3[Cu_{0.8-x}Zn_xMn_{0.2}]_2Fe_{24}O_{41}$ ($x = 0.0$–0.8) samples.

x	M_s (emu/g)	H_c (Oe)	M_r (emu/g)	M_{rs}	H_a (kOe)	χ_i	μ_i
0	45	735	11	0.24	11.18	0.0944	2.19
0.2	50	35	1.5	0.030	8.58	0.1678	3.11
0.4	48	28	0.99	0.021	5.61	0.1874	3.35
0.6	54	38	1.7	0.031	6.05	0.1805	3.27
0.8	53	107	4.3	0.081	7.92	0.1914	3.40

values of 58-59 emu/g [38, 48, 54]. Considering that the Zn-rich samples consist of a mixture of stoichiometric ratios of Zn_2Z, Zn_2Y, and M-type hexaferrites, with $Zn_2Z = Zn_2Y + M$,

TABLE 4: Crystallographic sites of small metal ions in Z-type baruim ferrite, their coordination, position in the unit cell, occupancy, and spin orientation of magnetic ions in the site [52].

Site	Coordination	Block	Number of ions/formula	Spin orientation
$12k_{VI}$	Octahedral	R-S	6	Up
$2d_V$	Five fold	R	1	Up
$4f_{VI}$	Octahedral	R	2	Down
$4e_{VI}$	Octahedral	T	2	Down
$4e_{IV}$	Tetrahedral	S	2	Down
$4f_{IV}$	Tetrahedral	S	2	Down
$4f*_{IV}$	Tetrahedral	T	2	Down
$4f*_{VI}$	Octahedral	S	2	Up
$12k*_{VI}$	Octahedral	T-S	6	Up
$2a_{VI}$	Octahedral	T	1	Up

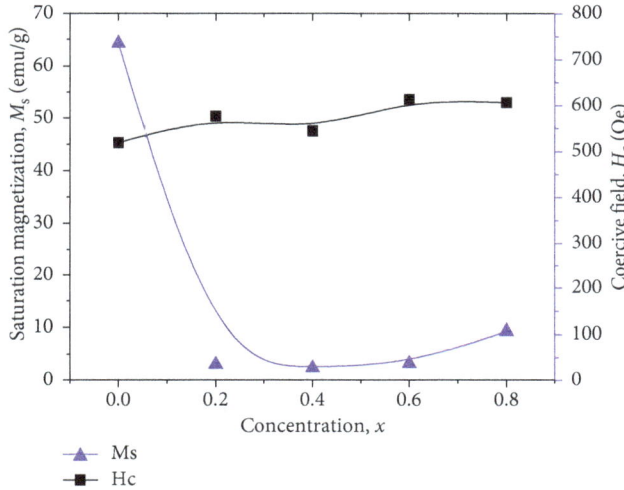

FIGURE 6: The saturation magnetization and coercivity as a function of concentration (x).

the saturation magnetization of the sample can be calculated from the following formula:

$$M_s = \left[\frac{M_M}{M_Z}\right](M_s)_M + \left[\frac{M_Y}{M_Z}\right](M_s)_Y, \qquad (4)$$

where M_M, M_Y, and M_Z are the molecular weights of BaM, Zn_2Y, and Zn_2Z hexaferrites, respectively, and $(M_s)_M$ and $(M_s)_Y$ are the saturation magnetizations of BaM and Zn_2Y hexaferrites, respectively. Using typical values of the saturation magnetization of 72 emu/g for BaM [55] and 35.2 emu/g for Zn_2Y [56], the calculated value of the saturation magnetization of the sample is ~51 emu/g. The fluctuations (of ~5%) of the observed values of Zn-substituted samples about the calculated value can be attributed to the presence of Cu and Mn ions in the lattice.

Figure 6 also shows that the coercivity decreased sharply at all levels of Zn substitution for Cu. In light of the XRD results, the magnetically hard BaM hexaferrite was one of the observed structural components of the sample with $x = 0$, which could be responsible for the relatively high coercivity

of the sample. The significant reduction of the coercivity of the Zn-substituted samples could therefore be associated with the transformation of the hard BaM magnetic phase to the soft Z-type phase, whereas the fluctuations of the coercivity of the Zn-substituted samples could be due to differences of the microstructure and particle size as demonstrated by SEM imaging.

Figure 7 shows the behavior of the anisotropy field (H_a) with increasing Zn concentration (x). The value of H_a demonstrated decreasing tendency with zinc substitution, recording values of 5.61 and 6.05 kOe at $x = 0.4$ and $x = 0.6$, respectively. These values are in good agreement with the reported value of 5.6 kOe for Zn_2Z hexaferrites [54]. Also, the values of 8.58 kOe and 7.92 kOe for the samples with $x = 0.2$ and 0.8 are in reasonable agreement with previously reported values of 6.66 and 7.0 kOe for Zn_2Z hexaferrites [45].

The initial susceptibility (χ_i) and magnetic permeability (μ_i) of each sample was determined from the initial magnetization curve (not shown for brevity). The obtained values of these magnetic parameters are also listed in Table 3 for all samples under investigation. The data indicated that the magnetic permeability improved by zinc substitution for Cu, and the improvement reached 55% at $x = 0.8$. This improvement could be potentially important for microwave applications of these materials.

3.3.2. Thermomagnetic Measurements.

Figure 8 shows thermomagnetic curves for $Ba_3[Cu_{0.8-x}Zn_xMn_{0.2}]_2Fe_{24}O_{41}$ samples at constants applied field of 100 Oe. A homogeneous sample with sharp magnetic phase-transition at the Curie temperature (T_c) should give a sharp peak in the derivative of the thermomagnetic curve at that temperature. The thermomagnetic curve for the sample with $x = 0.0$ exhibited magnetic transition at 336°C, and a second magnetic transition at 440°C, preceded by a Hopkinson peak at $T = 424$°C, which is an indicator of the existence of some small particles with superparamagnetic behavior near the Curie temperature [57]. In light of the reported values for the Curie temperatures of various pure hexaferrite phases (Table 5), the phase transition at 440°C is associated with the ferromagnetic-to-paramagnetic transition of the Cu_2Z phase. On the contrary, the magnetic transition associated with the peak in the derivative curve at 336°C is close to the reported magnetic transition temperature of Mn_2Y, where the difference between our value and those reported for the Mn_2Y phase could be due to the insufficiency of Mn and Cu content to form pure Mn_2Y. In addition, since the substitution of Fe by Cu in BaM was found to result in a decrease of the Curie temperature [32], the peak structure around 440°C could also be associated with ferromagnetic-to-paramagnetic transition in Cu-partially substituted BaM.

The derivative curve for the sample with $x = 0.2$ exhibited a relatively strong and sharp peak at 370°C, which is associated with the Curie temperature of the Zn_2Z phase. The observed higher transition temperature (compared to reported value of 360°C), and the occurrence of another small peak at 402°C, could be associated with sample

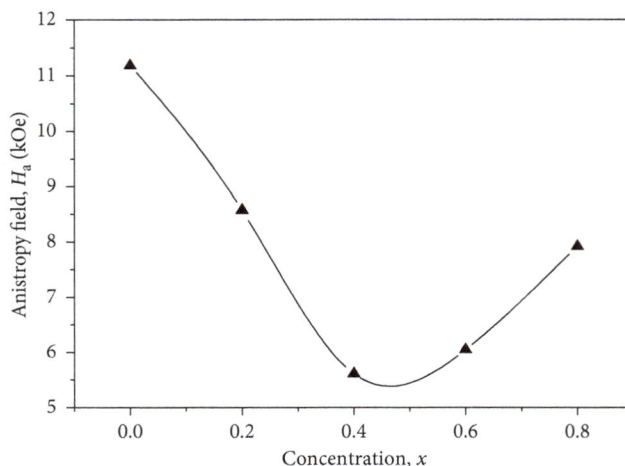

FIGURE 7: The anisotropy field as a function of Zn concentration (x).

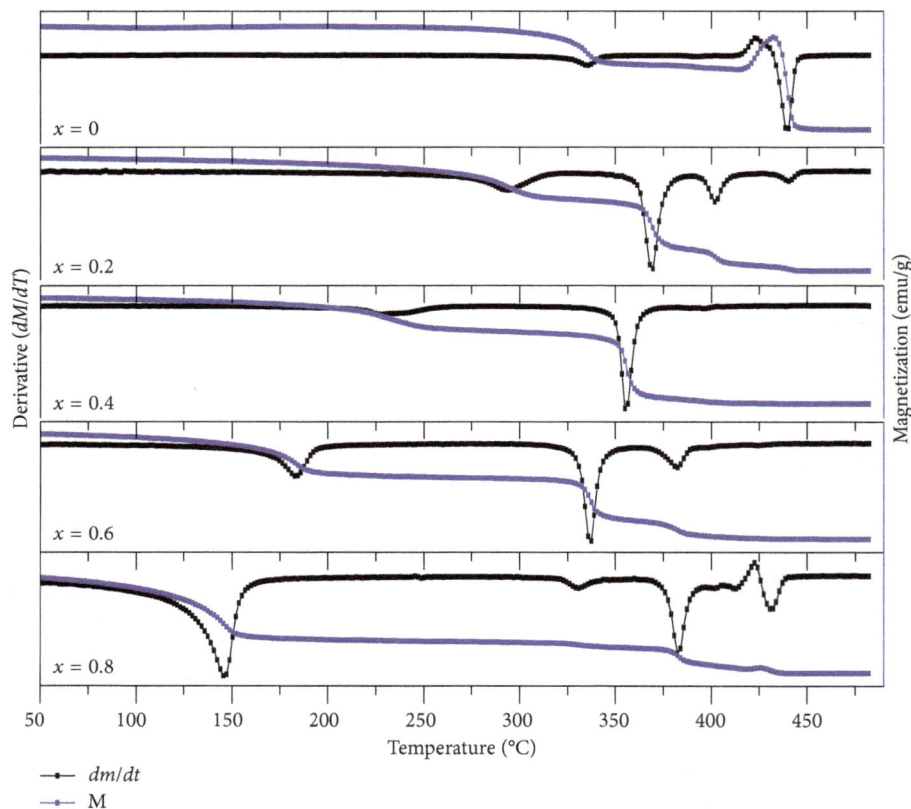

FIGURE 8: Thermomagnetic curves and their derivatives for the $Ba_3[Cu_{0.8-x}Zn_xMn_{0.2}]_2Fe_{24}O_{41}$.

TABLE 5: Curie temperature T_c (°C) for the hexaferrite phases observed in the samples [48].

Phase	BaM	Zn_2Z	Cu_2Z	Zn_2Y	Mn_2Y
T_c (°C)	450	360	440	130	290

inhomogeneity, where different phases of Cu-substituted Zn_2Z coexist with Curie temperatures between the characteristics of Zn_2Z and Cu_2Z. Furthermore, the small peak at

440°C could be associated with magnetic transitions of BaM and pure Cu_2Z phases, and the small peak at 294°C is in good agreement with the transition temperature of the Mn_2Y phase.

The derivative curve for the sample with $x = 0.4$ exhibited a strong peak at 356°C, which is associated with the transition temperature of the pure Zn_2Z phase. Also, the small peak at 396°C is associated with the transition temperature of the $(Cu,Zn)_2Z$ (Zn-substituted Cu_2Z) phase. On the

TABLE 6: Curie temperatures of the magnetic phases in $Ba_3[Cu_{0.8-x}Zn_xMn_{0.2}]_2Fe_{24}O_{41}$ samples.

x	Curie temperature T_c (°C)				Magnetic phases
0.0	336	—		440	Mn_2Y, BaM, Cu_2Z
0.2	294	370	402	440	Mn_2Y, Zn_2Z, Cu_2Z
0.4	227	356		396	$(Mn,Zn)_2Y$, Zn_2Z, $(Cu,Zn)_2Z$
0.6	183	338		383	$(Mn,Zn)_2Y$, Zn_2Z, $(Cu,Zn)_2Z$
0.8	146	329	383	431	Zn_2Y, Zn_2Z, BaM

contrary, the transition temperature at 227°C lies between the transition temperatures of Zn_2Y (130°C) and Mn_2Y (290°C) and could therefore be associated with the $(Mn, Zn)_2Y$ phase.

The magnetic transition at 338°C for the sample with $x = 0.6$ could be associated with the Zn_2Z phase, while the magnetic transition at 383°C is associated with the $(Cu, Zn)_2Z$ phase. Also, the thermomagnetic curve exhibited a small peak at 183°C, which can be associated with the Curie temperature of the $(Mn,Zn)_2Y$ phase. Finally, the thermomagnetic curve for the sample with $x = 0.8$ exhibited a strong peak at 146°C, which is associated with the transition temperature of the Zn_2Y phase. The phase transitions at 329°C and 383°C could be associated with Zn_2Z phases with slightly different chemical stoichiometry. The transition temperature of 431°C, however, could be associated with BaM with partial substitution for Fe. The observed Curie temperatures of the different magnetic phases in the samples are listed in Table 6.

The above results indicated that the Curie temperature of the magnetic phases generally decreased with the increase of Zn content as a consequence of the reduction of the superexchange interaction. The absence of Hopkinson peaks in the thermomagnetic curves of the Zn-substituted samples is indicative of the disappearance of superparamagnetic particles due to particle growth, in agreement with SEM imaging results.

4. Conclusion

The double-sintering process was successful in producing high-density $[Cu_{0.8-x}Zn_xMn_{0.2}]_2Z$ hexaferrites at a relatively low temperature of 1200°C. Intermediate levels of Zn substitution for Cu $(0.2 \leq x \leq 0.6)$ was found to be suitable for the production of a highly pure Z-type phase with increased saturation magnetization and reduced coercivity and magnetocrystalline anisotropy. All samples were highly dense, with a record density of 97% observed for the sample with $x = 0.2$. DGG, and topotactical reactions were responsible for growth of large particles and nonparticulate masses in the samples. The thermomagnetic curves demonstrated the complexity of the fabricated materials and indicated that the samples consisted of different magnetic phases with Curie temperatures generally decreasing with the increase of Zn concentration, which leads to the reduction of the superexchange interactions.

Conflicts of Interest

The authors declare that they have no conflicts of interest.

References

[1] P. Chang, L. He, D. Wei, and H. Wang, "Textured Z-type hexaferrites $Ba_3 Co_2 Fe_{24} O_{41}$ ceramics with high permeability by reactive templated grain growth method," *Journal of the European Ceramic Society*, vol. 36, no. 10, pp. 2519–2524, 2016.

[2] R. C. Pullar, "Hexagonal ferrites: a review of the synthesis, properties and applications of hexaferrite ceramics," *Progress in Materials Science*, vol. 57, no. 7, pp. 1191–1334, 2012.

[3] V. G. Harris, A. Geiler, Y. Chen et al., "Recent advances in processing and applications of microwave ferrites," *Journal of Magnetism and Magnetic Materials*, vol. 321, no. 14, pp. 2035–2047, 2009.

[4] S. Mahmood, "Ferrites with high magnetic parameters," in *Hexaferrite Permanent Magnetic Materials*, I. A.-A. S. H. Mahmood, Ed., pp. 111–152, Materials Research Forum LLC, Millersville, PA, USA, 2016.

[5] S. H. Mahmood, "High performance permanent magnets," in *Hexaferrite Permanent Magnetic Materials*, I. A.-A. S. H. Mahmood, Ed., pp. 47–73, Materials Research Forum LLC, Millersville, PA, USA, 2016.

[6] S. H. Mahmood, "Permanent magnet applications," in *Hexaferrite Permanent Magnetic Materials*, I. A.-A. S. H. Mahmood, Ed., pp. 153–165, Materials Research Forum LLC, Millersville, PA, USA, 2016.

[7] I. Abu-Aljarayesh, "Magnetic recording," in *Hexaferrite Permanent Magnetic Materials*, I. A.-A. S. H. Mahmood, Ed., pp. 166–181, Materials Research Forum LLC, Millersville, PA, USA, 2016.

[8] Ü. Özgür, Y. Alivov, and H. Morkoç, "Microwave ferrites, part 2: passive components and electrical tuning," *Journal of Materials Science: Materials in Electronics*, vol. 20, no. 10, pp. 911–952, 2009.

[9] P. Slick, "Ferrites for non-microwave applications," in *Handbook of Ferromagnetic Materials*, E. P. Wohlfarth, Ed., pp. 189–241, North-Holland Publishing Company, New York, NY, USA, 1980.

[10] J. Nicolas, "Microwave ferrites," in *Ferromagnetic Materials*, E. P. Wohlfarth, Ed., pp. 243–296, North-Holland Publishing Company, New York, NY, USA, 1980.

[11] J. Temuujin, M. Aoyama, M. Senna et al., "Preparation and properties of ferromagnetic Z-type hexaferrite from wet milled mixtures of intermediates," *Journal of Magnetism and Magnetic Materials*, vol. 311, no. 2, pp. 724–731, 2007.

[12] Y. Bai, F. Xu, L. Qiao, and J. Zhou, "High frequency magnetic mechanism of Ni-substituted Co_2Z hexagonal ferrite," *Materials Research Bulletin*, vol. 44, no. 4, pp. 898–900, 2009.

[13] X. Zhang, Z. Yue, S. Meng, B. Peng, and L. Yuan, "Magnetic and electrical properties of Z-type hexaferrites sintered in different atmospheres," *Materials Research Bulletin*, vol. 65, pp. 238–242, 2015.

[14] M. Aoyama, J. Temuujin, M. Senna, T. Masuko, C. Ando, and H. Kishi, "Preparation and characterization of Z-type hexaferrites, $Ba_{3(1-x)} Sr_{3x} Co_2Fe_{24}O_{41}$ with $x = 0$–0.5, via a two-step calcination with an intermediate wet milling," *Journal of Electroceramics*, vol. 17, no. 1, pp. 61–64, 2006.

[15] V. d. R. Caffarena, J. L. Capitaneo, T. Ogasawara, and M. S. Pinho, "Microwave absorption properties of Co, Cu, Zn: substituted hexaferrite polychloroprene nanocomposites," *Materials Research*, vol. 11, no. 3, pp. 335–339, 2008.

[16] J. Xu, G. Ji, H. Zou, Y. Zhou, and S. Gan, "Structural, dielectric and magnetic properties of Nd-doped Co_2Z-type hexaferrites," *Journal of Alloys and Compounds*, vol. 509, no. 11, pp. 4290–4294, 2011.

[17] F. Lotgering, "Topotactical reactions with ferrimagnetic oxides having hexagonal crystal structures—I," *Journal of Inorganic and Nuclear Chemistry*, vol. 9, no. 2, pp. 113–123, 1959.

[18] N. Solanki and R. B. Jotania, "Influence of Ca-doping on structural, magnetic and dielectric properties of $Ba_3Co_{2-x}Ca_xFe_{24}O_{41}$ hexaferrite powders," *Solid State Phenomena*, vol. 241, pp. 226–236, 2015.

[19] W. H. Dong and H. H. Yong, "Co_2Z type hexagonal ferrite prepared by sol-gel processes," *Materials Chemistry and Physics*, vol. 95, no. 2-3, pp. 248–251, 2006.

[20] H. Zhang, J. Zhou, Y. Wang, L. Li, Z. Yue, and Z. Gui, "The effect of Zn ion substitution on electromagnetic properties of low-temperature fired Z-type hexaferrite," *Ceramics International*, vol. 28, no. 8, pp. 917–923, 2002.

[21] H. Zhang, J. Zhou, Y. Wang et al., "Investigation on physical characteristics of novel Z-type Ba_3 $Co_{2(0.8-x)}$ $Cu_{0.40}Zn_{2x}$ Fe_{24} O_{41} hexaferrite," *Materials Letters*, vol. 56, no. 4, pp. 397–403, 2002.

[22] J. Bao, J. Zhou, Z. Yue, L. Li, and Z. Gui, "Dielectric behavior of Mn-substituted Co_2Z hexaferrites," *Journal of Magnetism and Magnetic Materials*, vol. 250, pp. 131–137, 2002.

[23] T. Tachibana, T. Nakagawa, Y. Takada et al., "X-ray and neutron diffraction studies on iron-substituted Z-type hexagonal barium ferrite: $Ba_3Co_{2-x}Fe_{24+x}O_{41}$ (x = 0–0.6)," *Journal of Magnetism and Magnetic materials*, vol. 262, pp. 248–257, 2003.

[24] S. Sharma, K. Daya, S. Sharma, K. M. Batoo, and M. Singh, "Sol–gel auto combustion processed soft Z-type hexa nanoferrites for microwave antenna miniaturization," *Ceramics International*, vol. 41, no. 5, pp. 7109–7114, 2015.

[25] X. Wang, L. Li, S. Su, Z. Gui, Z. Yue, and J. Zhou, "Low-temperature sintering and high frequency properties of Cu-modified Co_2Z hexaferrite," *Journal of the European Ceramic Society*, vol. 23, no. 5, pp. 715–720, 2003.

[26] V. da Rocha Caffarena and T. Ogasawara, "Characterization of Co-Zn-doped Z-type barium hexaferrite produced by co-precipitation method," *Journal of Metastable and Nanocrystalline Materials*, vol. 20-21, pp. 705–710, 2004.

[27] R. S. Alam, M. Moradi, H. Nikmanesh, J. Ventura, and M. Rostami, "Magnetic and microwave absorption properties of $BaMg_{x/2}Mn_{x/2}Co_xTi_{2x}Fe_{12-4x}O_{19}$ hexaferrite nanoparticles," *Journal of Magnetism and Magnetic Materials*, vol. 402, pp. 20–27, 2016.

[28] S. H. Mahmood, A. A. Ghanem, I. Bsoul, A. Awadallah, and Y. Maswadeh, "Structural and magnetic properties of $BaFe_{12-2x}Cu_xMn_xO_{19}$ hexaferrites," *Materials Research Express*, vol. 4, no. 3, article 036105, 2017.

[29] H. Sözeri, Z. Mehmedi, H. Kavas, and A. Baykal, "Magnetic and microwave properties of $BaFe_{12}O_{19}$ substituted with magnetic, non-magnetic and dielectric ions," *Ceramics International*, vol. 41, no. 8, pp. 9602–9609, 2015.

[30] S. H. Mahmood, A. Awadallah, Y. Maswadeh, and I. Bsoul, "Structural and magnetic properties of Cu-V substituted M-type barium hexaferrites," *Proceedings of IOP Conference Series: Materials Science and Engineering*, vol. 92, article 012008, 2015.

[31] D. Vinnik, A. Y. Tarasova, D. Zherebtsov et al., "Cu-substituted barium hexaferrite crystal growth and characterization," *Ceramics International*, vol. 41, no. 7, pp. 9172–9176, 2015.

[32] A. Awadallah, S. H. Mahmood, Y. Maswadeh et al., "Structural, magnetic, and Mossbauer spectroscopy of Cu substituted M-type hexaferrites," *Materials Research Bulletin*, vol. 74, pp. 192–201, 2016.

[33] Y. Bai, F. Xu, L. Qiao, and J. Zhou, "Effect of Mn doping on physical properties of Y-type hexagonal ferrite," *Journal of Alloys and Compounds*, vol. 473, no. 1-2, pp. 505–508, 2009.

[34] M. Chua, Z. Yang, and Z. Li, "Structural and microwave attenuation characteristics of ZnCuY barium ferrites synthesized by a sol–gel auto combustion method," *Journal of Magnetism and Magnetic Materials*, vol. 368, pp. 19–24, 2014.

[35] I. Ali, M. Islam, M. N. Ashiq, M. A. Iqbal, M. Awan, and S. Naseem, "Role of Tb–Mn substitution on the magnetic properties of Y-type hexaferrites," *Journal of Alloys and Compounds*, vol. 599, pp. 131–138, 2014.

[36] S. Bierlich and J. Töpfer, "Zn-and Cu-substituted Co_2Y hexagonal ferrites: sintering behavior and permeability," *Journal of Magnetism and Magnetic Materials*, vol. 324, no. 10, pp. 1804–1808, 2012.

[37] R. A. Nandotaria, C. C. Chauhan, and R. B. Jotania, "Effect of non-ionic surfactant concentration on microstructure, magnetic and dielectric properties of strontium-copper hexaferrite powder," *Solid State Phenomena*, vol. 232, pp. 93–110, 2015.

[38] Z. Li, L. Guoqing, N.-L. Di, Z.-H. Cheng, and C. Ong, "Mössbauer spectra of CoZn-substituted Z-type barium ferrite $Ba_3Co_{2-x}Zn_xFe_{24}O_{41}$," *Physical Review B*, vol. 72, no. 10, article 104420, 2005.

[39] X. Wang, L. Li, Z. Yue, S. Su, Z. Gui, and J. Zhou, "Preparation and magnetic characterization of the ferroxplana ferrites $Ba_3Co_{2-x}Zn_xFe_{24}O_{41}$," *Journal of Magnetism and Magnetic Materials*, vol. 246, no. 3, pp. 434–439, 2002.

[40] J. Li, H.-F. Zhang, G.-Q. Shao et al., "Synthesis and properties of new multifunctional hexaferrite powders," *Procedia Engineering*, vol. 102, pp. 1885–1889, 2015.

[41] H.-I. Hsiang and R.-Q. Yao, "Hexagonal ferrite powder synthesis using chemical coprecipitation," *Materials Chemistry and Physics*, vol. 104, no. 1, pp. 1–4, 2007.

[42] A. P. Daigle, M. Geiler, A. Geiler et al., "Permeability spectra of Co_2Z hexaferrite compacts produced via a modified aqueous co-precipitation technique," *Journal of Magnetism and Magnetic Materials*, vol. 324, no. 22, pp. 3719–3722, 2012.

[43] R. C. Pullar, I. K. Bdikin, and A. K. Bhattacharya, "Magnetic properties of randomly oriented BaM, SrM, Co_2Y, Co_2Z and Co_2W hexagonal ferrite fibres," *Journal of the European Ceramic Society*, vol. 32, no. 4, pp. 905–913, 2012.

[44] T. Nakamura and E. Hankui, "Control of high-frequency permeability in polycrystalline (Ba, Co)-Z-type hexagonal ferrite," *Journal of Magnetism and Magnetic Materials*, vol. 257, no. 2-3, pp. 158–164, 2003.

[45] E. S. Alhwaitat, S. H. Mahmood, M. Al-Hussein et al., "Effects of synthesis route on the structural and magnetic properties of $Ba_3Zn_2Fe_{24}O_{41}$ (Zn_2Z) nanocrystalline hexaferrites," *Ceramics International*, vol. 44, no. 1, pp. 779–787, 2018.

[46] T. Kato, H. Mikami, and S. Noguchi, "Performance of Z-type hexagonal ferrite core under demagnetizing and external static fields," *Journal of Applied Physics*, vol. 108, no. 3, article 033903, 2010.

[47] R. D. Shannon, "Revised effective ionic radii and systematic studies of interatomic distances in halides and chalcogenides," *Acta Crystallographica Section A*, vol. 32, no. 5, pp. 751–767, 1976.

[48] J. Smit and H. P. J. Wijn, *Ferrites*, Wiley, New York, NY, USA, 1959.

[49] B. D. Cullity and C. D. Graham, *Introduction to Magnetic Materials*, John Wiley & Sons, Hoboken, NJ, USA, 2nd edition, 2011.

[50] S. H. Mahmood, G. H. Dushaq, I. Bsoul et al., "Magnetic properties and hyperfine interactions in M-type $BaFe_{12-2x}Mo_xZn_xO_{19}$ hexaferrites," *Journal of Applied Mathematics and Physics*, vol. 2, no. 5, pp. 77–87, 2014.

[51] A. Awadallah, S. H. Mahmood, Y. Maswadeh, I. Bsoul, and A. Aloqaily, "Structural and magnetic properties of vanadium doped M-type barium hexaferrite ($BaFe_{12-x}V_xO_{19}$)," *IOP Conference Series: Materials Science and Engineering*, vol. 92, article 012021, 2015.

[52] A. Sharbati, S. Choopani, A. Ghasemi, I. Al-Amri, C. Machado, and A. Paesano Jr., "Synthesis and magnetic properties of nanocrystalline $Ba_3Co_{2(0.8-x)}Mn_{0.4}Ni_{2x}Fe_{24}O_{41}$ prepared by citrate sol-gel method," *Digest Journal of Nanomaterials and Biostructures*, vol. 6, pp. 187–198, 2011.

[53] M. Rasly and M. Rashad, "Structural and magnetic properties of Sn–Zn doped BaCo₂Z-type hexaferrite powders prepared by citrate precursor method," *Journal of Magnetism and Magnetic Materials*, vol. 337-338, pp. 58–64, 2013.

[54] Z. Li, Y. Wu, G. Lin, and L. Chen, "Static and dynamic magnetic properties of CoZn substituted Z-type barium ferrite $Ba_3Co_xZn_{2-x}Fe_{24}O_{41}$ composites," *Journal of Magnetism and Magnetic Materials*, vol. 310, no. 1, pp. 145–151, 2007.

[55] S. H. Mahmood, A. N. Aloqaily, Y. Maswadeh et al., "Effects of heat treatment on the phase evolution, structural, and magnetic properties of Mo-Zn doped M-type hexaferrites," *Solid State Phenomena*, vol. 232, pp. 65–92, 2015.

[56] I. Odeh, H. M. El Ghanem, S. H. Mahmood, S. Azzam, I. Bsoul, and A.-F. Lehlooh, "Dielectric and magnetic properties of Zn-substituted Co₂Y barium hexaferrite prepared by sol-gel auto combustion method," *Physica B: Condensed Matter*, vol. 494, pp. 33–40, 2016.

[57] S. H. Mahmood and I. Bsoul, "Hopkinson peak and superparamagnetic effects in $BaFe_{12-x}Ga_xO_{19}$ nanoparticles," *EPJ Web of Conferences*, vol. 29, article 00039, 2012.

6

The Impact of Magnetic Materials in Renewable Energy-Related Technologies in the 21st Century Industrial Revolution

Wallace Matizamhuka (ID)

Vaal University of Technology, Department of Metallurgical Engineering, Andries Potgieter Blvd, Vanderbijlpark, South Africa

Correspondence should be addressed to Wallace Matizamhuka; wallace@vut.ac.za

Academic Editor: Andres Sotelo

Magnetic materials specifically permanent magnets are critical for the efficient performance of many renewable energy technologies. The increased reliance on renewable energy sources has accelerated research in energy-related technologies the world over. The use of rare-earth (RE) metals in permanent magnets continues to be a source of greater concern owing to the limited RE supply coupled with dwindling reserves on the globe. This review focuses on how this has impacted on the state-of-the-art magnetic materials that continue to play a pivotal role in driving renewable energy technologies. Magnetic materials are perceived as key in driving the 21st century industrial revolution, and the participation of South Africa in this energy paradigm is critical in driving a new industrial revolution within the African continent. A number of opportunities are highlighted, and clarity is given on the several ubiquitous misconceptions and the risks on the heavy reliance on a single source for RE magnetic materials.

1. Introduction

In recent years, technology advancement focus has shifted towards renewables as the new energy source frontiers. Magnetic materials play a pivotal role in the efficient performance of devices in a wide range of applications such as electric power generation, transportation, air-conditioning, and telecommunications. The drive towards improving electricity transmission efficiency and the replacement of oil-based fuels by electric motors in transportation technologies has motivated researchers to focus on magnetic material technologies [1]. The increased demand for electricity in the past few decades will require a strong investment in energy-efficient power generation methods in some instances; lightweight and smaller sized devices are preferred such as in transportation and wind power [1]. The historical evolution of permanent magnetic materials spans over a 100-year period [1]. Manufacturing techniques of these magnets are well established, and the energy densities (a key figure of merit for permanent magnets) have been enhanced from ~1 MGOe for steels, increasing to ~3 MGOe for hexagonal ferrites, and peaking at ~56 MGOe for neodymium-iron-boron (Nd-Fe-B) magnets in the early 2000s [1]. The need for maximised energy densities at various operating temperatures has directed the research and development of rare-earth (RE) permanent magnets (RPMs) possessing improved temperature stability for electromotor applications [1]. However, due to the scarcity of RE magnetic metals such as Dy-dysprosium, Pr-praseodymium, and Sm-samarium, a more practical approach, which seems to be gaining more ground, is manipulation on the structure of grain boundary phases and internal interfaces, which enable better understanding of relevant coercivity mechanisms. Another approach is the development of textured nanocomposites, which may lead to the next generation of permanent magnets.

Although the relevance of this sector may be of little significance to the South African industry currently, it is indeed a reality that the world is moving towards cleaner and more efficient energy sources. As such, it is anticipated that a new industrial revolution will be ignited by such need, which will bring online specialised industries to meet the demand. In 2015, Stegan published a warning article on the concerns over the "rare-earth crisis" [2]. This was meant to serve as a wake-up call for decision-makers on the need to develop alternative supply chains for RE-based magnetic materials around the world. The present review, therefore, seeks to re-emphasise this need and make the South African policymakers, science and engineering community, and interested parties aware of the opportunities that lie ahead, which may require special funding, especially in research and development. The review provides a background on magnetic materials and further gives some highlights regarding the world market for magnetic materials. Special attention is given to RE-based permanent magnets as a key ingredient to the sustenance of the 21st century industrial revolution. The author also intends to clarify the several ubiquitous misconceptions and the risks of the heavy reliance on a single source for RE magnetic materials. The review concludes with an assessment of the available alternatives to address the shortages and the role South Africa can play in this rare-earth crisis.

2. Historical Background of Magnetic Materials

Over the years, society requirements have become more advanced and magnetic materials have become pivotal in the advancement of human civilisation. Magnetic product applications have evolved from the simple magnet distribution needs in the early 1930s to the more advanced high-performance motor applications in today's electric vehicles. A number of magnetic materials are available, ranging from the low-cost and low-energy ferrites to the more expensive and high-performance RE materials. Magnetic materials are generally classified in terms of their magnetic properties and uses. For instance, a material that is easily magnetised and demagnetised is referred to as a soft magnetic material, whereas a material that is difficult to demagnetise is referred to as a hard (permanent) magnetic material [3].

Since the 1930s, Alnico magnets have been widely used, initially in military electronic applications and later in civilian versions such as automotive and aircraft sensor applications. The development of Alnico magnets marked the beginning of a new way of thinking about magnetic materials where composite materials with multiple phases produced attribute superior to those of the individual components [4]. Alnico magnets are alloys mainly based on nickel, cobalt, and iron with smaller amounts of aluminium, copper, and titanium (typical composition in wt.%: Fe-35; Co-35; Ni-15; Al-7; Cu-4; Ti-4) [4]. They possess a fine microstructure consisting of micron- or submicron-scale ferromagnetic particles dispersed in a weak magnetic matrix [4]. They derive their magnetic strength by virtue of a phase separation in the alloy into ferromagnetic FeCo-rich and weakly magnetic NiAl-rich phases precipitated from the high-temperature homogenous composition [4]. To this date, the so-called *supermagnets* are based on this very principle [4].

In 1952, the Phillips Company (Eindhoven, Netherlands) announced the successful commercialisation of the first ceramic magnets [5]. These complex oxides are based on the prototypical composition $MO.6Fe_2O_3$ or equivalently $MFe_{12}O_{16}$ where M represents the divalent metals Ba, Sr, or Pb [4]. The most popular of these ceramic magnets is barium ferrite or barium hexaferrite ($BaFe_{12}O_{19}$) [4]. These magnets have commercial significance owing to their low cost and chemical inertness and because they are easy to process. They are classified as ferrimagnetics with both ferromagnetic (FM) and antiferromagnetic (AF) coupling between atomic moments, and the magnetic coupling depends on the specific crystallographic position of Fe ions [4]. However, the major drawback of these magnets is the decrease in magnetisation values with increasing temperatures, brittle behaviour, and low magnetisation values at room temperature [4]. This is offset by the high Curie temperature $T_c \sim 1223$–$1248°C$ (defined as the transition temperature from ferromagnetism to paramagnetism) making them suitable in areas of spintronic materials, battery cathodes, microwave communication, electric motors, and high-T_c superconductors [6–9].

In the mid-1960s, under the direction of Dr. Karl J. Strnat at the US Air Force Materials Laboratory later at the University of Dayton, Ohio, large magnetic products were reported in intermetallic compounds based on samarium-cobalt typically 5.1 MGOe ($40.6 kJ/m^3$) and later optimised to 18 MGOe ($143.2 kJ/m^3$ [10]. This family of compounds consisted of the general formula $RE(TM)_5$ containing RE metals Y-yttrium, Ce-cerium, Pr-praseodymium, Sm-samarium, and transition metal (TM) cobalt [4]. In 1972, further exploration resulted in the discovery of a new compound $RE_2(TM)_{17}$ the so-called "2-17" compounds [4]. $Sm_2(Co,Fe)_{17}$ was reported to possess a theoretical maximum energy product of up to 60 MGOe ($477.5 kJ/m^3$) [11]. The SmCo magnets were later commercialised with typical energy products in the range 22–32 GMOe (175–$255 kJ/m^3$) depending on the composition combined with an attractive Curie temperature ($\sim750°C$) making them suitable for high-temperature applications [12].

The drive towards neodymium-iron-boron ($Nd_2Fe_{14}B$) magnets was a result of the increased cost of Co in the late 1970s, a critical ingredient in the SmCo magnets [4]. Political instability in the DRC (former Zaire: source of 60 percent of Co world supply) in 1978 jeopardised the global supply of cobalt [4]. In 1982, the US Budget Office published a strategic policy option to minimise the US reliance on cobalt and to focus on cobalt substitutes for the manufacture of high-energy magnets [4]. In the mid-1980s, a new iron-based supermagnet $Nd_2Fe_{14}B$ (also known as 'Neo' or 2-14-1) was produced simultaneously at General Motors US through a rapidly solidified synthesis method and at Sumitomo through a liquid phase sintering method [13–15]. Today the commercialised supermagnets based on RE intermetallic compounds $Nd_2Fe_{14}B$ have typical maximum energy products on the order of 56MGOe ($\sim445.7 kJ/m^3$) with remanence of $B_r \sim 14 kG$ (1.4 T) and intrinsic coercivity of $H_{ci} \sim 10 kOe$ (796 kA/m) [16]. To put this into perspective, most magnets used on souvenirs such as displays on refrigerators possess typical energy products of <1MGOe ($8 kJ/m^3$) [4]. Although these magnets achieved high-performance scores, they possess a marginal Curie temperature in the range of 300–400°C with an operating temperature restricted to ~150°C plagued by both brittleness and a large propensity to corrode [4]. Figure 1 shows the development cycle of magnetic materials as described above.

2.1. World Market for Permanent Magnets.

In recent years, the choice of a permanent magnetic material for a given application is mainly based on a balanced consideration of price and performance [4]. The design goal for lightweight devices and smaller sizes has enabled NdFeB to be the magnet of choice for higher-end applications [4]. It is noteworthy to mention here that the fastest growing market for permanent magnetic materials is the energy-related applications [1]. The production of sintered NdFeB magnets has experienced a phenomenal growth from ~6000 t in 1996 to ~63,000 t in 2008 with the bulk of this (~80%) being produced in China [18]. The driving force behind the growth in demand for permanent magnets is the high-energy consumer electronic products such as DVDs, iPods, cameras, sensors, and cellphones [1]. The use of high-energy magnets enables the miniaturisation of these devices, which drastically reduces the electrical power requirements of such devices [1]. There are also other applications where NdFeB magnets have been used in large quantities such as in electric-assisted/electric vehicles (EAVs/EVs), speakers, magnetic separation units, windmill generators, magnetic resonance imaging (MRI), and electric bicycles [1].

Hexaferrite magnets have the largest global market share on a tonnage basis accounting to ~85 pct (by wt) of total sales by virtue of their much lower price [1, 4]. However, the NdFeB family of magnets remains the permanent magnet of choice for high-end applications and represents over 50 percent of magnet sales on a dollar basis (Figure 2) [19, 20]. A projection on the permanent magnet sales for the four major magnet types: Alnico, SmCo, ferrite, and NdFeB shows a quasiexponential growth from 1985 to 2020 with NdFeB magnet sales projected to be over $17bn by 2020 (Figure 2) [19, 20].

The use of hard magnetic material NdFeB has offered quite significant performance benefits, which has enabled the development of highly efficient traction motors not possible with other technologies [21]. However, hard NdFeB magnets contain the RE Nd whose supply together with that of other RE metals (Sm-samarium, Dy-dysprosium, Gd-gadolinium, Pr-praseodymium, Pm-promethium, and Er-erbium) is environmentally unsustainable [21]. This has resulted in the prices of such RE metals soaring during the 2011–2012 period (Figure 3), raising a lot of concern over their continued use as ingredients for hard magnetic materials [21].

The discussion below provides insight on the current status and the efforts made so far in the search for alternative materials to replace the RE magnets.

2.2. Magnetic Materials in Renewable Energy Applications.

Historically, the drive behind the development of permanent magnets emanates from the need to obtain high magnetic energy product over smaller volumes of magnets, which could be utilised in a number of technological applications such as clean energy technologies (wind turbine generators and hybrid regenerative motors), transportation components, and

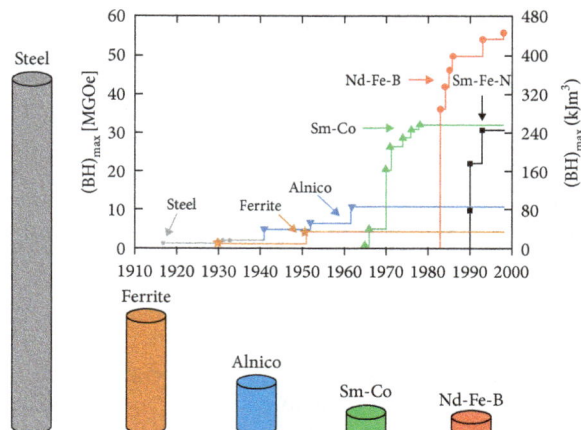

FIGURE 1: A schematic representation of the development cycle for permanent magnetic materials and a representation of different types of materials with comparable energy densities [17].

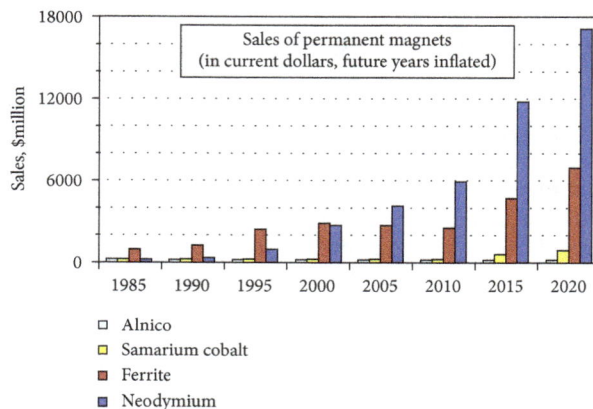

FIGURE 2: Projected growth in demand for permanent magnets from 1985 to 2020 [19].

consumer products [4]. Magnetic materials play a pivotal role in modern society owing to their unique ability to perform a number of tasks as follows:

(i) Convert mechanical to electrical energy

(ii) Transmit and distribute electric power

(iii) Facilitate microwave communications

(iv) Provide basis for data storage systems

Theoretically, a strong permanent magnet is characterised by a large remnant magnetic flux (remnant or B_r), which must be maintained in the absence of a magnetic field coupled with a large resistance to demagnetisation (coercivity of H_c or intrinsic coercivity of $_iH_c$). Magnetic properties can be either intrinsic or extrinsic. Intrinsic magnetic properties are those determined by the crystal structure and composition of the material and are ideally insensitive to the material's microstructure. Such properties include saturation magnetisation, M_s, and magnetic ordering temperatures, i.e., ferromagnetic Curie temperature (T_c) and antiferromagnetic Neel temperature (T_N) [4]. T_c and T_N define the temperatures at which ambient thermal

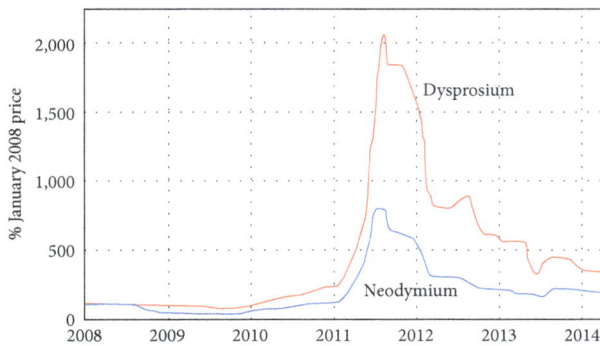

FIGURE 3: Price spike of RE metals experienced during the 2011–12 period as a result of controlled supplies [19].

energy becomes sufficiently large to destroy effective magnetic ordering. Key to the high performance of permanent magnets is that all the parameters mentioned above should be insensitive to temperature in order to maintain their integrity under elevated temperature operating environments [4]. A Curie temperature well above room temperature is more suitable, and the constituent materials should be preferably inexpensive, easy to process, lightweight, nontoxic, and corrosion resistant [22].

There is no doubt that policymakers, scientists, and other interested stakeholders around the globe are focused on reducing the reliance on hydrocarbon energy sources in favour of renewables [2]. A number of reasons have been cited, which include oil price volatility and economic vulnerability, concerns over global warming, and the general need for diversification in energy portfolios [2]. Permanent magnets find use in many renewable energy technologies and are key to the success of the renewable energy industry. An assessment by the US Department of Energy in 2011 on the criticality of REs to clean energy applications for both short-term (0–5 yrs) and medium-term (5–15 yrs) periods clearly indicates the importance of REs in the sustenance of renewable energy technologies [23]. A selected few REs relevant to the present review are summarised in the Table 1. Rare-earth elements (REEs) with the greatest severity of supply risk are considered critical, and those at medium or low risk are deemed near critical or not critical, respectively.

According to the World Wind Energy Association (WWEA) statistics, wind power is the fastest growing sector within the renewable energy sector [24]. Wind power capacity is expected to reach 1.9 mln MW in 2020. Permanent magnet is a key component in the construction of wind turbine generators used for transforming mechanical power into electrical power [1]. The design of wind turbine generator drive system (Figure 4) has evolved over the years to meet higher demands for greater energy yield, reliability, and lower maintenance requirements [25]. NdFeB permanent magnets enable the replacement of mechanical gearboxes in wind turbines with direct-drive (DD) permanent magnet generators thus reducing the overall turbine weight, cost of other components such as concrete and steel required to support heavy gearboxes, and a reduction in the number of moving parts which basically allows for greater reliabilities and efficiencies [2, 26].

The new design offers benefits such as lower volume and weight, higher operating efficiency, higher torque density, easiness to assemble and maintain, and 50% lower internal heat generation [27–29].

Since the invention of the "mixed drive autovehicle" by the German inventor Henri Pieper in 1905, there has been phenomenal progress made to the rudimentary design patented then [1]. Today, electric/hybrid electric vehicles (EVs/HEVs) have re-emerged as a realistic alternative to gasoline internal combustion vehicles [1]. The success of EVs is owed to the highly efficient permanent magnet motors used to run the power train of the EV. In 1997, HEV became mainstream with the launch of the Toyota Prius [21]. Today, other automobile manufacturers have launched their own EV/HEV brands such as BMW i3 and Nissan Leaf owing to improvements made in the technology over the years [21]. The use of hard magnetic material NdFeB offers quite significant benefits, which has enabled the development of highly efficient traction motors not possible with other technologies. Typical composition of NdFeB magnets used in traction motors is around $Nd_{22}Dy_{11}Fe_{6.5}B_1Cu_{0.1}$ by wt.% implying ~33wt.% comprises the precious RE elements [30].

The magnetocaloric effect (MCE) is an alternative refrigeration method, which makes use of adiabatic magnetisation [1]. The working principle of MCE is based on the concept that the temperature of a suitable material changes when magnetised or demagnetised [31]. Magnetisation of a magnetocaloric material is equivalent to the compression of gas (heating), while demagnetisation is equivalent to expansion of gas (cooling) [32]. The MCE is fast becoming the preferred refrigeration method of the future owing to a number of benefits in comparison with the compressor-based refrigeration method. The most prominent benefits of MCE refrigeration include absence of harmful gases, generation of much less noise, and that it can be built more compactly because the working material is solid (Figure 5) [1]. Moreover, it has been demonstrated that the cooling efficiency in magnetic refrigerators containing gadolinium (Gd) can reach 60 percent of the theoretical efficiency limit compared to only ~45 pct in the best gas-compression refrigerators [33].

Lastly, the use of soft magnetic materials in transformers for power generation and conversion for the electrical grid plays a pivotal role in electricity generation. The performance of soft magnets is material specific and is dominated by properties such as low coercivity and core losses, high saturation magnetisation, resistivity, and permeability, which makes these materials more attractive for the efficient transmission and distribution of electricity [1]. There are efforts to revolutionise the way power is delivered by designing advanced electric storage systems, smart controls, and power electronics for AC-DC conversion, referred to as "smart grids," using a number of advanced materials and devices to provide greater efficiencies and more affordable and sustainable energy use for the long term [22, 34].

2.3. Sustainability of Permanent Magnets Supply in the 21st Century. The RE metals are pivotal in the production of RE supermagnets owing to their superior properties from the

TABLE 1: A summary of selected rare-earth elements, applications, and criticality to clean energy [2].

Atomic no.	Name	Type	Selected applications	Crustal abundance (ppm)	Criticality to clean energy: short/medium term
57	Lanthanum	Light	Battery alloys, lasers, phosphors	31	Near critical/not critical
58	Cerium	Light	Ni-metal hydride (NiMH) batteries for hybrid/electric vehicles, phosphor powders	63	Near critical/not critical
59	Praseodymium	Light	Permanent magnets, NiMH batteries, photographic filters	7.1	Not critical/not critical
60	Neodymium	Light	Permanent magnets, lasers, astronomical instruments	27	Critical/critical
62	Samarium	Light	Permanent magnets, reactor control rods	4.7	Not critical/not critical
65	Terbium	Heavy	Permanent magnets, lighting and display phosphors	0.7	Critical/critical
66	Dysprosium	Heavy	Permanent magnets, lasers, lighting	3.9	Critical/critical
67	Holmium	Heavy	Magnets	0.83	N/A/N/A
69	Thulium	Heavy	Magnets	0.3	N/A/N/A

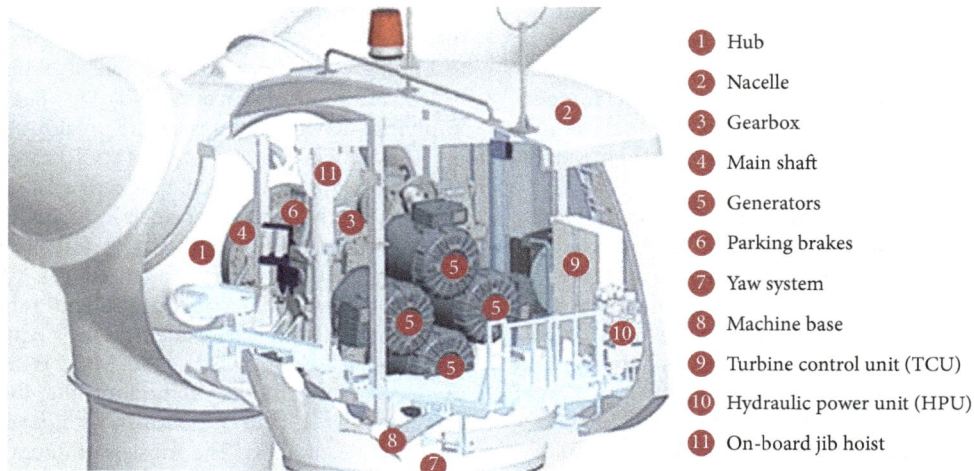

1 Hub
2 Nacelle
3 Gearbox
4 Main shaft
5 Generators
6 Parking brakes
7 Yaw system
8 Machine base
9 Turbine control unit (TCU)
10 Hydraulic power unit (HPU)
11 On-board jib hoist

FIGURE 4: Exploded view of a typical 2.5 MW clipper wind turbine showing the position of four permanent magnet generators (5) [25].

high magnetisation provided by the 3d transition metal crystalline sublattice to the extremely strong magneto-crystalline anisotropy field provided by the 4f electrons [22]. RE elements consist of 17 chemical elements in the periodic table, namely, scandium (Sc), yttrium (Y), and 15 lanthanides. The lanthanides are categorised as light REs composed of elements with atomic numbers $Z = 57$ (lanthanum, La) to 61 (promethium, Pm), the medium RE elements ranging from $Z = 62$ (samarium, Sm) to $Z = 64$ (gadolinium, Gd), and lastly, the heavy RE elements include $Z = 65$ (terbium, Tb) through to $Z = 71$ (lutetium, Lu) [4]. The elements, which are key to the production of permanent magnets are Pr-praseodymium ($Z = 59$), Nd-neodymium ($Z = 60$), Sm-samarium ($Z = 62$), and Gd-gadolinium ($Z = 64$) for specialised applications and extremely important RE elements are Tb-terbium ($Z = 65$) and Dy-dysprosium ($Z = 66$) [4]. It is fascinating to note that the rarest elements in the earth's crust are not particularly the rare elements. For instance, cerium, which is part of the REs, is the 25th most abundant element of the 78 common elements in the earth's crust (~ 60 ppm) with the elements thulium and lutetium being the least abundant REs (~ 0.5 ppm) but are still more plentiful than the precious metals such as gold and platinum [1]. Of note is that a century

ago, the RE elements were referred to as rare mainly because of their rarity, but nowadays, rare refers more to the difficulty in isolating single elements from their ores owing to their extremely similar physicochemical properties [2]. Moreover, the global distribution of REs is uneven with major world geologic supplies originating from a handful of sources (Table 2 and Figure 6) [4].

A recent analysis by the US Geological Survey indicates that China holds approximately 39 percent of the world's total reserves of REs accounting to 55,000,000 t with the rest distributed as follows: Brazil (22,000,000 t), the Commonwealth of Independent States (CIS) (19,000,000 t), the US (13,000,000 t), India (3,100,000 t), and Australia (2,100,000 t) with the remainder 25,800,000 t distributed among smaller reserves in Malaysia, Vietnam, and other countries [36, 37]. The RE global production stands at about 110,000 tpy of which China supplies 90 pct with the remaining 10 pct spread over smaller suppliers as follows: the US (~ 4000 tpy), India (~ 2900 tpy), Russia (~ 2400 tpy), and Australia (~ 2000 tpy) with smaller amounts from Brazil, Malaysia, and Vietnam [36, 37]. Currently, China is the only country in the world with the capacity to process heavy REs with its integrated supply chain developed over the past few decades [38].

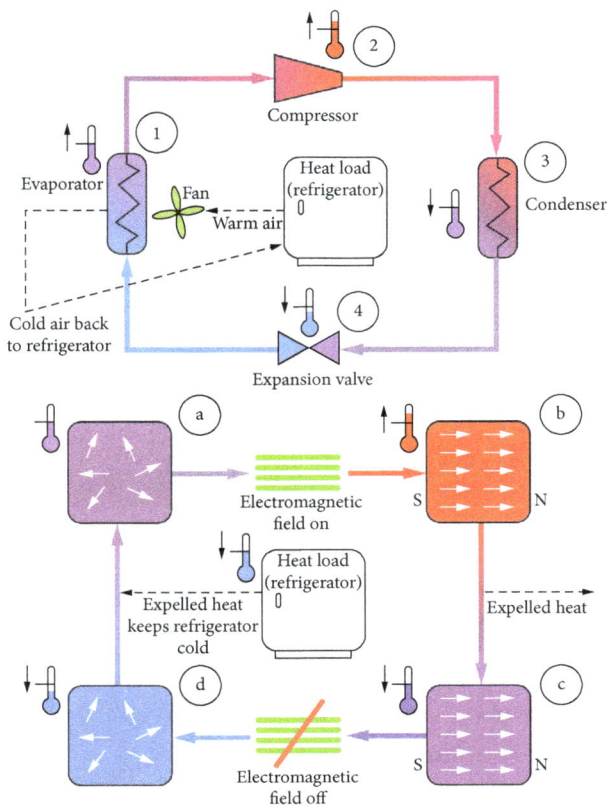

FIGURE 5: A comparison between the conventional vapour compression refrigeration (a) and magnetic refrigeration (b) [32].

TABLE 2: A summary of the world production and reserves of rare-earth elements in 2010 [4, 20].

Country	Mine production (metric tons)	Percent of total	Reserves (million metric tons)	Percent of total
China	130,000	97.3	55.0	50
United States	None	—	13.0	13
Russia	Not reported	Not reported	19.0	17
Australia	Not reported	Not reported	1.6	1.5
India	2700	2	3.1	2.8
Brazil	550	0.42	Small	—
Malaysia	250	0.27	Small	—
Others	N/A	22	—	20
Total	133,600	—	1100	

In 1995, a Chinese-based consortium acquired General Motors Magnequench-Delco Remy Division, created in 1986 to commercialise $Nd_2Fe_{14}B$ magnetic materials [39]. A mass exodus of RE magnet manufacturing capability from the US to China and subsequently an effective 'brain drain' of engineers and scientists [4] followed this.

It is estimated that the global RE industry was worth $1,3 bn in 2010, whereas the end-user industries requiring REs are worth a lucrative ~$4.8 trillion the same year [40]. This prompted the Chinese government to limit world exports of REs in a bid to attract more end use manufacturing industries to operate in China [40]. This spiralled the so-called rare-earth crisis in the late 2000s. In the year 2011, the US, EU, and Japan filed a WTO complaint against China of which China was found to have violated the international trade law by restricting overseas sales of REs [41].

Although the RE crisis is expected to subside over the years, the supply of REs for high-end applications will strategically remain important. Owing to the high demand for advanced permanent magnets to power high-end technologies, there is a clear shift in the future design paradigms towards low or zero RE content [4]. However, the complete substitution of REs in permanent magnets has proven problematic, and manufacturers and scientists are rather more focused on reducing RE content [2]. On the contrary, the complete elimination of REs in renewable technologies is a lengthy process, which can easily take a decade before complete replacement can be realised.

In recent years, the re-evaluation of mines outside China has taken the centre stage. With the currently prevailing attractive prices for REs, reopening mines has appeared attractive again [2]. One such mine is the Steenkampskraal mine in South Africa, which came online in the late 1950s, primarily producing thorium from an ore containing REs [2, 42]. The mine is currently being refurbished as an RE mine, and in 2014, it was reported that a pilot plant testing the extraction of REs has been preliminarily successful [42]. However, the US Department of Energy projected that the increase in demand for RE supplies in sufficient quantities will be required to offset the heavy reliance on Chinese mines [23]. In its 2013 survey, the US Geological Survey presented additional surveys for RE reserves within the African continent, which include Mozambique, Malawi, Madagascar, South Africa, and Tanzania [35].

2.4. Innovation of Alternative Permanent Magnetic Materials. It is no doubt that the use of RE metals Nd and Dy imparts high maximum energy product, resistance to demagnetisation, and high-temperature stability in $(Nd,Dy)_2(Fe, Co)_{14}B$ magnets in comparison with other magnetic materials [2, 21]. The maximum energy product is defined as a measure of the magnetic energy, which can be stored per unit volume by a magnetic material. Mathematically, it can be expressed as the product of the residual magnetic flux density (degree of magnetisation, M) and its coercivity (ability to resist demagnetisation once magnetised, H) [21]. It has been demonstrated that NdFeB magnets allow a very strong magnetic field to be generated in a very small volume. To put this into perspective, about five times less the cross-sectional area of NdFeB is required to produce the same magnetic field as an electromagnetic coil [21]. Moreover, an electromagnetic coil produces more losses in the winding arising due to electrical resistance of the conductor [21].

Although RE magnetic materials offer a number of benefits, it has been demonstrated that their replacement does offer some attraction in terms of cost, environmental footprint, and some aspects of performance [21]. There has been growing concern worldwide over the security of the

Canada

Russia *and other former Soviet Union countries*

Russia	
Mine Production (mt)	-
Reserves (mmt)	19.0
Reserve Base (mmt)	21.0

United States

United States	
Mine Production (mt)	none
Reserves (mmt)	13.0
Reserve Base (mmt)	14.0

1% *from Russia*
3% *from France*
91% *from China*
3% *from Japan*
2% *from Other*

China

China	
Mine Production (mt)	105,000
Reserves (mmt)	55
Reserve Base (mmt)	89

India

India	
Mine Production (mt)	2,800
Reserves (mmt)	2.7
Reserve Base (mmt)	1.3

Brazil

Brazil	
Mine Production (mt)	250
Reserves (mmt)	small
Reserve Base (mmt)	-

Malaysia

Malaysia	
Mine Production (mt)	280
Reserves (mmt)	small
Reserve Base (mmt)	-

Australia

Australia	
Mine Production (mt)	-
Reserves (mmt)	1.6
Reserve Base (mmt)	5.8

S. Africa

Mine Production (2011): 108,330 mt
- China (105,000) 96.9%
- India (2,800) 2.6%
- Malaysia (280) .26%
- Brazil (250) .23%

Reserves (2010): 110 mmt
- China (55) 50%
- Other (18.7) 17%
- Russia (19) 17%
- U.S. (13) 12%
- Australia (1.6) 1.5%
- India (2.7) 2.5%

Reserve Base (2008): 154 mmt
- China (89) 59.3%
- Russia (21) 14%
- U.S. (14) 9.3%
- Australia (5.8) 3.9%
- India (1.3) 1%
- Other (23) 12.5%

Other: *Brazil, Malaysia, Canada, S. Africa, etc.*

U.S. Imports
Countries with significant REE production, reserves, or reserve base
mt: *metric tons*
mmt: *million metric tons*

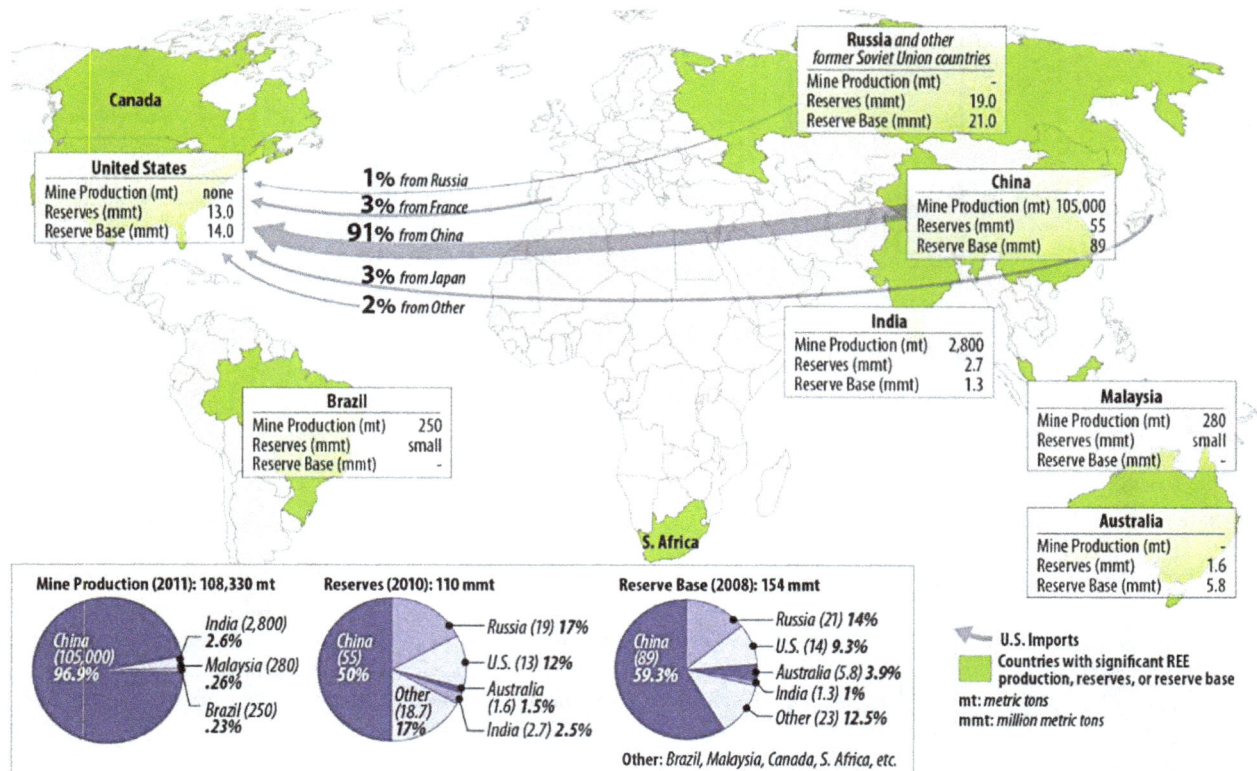

FIGURE 6: A representation of rare-earth world production, reserves, and US imports [35].

supply of REs critical to the manufacture of NdFeB magnets [43]. It must be noted, however, that a shift from REs may also trigger increased costs in the replacement metals, which could easily offset any benefits of eliminating REs [21]. A number of approaches are being examined to develop new RE-free magnet alloys and processing concepts to design structures that can match the magnetic properties of NdFeB permanent magnets.

Hitachi Metals developed a process, which involves the diffusion of dysprosium (Dy) into magnetic materials as opposed to direct alloying [44]. This effectively reduces the amount of Dy required in the magnetic products. There are reports on technologies based on size reduction into the nanorange to enhance the maximum energy product [45]. Design engineers have also used computer modelling to optimise electrical machine geometries, which in turn maximises power output whilst maintaining a low amount of REs required [44].

Lastly, it has been demonstrated that approximately 25 pct scrap is generated during machining of sintered magnets, and an enormous amount of this waste is discarded without attempting to recycle [46]. The current high prices of RE magnets have compelled producers to consider recycling, and a number of attempts are underway to develop technologies, which are efficient for the production of not only RE magnets from scrap but also disused magnets from old devices. There are two methods that have been developed to obtain powders from NdFeB waste-sintered magnets, namely, the hydrogen decrepitation (HD) process

and the hydrogen decomposition desorption recombination (HDDR) process [18, 47, 48]. However, the major drawback is to develop high-coercivity anisotropic sintered magnets. Furthermore, the highly oxidised Nd-rich phases obtained during the recycling process are not well suited for the production of fully densified NdFeB magnets without some form of blending with purer phases [49]. In 2003, Kawasaki et al. reported that sintered magnets were successfully reproduced by blending jet-milled NdFeB with a $Nd_{80}Fe_{20}$ binary alloy [50]. Zakotnik et al. [51] added 1.0 wt.% Nd to recycled jet-milled NdFeB powders, which was sufficient to recover the magnetic properties through a hydrogen decrepitation (HD) method. Li et al. blended recycled milled NdFeB magnets with 24 wt.% $Nd_{22}PrFe_{14}B$ powder, which restored B_r to 99.2 pct, $_iH_c$ to 105.65 pct, and $(BH)_{max}$ to 98.65 pct [52, 53]. In two separate studies, the use of DyF_3 salt to produce $(Nd,Dy)_2Fe_{14}B$ shell on the surface of recycled powders was proposed [53, 54]. Recently, Sepehri-Amin et al. [49] reported a recycled magnet with a nominal composition $Nd_{21.63}Pr_{6.43}Dy_{3.42}Fe_{64.75}Ga_{0.1}Zr_{0.11}Al_{0.28}Co_{1.74}Cu_{0.32}B_{0.97}C_{0.12}O_{0.13}$ wt.% produced through a grain boundary modifier $Nd_2Fe_{14}B$ in powder form. The magnetic properties of the recycled magnet were reported to be superior to those of commercial NMX-43SH-grade sintered magnet. This was attributed to the formation of the distinct grain boundary phase and enrichment of ~0.8% at Dy in the shell of $Nd_2Fe_{14}B$ grains [49]. In all these studies, it is clear that the composition selection is quite critical to recover comparable hard magnetic properties.

3. Outlook

It is no doubt that the demand for RE permanent magnets is on the increase and there are a number of arguments that RE magnets may not offer the best long-term solution owing to the high cost and risk in the supply of critical rare-earth elements (REEs) used in permanent magnets. Moreover, some researchers have indicated that the influence of permanent magnets in motors may act as a source of inefficiency. However, in the author's opinion there exist opportunities in the supply chain development of RE magnets within South Africa in the short to medium term, provided the bigger players do not resort to predatory pricing behaviour to scare off emerging players as witnessed during the 2011-12 period. Furthermore, South African higher learning institutions need to step up their support by formulating specialised training for engineers and technologists in the area of magnetic materials including the research and development of such. There is indeed a critical skill shortage in this area world over. Research capability needs to be geared up towards finding alternative hard magnetic materials containing low or no REs, design of alternative technologies, and developing recycling technologies. Through such initiatives, some automotive manufacturers such as Renault and Tesla have successfully designed wound rotor and induction motor technologies, respectively, in their quest to finding alternatives to RE permanent magnets. In the long term, there is indeed potential of completely replacing RE magnets with low-cost ferrites with even higher traction. It is simply a matter of time and investing enough funding to achieve a level of design maturity commensurate with performance targets sought. In conclusion, South Africa can take advantage of this opportunity by collaborating with organisations in the know to enable the establishment of a niche-level 21st century industry within Africa.

Conflicts of Interest

The author declares that there are no conflicts of interest.

References

[1] O. Gutfleisch, M. A. Willard, E. Brück, C. H. Chen, S. G. Sankar, and J. P. Liu, "Magnetic materials and devices for the 21st century: stronger, lighter, and more energy efficient," *Advanced Materials*, vol. 23, no. 7, pp. 821–842, 2011.

[2] K. S. Stegen, "Heavy rare earths, permanent magnets, and renewable energies: an imminent crisis," *Energy Policy*, vol. 79, pp. 1–8, 2015.

[3] I. R. Harris and A. J. Williams, *Material Science and Engineering*, EOLSS, Vol. II, EOLSS, Oxford, UK, 2000.

[4] L. H. Lewis and F. Jimenez-Villacorta, "Perspectives on permanent magnetic materials for energy conversion and power generation," *Metallurgical and Materials Transactions A*, vol. 44A, no. S1, pp. S2–S20, 2013.

[5] J. Went, G. W. Rathenau, E. W. Gorter, and G. W. van Oosterhout, "Hexagonal iron-oxide compounds as permanent-magnet materials," *Physical Review*, vol. 86, no. 3, pp. 424-425, 1952.

[6] J. Goodenough, *Magnetism and the Chemical Bond, Interscience Monographs on Chemistry: Inorganic Chemistry Section*, Interscience, Geneva, Switzerland, 1963.

[7] A. Mann, "Still in suspense: a quarter of a century after the discovery of high-temperature superconductivity, there is still heated debate about how it works," *Nature*, vol. 475, no. 7356, pp. 280–282, 2011.

[8] J. Heber, "Materials science: enter the oxides," *Nature News*, vol. 459, no. 7243, pp. 28–30, 2009.

[9] S. A. Wolf, "Spintronics: a spin-based electronics vision for the future," *Science*, vol. 294, no. 5546, pp. 1488–1495, 2001.

[10] K. Strnat, G. Hoffer, J. Olson, W. Ostertag, and J. J. Becker, "A family of new cobalt-based permanent magnet materials," *Journal of Applied Physics*, vol. 38, no. 3, pp. 1001-1002, 1967.

[11] K. Strnat, "The hard-magnetic properties of rare earth-transition metal alloys," *IEEE Transactions on Magnetics*, vol. 8, no. 3, pp. 511–516, 1972.

[12] J. Zhou, R. Skomski, C. Chen, G. C. Hadjipanayis, and D. J. Sellmyer, "Sm–Co–Cu–Ti high-temperature permanent magnets," *Applied Physics Letters*, vol. 77, no. 10, pp. 1514–1516, 2000.

[13] J. Croat, "Bonded rare earth-iron magnets," US Patent 4902361 A, 1986.

[14] Y. Matsuura, "Process for producing permanent magnet materials," US Patent 4597938, 1986.

[15] N. C. Koon, "Hard magnetic alloys of a transition metal and lanthanide," US Patent 4402770, 1983.

[16] J. M. D. Coey, *Magnetism and Magnetic Materials: Principles and Applications*, Cambridge University Press, New York, NY, USA, 2010, ISBN-13: 978-0521816144.

[17] O. Gutfleisch, "Controlling the properties of high energy density permanent magnetic materials by different processing routes," *Journal of Physics D: Applied Physics*, vol. 33, no. 17, pp. R157–R172, 2000.

[18] O. Gutfleisch, K. Güth, T. G. Woodcock, and L. Schultz, "Recycling used Nd-Fe-B sintered magnets via a hydrogen-based route to produce anisotropic resin bonded magnet," *Advanced Energy Materials*, vol. 3, pp. 151–155, 2013.

[19] Arnold Magnetic Technologies, http://www.arnoldmagnetics.com/.

[20] M. Humphries, *Rare Earth Elements: The Global Supply Chain Congressional Research Service Report 7-5700*, 2013, http://www.crs.gov.pp.R41347.

[21] J. D. Widmer, R. Martin, and M. Kimiabeigi, "Electric vehicle traction motors without rare earth magnets," *Sustainable Materials and Technologies*, vol. 3, pp. 7–13, 2015.

[22] Litos Strategic Communication under DOE contract DE-AC26-04NT41817. The Smart Grid: An Introduction, http://www.oe.energy.gov/DocumentsandMedia/DOE_SG_Book_Single_Pages(1).pdf.

[23] US DOE (US Department of Energy), *Critical Materials Strategy*, US Department of Energy, Washington DC, USA, 2011, http://energy.gov/sites/prod/files/DOE_CMS2011_FINAL_Full.pdf.

[24] World Wind Energy Association, http://www.windea.rg, report from March 2010.

[25] Clipper Windpower, http://www.clipperwind.com/pdf/liberty_brochure.pdf.

[26] R. Kleijn, "Materials and energy: a story of linkages. Materials requirement of new energy technologies, resource capacity, and interconnected materials flow," PhD Thesis, Leiden University, Netherlands, 2012, https://openaccess.leidenuniv.nl/handle/1887/19740.

[27] S. Makaremi, *Clipper's Design Approach to Improving Reliability*, Clipper Windpower, Inc, Cedar Rapids, IA, USA, 2007.

[28] G. Bywaters, V. John, J. Lynch, P. Mattila, G. Norton, and J. Stowell, *Northern Power Systems Wind PACT Drive Train Alternative Design Study Report (2005)*, National Renewable Energy Laboratory, Golden, CO, USA, 2005, http://www.nrel.gov/docs/fy05osti/35524.pdf.

[29] E. De Vries, *Danes Hail Lidar Breakthrough in Wind*, Renewable Energy World Int. Magazine, 2010, https://www.renewableenergyworld.com/articles/2010/03/danes-hail-lidar-breakthrough-in-wind.html.

[30] K. Hono and H. Sepehri-Amin, "Strategy for high-coercivity Nd-Fe-B magnets," *Scripta Materialia*, vol. 67, pp. 530–536, 2012.

[31] Toyota Motor Corporation, *Public Affairs Division, "Toyota Hybrid System"*, Toyota Motor Corporation, Japan, 2003.

[32] H. C. Singal, A. Mahajan, and R. Singh, "Magnetic refrigeration-a review-a boon for the coming generations," *International Journal of Mechanical Engineering*, vol. 3, no. 5, pp. 46–52, 2016.

[33] C. Zimm, A. Jastrab, A. Sternberg et al., "Description and performance of a near-room temperature magnetic refrigerator," *Advances in Cryogenic Engineering*, vol. 43, pp. 1759–1766, 1998.

[34] US Department of Energy, *Grid 2030-A National Vision for Electricity's Second 100 Years*, Office of Electric Transmission and Distribution, US Department of Energy, USA, 2003.

[35] USGS (US Geological Survey), *2011 Minerals Yearbook: rare earths*, US Department of Interior, Reston, Virginia, USA, 2013, http://minerals.usgs.gov/minerals/pubs/commodity/rare_earths/.

[36] USGS (US Geological Survey), *The Rare Earth Elements-Vital to Morden Technologies and Lifestyles. Fact Sheet 2014-3078*, US Department of Interior, Reston, Virginia, USA, 2014, http://pubs.usgs.gov/fs/2014/3078/.

[37] USGS (US Geological Survey), *Mineral Commodity Summaries 2014*, US Department of the Interior, Reston, Virginia, USA, 2014, http://minerals.usgs.gov/minerals/pubs/commodity/rare_earths/mcs-2014-raree.pdf.

[38] J. Lifton, "Where are the non-Chinese heavy rare earths going to come from and who's going to buy them?," *News Analysis, Rare Earths*, 2012.

[39] Neo Materials Technologies, http://www.magnequench.com.

[40] J. Seaman, *Rare Earth and Clean Energy: Analysing China's Upper Hand. IFRI Working Paper*, L'Institut Français des Relations Internationals, Paris, France, 2010, http://www.ifri.org/downloads/noteenergieseaman.pdf.

[41] D. Jolly, *China Export Restrictions on Metals Violate Global Trade Law, Panel Finds*, The New York Times, New York, NY, USA, http://nytimes.com/2014/03/27/business/international.

[42] GWMG (Great Western Minerals Group), *Great Western Minerals Successfully Produces Rare Earth Carbonate from Mini Pilot Plant Testing*, GWMG (Great Western Minerals Group), Saskatoon, Canada, 2014, http://www.gwmg.ca/investors/news.

[43] K. Bourzac, *The Rare-earth Crisis. MIT Technology Review*, vol. 114, no. 3, pp. 58–63, 2011, http://www.technologyreview.com/featuredstory/423730/the-rare-earth-crisis/.

[44] R. Gehm, *Hitachi Metals Reduces Rare-earth Dysprosium in Electric-Motors Magnets*, SAE International, Pittsburgh, PA, United States, 2013, http://articles.sae.org/11988.

[45] S. Kozawa, *Trends and Problems in Research of Permanent Magnets for Motors-Addressing Scarcity Problem of Rare Earth Elements*, 2010, NISTEP Science & Technology Foresight Center, http://data.nistep.go.jp/dspace/bitstream/11035/2854/1/NISTEP-STT038-40.pdf.

[46] K. Miura, M. Itoh, and K. I. Machida, "Extraction and recovery characteristics of Fe element from Nd–Fe–B sintered magnet powder scrap by carbonylation," *Journal of Alloys and Compounds*, vol. 466, no. 1-2, pp. 228–232, 2008.

[47] E. A. Perigo, S. C. da Silva, R. V. Martin, H. Takiishi, and F. J. G. Landgraf, "Properties of hydrogenation-disproportionation-desorption-recombination NdFeB powders prepared from recycled sintered magnets," *Journal of Applied Physics*, vol. 111, no. 7, article 07A725, 2012.

[48] R. S. Sheridan, R. Sillitoe, M. Zakotnik, I. R. Harris, and A. J. Williams, "Anisotropic powder from sintered NdFeB magnets by HDDR processing route," *Journal of Magnetism and Magnetic Materials*, vol. 324, pp. 63–67, 2012.

[49] H. Sepehri-Amin, T. Ohkubo, M. Zakotnik et al., "Microstructure and magnetic properties of grain boundary modified recycled Nd-Fe-B sintered magnets," *Journal of Alloys and Compounds*, vol. 694, pp. 175–184, 2017.

[50] T. Kawasaki, M. Itoh, and K. Machida, "Reproduction of Nd-Fe-B sintered magnet scraps using a binary alloy blending technique," *Materials Transactions*, vol. 44, no. 9, pp. 1682–1685, 2003.

[51] M. Zakotnik, I. R. Harris, and A. J. Williams, "Multiple recycling of NdFeB-type sintered magnets," *Journal of Alloys and Compounds*, vol. 469, no. 1-2, pp. 314–321, 2009.

[52] X. T. Li, M. Yue, W. Q. Liu et al., "Large batch recycling of waste Nd–Fe–B magnets to manufacture sintered magnets with improved magnetic properties," *Journal of Alloys and Compounds*, vol. 649, pp. 656–660, 2015.

[53] T. S. Jang, D. H. Lee, A. S. Kim, S. Namgung, H. W. Kwon, and D. H. Hwang, "Recovery of high coercivity of the powders obtained by crushing Nd–Fe–B sintered magnet scraps," *Physica Status Solidi (A)*, vol. 201, no. 8, pp. 1794–1797, 2004.

[54] H. W. Kwon, I. C. Jeong, A. S. Kim et al., "Restoration of coercivity in crushed Nd-Fe-B magnetic powder," *Journal of Magnetism and Magnetic Materials*, vol. 304, no. 1, pp. e219–e221, 2006.

7

Effect of High-Temperature Curing Methods on the Compressive Strength Development of Concrete Containing High Volumes of Ground Granulated Blast-Furnace Slag

Wonsuk Jung[1] and Se-Jin Choi[2]

[1]*School of Mechanical Engineering, Chungnam National University, Daejeon, Republic of Korea*
[2]*Department of Architectural Engineering, Wonkwang University, 460 Iksan-daero, Iksan 54538, Republic of Korea*

Correspondence should be addressed to Se-Jin Choi; csj2378@wku.ac.kr

Academic Editor: Xiao-Yong Wang

This paper investigates the effect of the high-temperature curing methods on the compressive strength of concrete containing high volumes of ground granulated blast-furnace slag (GGBS). GGBS was used to replace Portland cement at a replacement ratio of 60% by binder mass. The high-temperature curing parameters used in this study were the delay period, temperature rise, peak temperature (PT), peak period, and temperature down. Test results demonstrate that the compressive strength of the samples with PTs of 65°C and 75°C was about 88% higher than that of the samples with a PT of 55°C after 1 day. According to this investigation, there might be optimum high-temperature curing conditions for preparing a concrete containing high volumes of GGBS, and incorporating GGBS into precast concrete mixes can be a very effective tool in increasing the applicability of this by-product.

1. Introduction

About 14 million tons of blast-furnace slag, the by-product of steel industry, is produced annually in Korea [1]. Ground granulated blast-furnace slag (GGBS), either as a constituent of cement or as a mineral admixture, is widely used to make not only traditional concrete but also high-performance concrete, which has several advantages including workability, long-term strength, and durability [2]. Recently, to reduce CO_2 production in the cement and concrete industry, many studies have been conducted on environmentally friendly concrete with high-volume supplementary cementitious materials [3–9]. It was reported [10–12] that concrete using GGBS has the advantage of an earlier strength development under high-temperature curing conditions owing to the temperature-dependent characteristic of GGBS. Owing to this, GGBS might be effectively used in preparing precast concrete manufactured with high-temperature curing.

While significant literature is available on supplementary cementitious materials in concrete, such as GGBS and fly ash, few works have studied the effect of high-temperature curing conditions on the strength properties of concrete containing high-volume supplementary cementitious materials [11, 13–15]. Miura and Iwaki [11] evaluated the strength development of concrete incorporating high levels of GGBS at low temperatures, demonstrating that GGBS concrete with a specific surface area of $400 \, m^2/kg$ faces serious strength development problems at early ages and with low curing temperatures. Aldea et al. [13] investigated the effects of curing conditions on the properties of concrete prepared with slag replacement. Test results indicated that slag replacement of up to 50% by mass had little effect on strength and steam curing reduced the compressive strength as compared to other curing types (such as autoclaving and normal curing). Yazici et al. [14] investigated the effect of steam curing on class C high-volume fly ash concrete mixtures, indicating that steam curing accelerated the 1-day strength development but the long-term strength was greatly reduced. Liu et al. [15] evaluated the influence of steam curing on the compressive strength of concrete containing supplementary cementing

FIGURE 1: Autotemperature-controlled curing chamber.

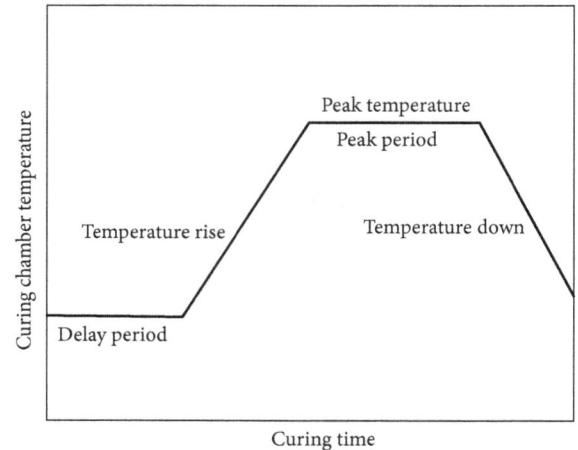

FIGURE 2: High-temperature curing cycle.

materials. Test results indicated that the concrete containing ultrafine fly ash (UFA) had a low early strength after 13 hours of steam curing and the difference between the 28-day compressive strength of 13-hour steam-cured concrete and that of moist-cured concrete was large.

This study examines how the compressive strength of concrete containing high volumes of GGBS is affected by various high-temperature curing methods. To date, the compressive strength development of high-volume GGBS concrete according to various high-temperature curing conditions—including delay period, temperature rise, peak temperature and period, and temperature down—has not been reported. The goal of this work is to demonstrate how GGBS can be used more efficiently and effectively in precast concrete using high-temperature curing methods.

2. Materials and Methods

2.1. Materials. ASTM Type I Portland cement with a specific gravity of 3.15 and crushed gravel with a specific gravity of 2.67 and fineness modulus of 6.75 were used. The fine aggregates used were crushed sand and washed sea sand (specific gravities of 2.62 and 2.66, resp.), which are commonly used in Korea. GGBS with a specific gravity of 2.89 and Blaine fineness of 4490 cm^2/g was obtained from Po-hang, South Korea. The chemical composition of the cement and GGBS, determined by X-ray fluorescence, is shown in Table 1.

2.2. Mix Proportions and Specimen Preparation. In this study, GGBS was used to replace Portland cement at a replacement ratio of 60% by binder mass. A constant water-to-binder ratio (w/b) of 0.35 was used in this investigation. The mix proportions are given in Table 2.

Concrete was mixed in a twin shaft mechanical mixer. Specimens were cast in a cylindrical mold (100 mm diameter, 200 mm length) for the compressive strength test. After casting, the cylinder molds were moved to the autotemperature-controlled curing chamber with a 100% relative humidity, shown in Figure 1, and cured according to various high-temperature curing methods. The samples were then removed from the molds and cured in a water-curing tank before the compressive strength test.

Figure 2 shows the common high-temperature curing cycle used in manufacturing precast concrete. The high-temperature curing methods used in this investigation involved delay periods (DPs) of 2, 3, and 4 h at 20°C; temperature rises (TRs) of 10, 15, and 20°C/h; peak temperatures (PTs) of 55, 65, and 75°C; peak periods (PPs) of 3, 4, and 5 h; and temperature downs (TDs) of 5, 10, and 15°C/h. The high-temperature curing methods are detailed in Figure 3 and Table 3. The compressive strength of the concrete was tested after 1, 14, and 28 days, in accordance with ASTM C39. Each strength is the average value of three samples.

3. Results and Discussion

3.1. Delay Period and Compressive Strength. The variation in the compressive strength with DP is shown in Figure 4. The 1-day compressive strength of concrete mixes with DPs of 2, 3, and 4 h was 30, 32.5, and 31 MPa, respectively. After 14 days, the values for the mixes with a DP of 3 h were about 4.8–8.8% higher than the other mixes. In addition, the compressive strength of the concrete mix with a DP of 3 h was greater than the other mixes after 28 days, with a value of 49.5 MPa. In contrast, the DP of 2 h gave the lowest values for each test day. After 28 days, the compressive strength of the 3 h DP concrete was about 10% higher than that with a DP of 2 h.

3.2. Temperature Rise and Compressive Strength. The compressive strength variation with TR is given in Figure 5 at different ages. The 1-day compressive strength of the sample with a TR of 15°C/h was 32.5 MPa, and this value was higher than that of other samples with a compressive strength of 27-28 MPa. The compressive strength of the 20°C/h TR concrete was 28 MPa after 1 day, about 3.7% higher than that for the 10°C/h TR concrete. However, the later compressive strength of the 20°C/h TR concrete was 41 MPa (14 days) and 47 MPa (28 days), about 2.1–5.0% lower than the corresponding values of 10°C/h TR sample. This test results demonstrate that high heating rate might result in an advantageous high early strength development, but a disadvantageous long-term

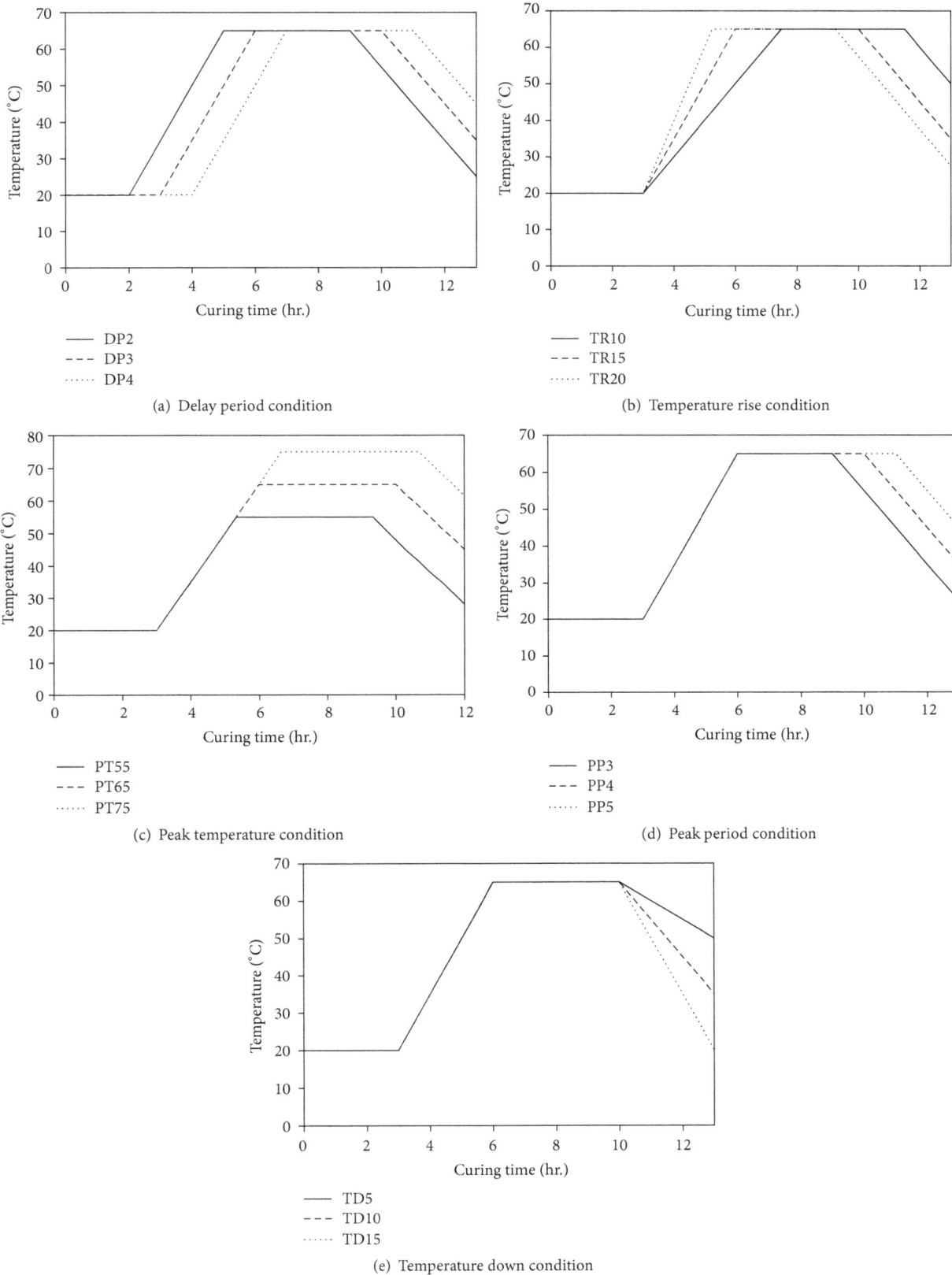

(a) Delay period condition

(b) Temperature rise condition

(c) Peak temperature condition

(d) Peak period condition

(e) Temperature down condition

FIGURE 3: Various high-temperature curing methods used in this study.

TABLE 1: Chemical composition of Portland cement and GGBS.

Materials	SiO_2	Al_2O_3	Fe_2O_3	K_2O	CaO	MgO	Na_2O	SO_3
Cement	21.20	4.64	2.91	1.22	61.90	1.87	0.29	2.31
GGBS	34.00	16.40	0.50	0.45	37.20	6.29	1.33	2.71

TABLE 2: Mixture proportions of concrete.

W/B %	Cement kg/m^3	GGBS kg/m^3	Water kg/m^3	Gravel kg/m^3	Crushed sand kg/m^3	Sea sand kg/m^3
35	168	252	140	1012	406	413

TABLE 3: Steam curing methods.

Curing methods	DP, hours	TR, °C/h	PT, °C	PP, hours	TD, °C/h
Delay period (DP)	2, 3, 4	15	65	4	10
Temperature rise (TR)	3	10, 15, 20	65	4	10
Peak temperature (PT)	3	15	55, 65, 75	4	10
Peak period (PP)	3	15	65	3, 4, 5	10
Temperature down (TP)	3	15	65	4	5, 10, 15

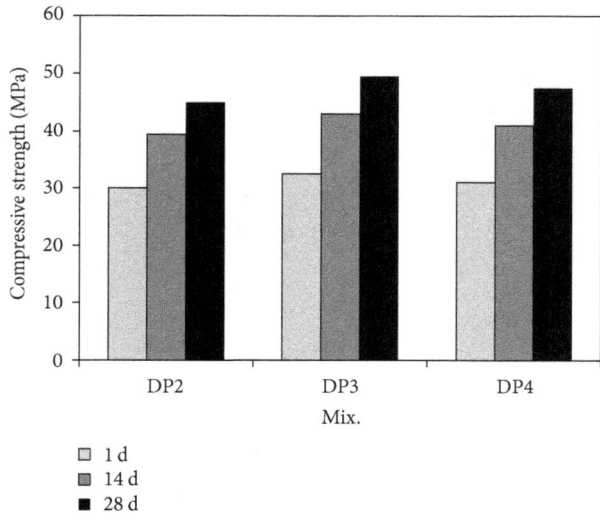

FIGURE 4: Compressive strength versus delay period.

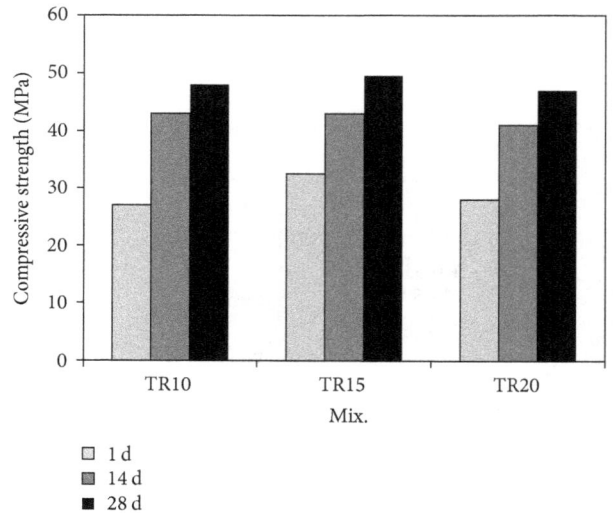

FIGURE 5: Compressive strength versus temperature rise.

strength development. An earlier study [11] reported that heat curing seems to have an adverse influence on strength development, particularly at later ages.

3.3. Peak Temperature and Compressive Strength. Figure 6 shows the compressive strength variation of concrete mixes with different PTs at different ages. The 1-day compressive strength of the 65°C and 75°C PT samples was 32.5 and 32 MPa, respectively. These values are about 88–91% higher than that of the sample with the lowest PT of 55°C (17 MPa). This result is similar to that of a previous study [10]. Barnett et al. reported that the early-age strength is much more sensitive to temperature for samples with higher levels of GGBS. The

lowest 14-day compressive strength was for the sample with a PT of 55°C (36 MPa). After 28 days, the compressive strength of the 65°C PT concrete was the greatest (49.5 MPa), being about 10–18% higher than the samples with PTs of 55°C and 75°C.

3.4. Peak Period and Compressive Strength. The variation in compressive strength with PP is given in Figure 7 for different ages. The 1-day compressive strength was highest for the concrete with a PP of 4 h. The 1-day compressive strength values of the samples with PPs of 3, 4, and 5 h were 25, 32.5, and 28 MPa, respectively. In addition, after 14 days, the compressive strength of the 4 h PP condition was 4.8% higher

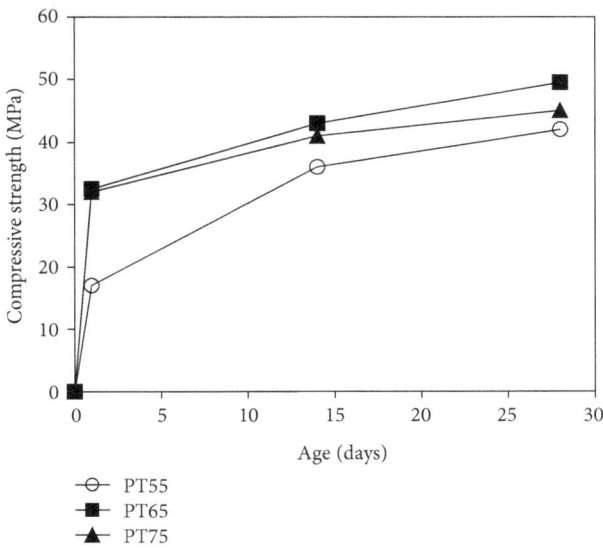

FIGURE 6: Compressive strength versus peak temperature.

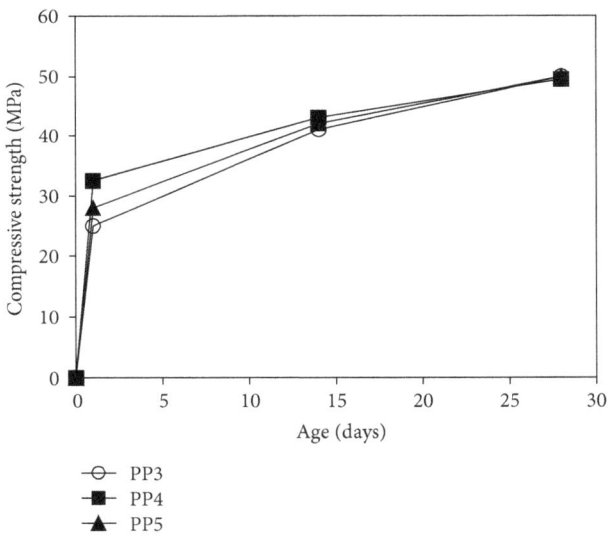

FIGURE 8: Compressive strength versus temperature down.

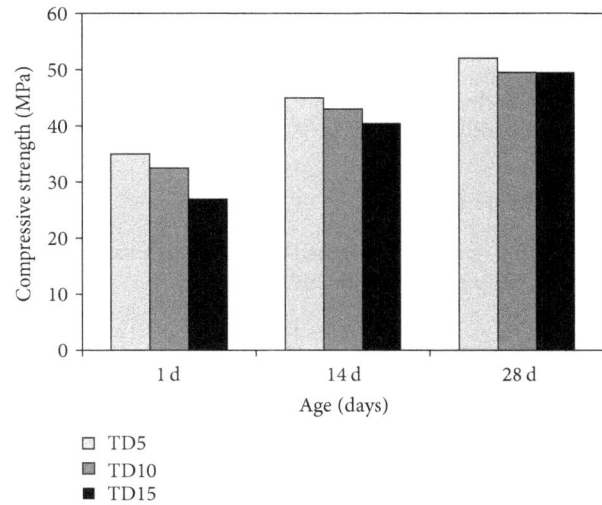

FIGURE 7: Compressive strength versus peak period.

than that of the concrete with a PP of 3 h. After 28 days, the compressive strengths of all the samples were the same with values ranging from 49.5 to 50 MPa.

3.5. Temperature Down and Compressive Strength. Figure 8 shows the variation in the compressive strength for the concrete with different TD values at different ages. The 1-day compressive strength of the concrete with a TD of 5°C/h (35 MPa) was about 30% higher than that with a TD of 15°C/h. It is assumed that the rapid cooling during steam curing might have a negative influence on the concrete compressive strength at an early age. The compressive strength of the 15°C/h TD condition was 40.5 MPa after 14 days, around 6–11% lower than the other mixes. However, the compressive

strength of the concrete with a TD of 15°C/h was similar to that for the 10°C/h TD condition after 28 days.

4. Conclusions

The following conclusions were obtained from the present investigation.

(1) The compressive strength of the concrete mix with a DP of 3 h (49.5 MPa) was greater than that of the other mixes after 28 days. In contrast, the compressive strength of the mix with a DP of 2 h was the lowest value after 28 days.

(2) The compressive strength of the sample with a TR of 20°C/h was about 3.7% higher than that of the 10°C/h TR sample after 1 day. However, the compressive strength of the sample with 20°C/h TR was lower than that of the sample with a TR of 10°C/h. This demonstrates that the high heating rate might result in an advantageous high early strength development, but a disadvantageous long-term strength development.

(3) The 1-day compressive strength of the samples with PTs of 65°C and 75°C was about 32 MPa, about 88% higher than the 55°C PT condition. After 28 days, the value for the sample with a PT of 65°C was the greatest.

(4) The compressive strength of the sample with a PP of 4 h was 4.8% higher than that of the 3 h PP sample after 14 days. After 28 days, the compressive strength of all the samples was the same.

(5) The 1-day compressive strength of the sample with a TD of 5°C/h was about 30% higher than that with a TD of 15°C/h. It is assumed that the rapid cooling during steam curing might adversely affect the compressive strength of the concrete at an early age.

The results of this investigation suggest that there might be optimum high-temperature curing conditions for the preparation of concrete containing high volumes of GGBS. Incorporating GGBS into precast concrete mixes can be a very effective tool in reducing CO_2 production in the cement industry and increasing the applicability of GGBS. Further studies are needed to establish the relationships between the strength properties of concrete containing high volumes of supplementary cementitious materials and the water-to-binder ratio, mineral admixture type, and so on.

Conflicts of Interest

The authors declare that there are no conflicts of interest regarding the publication of this paper.

Acknowledgments

This paper was supported by Wonkwang University in 2015.

References

[1] Korea Concrete Institute, "Concrete & Environment," vol.2, Kimoondang, Seoul, Korea, 2016.

[2] K. M. Lee, H. K. Lee, S. H. Lee, and G. Y. Kim, "Autogenous shrinkage of concrete containing granulated blast-furnace slag," *Cement and Concrete Research*, vol. 36, no. 7, pp. 1279–1285, 2006.

[3] S.-J. Choi and Y.-S. Jeon, "The Fluidity and Hardened Properties of Eco-Friendly Low Cement Concrete with 3 Types of Binders," *Journal of the Architectural Institute of Korea Structure & Construction*, vol. 31, no. 11, pp. 63–70, 2015.

[4] H. Yazıcı, "The effect of curing conditions on compressive strength of ultra high strength concrete with high volume mineral admixtures," *Building and Environment*, vol. 42, no. 5, pp. 2083–2089, 2007.

[5] M.-H. Zhang and J. Islam, "Use of nano-silica to reduce setting time and increase early strength of concretes with high volumes of fly ash or slag," *Construction and Building Materials*, vol. 29, no. 4, pp. 573–580, 2012.

[6] S.-J. Choi, J.-S. Mun, K.-H. Yang, and S.-J. Kim, "Compressive fatigue performance of fiber-reinforced lightweight concrete with high-volume supplementary cementitious materials," *Cement and Concrete Composites*, vol. 73, pp. 89–97, 2016.

[7] G. Hannesson, K. Kuder, R. Shogren, and D. Lehman, "The influence of high volume of fly ash and slag on the compressive strength of self-consolidating concrete," *Construction and Building Materials*, vol. 30, pp. 161–168, 2012.

[8] K. H. Mo, U. Johnson Alengaram, M. Z. Jumaat, and S. P. Yap, "Feasibility study of high volume slag as cement replacement for sustainable structural lightweight oil palm shell concrete," *Journal of Cleaner Production*, vol. 91, pp. 297–304, 2015.

[9] I. Papayianni and E. Anastasiou, "Production of high-strength concrete using high volume of industrial by-products," *Construction and Building Materials*, vol. 24, no. 8, pp. 1412–1417, 2010.

[10] S. J. Barnett, M. N. Soutsos, S. G. Millard, and J. H. Bungey, "Strength development of mortars containing ground granulated blast-furnace slag: effect of curing temperature and determination of apparent activation energies," *Cement and Concrete Research*, vol. 36, no. 3, pp. 434–440, 2006.

[11] T. Miura and I. Iwaki, "Strength development of concrete incorporating high levels of ground granulated blast-furnace slag at low temperature," *Materials Journal*, vol. 97, no. 1, pp. 66–70, 2000.

[12] J. I. Escalante, L. Y. Gómez, K. K. Johal, G. Mendoza, H. Mancha, and J. Méndez, "Reactivity of blast-furnace slag in Portland cement blends hydrated under different conditions," *Cement and Concrete Research*, vol. 31, no. 10, pp. 1403–1409, 2001.

[13] C.-M. Aldea, F. Young, K. Wang, and S. P. Shah, "Effects of curing conditions on properties of concrete using slag replacement," *Cement and Concrete Research*, vol. 30, no. 3, pp. 465–472, 2000.

[14] H. Yazici, S. Aydin, H. Yiğiter, and B. Baradan, "Effect of steam curing on class C high-volume fly ash concrete mixtures," *Cement and Concrete Research*, vol. 35, no. 6, pp. 1122–1127, 2005.

[15] B. Liu, Y. Xie, and J. Li, "Influence of steam curing on the compressive strength of concrete containing supplementary cementing materials," *Cement and Concrete Research*, vol. 35, no. 5, pp. 994–998, 2005.

8

Electrical and Vibrational Studies of $Na_2K_2Cu(MoO_4)_3$

Wassim Dridi,[1] Mohamed Faouzi Zid,[1] and Miroslaw Maczka[2]

[1]Laboratory of Materials, Crystal Chemistry and Applied Thermodynamics, Faculty of Sciences of Tunis,
 University of Tunis El Manar, El Manar II, 2092 Tunis, Tunisia
[2]Institute of Low Temperature and Structure Research, Polish Academy of Sciences, P.O. Box 1410, 50-950 Wrocław 2, Poland

Correspondence should be addressed to Mohamed Faouzi Zid; faouzi.zid@fst.rnu.tn

Academic Editor: Pascal Roussel

The complex impedance of $Na_2K_2Cu(MoO_4)_3$ material has been investigated in the temperature range of 653–753 K and in the frequency range of 40 Hz–5 MHz. Electrical behavior of the studied material is explained through an equivalent circuit model which takes into account the contributions of grains and grains boundaries. The number of vibrational modes was calculated using group theoretical approach. The infrared and Raman spectra have also been measured and vibrational assignment has been proposed.

1. Introduction

A great deal of interest has been devoted to the chemistry of molybdenum; a significant number of new molybdates have been synthesized and characterized. Molybdate chemistry has developed rapidly and this development can be explained by several factors, especially the improvement of the structural X-ray diffraction analysis, which has been a fundamental tool used for determination of crystal structures. But this renewed interest is also explained by the fact that many molybdates are suitable materials for technological applications. Molybdates exhibit various physicochemical properties, which are related to both the nature of the elements associated with the molybdate groups and the degree of opening of the formed framework. In these materials, the anionic framework is usually built from MoO_4 tetrahedra linked to the transition metal polyhedra, leading to a large variety of crystal structures with a high capacity for cationic substitution. The chemistry of inorganic molybdate materials has been significantly advanced thanks to their valuable electrical and optical properties, which make them promising for various applications such as photoluminescence [1], ionic conductivity [2–4], laser materials [5, 6], and piezoelectrics [7]. The high-temperature superconductivity present in the copper-oxygen ceramic systems resulted in an increasing structural and physicochemistry interest of materials containing Cu-O [8]. Among these materials we can mention

the copper molybdate $Cu_3Mo_2O_9$ doped with lithium, which displays high coulombic efficiency in lithium-ion batteries and excellent charge-discharge stability [9]. Another example is $Li_2Cu_2(MoO_4)_3$ material, which presents a high ionic conductivity ($\sigma = 5.810^{-3}$ $Ohm^{-1} \cdot cm^{-1}$ at 400 K, $E_a = 0.33$ eV) [10].

The vibrational spectroscopic studies of molybdates have attracted particular attention of a large number of researchers [11–16]. This attention is due to the catalytic activity of (MoO_4^{2-}) groups in hydrocarbons oxidations [17–20] and due to negative thermal expansion, ferroelasticity, and pressure-induced amorphization [21]. Furthermore, the interpretation of laser properties needs knowledge of vibrational level distribution [22]. According to this approach, we have decided to explore system A-Mo-Cu-O (A = alkali metal). The purpose of this study is to analyze the electrical response of the grain and grains boundaries effects, which greatly influence the electrical properties, and to understand the molecular structure at microscopic level of novel $Na_2K_2Cu(MoO_4)_3$ compound. This study can provide important information on the conductivity, which is very important for practical applications. In this paper, we will describe the synthesis method and the characterization of $Na_2K_2Cu(MoO_4)_3$ by Infrared, Raman, and complex impedance spectroscopies. Raman and IR selection rules will be also analyzed using factor group analysis.

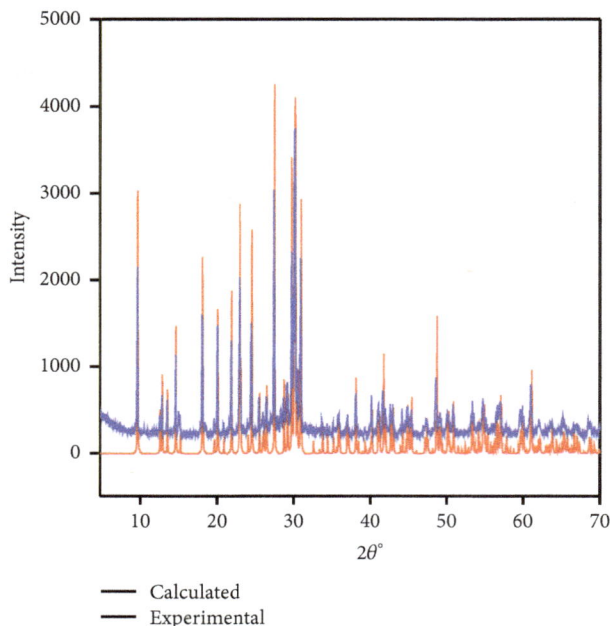

FIGURE 1: Calculated and experimental powder X-ray diffraction patterns of $Na_2K_2Cu(MoO_4)_3$.

2. Experimental

2.1. Polycrystalline Powder Synthesis of $Na_2K_2Cu(MoO_4)_3$. The $Na_2K_2Cu(MoO_4)_3$ polycrystalline powder was prepared by a conventional solid-state reaction from high-purity starting reagents of Na_2CO_3, K_2CO_3, $Cu(CO_2CH_3)\cdot H_2O$, and $(NH_4)_6Mo_7O_{24}\cdot 4H_2O$. These reagents were weighted according to the stoichiometric ratio. They were mixed and ground together in an agate mortar and heated progressively to 773 K in porcelain crucible with intermittent cooling and regrindings. The powder was analyzed by X-ray powder diffraction, using a PAN-analytical X'Pert PRO X-ray diffractometer equipped with copper anticathode ($\lambda K_\alpha = 1.5406$ Å). The unit cell parameters were refined using Celref 3.0 program and calculated to be as follows: $a = 7.5010(7)$ Å, $b = 9.3411(4)$ Å, and $c = 9.3670(7)$ Å and $\alpha = 92.59(3)°$, $\beta = 105.32(9)°$, and $\gamma = 105.44(6)°$. The powder X-ray diffraction pattern was in agreement with single-crystal structure (Figure 1).

2.2. Complex Impedance Spectroscopy. Pellet was prepared by isostatic pressing at 4 kbar and sintering at 773 K for 12 h in air with 10 $Kmin^{-1}$ heating and cooling rates. The thickness and surface of pellet were about 0.22 cm and 1.25 cm^2 having a geometric factor of $e/s = 0.176$ cm^{-1}. Electrical measurements were carried out in air by complex impedance spectroscopy using Agilent 4294A Precision Impedance Analyzer in the 40 Hz–5 MHz frequency range and 653–753 K temperature range. The sinusoidal AC voltage applied is of 0.5 V. The measuring cell having the sample inserted between two platinum discs used as ion-blocking electrodes is heated in an electric oven under dry air. The measurements were

carried out after temperature stabilization of the device every 30 min with a pitch of 10 K. The advantage of this method lies in the fact that it is possible to separate the physical phenomena according to their speed; the fast phenomena will take place at high frequencies and the slow ones at low frequencies. Analysis of impedance spectroscopy data can provide information on charge carrier dynamics in ionic conductors [23].

2.3. Vibrational Spectroscopies. The infrared spectrum was recorded at room temperature using Thermo Scientific Nicolet iS50 FT-IR Spectrometer. The frequency range was from 400 to 4000 cm^{-1} and the spectral resolution was 3 cm^{-1}. We are interested only in the domain 400–1100 cm^{-1} containing the most significant solid-state absorption bands. To obtain the Raman spectrum of the powdered sample, LAB RAMAN HR800 spectrometer was used. The frequency range was from 100 to 1100 cm^{-1} and the spectral resolution was 2 cm^{-1}. The measurement was carried out on a thin pellet. The sample was analyzed with an excitation wavelength of 632.81 nm and a power was adjusted to 1 mW in order to avoid any degradation. Spectroscopic studies are used to obtain the distribution of vibrational levels and assignment to the respective normal modes of $Na_2K_2Cu(MoO_4)_3$.

3. Results and Discussion

3.1. Structure Description. $Na_2K_2Cu(MoO_4)_3$ crystallizes in the triclinic space group P-1 with $a = 7.4946(8)$ Å, $b = 9.3428(9)$ Å, $c = 9.3619(9)$ Å, $\alpha = 92.591(7)°$, $\beta = 105.247(9)°$, $\gamma = 105.496(9)°$, $V = 604.7(Å^3)$, $Z = 2$, $R(F^2) = 0.022$, and $R_w(F^2) = 0.056$. Both cations K1 and K3 are located in the center of inversion, and all other atoms are at general positions. The structure of $Na_2K_2Cu(MoO_4)_3$ can be described as a one-dimensional framework formed by ribbons arranged in parallel to a axis with interribbons spaces containing Na^+ and K^+ monovalent cations directed to the free vertices of the tetrahedra MoO_4 (Figure 2). These structural characteristics encouraged us to study the electrical properties. CIF file containing complete information on the studied structure was deposited with FIZ Karlsruhe, 76344 Eggenstein-Leopoldshafen, Germany; e-mail: crysdata(at)fiz-karlsruhe(dot)de, deposition number CSD-430379).

3.2. Electrical Properties. The Nyquist plots in the temperature range from 653 K to 753 K are shown in Figure 3. When temperature increases, the radius of semicircles decreases and consequently the ionic conductivities increase with the temperature. We notice the presence of two hardly distinguishable semicircles, which proves the presence of two relaxation phenomena. The first arc existing towards higher frequencies corresponds to the movement of ions across the grain (bulk), which represents intrinsic conduction and gives rise to an intragranular resistance. The second arc, observed at lower-frequency, corresponds to movement of ions through the grain boundaries [24, 25]. The electrical behavior of $Na_2K_2Cu(MoO_4)_3$ is interpreted through an equivalent electrical circuit formed by two cells arranged

FIGURE 2: Projection of $Na_2K_2Cu(MoO_4)_3$ structure according to a axis.

in series, constituted by the parallel combination of the following: $R_g \parallel C1$ and $R_{gb} \parallel CPE1$ corresponding to the contributions of grains and grains boundaries, respectively. R_g and R_{gb} are the resistances of grains and grains boundaries, respectively. $C1$ is the pure capacitance of grain and $CPE1$ is the fractal capacitance constant phase element according to grains boundary. Electrical parameters were measured as a function of temperature. The intercepts of the semicircular arcs with the real axis give an estimation of the resistance of the studied material. Zview software [26] was used to fit these curves. The total resistance, R_{total}, follows the relation $R_g + R_{gb} = R_{total}$. The conformity between the experimental and calculated curves (fit) on the whole temperature range proves the validity of the proposed equivalent circuit. Electrical parameters are represented in Table 1.

In order to determine the direct conductivity for the grain interior σ_g, grain boundary σ_{gb}, and total conductivity σ_{tot}, we used the following equation:

$$\sigma_i = \frac{e}{R_i} \cdot \frac{1}{s}. \tag{1}$$

Values of ionic conductivities in $Na_2K_2Cu(MoO_4)_3$ material are represented in Table 2.

The activation energies was obtained by linear fitting of the ionic conductivities values at different temperatures by applying the Arrhenius equation:

$$T = \sigma_0 \exp\left(-\frac{E_a}{k_b T}\right), \tag{2}$$

where σ is the temperature dependent ionic conductivity, σ_0 is the ionic conductivity at absolute zero temperature,

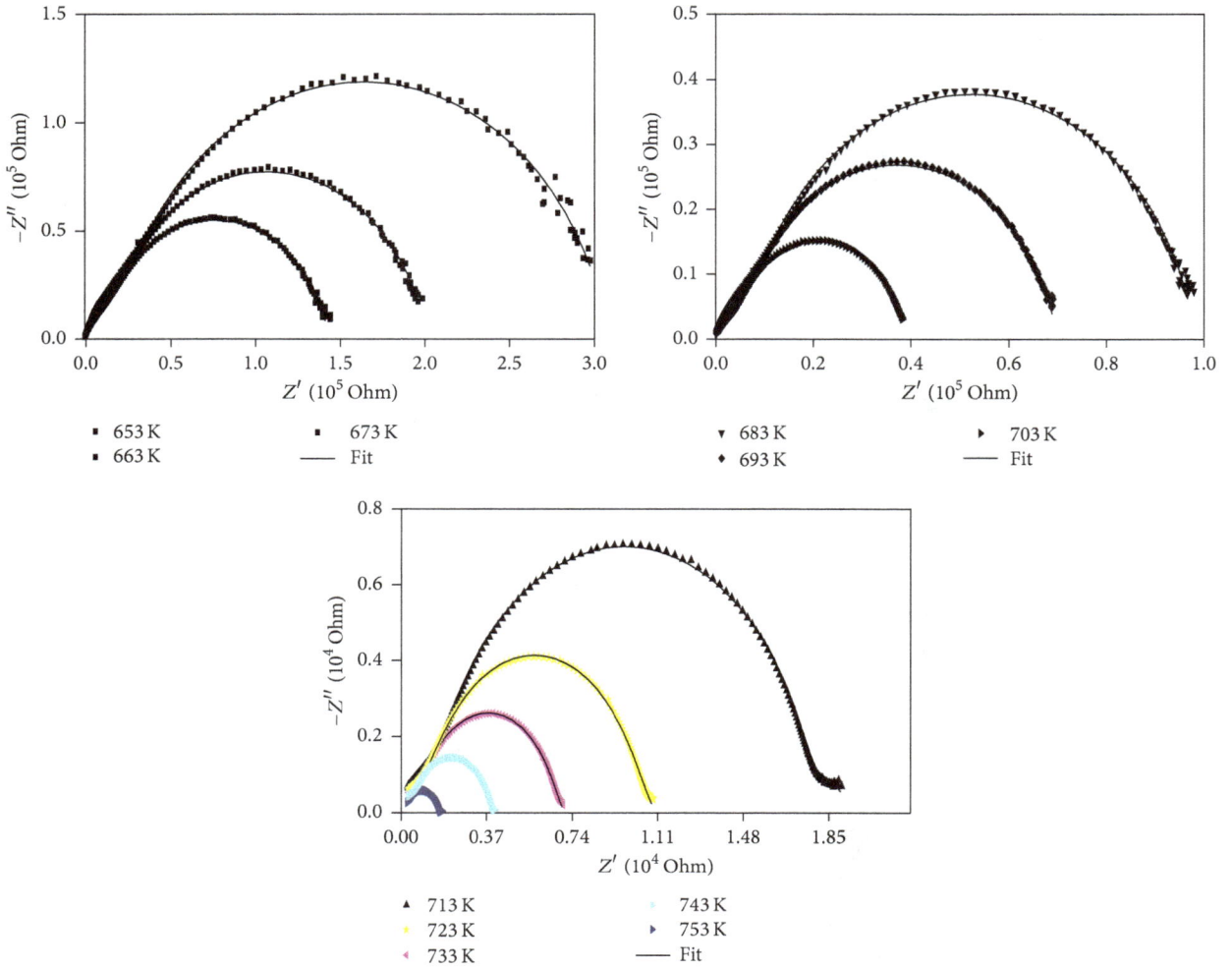

FIGURE 3: Complex impedance spectrum of $Na_2K_2Cu(MoO_4)_3$ over temperature range 653 and 753 K.

TABLE 1: Simulated data of the electrical parameters for the proposed equivalent circuit.

T (K)	R_g (Ω)	C_g (F)	R_{gb} (Ω)	C_{gb} (F)	$\sum R_i$ (Ω)
653	$2.087(4)10^5$	$1.118(2)10^{-11}$	$2.874(1)10^6$	$8.683(2)10^{-11}$	$3.083(1)10^6$
663	$1.354(1)10^5$	$1.078(1)10^{-11}$	$1.877(2)10^6$	$9.442(1)10^{-11}$	$2.012(2)10^6$
673	$9.711(9)10^4$	$9.131(4)10^{-11}$	$1.339(2)10^6$	$9.908(4)10^{-11}$	$1.436(2)10^6$
683	$6.797(9)10^4$	$9.639(5)10^{-12}$	$9.159(1)10^5$	$1.032(1)10^{-10}$	$9.838(9)10^5$
693	$4.915(3)10^4$	$9.010(1)10^{-12}$	$6.499(8)10^5$	$1.078(3)10^{-10}$	$6.991(3)10^5$
703	$2.913(9)10^4$	$7.893(2)10^{-12}$	$3.610(6)10^5$	$1.145(4)10^{-10}$	$3.901(9)10^5$
713	$1.283(3)10^4$	$6.905(2)10^{-12}$	$1.677(1)10^5$	$1.349(2)10^{-10}$	$1.805(3)10^5$
723	$8.670(1)10^3$	$6.781(3)10^{-12}$	$9.794(8)10^4$	$1.518(1)10^{-10}$	$1.066(1)10^5$
733	$6.670(2)10^3$	$6.793(4)10^{-12}$	$6.197(5)10^4$	$1.578(2)10^{-10}$	$6.864(5)10^4$
743	$4.831(3)10^3$	$7.066(1)10^{-12}$	$3.464(6)10^4$	$1.803(2)10^{-10}$	$3.947(7)10^4$
753	$2.715(2)10^3$	$7.954(2)10^{-12}$	$1.377(1)10^4$	$1.984(2)10^{-10}$	$1.648(6)10^4$

E_a is the activation energy of cations migration, k_b is the Boltzmann constant, and T is the absolute temperature. The variation of $\log(\sigma(S \cdot K \cdot cm^{-1}))$ versus $1000/T$ (K^{-1}) is represented in Figure 4. Activation energies values are represented in Table 3.

We note that the total conductivity of our compound is less than the bulk conductivity but higher than the grain boundary one. This reveals the existence of a partial blockage of the charge carriers by the grain boundaries [27]. Therefore, the conductivity of our material is limited by the

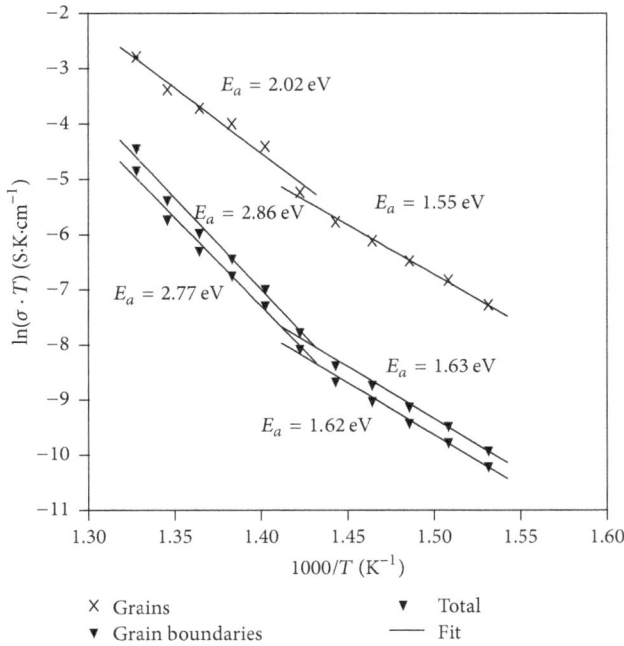

FIGURE 4: Arrhenius plot of conductivity of $Na_2K_2Cu(MoO_4)_3$.

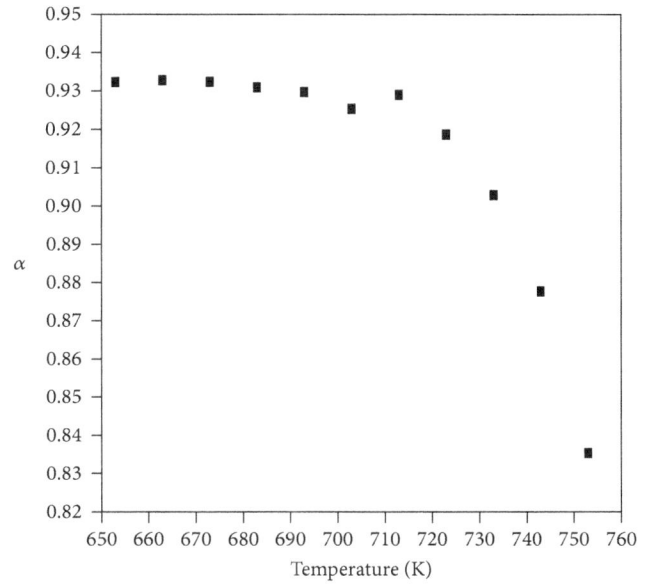

FIGURE 5: Variation of the blockage factor α with temperature.

TABLE 2: Ionic conductivities measurements value as a function of temperature in $Na_2K_2Cu(MoO_4)_3$ material.

T (K)	Conductivities ($\Omega\,cm^{-1}$)		
	σ_g	σ_{gb}	σ_{total}
653	1.053^{-6}	7.652^{-8}	5.707^{-8}
663	1.624^{-6}	1.172^{-7}	8.745^{-8}
673	2.265^{-6}	1.642^{-7}	1.225^{-7}
683	3.236^{-6}	2.401^{-7}	1.788^{-7}
693	4.475^{-6}	3.384^{-7}	2.517^{-7}
703	7.550^{-6}	6.093^{-7}	4.510^{-7}
713	1.714^{-5}	1.311^{-6}	9.748^{-7}
723	2.537^{-5}	2.246^{-6}	1.650^{-6}
733	3.298^{-5}	3.549^{-6}	2.563^{-6}
743	4.553^{-5}	6.349^{-6}	4.458^{-6}
753	8.103^{-5}	1.597^{-5}	1.067^{-5}

TABLE 3: Activation energies in $Na_2K_2Cu(MoO_4)_3$ material.

Contribution	Temperature range (K)	E_a (eV)
Grains	653–673	1.55
	673–753	2.02
Grains boundaries	653–673	1.63
	673–753	2.86
Total	653–673	1.62
	673–753	2.77

low conductivity of the grain boundaries. The influence of the grains boundaries conductivity on the total conductivity can be evaluated quantitatively by the blocking factor α. This parameter characterizes the fraction of the load carriers blocked in the case where a direct current flows through the sample. It can also be calculated using the following equation [28, 29]:

$$\alpha = \frac{R_{gb}}{R_{tot}}. \tag{3}$$

Figure 5 shows the variation of the blocking factor as a function of the temperature. It is found that the blocking factor decreases with the temperature. Therefore, the increase in temperature causes a decrease in the blocking effect by the limits of the grains.

3.3. Vibrational Study. $Na_2K_2Cu(MoO_4)_3$ crystallizes in the triclinic space group P-1 which corresponds to C_i factor group. There are two molecules per unit-cell, so there are also two molecules per Bravais cell. Mo1, Mo2, Mo3, Cu1, K2, Na1, Na1, and O atoms occupy C_1 symmetry whereas K1 and K3 atoms occupy C_i symmetry. The Bravais cell comprises 40 atoms that have 120 zone center degrees of freedom. The structure of $Na_2K_2Cu(MoO_4)_3$ compound is centrosymmetric; a complete assignment of the crystal modes requires both IR and Raman spectra [30]. The crystal vibrational modes are obtained by group theoretical calculations developed by Fateley et al. [31]. Factor group analysis of $Na_2K_2Cu(MoO_4)_3$ is represented in Table 4. The vibrational irreducible representation for the triclinic phase at the center of the Brillouin zone ($k = 0$) is

$$\Gamma_{vibr} = 57A_g + 63A_u, \tag{4}$$

3 acoustic and ($3N - 3$) optic modes, where N is the number of atoms in the unit cell [32]:

$$\Gamma_{acoustic} = 3A_u$$

$$\Gamma_{optic} = 57A_g + 60A_u. \tag{5}$$

TABLE 4: Factor group analysis of the triclinic phase $Na_2K_2Cu(MoO_4)_3$.

	Atoms	Site symmetry	n_s	t_γ	γ	f_γ	ζ	C_ζ	a_ζ	$\sum \Gamma_{atom}$
P-1 $Z = 2$	Γ_{transl} Mo1, Mo2, Mo3, Cu1, K2, Na1, Na2, 12 O	C_1	2	3	$A(T_x, T_y, T_z)$	6	A_g A_u	1 1	3 3	$19(3A_g + 3A_u)$
$Z^B = 2$	Γ_{transl} K1, K3	C_i	1	3	$A_u(T_x, T_y, T_z)$	3	A_u	1	3	$2(3A_u)$

Z^B: number of molecules per primitive Bravais cell, n_s: number of positions, t_γ: number of translations of a site species γ, and f_γ: the degree of vibrational freedom present in each site species γ.
C_ζ represents the degeneracy of the species ζ of the factor group, whereas a_ζ is the number of lattice vibrations of the equivalent set of atoms in species ζ of the factor group.

Infrared and Raman active modes are as follows:

$$\Gamma_{raman} = 57A_g$$
$$\Gamma_{infrared} = 60A_u. \quad (6)$$

In the free state tetrahedral MoO_4 ion has T_d symmetry and four vibrational modes with the following wavenumbers: $\nu_1(A_1)$ nondegenerate symmetric stretching at 936 cm^{-1}, $\nu_2(E)$ doubly degenerate symmetric bending at 220 cm^{-1}, $\nu_3(F_2)$ triply degenerate asymmetric stretching at 895 cm^{-1}, and $\nu_4(F_2)$ triply degenerate asymmetric bending at 365 cm^{-1} [33]. Moreover, all four vibrational modes are active in the Raman spectra, but only F_2 stretching and bending vibrations are active in the IR spectra. However, when this ion is located in the crystal lattice, its symmetry is lowered due to the constraints imposed by the lattice. So, the local symmetry of the three MoO_4 tetrahedra decreases to C_1. Because of this lowering of symmetry, all modes become active in Raman and in infrared and degenerate modes raise their degenerations. Therefore, ν_3 and ν_4 are split into three bands $3A$ and ν_2 into two $2A$. The correlation between the point group of T_d symmetry of the free anion MoO_4, its site-symmetry C_1, and its factor group C_i is represented in Scheme 1. According to Basiev et al. [34], the vibrational modes observed in Raman spectra of molybdates can be classified into two groups, internal and external modes. The internal vibrational modes of each type of MoO_4 derived from the correlation scheme are equal to $Z(3n - 6) = 18$, where n is the number of atoms in the molecular MoO_4:

$$\Gamma_{MoO_4} = 9A_g^{(R)} + 9A_u^{(IR)}$$
$$\Gamma\nu_1 = A_g^{(R)} + A_u^{(IR)}$$
$$\Gamma\nu_2 = 2A_g^{(R)} + 2A_u^{(IR)} \quad (7)$$
$$\Gamma\nu_3 = 3A_g^{(R)} + 3A_u^{(IR)}$$
$$\Gamma\nu_4 = 3A_g^{(R)} + 3A_u^{(IR)}.$$

The external vibrational modes of MoO_4 are divided into translational modes which includes acoustic and lattice modes and librational modes [35], presented as follows:

$$\Gamma_{translation} = \Gamma_{acoustic} + \Gamma_{lattice} = 3A_g^{(R)} + 3A_u^{(IR)}$$
$$\Gamma_{libration} = 3A_g^{(R)} + 3A_u^{(IR)}. \quad (8)$$

Point group T_d	Site group C_1	Factor group C_i
$A_1(\nu_1)$ (R)	A	$A_g^{(R)} + A_u^{(IR)}$
$E(\nu_2)$ (R)	A	$A_g^{(R)} + A_u^{(IR)}$
	A	$A_g^{(R)} + A_u^{(IR)}$
$F_2(\nu_3)$ (R and IR)	A	$A_g^{(R)} + A_u^{(IR)}$
	A	$A_g^{(R)} + A_u^{(IR)}$
	A	$A_g^{(R)} + A_u^{(IR)}$
$F_2(\nu_4)$ (R and IR)	A	$A_g^{(R)} + A_u^{(IR)}$
	A	$A_g^{(R)} + A_u^{(IR)}$
	A	$A_g^{(R)} + A_u^{(IR)}$

SCHEME 1: Correlation scheme for the internal modes of MoO_4 in $Na_2K_2Cu(MoO_4)_3$ structure.

The comparison of the infrared and Raman bands positions shows that the majority of these bands do not coincide. Indeed, the observed IR bands appear at wavenumbers different from those in the Raman spectrum (Figure 6). This is in agreement with the centrosymmetric character of $Na_2K_2Cu(MoO_4)_3$ structure [36]. The Raman spectrum can be separated into two parts with a wide empty gap in the range 500–700 cm^{-1} that is commonly observed in molybdates containing MoO_4 tetrahedra [22, 37–43]. The proposed assignment of the vibrational spectra of MoO_4 in $Na_2K_2Cu(MoO_4)_3$ is realised by considering the following criteria: ν_1 bands are generally very strong in the Raman and weaker in the infrared spectra, whereas an opposite behavior is usually observed for ν_3 bands. ν_2 bands are usually stronger in the Raman spectra than those corresponding to ν_4 modes but in the infrared spectra ν_4 band is generally more intense [44]. The Mo-O stretching modes are located in the range 720–930 cm^{-1} whereas the bending modes are situated in the range 380–330 cm^{-1}. Wavenumbers and assignment of the internal vibrational modes of MoO_4 tetrahedron are listed in Table 5.

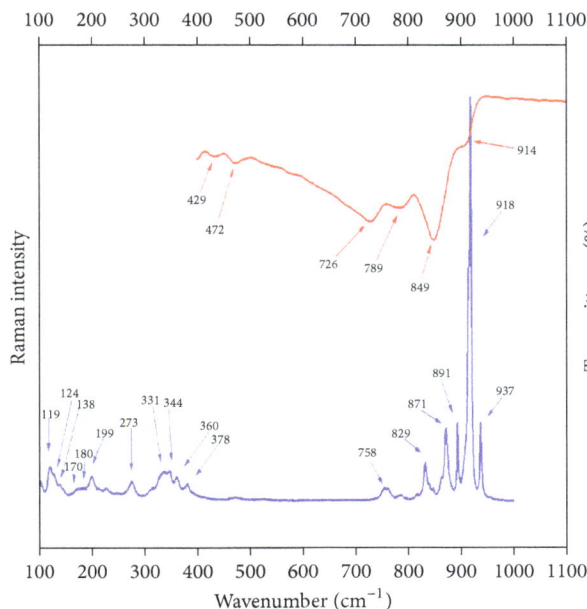

FIGURE 6: Infrared and Raman spectra of $Na_2K_2Cu(MoO_4)_3$.

TABLE 5: Assignment of the internal vibrational modes frequencies of MoO_4 tetrahedron.

Assignment	Infrared (ν/cm^{-1})	Raman (ν/cm^{-1})
$\nu_1(MoO_4)$	914	937, 918, 891
$\nu_2(MoO_4)$		344, 331, 273
$\nu_3(MoO_4)$	849, 789, 726	871, 829, 758
$\nu_4(MoO_4)$	429, 472	378, 360

In the Raman spectrum, bands located below $300\,cm^{-1}$ are attributed to external vibrations involving the librational and translational modes of the MoO_4 anions and translational modes of cations; the distinguishing between librational and translational modes is difficult. But in general librational modes have higher wavenumbers and intensities than the translational modes [45]. Furthermore, since the atomic mass of molybdenum is larger than that of copper, potassium, and sodium, translations of the MoO_4^{-2} ions should be observed at lower wavenumbers than translations of Cu^{2+}, K^+, and Na^+ [13]. Based on these rules, we propose assignment of the $273\,cm^{-1}$ band to $T'(Na^+)$ modes, those at 119 and $124\,cm^{-1}$ to $L(MoO_4)$ modes, and the remaining bands in the $138-199\,cm^{-1}$ range to the coupled modes involving translational motions of the molybdate, potassium, and copper ions.

4. Conclusion

Polycrystalline powder of $Na_2K_2Cu(MoO_4)_3$ was obtained by standard solid-state reaction at 773 K. X-ray diffraction studies show that this compound crystallizes in the triclinic symmetry with the P-1 space group. Ionic conductivity of the investigated material is characterized by the existence of a partial blockage of the charge carriers by the grain boundaries. The blocking effect generated by the limits of the grains decreases with temperature. The centrosymmetric space group P-1 of our structure is confirmed by the noncoincidence of majority of Raman and IR bands. Vibrational study indicates the lowering of symmetry of molybdate anion from T_d to C_1 symmetry.

Conflicts of Interest

The authors declare that there are no conflicts of interest regarding the publication of this paper.

References

[1] R. Grasser, E. Pitt, A. Scharmann, and G. Zimmerer, "Optical properties of CaWO4 and CaMoO4 crystals in the 4 to 25 eV region," *physica Status Solidi (b)*, vol. 69, no. 2, pp. 359–368, 1975.

[2] A. A. Savina, S. F. Solodovnikov, D. A. Belov et al., "Synthesis, crystal structure and properties of alluaudite-like triple molybdate Na25Cs8Fe5(MoO4)24," *Journal of Solid State Chemistry*, vol. 220, pp. 217–220, 2014.

[3] M. Hartmanova, M. T. Le, M. Jergel, V. Šmatko, and F. Kundracik, "Structure and electrical conductivity of multicomponent metal oxides having scheelite structure," *Russian Journal of Electrochemistry*, vol. 45, no. 6, pp. 621–629, 2009.

[4] N. I. Sorokin, "Ionic conductivity of double sodium-scandium and cesium-zirconium molybdates," *Physics of the Solid State*, vol. 51, no. 6, pp. 1128–1130, 2009.

[5] Y. Zhang, H. Cong, H. Jiang, J. Li, and J. Wang, "Flux growth, structure, and physical characterization of new disordered laser crystal LiNd(MoO4)2," *Journal of Crystal Growth*, vol. 423, pp. 1–8, 2015.

[6] N. M. Kozhevnikova and O. A. Kopylova, "Synthesis and X-ray diffraction and IR spectroscopy studies of ternary molybdates Li3Ba2R3(MoO4)8 (R = La-Lu, Y)," *Russian Journal of Inorganic Chemistry*, vol. 56, no. 6, pp. 935–938, 2011.

[7] A. E. Sarapulova, B. Bazarov, T. Namsaraeva et al., "Possible piezoelectric materials CsMZr0.5(MoO4)3 (M = Al, Sc, V, Cr, Fe, Ga, In) and CsCrTi0.5(MoO4)3: structure and physical properties," *Journal of Physical Chemistry C*, vol. 118, no. 4, pp. 1763–1773, 2014.

[8] J. Hanuza, M. Andruszkiewicz, Z. Bukowski, R. Horyń, and J. Klamut, "Vibrational spectra and internal phonon calculations for the M2Cu2O5 binary oxides (M = In, Sc, Y or from Tb to Lu)," *Spectrochimica Acta Part A: Molecular Spectroscopy*, vol. 46, no. 5, pp. 691–704, 1990.

[9] J. Xia, L. X. Song, W. Liu et al., "Highly monodisperse Cu3Mo2O9 micropompons with excellent performance in photocatalysis, photocurrent response and lithium storage," *RSC Advances*, vol. 5, no. 16, pp. 12015–12024, 2015.

[10] S. F. Solodovnikov, R. F. Klevtsova, and P. V. Klevtsov, "A correlation between the structure and some physical properties of binary molybdates (Tungstates) of uni- and bivalent metals," *Journal of Structural Chemistry*, vol. 35, no. 6, pp. 879–889, 1994.

[11] S. S. Saleem, G. Aruldhas, and H. D. Bist, "Raman and infrared spectra of GdTb (MoO4)3 single crystal in the region 250–1000 cm-1," *Spectrochimica Acta Part A: Molecular Spectroscopy*, vol. 39, no. 12, pp. 1049–1053, 1983.

[12] B. G. Bazarov, R. F. Klevtsova, T. T. Bazarova et al., "Double molybdate Tl2Mg4(MoO4)3: synthesis, structure, and properties," *Russian Journal of Inorganic Chemistry*, vol. 51, no. 10, pp. 1577–1580, 2006.

[13] M. MacZka, A. Pietraszko, W. Paraguassu et al., "Structural and vibrational properties of $K_3Fe(MoO_4)_2(Mo_2O_7)$—a novel layered molybdate," *Journal of Physics Condensed Matter*, vol. 21, no. 9, Article ID 095402, 2009.

[14] M. Isaac, V. U. Nayar, D. D. Makitova, V. V. Tkachev, and L. O. Atovmjan, "Infrared and polarized Raman spectra of $LiNa_3(MoO_4)_2 \cdot 6H_2O$," *Spectrochimica Acta. Part A: Molecular and Biomolecular Spectroscopy*, vol. 53, no. 5, pp. 685–691, 1997.

[15] N. M. Kozhevnikova, "Synthesis and phase formation study in K_2MoO_4-$SrMoO_4$-$R_2(MoO_4)_3$ systems (where R = Pr, Nd, Sm, Eu, and Gd)," *Russian Journal of Inorganic Chemistry*, vol. 57, no. 5, pp. 646–649, 2012.

[16] R. L. Frost, J. Bouzaid, and I. S. Butler, "Raman spectroscopic study of the molybdate mineral Szenicsite and comparison with other paragenetically related molybdate minerals," *Spectroscopy Letters*, vol. 40, no. 4, pp. 603–614, 2007.

[17] S. S. Saleem, "Infrared and Raman spectroscopic studies of the polymorphic forms of nickel, cobalt and ferric molybdates," *Infrared Physics*, vol. 27, no. 5, pp. 309–315, 1987.

[18] Y. S. Yoon, W. Ueda, and Y. Moro-oka, "Selective conversion of propane to propene by the catalytic oxidative dehydrogenation over cobalt and magnesium molybdates," *Topics in Catalysis*, vol. 3, no. 3-4, pp. 265–275, 1996.

[19] W. Ueda, Y.-S. Yoon, K.-H. Lee, and Y. Moro-oka, "Catalytic oxidation of propane over molybdenum-based mixed oxides," *Korean Journal of Chemical Engineering*, vol. 14, no. 6, pp. 474–478, 1997.

[20] J. D. Pless, B. B. Bardin, H.-S. Kim et al., "Catalytic oxidative dehydrogenation of propane over Mg-V/Mo oxides," *Journal of Catalysis*, vol. 223, no. 2, pp. 419–431, 2004.

[21] M. MacZka, A. G. Souza Filho, W. Paraguassu, P. T. C. Freire, J. Mendes Filho, and J. Hanuza, "Pressure-induced structural phase transitions and amorphization in selected molybdates and tungstates," *Progress in Materials Science*, vol. 57, no. 7, pp. 1335–1381, 2012.

[22] J. Hanuza and L. Macalik, "Polarized i.r. and Raman spectra of orthorhombic $KLn(MoO_4)_2$ crystals (Ln = Y, Dy, Ho, Er, Tm, Yb, Lu)," *Spectrochimica Acta Part A: Molecular Spectroscopy*, vol. 38, no. 1, pp. 61–72, 1982.

[23] A. Lasia, *Electrochemical Impedance Spectroscopy and Its Applications*, vol. 32 of *Modern Aspects of Electrochemistry*, Springer US, 2002.

[24] B. Louati and K. Guidara, "Dielectric relaxation and ionic conductivity studies of $LiCaPO_4$," *Ionics*, vol. 17, no. 7, pp. 633–640, 2011.

[25] H. Mahamoud, B. Louati, F. Hlel, and K. Guidara, "Impedance and modulus analysis of the $(Na_{0.6}Ag_{0.4})_2PbP_2O_7$ compound," *Journal of Alloys and Compounds*, vol. 509, no. 20, pp. 6083–6089, 2011.

[26] D. Johnson, *Zview Version 3.1c*, Scribner Associates, 1990–2007.

[27] R. Ben Said, B. Louati, and K. Guidara, "Conductivity behavior of the new pyrophosphate $NaNi_{1.5}P_2O_7$," *Ionics*, vol. 22, no. 2, pp. 241–249, 2016.

[28] R. Gerhardt and A. S. Nowick, "Grain-boundary effect in ceria doped with trivalent cations: I, electrical measurements," *Journal of the American Ceramic Society*, vol. 69, no. 9, pp. 641–646, 1986.

[29] M. J. Verkerk, B. J. Middelhuis, and A. J. Burggraaf, "Effect of grain boundaries on the conductivity of high-purity ZrO_2—Y_2O_3 ceramics," *Solid State Ionics*, vol. 6, no. 2, pp. 159–170, 1982.

[30] J. Hanuza, "Raman scattering and infra-red spectra of tungstates $KLn(WO_4)_2$ family (Ln = La÷Lu)," *Journal of Molecular Structure*, vol. 114, pp. 471–474, 1984.

[31] W. G. Fateley, F. R. Dollish, N. J. McDevitt, and F. F. Bentley, *Infrared and Raman Selection Rules for Molecular and Lattice Vibrations: The Correlation Method*, John Wiley & Sons, New York, NY, USA, 1972.

[32] E. D. Palik, *Handbook of Optical Constants of Solids*, vol. 2, Academic Press, 1991.

[33] K. Nakamoto, *Infrared and Raman Spectra of Inorganic and Coordination Compounds*, Mir, Moscow, Russia, 1966.

[34] T. T. Basiev, A. A. Sobol, P. G. Zverev, L. I. Ivleva, V. V. Osiko, and R. C. Powell, "Raman spectroscopy of crystals for stimulated Raman scattering," *Optical Materials*, vol. 11, no. 4, pp. 307–314, 1999.

[35] L. Nalbandian and G. N. Papatheodorou, "Raman spectra and molecular vibrations of Au_2Cl_6 and $AuAlCl_6$," *Vibrational Spectroscopy*, vol. 4, no. 1, pp. 25–34, 1992.

[36] S. Kaoua, S. Krimi, S. Péchev et al., "Synthesis, crystal structure, and vibrational spectroscopic and UV-visible studies of $Cs_2MnP_2O_7$," *Journal of Solid State Chemistry*, vol. 198, pp. 379–385, 2013.

[37] J. Hanuza, L. Macalik, and K. Hermanowicz, "Vibrational properties of $KLn(MoO_4)_2$ crystals for light rare earth ions from lanthanum to terbium," *Journal of Molecular Structure*, vol. 319, pp. 17–30, 1994.

[38] V. V. Atuchin, O. D. Chimitova, T. A. Gavrilova et al., "Synthesis, structural and vibrational properties of microcrystalline $RbNd(MoO_4)_2$," *Journal of Crystal Growth*, vol. 318, no. 1, pp. 683–686, 2011.

[39] V. V. Atuchin, V. G. Grossman, S. V. Adichtchev, N. V. Surovtsev, T. A. Gavrilova, and B. G. Bazarov, "Structural and vibrational properties of microcrystalline $TlM(MoO_4)_2$ (M = Nd, Pr) molybdates," *Optical Materials*, vol. 34, no. 5, pp. 812–816, 2012.

[40] V. V. Atuchin, O. D. Chimitova, S. V. Adichtchev et al., "Synthesis, structural and vibrational properties of microcrystalline β-$RbSm(MoO_4)_2$," *Materials Letters*, vol. 106, pp. 26–29, 2013.

[41] L. Macalik, "Comparison of the spectroscopic and crystallographic data of Tm^{3+} in the different hosts: $KLn(MoO_4)_2$ where Ln=Y,La,Lu and M=Mo,W," *Journal of Alloys and Compounds*, vol. 341, no. 1-2, pp. 226–232, 2002.

[42] L. Macalik, E. Tomaszewicz, M. Ptak, J. Hanuza, M. Berkowski, and P. Ropuszynska-Robak, "Polarized Raman and IR spectra of oriented $Cd_{0.9577}Gd_{0.0282}\square_{0.0141}MoO_4$ and $Cd_{0.9346}Dy_{0.0436}\square_{0.0218}MoO_4$ single crystals where $\square$ denotes the cationic vacancies," *Spectrochimica Acta Part A: Molecular and Biomolecular Spectroscopy*, vol. 148, pp. 255–259, 2015.

[43] V. Dmitriev, V. Sinitsyn, R. Dilanian et al., "In situ pressure-induced solid-state amorphization in $Sm_2(MoO_4)_3$, $Eu_2(MoO_4)_3$ and $Gd_2(MoO_4)_3$ crystals: chemical decomposition scenario," *Journal of Physics and Chemistry of Solids*, vol. 64, no. 2, pp. 307–312, 2003.

[44] E. J. Baran, M. B. Vassallo, C. Cascales, and P. Porcher, "Vibrational spectra of double molybdates and tungstates of the type $Na_5Ln(XO_4)_4$," *Journal of Physics and Chemistry of Solids*, vol. 54, no. 9, pp. 1005–1008, 1993.

[45] S. S. Saleem and G. Aruldhas, "Raman and infrared spectra of lanthanum molybdate," *Journal of Solid State Chemistry*, vol. 42, no. 2, pp. 158–162, 1982.

9

Carbon Nanofoam by Pulsed Electric Arc Discharges

David Saucedo-Jimenez, Isaac Medina-Sanchez, and Carlos Couder Castañeda ⓘ

Centro de Desarrollo Aeroespacial, Instituto Politécnico Nacional, Belisario Dominguez 22, Centro, 06610 Ciudad de México, Mexico

Correspondence should be addressed to Carlos Couder Castañeda; ccouder@ipn.mx

Academic Editor: Marco Cannas

The aim of this article was to report the carbon nanofoam synthesis by a new method and a new catalytic mixture. Using the pulsed electric arc discharge method, carbon nanofoam was synthesized. The synthesis was carried out in a controlled atmosphere at 200 torr of hydrogen pressure. The pulsed electric arc discharge was established between two graphite electrodes with 22.8 kVA of power and 150 A DC current; the cathode was relatively motionless and was made of a pure carbon rod of 6 mm diameter, and the spinner anode was a pure carbon disc spinning at 600 rpm; over the disc was an annular cavity where the new catalytic mixture of 93.84/2.56/1.43/0.69/1.48 of C/Ni/Fe/Co/S molar fraction was deposited in a geometrically fixed way by 8 catalytic mixture blocks and 8 empty spaces, and the discharge frequency was 80 Hz. After the synthesis was made, the resulting products were deposited on the electrodes, proving that our method can synthesize different carbon nanostructures easily and at low cost.

1. Introduction

Synthesized materials include many further distinct forms of carbon such as fullerenes, multiwalled carbon nanotubes, and single-walled carbon nanotubes, which are based on a mixture of sp^2 and sp^3 hybridized carbon atoms. The structural phase space of carbon between graphite-like and diamond-like hybridization states now has many occupants.

There is even more variety in the electronic properties of various carbon allotropes, which range from superconductivity to ferromagnetism and tunable electrical conductivity [1].

Nanofoams are of considerable current interest due to their unique structure, which lies between two and three dimensions, allowing for many new types of materials with promising new functions for future technologies.

Since many material functions rely on molecule surface interactions and low-dimensional properties, materials with large surface areas and a quantum-confined nanoscale nature are of particular interest. Such conditions are provided by porous materials with nanometer dimensions [2].

Nanofoams of different materials have been used for specific applications, varying from medicinal applications by

using nickel nanofoam [3] to semiconductors, insulators, and acoustic-optic materials based on silicon [4, 5, 6] or to simply characterize other nanofoam structures such as gold and titanium nanofoams using hydrogen nanofoams [7].

The possibility to produce foams with controlled and reproducible mean density, area, and thickness, with a satisfactory adhesion on a solid substrate, might be significant to achieve desired material properties and/or for specific applications. Carbon represents a particularly suitable choice as foam constituent since, being a light element, it makes it easier to reach low densities, and because it can in principle lead to a monoelemental film due to its volatile oxides [8].

Pulsed laser deposition (PLD) has been proven as one of the most popular ways of synthesizing carbon nanofoams since it creates highly uniform samples, good adhesion to the substrate, and high porosity, with the possibility of reducing the density range by varying the forming conditions.

While many researchers focus on improving the uniformity and reducing the density range, few have proposed new methods for the formation of carbon nanofoams. Such is the case of Mitchell et al. [9], which proposal is to form high-quality carbon nanofoam via low-temperature hydrothermal carbonization of sucrose. Their fundamental idea for this was to develop an environmentally friendly method, without the use of any chemical. Using this technique, the foams tend to be composed of micrometer-sized spheres of predominantly sp^2 carbon that forms a three-dimensional open scaffold. These so-called micropearls are usually detected as individual units which are weakly connected, forming the foam structure. Their interaction is not strong enough for coalescence to occur. The hydrothermal process allows for the variation of growth parameters, which may lead to further foam morphologies. The study of the parameter-morphology relationship can help to better understand the hydrothermal carbonization process, and in addition, to tune the growth toward particular material structures.

Some methods for improving the formation of carbon nanofoam include the one mentioned by Zel'dovich et al. [10] that has been used since then with many variations of gas density or laser power, where it is stated that the presence of a gas atmosphere during the PLD process may have two different objectives: the use of a chemically reactive gas is necessary either to correct incongruent ablation or to introduce in the deposited film atomic species that are lacking in the target material.

Typically, the pressure of the ambient gas is kept low to preserve the energy of the atomic species impinging onto the surface, thus allowing compact, uniform films to grow. In another scenario, inert gases at high pressure are mostly used. When laser-generated plasma becomes spatially confined, most of its initial kinetic energy is converted into thermal energy, leading to a rise of the temperature up to several thousands of Kelvin degrees. Under such conditions, mutual aggregation of plasma constituents is favored, thus opening the way to the synthesis of nanoparticles during plasma propagation and possibly to the growth of nanostructured materials by PLD [11].

Also, Filipescu et al. [12] have proposed radiofrequency-assisted pulsed laser deposition for the creation of WO$_x$ (tungsten oxide) films. This favors the formation of nanostructured agglomerates that uniformly cover the substrate surface for this particular sample.

In this paper, a new method for synthesizing carbon nanofoam is proposed by using pulsed arch electric discharges in a low-frequency controlled scenario and a new proposal for the catalytic mixture. This is a modification of previous reported analysis that has been proven to work under the circumstances described throughout this work.

2. Experimental Section

The synthesis of carbon nanofoam by pulsed electric arc discharge with spinner anode was located inside the chamber developed by Saucedo-Jimenez et al. [13]. The chamber is made of 304s stainless steel, which can be easily manipulated to perform the experimental setup, and the body can be removed from the base, likewise, the upper part of the body has a cover that can be removed (Figure 1(a)). At the side of the chamber body, there is a boron-silicate observation window transversely to access and do any possible

adjustment not contemplated during the experimental setup and to make visual contact during the development of the experiment; this window can also be removed. The chamber temperature is controlled by a cold-water circulation cooling system. In order to perform the pulsed and periodic electric arc discharge, two electrodes with a potential difference between them were considered; these electrodes have a 3 mm minimum gap to generate the electric arch discharge; however, this setup is made in such a way that this gap varies in time, so that the electric arc cannot be established continuously. Figure 1(b) shows the configuration of the components inside the chamber.

The proximity and separation of the electrodes have been made by a mechanical means, by using a discontinuous circular anode, formed by catalytic mixture blocks and empty blocks. The cathode was located in a relatively motionless position, thus maintaining the specified gap, and can only move a few millimeters in a vertical way. The anode is a spinning carbon disc of 10 cm diameter with a constant circular motion, and over this disc, there is a channel of 3.5 cm of radius, 5 mm depth, and 1.5 cm width, as shown in Figure 2(a). When a catalytic mixture block approaches the cathode, the gap reduces and the discharge is established. At a subsequent time, when an empty sector of the anode reaches the cathode, the gap between the electrodes increased and the arc discharge was suspended. With this mechanism, a pulsed and periodic electric arc discharge is generated. The anode motion was constant considering its geometrical configuration, and a periodical catalytic mixture deposition was done; hence, the pulsed discharge was periodical too. The cathode is a pure sharpened graphite rod of 6 mm diameter to favor the emission of electrons by the tip effect, as shown in Figure 2(b).

Within the channel, a catalytic mixture is deposited, forming compacted catalytic mixture blocks distributed in a special geometric design, establishing sectors with catalytic mixture (teeth) and sectors without this; the arrangement of this mixture was forming 16 sectors of the same dimensions, thus forming 8 sectors of catalytic mixture and 8 sectors without mixing.

The constant angular velocity is set at 600 rpm, so the discharge time is of 6 ms approximately, which is a reasonable time to allow a short growth of the nanotube before an abrupt change in the environment, where it is immersed as reported by Puretzky et al. [14].

The direction of the rotation was set counterclockwise; this rotation generates a drag of the jet plasma, and this effect increases the plasma volume, thus having a greater space for the nucleation and growth of the nanostructures. This growth is achieved due to the angle of the cathode that is located relatively normal to the anode and is of about 60°. The catalytic mixture employed was as follows: 93.84/2.56/1.43/0.69/1.48 of C/Ni/Fe/Co/S in molar fractions. This catalytic mixture is a modification to that used by Liu et al. [15] and Tibbetts et al. [16], by using this, it is intended to induce topological defects as pentagonal rings by increasing the sulfur concentration to a double proportion; also, this mixture adds the sulfur in its elemental stage, while in the original, the sulfur is added as FeS.

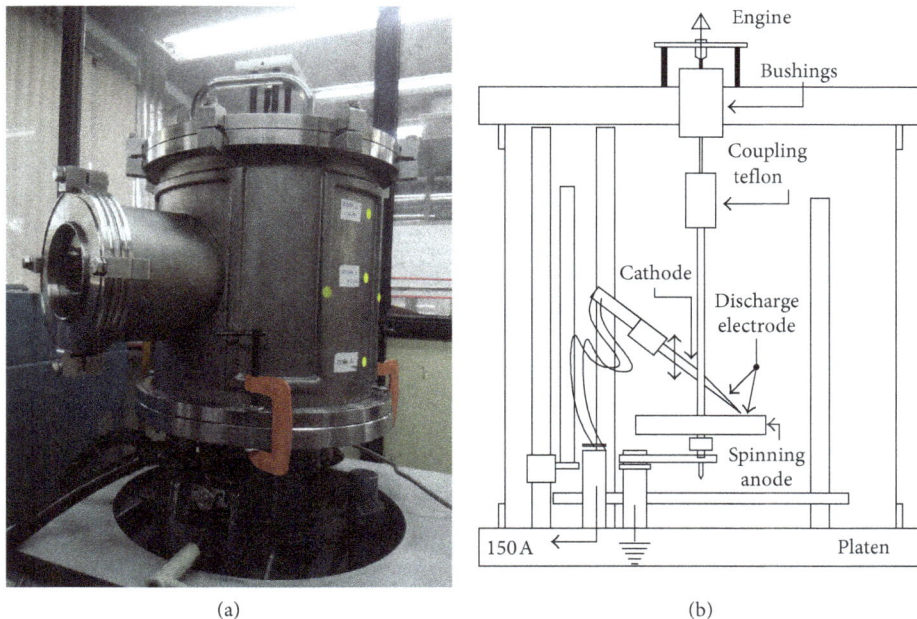

FIGURE 1: Carbon nanostructures growth chamber. (a) Growth chamber. (b) Growth chamber CAD design.

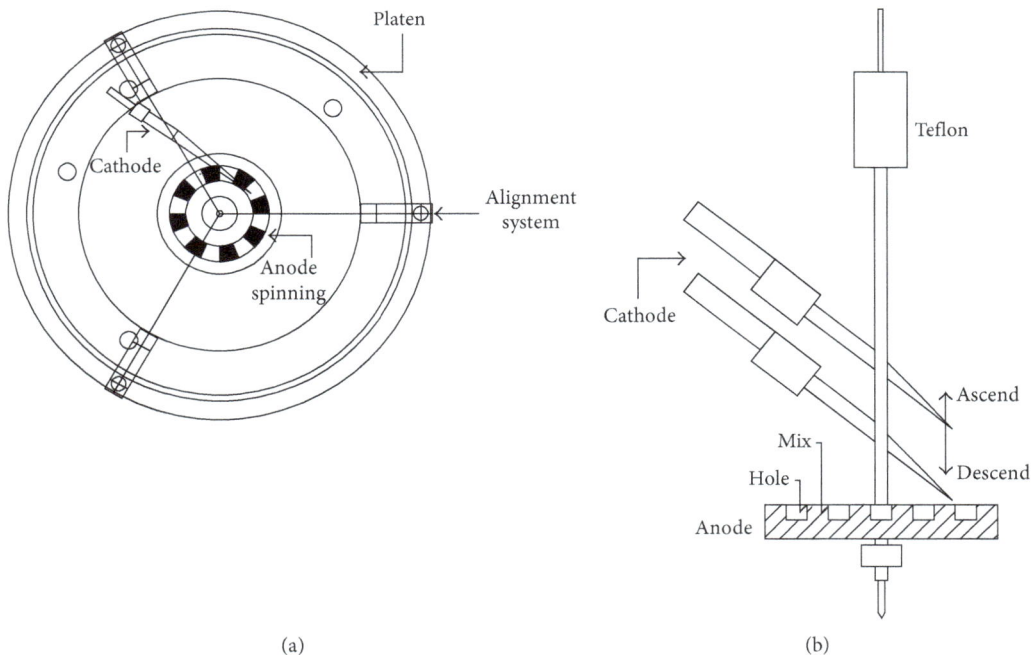

FIGURE 2: Carbon nanostructures growth chamber interior design. (a) Growth chamber top view. (b) Ascendant and descendant moving tip mechanism.

3. Results and Discussions

The first result after the experiment showed no deposition of nanotubes in spiderweb form; as a result, there was no growth of single wall or few wall carbon nanotubes, possibly due to excess sulfur concentration; nevertheless, carbon nanostructures were deposited on the walls of the chamber and on the surface of the electrodes. The material deposited on the walls of the chamber was characterized by scanning electron microscopy and was found mostly as amorphous carbon. The material deposited in the electrodes was characterized by the same microscopy, and a different material was found; this material showed very porous nanostructures, similar to fractals. This material was characterized by scanning and transmission electron microscopy as well as by Raman spectroscopy.

3.1. Scanning Electron Microscopy. The samples collected at the electrodes (cathode and anode) were characterized by scanning electron microscopy using the microscope mark FEI model Nova 200 NanoLab dual beam, focused ion beam (FIB), finding similar structures in both electrodes with the difference that the structures deposited in the cathode are free of amorphous carbon and catalyst particles than those deposited in the anode.

Figures 3 and 4 show the synthesized product that was observed and deposited on the surface of the anode. Structures highly contaminated by a large amount of amorphous carbon were found to be 35.66%, while the number of carbon nanofoam was 11.94%, and the proportion of catalyst particles was 52.4%. Figure 3 (3000x magnification) shows large rope formations with dimensions around 1 micron in width, and the ropes observed could be formed by carbon nanotubes. By the scanning electron microscopy characterization, we observed the rope surface, finding what could possibly be carbon nanotubes with diameters of about 30 nm. Figure 4 (6000x magnification) shows formation consisting of carbon nanotubes and catalyst particles, and the nanotube diameters are approximately 35 nm, while the particle diameters are on average 380 nm.

The morphology of the synthesized and deposited nanostructures on the surface of the cathode is shown in Figures 5 and 6; those nanostructures are similar to the ones shown in Figures 3 and 4, with the difference that they are not excessively contaminated by amorphous carbon, with an amorphous carbon proportion of 15.46%, while the proportion of nanofoam carbon is of 81.82%, and the proportion of catalyst particles is 2.71%. The structure of the sample resembles a fractal, a typical shape of the carbon nanofoam; the dimensions of the nanotube-like structures that shape the carbon nanofoam are 2.7 nm in diameter and approximately 100 nm in length, before breaking translation symmetry and changing its direction.

The proportion of amorphous carbon was calculated by making a very fine mesh on the micrograph.

3.2. Raman Spectroscopy. Resonance Raman spectra from samples collected at the electrodes (cathode and anode) are acquired using the confocal micro-Raman spectrometer Horiba Jobin Yvon (LabRam HR800), with CCD detector, power resolution 1024×256 pixels, variable spot size in range from 0.86 to 3.1 μm, optimized spectral range of 200–1100 nm (100–4500 cm^{-1}), with laser lines of 532.07 nm (2.34 eV green with power 43.4 mW), 632.79 nm (1.97 eV red with power 6.3 mW), and 784.12 nm (1.48 eV near infrared with power of 56.7 mW).

The spectrometer uses a confocal microscope (Olympus Bx41) with three objective lens of 10x, 50x and 100x magnification. The spectrometer have gratings of 300 gr/mm, 600 gr/mm, 950 gr/mm, and 1800 gr/mm. For 100x

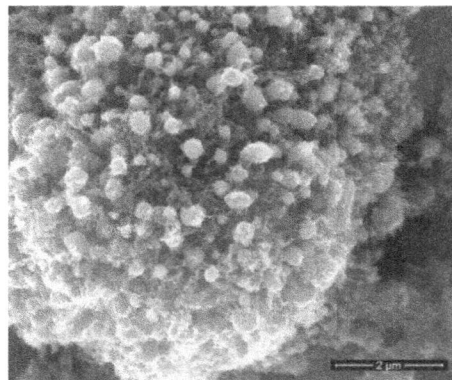

FIGURE 3: Anode magnification, 3000x.

FIGURE 4: Cathode magnification, 6000x.

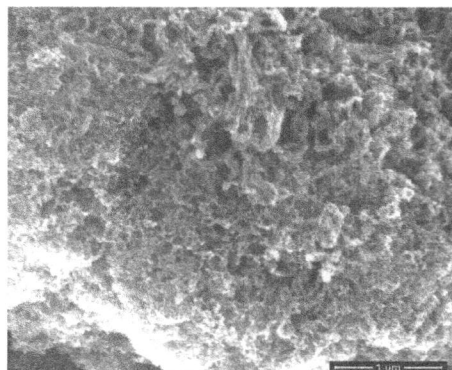

FIGURE 5: Cathode magnification, 6000x.

magnification in high resolution mode, the spectral resolution is 0.3 cm^{-1}. The process of measuring the Raman spectra for carbon nanofoam is made on a soda lime silicate safety glass substrate.

In the scanning electron microscopy, it is observed that this carbon nanofoam is conformed by structures similar to carbon nanotubes; therefore, the Raman spectroscopy characterization is proposed, to analyze the spectrum with the theory developed for carbon nanotubes [17, 18], and

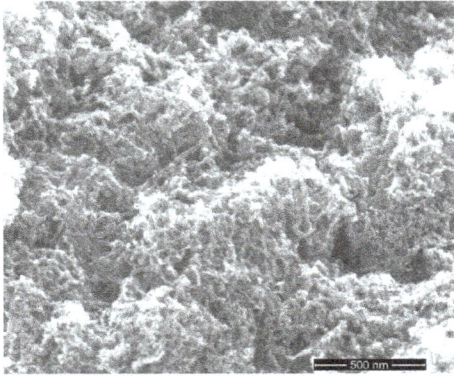

FIGURE 6: Cathode magnification, 12000x.

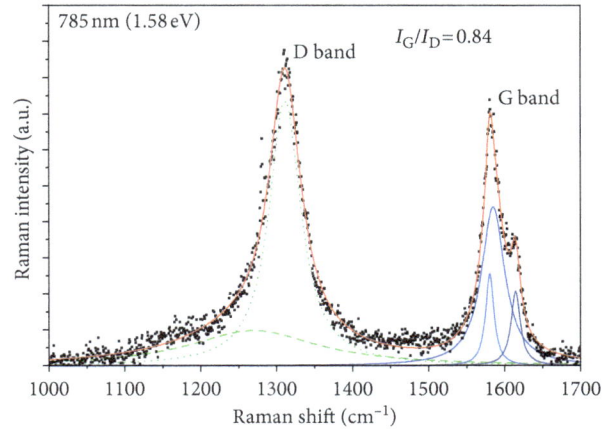

FIGURE 7: Cathode Raman spectra, 1.58 eV.

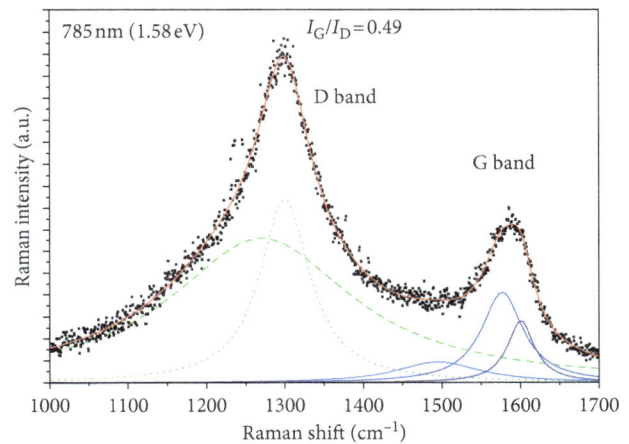

FIGURE 8: Anode Raman spectra, 1.58 eV.

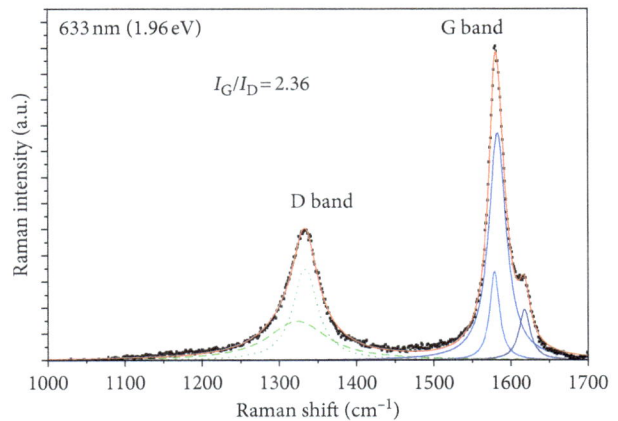

FIGURE 9: Cathode Raman spectra, 1.96 eV.

a possible interpretation to the signals obtained by this technique is stated.

By means of this characterization, it is determined that the structure of the carbon nanofoam is not formed by carbon nanotubes of few walls since no response was obtained in the region RBM 50–400 cm^{-1}, characteristic of radial breathing mode vibrations of the carbon atoms; however, a response is obtained in the region of 1200–1700 cm^{-1}, corresponding to the tangential vibrations of the carbon atoms that have the nanotubes. In this same region, there is the response of disorder band corresponding to heteroatoms, vacancies, heptagon-pentagon pairs, or even the presence of impurities, and so on. Considering that the possible carbon nanotubes in the sample are multiwalled carbon nanotubes with lengths of 100 nm approximately, before breaking the translation symmetry, it is clear that the response in the region 1450 cm^{-1} corresponds to disorder-induced D band due to amorphous carbon and heptagon-pentagon pairs present in the sample, as was mentioned in the scanning electron microscopy. Figures 7–12 give a general view of the Raman spectra from samples of carbon nanofoam. The signal shows the Raman features corresponding to tangential modes (G band) and disorder-induced D band. The D-band intensity usually is 100 times smaller than that of the G band; large D-band peaks compared with the G-peak intensity usually indicate the presence of amorphous carbon [19]. For the Raman spectra analysis, usually there are two characteristics that differentiate the D band: defects in the structure carbon nanotubes and amorphous carbon. The Raman spectra from carbon nanotubes are composed of a broad peak upon which is superimposed a sharper peak; the broad feature refers to amorphous carbon, and the sharper feature refers to the carbon nanotube defects [20].

The fit parameters of the experimental data obtained from the Raman characterization are shown in Tables 1–6.

3.2.1. Fit Parameters for Raman Spectra Data in Anode.
Table 1 shows the adjustment of the G region or tangential vibrations; the first two signals correspond to defects in the nanotubes; of these, the narrowest signal, 1338.13 cm^{-1}, corresponds to bent nanotubes that in this case correspond

to carbon nanofoam. The widest signal, 1346.26 cm^{-1}, corresponds to amorphous carbon. The last three signals correspond to tangential vibrations of the carbon atoms, the signal in 1549.20 cm^{-1} corresponds to the circumferential

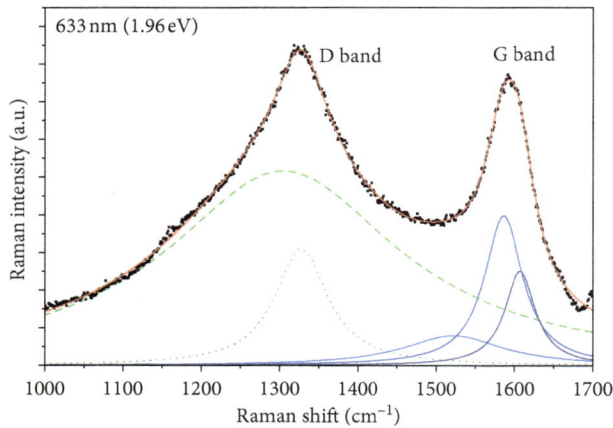

FIGURE 10: Anode Raman spectra, 1.96 eV.

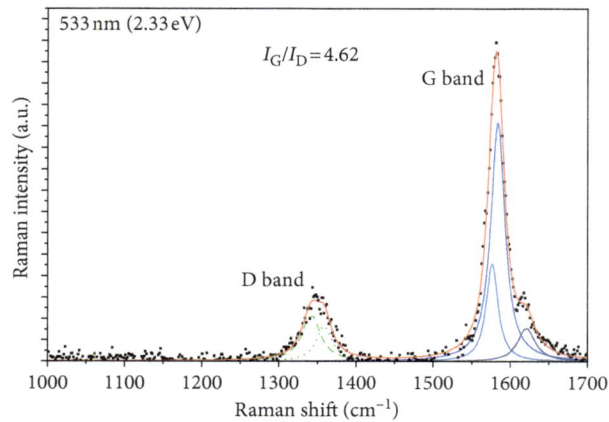

FIGURE 11: Cathode Raman spectra, 2.33 eV.

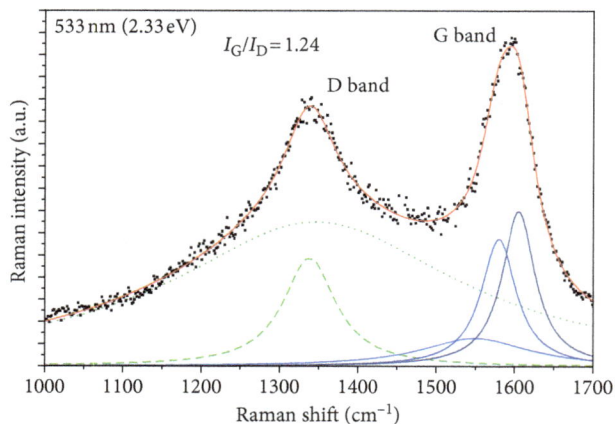

FIGURE 12: Anode Raman spectra, 2.33 eV.

nanotubes; of these, the narrowest signal, $1327.73\,\mathrm{cm}^{-1}$, corresponds to bent nanotubes that in this case correspond to carbon nanofoam. The widest signal, $1304.29\,\mathrm{cm}^{-1}$, corresponds to amorphous carbon. The last three signals correspond to tangential vibrations of the carbon atoms, the signal in $1521.89\,\mathrm{cm}^{-1}$ corresponds to the circumferential vibration, the second to the vibration in $1585.84\,\mathrm{cm}^{-1}$ corresponds to the carbon materials with sp^2 hybridization, and the last signal in $1606.99\,\mathrm{cm}^{-1}$ corresponds to the vibration in the direction parallel to the axis tube.

Table 3 shows the adjustment of the G region or tangential vibrations; the first two signals correspond to defects in the nanotubes; of these, the narrowest signal, $1299.98\,\mathrm{cm}^{-1}$, corresponds to bent nanotubes that in this case correspond to carbon nanofoam. The widest signal, $1270.23\,\mathrm{cm}^{-1}$, corresponds to amorphous carbon. The last three signals correspond to tangential vibrations of the carbon atoms, the signal in $1495.12\,\mathrm{cm}^{-1}$ corresponds to the circumferential vibration, the second to the vibration in $1576.89\,\mathrm{cm}^{-1}$ corresponds to the carbon materials with sp^2 hybridization, and the last signal in $1600.65\,\mathrm{cm}^{-1}$ corresponds to the vibration in the direction parallel to the axis tube.

Table 7 shows the proportions of amorphous carbon, and defects in nanostructures are summarized. Typical characteristics of the Raman spectra of carbon SP2 bonds are the G and D bands. The G band of tangential stretching is located approximately on $1580\,\mathrm{cm}^{-1}$ and corresponds to the carbon atoms vibrations in the tangential direction, that is, over the plane that defines the hexagons formed by the carbon atoms. The D band of induced disorder, located between 1250 and $1450\,\mathrm{cm}^{-1}$, describes the presence of substitutional heteroatoms, vacancies, grain boundaries, topological defects by finite size effects, and the presence of other carbon materials (amorphous carbon covering the nanotubes), among other defects. The intensities of these bands are known as IG and ID; the defect index (IG/ID) is useful for an approximation that indicates statistically the concentration of topological defects, rehybridization defects, incomplete links, and the presence of amorphous carbon.

3.2.2. Fit Parameters for Raman Spectra Data in Cathode. Table 4 shows the adjustment of the G region or tangential vibrations; the first two signals correspond to defects in the nanotubes; of these, the narrowest signal, $1358.76\,\mathrm{cm}^{-1}$, corresponds to bent nanotubes that in this case correspond to carbon nanofoam. The widest signal, $1342.01\,\mathrm{cm}^{-1}$, corresponds to amorphous carbon. The last three signals correspond to tangential vibrations of the carbon atoms, the signal in $1576.10\,\mathrm{cm}^{-1}$ corresponds to the circumferential vibration, the second to the vibration in $1583.39\,\mathrm{cm}^{-1}$ corresponds to the carbon materials with sp^2 hybridization, and the last signal in $1621.05\,\mathrm{cm}^{-1}$ corresponds to the vibration in the direction parallel to the axis tube.

Table 5 shows the adjustment of the G region or tangential vibrations; the first two signals correspond to defects in the nanotubes; of these, the narrowest signal, $1323.10\,\mathrm{cm}^{-1}$, corresponds to bent nanotubes that in this

vibration, the second to the vibration in $1579.64\,\mathrm{cm}^{-1}$ corresponds to the carbon materials with sp^2 hybridization, and the last signal in $1605.03\,\mathrm{cm}^{-1}$ corresponds to the vibration in the direction parallel to the axis tube.

Table 2 shows the adjustment of the G region or tangential vibrations; the first two signals correspond to defects in the

TABLE 1: Fitted parameters anode energy, 2.33 eV (533 nm).

Peak	Type	Amplitude (a.u.)	Center (cm^{-1})	FWHM (cm^{-1})	% area	Interpretation
1	Lorentz amp	97.95	1338.13	80.60	12.22	Nanotube defects
2	Lorentz amp	129.97	1346.26	443.91	61.67	Amorphous carbon
3	Lorentz amp	25.38	1549.20	166.85	5.60	G−
4	Lorentz amp	114.78	1579.64	54.88	9.62	G0
5	Lorentz amp	139.76	1605.03	51.73	10.90	G+

Laser characterization energy = 2.33 eV. The coefficient of determination $r^2 = 0.99$.

TABLE 2: Fitted parameters anode energy 1.96 eV (633 nm).

Peak	Type	Amplitude (a.u.)	Center (cm^{-1})	FWHM (cm^{-1})	% area	Interpretation
1	Lorentz amp	1030.99	1304.29	362.07	66.25	Amorphous carbon
2	Lorentz amp	623.54	1327.73	80.96	11.98	Nanotube defects
3	Lorentz amp	159.26	1521.89	153.65	5.17	G−
4	Lorentz amp	796.23	1585.84	59.84	11.02	G0
5	Lorentz amp	501.17	1606.99	47.79	5.57	G+

Laser characterization energy = 1.96 eV. The coefficient of determination $r^2 = 0.99$.

TABLE 3: Fitted parameters anode energy 1.58 eV (783 nm).

Peak	Type	Amplitude (a.u.)	Center (cm^{-1})	FWHM (cm^{-1})	% area	Interpretation
1	Lorentz amp	132.23	1270.23	280.15	56.72	Amorphous carbon
2	Lorentz amp	168.23	1299.98	77.28	24.73	Nanotube defects
3	Lorentz amp	18.83	1495.12	130.65	4.33	G−
4	Lorentz amp	82.05	1576.89	64.56	9.78	G0
5	Lorentz amp	56.05	1600.65	41.97	4.45	G+

Laser characterization energy = 1.58 eV. The coefficient of determination, $r^2 = 0.99$.

TABLE 4: Fitted parameters cathode energy 2.33 eV (533 nm).

Peak	Type	Amplitude (a.u.)	Center (cm^{-1})	FWHM (cm^{-1})	% area	Interpretation
1	Lorentz amp	22.15	1342.01	28.31	13.27	Amorphous carbon
2	Lorentz amp	17.00	1358.76	23.44	8.46	Nanotube defects
3	Lorentz amp	46.02	1576.10	18.53	17.87	G−
4	Lorentz amp	112.04	1583.39	21.93	51.44	G0
5	Lorentz amp	15.56	1621.05	27.94	8.85	G+

Laser characterization energy = 2.33 eV. The coefficient of determination $r^2 = 0.99$.

TABLE 5: Fitted parameters cathode energy 1.96 eV (633 nm).

Peak	Type	Amplitude (a.u.)	Center (cm^{-1})	FWHM (cm^{-1})	% area	Interpretation
1	Lorentz amp	151.11	1323.10	93.29	22.77	Nanotube defects
2	Lorentz amp	357.76	1333.53	57.58	22.91	Amorphous carbon
3	Lorentz amp	339.74	1578.80	15.08	8.82	G−
4	Lorentz amp	872.04	1582.78	26.85	39.53	G0
5	Lorentz amp	194.28	1617.77	18.10	5.90	G+

Laser characterization energy = 1.96 eV. The coefficient of determination $r^2 = 0.99$.

TABLE 6: Fitted parameters cathode energy 1.58 eV (785 nm).

Peak	Type	Amplitude (a.u.)	Center (cm^{-1})	FWHM (cm^{-1})	% area	Interpretation
1	Lorentz amp	19.77	1269.26	205.75	22.77	Amorphous carbon
2	Lorentz amp	148.16	1311.23	50.28	48.07	Nanotube defects
3	Lorentz amp	51.43	1580.37	13.36	4.55	G−
4	Lorentz amp	88.41	1584.86	37.35	20.97	G0
5	Lorentz amp	41.68	1614.55	15.19	4.14	G+

Laser characterization energy = 1.58 eV. The coefficient of determination $r^2 = 0.98$.

TABLE 7: Anode amorphous carbon concentration.

Energy (ev)	Amorphous concentration (%)	Defect index
2.33	61.67	1.24
1.96	66.25	0.91
1.5	56.72	0.49

case correspond to carbon nanofoam. The widest signal, $1333.53\,cm^{-1}$, corresponds to amorphous carbon. The last three signals correspond to tangential vibrations of the carbon atoms, the signal in $1578.80\,cm^{-1}$ corresponds to the circumferential vibration, the second to the vibration in $1582.78\,cm^{-1}$ corresponds to the carbon materials with sp^2 hybridization, and the last signal in $1617.77\,cm^{-1}$ corresponds to the vibration in the direction parallel to the axis tube.

Table 6 shows the adjustment of the G region or tangential vibrations; the first two signals correspond to defects in the nanotubes; of these, the narrowest signal, $1311.23\,cm^{-1}$, corresponds to bent nanotubes that in this case correspond to carbon nanofoam. The widest signal, $1269.26\,cm^{-1}$, corresponds to amorphous carbon. The last three signals correspond to tangential vibrations of the carbon atoms, the signal in $1580.37\,cm^{-1}$ corresponds to the circumferential vibration, the second to the vibration in $1584.86\,cm^{-1}$ corresponds to the carbon materials with sp^2 hybridization, and the last signal in $1614.55\,cm^{-1}$ corresponds to the vibration in the direction parallel to the axis tube.

Table 8 shows the proportions of amorphous carbon, and defects in nanostructures are summarized.

The physical processes that describe Raman and SEM characterizations are different, so even though the characterized area by SEM is larger, of about $25\,\mu m^2$ (for the analyzed micrography), the punctual excitation is smaller, approximately of $0.05\,\mu m^2$, while Raman characterization excited area is of less than $4\,\mu m^2$, obtaining punctual characterizations. Raman spectroscopy characterization is a resonant process; therefore, by exciting the sample with different lasers wavelength, diverse structures are excited; this depends on the electronic state densities of the samples.

The defects detected by Raman characterization do not correspond only to amorphous carbon; these defects correspond to topological defects, interstitial atoms, and vacancies. In the absence of a crystallographic structure corresponding to the amorphous carbon, a wider line shape is obtained, involving the mentioned defects, thus showing a more general view and reporting higher concentrations of amorphous carbon.

3.2.3. Transmission Electron Microscopy. The JEOL transmission electron microscope JEM-220Fs works with a field emission electron source (FEG) with 200 KeV and a column energy filter (Omega Filter). The JEM-2200FS also uses a new rotation-free optical system to form the micrography, which facilitates the acquisition of TEM micrographies and diffraction patterns. It is possible to take high-contrast

images, tomography, and work without the need of a dark room with the LCD camera. It has extensive processing tools of micrographies and computational analysis.

Figure 13 shows a multiwalled carbon nanotube with a catalytic mixture particle; in Figures 14 and 15, we can see the enlargement of the zones z1 and z2 and could measure the interplanar distance, which is close to the interplanar distance of the graphite.

The product deposited over the chamber walls and the grid structure is amorphous carbon. The product deposited over the electrode surfaces has similar structures in both with the difference that the structures deposited in the cathode are more free of amorphous carbon and catalyst particles than those deposited on the anode surface.

The discrepancy of the amorphous carbon concentrations localized on the cathode and anode is due to two factors, being these the heat exposure of the electrodes and the mass difference between the generated atoms cluster after the discharge. The heat exposure among the electrodes is different because of the anode discontinuities, also described as the catalytic mixture section, as shown in Figure 2(b). The anode is formed by eight teeth (catalytic mixture sections) and the cathode by only one point electrode; this implies that the temperature increase at the cathode is more frequent than the increment at one of the anode teeth. There is a formation of jet plasma after every discharge occurs; since the rotation of the anode base is counterclockwise, the plasma is dragged toward the cathode and covers it. This constant heat increases the mobility on the surface of the cathode, therefore allowing the atoms to be arranged on the areas of stable geometry or minimal energy.

When discharges occur, some materials are released from the anode; these can be a small set of atoms or a big set of atoms of catalytic mixture. The heavier elements cannot reach the cathode surface and are deposited on the surface of the teeth and in surrounding areas of the anode, that is, carbon nanofoam contaminated with amorphous carbon. Considering that on the cathode surface there are smaller and lighter set of atoms, their arrangement is more efficient, compared with the anode, where less frequent arrangement lapses occur and bigger and heavier molecules are located. Thus, the carbon nanofoam deposited on the cathode is less contaminated by amorphous carbon.

The electrical arc discharge is a complex and chaotic process. During this carbon nanofoam synthesis, most of the carbon nanostructures are produced, being impossible to produce 100% carbon nanofoam by this method. The modifications to this method increase the synthesis of carbon nanofoam; however, to increase the carbon nanofoam production rate, a more precise control of all the involved variables is required, that is, an automation of the system is needed.

4. Conclusion

The carbon product synthesized in the experiment was not deposited in spiderweb form. The product obtained was deposited over the electrode surfaces, chamber walls, and over the structure of the collecting grid in a black cotton way.

TABLE 8: Cathode amorphous carbon concentration.

Energy (ev)	Amorphous concentration (%)	Defect index
2.33	19.78	4.62
1.96	29.1	2.36
1.5	22.77	0.84

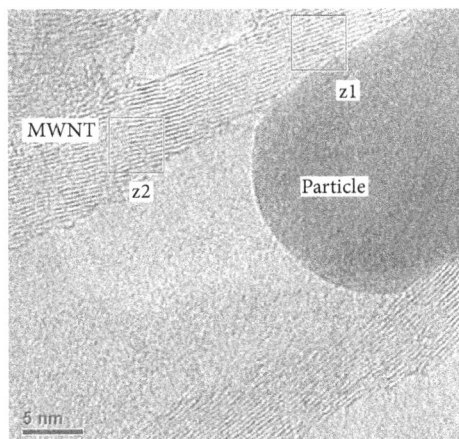

FIGURE 13: Multiwalled carbon nanotube.

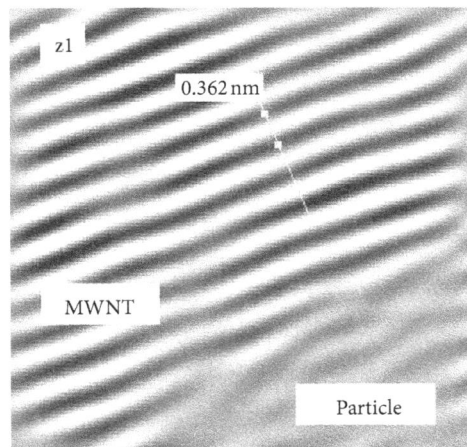

FIGURE 14: Zone z1 amplification.

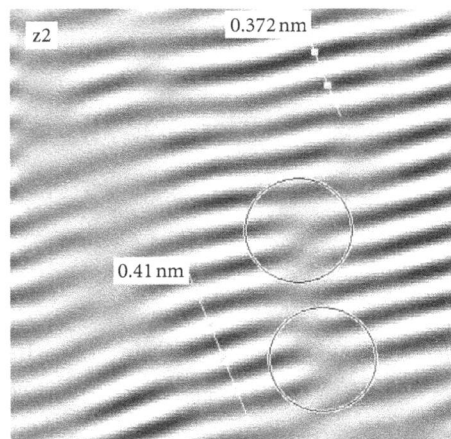

FIGURE 15: Zone z2 amplification.

In the characterization by scanning electron microscopy, it is found that the structures deposited over the anode surface have 35.66% amorphous carbon concentration, and the structures deposited over the cathode surface have 15.46% amorphous carbon concentration. The carbon nanofoam concentration over the anode surface is 11.94%, and the carbon nanofoam concentration over the cathode surface is 81.82%. The catalytic particle proportion on the anode is 52.4%, and the catalytic particle proportion on the cathode is 2.71%.

With the same technique, it is found that the carbon nanofoam structure is conformed by ropes with dimensions of 1 micron width and 100 nm length before breaking translation symmetry. The ropes are conformed by multi-walled carbon nanotubes, and the diameter is different in anode and cathode, the mean diameter of the carbon nanotube at the anode is 30 nm, and the mean diameter of the carbon nanotube at the cathode is 2.7 nm.

With Raman spectroscopy characterization, different amorphous carbon concentrations were found because this depends on the laser excitation. With the 2.33 eV laser energy, the proportion of amorphous carbon is 61.67%; with the 1.96 eV laser energy, the amorphous carbon proportion is 66.25%; and with the 1.53 eV laser energy, the amorphous carbon proportion is 56.72%.

With these results, it can be seen that the growth and deposit on the electrodes are different for the anode (catalytic mixture section) and the cathode because the cathode pulse discharge frequency is greater. When the pulsed discharge occurs at the cathode, the catalytic mixture section is present; at the next discharge, the same cathode and other catalytic mixture sections are present. The time for the next discharge in the first catalytic mixture is the same time the anode makes a full lap; because of this, the temperature at the cathode is higher and constantly increasing. The different cathode and anode gap height affect the material deposit, only lighter material can attach to the cathode, and the heavy material attaches to the anode, like amorphous carbon and catalytic particles. This depends on the anode angular velocity too.

Conflicts of Interest

The authors declare that there are no conflicts of interest regarding the publication of this paper.

Acknowledgments

The authors acknowledge the EDI grant by SIP/IPN. This research was partially supported by the projects 20181454, 20181441, and 20180472.

References

[1] A. V. Rode, E. G. Gamaly, A. G. Christy et al., "Unconventional magnetism in all-carbon nanofoam," *Physical Review B*, vol. 70, no. 5, article 054407, 2004.

[2] N. Frese, S. Taylor Mitchell, C. Neumann, A. Bowers, A. Gölzhäuser, and K. Sattler, "Fundamental properties of high-quality carbon nanofoam: from low to high density," *Beilstein Journal of Nanotechnology*, vol. 7, pp. 2065–2073, 2016.

[3] K. O. Iwu, A. Lombardo, R. Sanz, S. Scirè, and S. Mirabella, "Facile synthesis of ni nanofoam for flexible and low-cost non-enzymatic glucose sensing," *Sensors and Actuators B: Chemical*, vol. 224, pp. 764–771, 2016.

[4] T. Iino and K. Nakamura, "Acoustic and acousto-optic characteristics of silicon nanofoam," *Japanese Journal of Applied Physics*, vol. 48, no. 7, article 07GE01, 2009.

[5] A. Kurek, R. Xalter, M. Stürzel, and R. Mülhaupt, "Silica nanofoam (nf) supported single-and dual-site catalysts for ethylene polymerization with morphology control and tailored bimodal molar mass distributions," *Macromolecules*, vol. 46, no. 23, pp. 9197–9201, 2013.

[6] L. Liu, I. Botos, Y. Wang et al., "Structural basis of toll-like receptor 3 signaling with double-stranded rna," *Science*, vol. 320, no. 5874, pp. 379–381, 2008.

[7] L. T. Johnston, M. M. Biener, J. C. Ye, T. F. Baumann, and S. O. Kucheyev, "Pore architecture of nanoporous gold and titania by hydrogen thermoporometry," *Journal of Applied Physics*, vol. 118, no. 2, article 025303, 2015.

[8] A. Zani, D. Dellasega, V. Russo, and M. Passoni, "Ultra-low density carbon foams produced by pulsed laser deposition," *Carbon*, vol. 56, pp. 358–365, 2013.

[9] S. T. Mitchell, N. Frese, A. Gölzhäuser, A. Bowers, and K. Sattler, "Ultralight carbon nanofoam from naphtalene-mediated hydrothermal sucrose carbonization," *Carbon*, vol. 95, pp. 434–441, 2015.

[10] Y. B. Zel'dovich, Y. P. Raizer, W. D. Hayes, R. F. Probstein, and S. P. Gill, "Physics of shock waveshigh-temperature hydrodynamic phenomena, vol. 1," *Journal of Applied Mechanics*, vol. 34, no. 4, p. 1055, 1967.

[11] F. Neri, P. M. Ossi, and S. Trusso, "Propagation of laser generated plasmas through inert gases," *Laser and Particle Beams*, vol. 28, no. 1, pp. 53–59, 2010.

[12] M. Filipescu, P. M. Ossi, N. Santo, and M. Dinescu, "Radio-frequency assisted pulsed laser deposition of nanostructured wo x films," *Applied Surface Science*, vol. 255, no. 24, pp. 9699–9702, 2009.

[13] D. Saucedo-Jimenez, J. Ortiz-López, V. Garibay-Febles, and E. Palacios-González, "Sem and hrtem analysis of carbon nanostructures synthesized with a new pulsed electric arc discharge technique," *Acta Microscopica*, vol. 22, no. 3, pp. 289–299, 2013.

[14] A. A. Puretzky, H. Schittenhelm, X. Fan, M. J. Lance, L. F. Allard Jr., and D. B. Geohegan, "Investigations of single-wall carbon nanotube growth by time-restricted laser vaporization," *Physical Review B*, vol. 65, no. 24, article 245425, 2002.

[15] C. Liu, H. T. Cong, F. Li et al., "Semi-continuous synthesis of single-walled carbon nanotubes by a hydrogen arc discharge method," *Carbon*, vol. 37, no. 11, pp. 1865–1868, 1999.

[16] G. G. Tibbetts, C. A. Bernardo, D. W. Gorkiewicz, and R. L. Alig, "Role of sulfur in the production of carbon fibers in the vapor phase," *Carbon*, vol. 32, no. 4, pp. 569–576, 1994.

[17] M. S. Dresselhaus, G. Dresselhaus, R. Saito, and A. Jorio, "Raman spectroscopy of carbon nanotubes," *Physics reports*, vol. 409, no. 2, pp. 47–99, 2005.

[18] M. S. Dresselhaus, A. Jorio, A. G. Souza Filho, and R. Saito, "Defect characterization in graphene and carbon nanotubes using raman spectroscopy," *Philosophical Transactions of the Royal Society of London A: Mathematical, Physical and Engineering Sciences*, vol. 368, no. 1932, pp. 5355–5377, 2010.

[19] C. Andrea Ferrari, J. C. Meyer, V. Scardaci et al., "Raman spectrum of graphene and graphene layers," *Physical Review Letters*, vol. 97, no. 18, article 187401, 2006.

[20] S. Costa, E. Borowiak-Palen, M. Kruszynska, A. Bachmatiuk, and R. J. Kalenczuk, "Characterization of carbon nanotubes by raman spectroscopy," *Materials Science-Poland*, vol. 26, no. 2, pp. 433–441, 2008.

10

Effect of Alternative Wood Species and First Thinning Wood on Oriented Strand Board Performance

Fabiane Salles Ferro (ID),[1] **Amós Magalhães Souza**,[2] **Isabella Imakawa de Araujo**,[2] **Milena Maria Van Der Neut de Almeida**,[1] **André Luis Christoforo**,[3] **and Francisco Antonio Rocco Lahr**[2]

[1]*Department of Forest Engineering, State University of the Midwest of Paraná, PR 153 Km, 84500-000 Irati, PR, Brazil*
[2]*Department of Science and Engineering Materials, University of São Paulo, Av. Trabalhador São Carlense, 13566-590 São Carlos, SP, Brazil*
[3]*Department of Civil Engineering, Federal University of São Carlos, Rodovia Washington Luís, 13565-905 São Carlos, SP, Brazil*

Correspondence should be addressed to Fabiane Salles Ferro; fabi.salles.ferro@gmail.com

Academic Editor: Simon C. Potter

This study aimed to evaluate the feasibility of using and influence of alternative wood species such as Cambará, Paricá, Pinus, and wood from first thinning operations on oriented strand board (OSB) physical and mechanical properties. Besides that, an alternative resin, castor oil-based polyurethane, was used to bond the particles, due to the better environmental performance when compared to other resins commonly used worldwide in OSB production. Physical properties such as the moisture content, thickness swelling, and water absorption, both after 2 and 24 hours of water immersion, and mechanical properties such as the modulus of elasticity and resistance in static bending, in major and minor axes, and internal bonding were investigated. All tests were performed according to European code EN 300:2006. Results showed the influence of wood species on physical and mechanical properties. Panels made with higher density woods such as Cambará presented better physical performance, while those made with lower density woods such as Pinus presented better mechanical properties. Besides that, strand particle geometry was also influenced on all physical and mechanical properties investigated. Therefore, the feasibility of using alternative species and wood from first thinning and with castor oil-based polyurethane resin in OSB production was verified.

1. Introduction

Replacement of solid wood by composite panels for structural purposes, in several sectors of building construction, is often rising. Parameters concerning to ensure this context are, among others, definition of alternative raw materials (wood species and adhesives) as well as processing technologies that remain conferring physical and mechanical properties to the wood-based products, compatible with their specific applications [1–3].

Brazilian wood panel market, mostly constituted by producers of medium-density particleboards (MDPs), medium-density fiberboards (MDFs), and oriented strand boards (OSBs), is considerably growing in recent years [4]. Moreover, Brazil is among the countries with the most advanced manufacturing processes for wood panels from reforestation trees. According to data from the Brazilian Tree Industry (IBÁ), in 2015, the country was responsible for the production of 8 million m³ of reconstituted wood panels, being the seventh largest worldwide producer [5].

Oriented strand board (OSB) is a well-known kind of panel made of strands or wafers, usually oriented, bonded with waterproof synthetic resin, and consolidated under heat and pressure [6–8]. Particles of the surface layer are aligned and arranged in parallel direction to the length or width of the panel, while particles of the core layer are randomly oriented or aligned generally in the perpendicular direction to the surface layer particles [9].

OSB market is also growing worldwide, and it is expected a growth rate of ≈28% until 2022 [10]. This increasing consumption in different sectors (mainly construction, furniture, and packing) is related to improvements in panels' properties such as strength, workability, and versatility [10].

These panels have a diversified range of uses, like packages, pallets, stands for display, frames for furniture, fences, and formworks, among others, wherever they are mainly intended for structural application [11, 12].

OSB panel is generally manufactured with low-density wood species. In Brazil, for example, industry uses mainly *Pinus* sp. wood species (mainly *Pinus elliottii* and loblolly pine *Pinus taeda*) in OSB production. Consequently, aspects like easy adhesion but high values of water absorption and thickness swelling are commonly observed in these products, as pointed out by Nascimento and Morales [13] and de Souza et al. [14].

Environmental concerns throughout the world have been showing that OSB is really a very newsworthy product because inclusions of more dense materials are true possibilities to expand OSB production and uses. Document of the Canadian Forest Industries [15] evidences that, since that date, excellent quality OSB has been produced with up to 50% of medium- and high-density wood species addictions. But it is always necessary to establish new alternative inputs, once application of wood-based products remains growing in several developing countries like Brazil and South Africa [16–18].

As raw materials of low density for alternative inputs, it can be considered *Pinus* sp. from first thinning and Paricá (*Schizolobium amazonicum*) wood species from planted forest in the Brazilian Amazonian region [19, 20]. These inputs can represent an important percentage of proper wood to OSB production.

When OSB with high physical performance is required, it is convenient to suggest that medium-density wood species be employed, once water absorption and thickness swelling reach interesting values, turning viable the application of essences with expressive availability in Brazilian Tropical Forest as Amescla (*Trattinickia* sp.), Cajueiro (*Anacardium* sp.), and Cambará (*Erisma uncinatum*) as explained by Freitas et al. [17].

Considering Brazilian diversity of wood species, this study aimed to evaluate the feasibility of using and the influence of alternative wood species on the physical and mechanical properties of OSB. Besides that, an alternative resin, castor oil-based polyurethane resin, was used to bond particles, due to the better environmental performance when compared to phenol formaldehyde and methylene diphenyl diisocyanate (MDI), both commonly used in OSB production.

2. Materials and Methods

For the development of this study, three different wood species were used: *Pinus* sp. with a bulk density of 500 kg/m³, Cambará (*Erisma uncinatum*) with a bulk density of 720 kg/m³, Paricá (*Schizolobium amazonicum*) with a bulk density of 400 kg/m³, and *Pinus* sp. wood from first thinning, with a bulk density of 450 kg/m³. Each wood species is from a different region of Brazil. *Pinus* sp. wood species and *Pinus* sp. wood from first thinning came from São Carlos City, State of São Paulo (Brazil southeast region). Cambará wood species came

from Alta Floresta, State of Mato Grosso (Brazil midwest region), while Paricá came from Paragominas, State of Pará (Brazil north region).

Strands were generated from each wood species. For that, firstly, wood beams were cut into 90 mm wide and 35 mm thickness, which defined the strand's length and width, respectively. Strands were generated in a chipper disc, and they were obtained with a medium thickness of 0.7 mm as shown in Figure 1(a) [19–21].

OSBs were manufactured using an alternative resin, the castor oil-based polyurethane resin, of bicomponent type (Figure 1(b)). This resin is composed of polyol, a component derived from vegetable oil, and polyfunctional isocyanate (prepolymer), derived from crude oil. The ratio between polyol and isocyanate used was 1 : 1, and wax and other additives are not used in panel manufacturing.

For all experimental conditions, 12% of resin content was used, in relation to dry mass of particles. The use of castor oil-based polyurethane resin, as well as the ratio between its components, is justified by the excellent results obtained in previous studies with wood-based panels. Besides that, castor oil-based polyurethane resin presents better environmental performance when compared with other resins such as MDI (methylene diphenyl diisocyanate) and phenol formaldehyde, both commonly used worldwide by the OSB industries [19–22]. Table 1 shows the experimental design of this study.

OSBs were manufactured with a nominal density of 650 kg/m³ and in three layers (Figure 1(c)). Strands of each layer were manually distributed, with the strands of surface layers arranged in an oriented way (longitudinal direction of the panel), while particles of the core layer were randomly distributed. The face/core/face ratio of strands was 20 : 60 : 20, due to great results obtained in the literature such as Cloutier [23], Ferro et al. [24], and Nascimento et al. [25].

The formed mats were pressed for 10 minutes at a temperature of 100°C and specific pressure of 4 MPa (Figure 1(d)). For stabilization and complete cure of resin, panels were conditioned for 48 hours under environmental conditions. After this period, they were cut for subsequent removal of the specimens for physical and mechanical tests (Figure 1(e)) [26, 27].

Tests were performed according to European normative code EN 300 "*Oriented Strand Boards (OSB): Definitions, Classification and Specification*" [9] and complementary codes, due to the absence of Brazilian codes about OSB. Physical properties evaluated were the moisture content (MC), thickness swelling after 2 hours (TS2h) and 24 hours (TS24h), and water absorption after 2 hours (WA2h) and 24 hours (WA24h) of water immersion. Mechanical properties verified were the bending stiffness (MOEpar) and bending strength (MORpar) in the major axis and modulus of elasticity (MOEper) and resistance (MORper) in the minor axis and internal bonding (IB).

The Tukey test, at 5% of the significance level, was used to group the levels of the wood factor (Wood) (Cambará (Ca), Pinus (Pi), Paricá (Pa), and thinning (Wt)) used in the OSB manufacture. From Tukey's test, letter "a" denotes the level

FIGURE 1: Steps of OSB production.

TABLE 1: Experimental design.

Experimental condition	Wood species	Bulk density (kg/m³)
Ca	Cambará (*Erisma uncinatum*)	720
Pa	Paricá (*Schizolobium amazonicum*)	400
Pi	*Pinus* sp.	500
Wt	*Pinus* sp. wood from first thinning	450

TABLE 2: Number of determinations for each property and for each wood species used.

Properties	Cambará (Ca)	Pinus (Pi)	Paricá (Pa)	Thinning (Wt)
MC	5	12	14	30
TS2h	15	12	18	30
TS24h	15	12	18	30
WA2h	15	12	18	30
WA24h	15	12	18	30
MOEpar	15	12	18	28
MORpar	15	12	18	28
MOEper	5	4	5	5
MORper	5	4	5	5
IB	15	12	18	30

of the highest mean value factor, "b" the second highest mean value, and so on, and same letters imply levels with statistically equivalent means.

Anderson-Darling (AD) and Bartlett (Bt) variance homogeneity tests were used in the validation of Tukey's test. For the hypotheses formulated, P value (probability P) equal to or higher than the significance level implies normality and homogeneity of variances by property, which validates Tukey's test.

Table 2 shows the number of determinations for each property and for each wood species used in OSB production, which resulted in 620 experimental results.

3. Results and Discussion

The density mean values for all experimental conditions are in accordance with the nominal density of 650 kg/m³ initially defined for panel production. The values obtained were 680,

TABLE 3: Physical property results in %.

Properties	Cambará (Ca)	Pinus (Pi)	Paricá (Pa)	Thinning (Wt)	CV (%)
MC	8.67 bc	9.30 ab	8.13 c	9.57 a	3.25; 12.88
TS2h	5.74 b	15.34 a	13.84 a	12.11 a	28.51; 37.85
TS24h	8.78 c	25.90 a	28.74 a	19.54 b	12.92; 26.45
WA2h	8.21 c	34.92 a	21.31 b	20.88 b	23.10; 34.01
WA24h	21.21 c	52.75 a	57.63 a	29.96 b	13.62; 23.35

700, 640, and 620 kg/m³ for the OSB produced with Cambará, Pinus, Paricá, and wood from the thinning operations, respectively. Differences are associated with the manufacturing process of the panels.

Table 3 shows the mean values, the range of the coefficients of variation (CV), and Tukey's test results for physical properties evaluated.

The P values of the normality tests and homogeneity of variances for physical properties ranged in the intervals of 0.105 to 0.526 and 0.086 to 0.722, respectively, validating Tukey's test results.

Table 3 shows that, for moisture content (MC), the mean values for all evaluated treatments are within the ranges recommended by normative codes. According to European code EN 312 [28], the required moisture content must be between 5 and 13%, whereas according to the Brazilian normative code ABNT NBR 14810-2 [29], the MC for particleboards must be between 5 and 11%. Comparing the results obtained in this study with the related literature, the mean values are in conformity. Besides that, according to the technical report of the Technological Research Institute (IPT) [28], the moisture content for the OSB produced and marketed in Brazil is 7.8 ± 3%.

For thickness swelling after 2 h (TS2h) and 24 h (TS24h) of water immersion, the highest mean values were obtained for the OSB manufactured with *Pinus* sp. and Paricá wood species. EN 300 [9] normative code only mentions TS24h. Comparing the mean values shown in Table 2 with those recommended by normative code, it could be observed that the OSB made from *Pinus* sp. and Paricá wood species reached mean values of 25.9 and 28.74, respectively. These values are higher than the maximum permitted value (25%) for OSB type 1 (panels intended for application in dry conditions).

On the other hand, the OSB manufactured with Cambará wood species and wood from the thinning operations presented lower values than that recommended by normative code, being categorized as OSB type 4 (boards for use in humid conditions) and type 2 (boards for use in dry conditions), respectively.

Mean values for TS24h obtained in this study are consistent when confronted with related studies in which were evaluated resins from the same nature. Nascimento et al. [25] obtained TS24h of 14.4% for the OSB manufactured with *Piptadenia moniliformis* Benth. and castor oil-based polyurethane resin; Akrami et al. [1] for the OSB with *Populus tremula* and *Fagus sylvatica* wood species together with isocyanate resin (polymeric methylene diphenyl diisocyanate (pMDI)) obtained mean values of TS24h between 10.0% and 28.0%.

Due to the absence of properties such as TS2h and water absorption after 2 h (WA2h) and 24 h (WA24h) of water immersion in normative codes, the mean values resulting from this study were compared to the literature. Thus, the results obtained indicated physical performances consistent with those of Mendes et al. [30] for TS2h of 31.9%, WA2h of 91.5%, and WA24h of 102.4% for the OSB made with *Pinus oocarpa* wood species with phenol formaldehyde resin. Saldanha [31] obtained mean values for TS2h of 28.0%, WA2h of 58.6%, and WA24h of 74.2% for the OSB produced with *Pinus taeda*. Saldanha and Iwakiri [32] obtained mean values for TS2h of 31.3%, WA2h of 72.3%, and WA24h of 82.3% when panels were made with MUF (melamine-modified urea formaldehyde) resin.

As can be observed in Table 3, wood species influenced significantly on all OSB physical properties analyzed. For all of them, the panel made with Cambará wood species presented better performances, that is, lower mean values of thickness swelling and water absorption for the evaluated periods.

According to Hsu [33], thickness swelling and water absorption parameters are the sum of three components, that is, reversible swelling of the wood itself, spring back of compressed wood, and separation of furnish. Besides that, low-density woods tend to have greater porosity, consequently, greater water absorption in relation to the high-density woods. At the same target density, a lower wood density resulted on higher compression ratio; consequently, it will increase the water absorption and thickness swelling values. The stress inside the panel is particularly released when submitted to water immersion.

Comparing only low-density wood species, it could be observed that the OSB manufactured with thinning wood presented better performance in relation to the panels made with Pinus and Paricá wood species. This is mainly related to the strand's particle geometry obtained for each wood species. Strand generation from wood from the thinning operations resulted in a large amount of fines, which up to a certain amount improves the performance of these properties, once these assist in filling panel empty spaces, reducing the amount of water absorbed.

Table 4 shows the mean values, the range of the coefficients of variation (CV), and Tukey's test results for mechanical properties evaluated. P values of the normality tests and homogeneity of variances for mechanical properties varied between 0.291 and 0.876 and between 0.233 and 0.524, respectively, validating Tukey's test results.

It can be observed in Table 4 that the mean values of MOEpar ranged from 5463 for the OSB manufactured with Cambará wood species up to 8238 MPa for the OSB made

TABLE 4: Mechanical property results in MPa.

Properties	Cambará (Ca)	Pinus (Pi)	Paricá (Pa)	Thinning (Wt)	CV (%)
MOEpar	5463 c	8237 a	6932 b	6395 b	8.21; 17.74
MORpar	30.20 b	54.77 a	52.90 a	36.40 b	8.39; 35.23
MOEper	1638 bc	2437 a	1366 c	1818 b	2.88; 17.03
MORper	12.16 c	22.50 a	18.63 ab	13.94 bc	10.58; 22.47
IB	0.66 b	1.58 a	0.54 b	0.77 b	24.63; 0.22

with Pinus. For MOEpar property, all OSBs reached the minimum value (4800 MPa) recommended by EN 300 [9] for OSB type 4 (special panels for structural purposes). Regarding MORpar, mean values ranged from 31 MPa (Cambará) up to 55 MPa (Pinus). In addition, all conditions obtained the minimum value (30 MPa) recommended by EN 300 [9] for OSB categorized as type 4.

The mean values of all experimental conditions evaluated were consistent with those of 5080 MPa and 38.5 MPa for MOE and MOR in the major axis, respectively, reported in IPT technical report [34].

In Table 4, it also could be observed that, for MOEper, mean values varied between 1367 MPa for the OSB manufactured with Paricá and 2437 MPa for the OSB made with Pinus. In addition, only panels made with Pinus wood reached the minimum value of 1900 MPa recommended by normative code [9] for OSB type 4. Regarding the other treatments, it reached the minimum value of 1400 MPa recommended for OSB types 2 and 3 (both load-bearing boards). For MORper property, mean values ranged from 12 MPa (Cambará) up to 22 MPa (Pinus). According to EN 300 [9], panels made with Cambará and thinning wood reached the minimum value of 11 MPa required for OSB type 3, while panels made with Pinus and Paricá reached the required value of 16 MPa for categorization as type 4. However, both panels are intended for structural application.

Comparing the mean values of MOEper and MORper obtained in this study with results of the literature, it was observed that there is consistency between them. Surdi et al. [3], for the OSB produced with hybrids of Pinus and phenol formaldehyde resin, obtained mean values of 992 MPa and 15.3 MPa for MOEper and MORper, respectively. Neimsuwan [35], for the OSB made with MDI resin, obtained mean values of 1381 MPa for MOEper and 13.5 MPa for MORper.

From Table 4, it could also be observed that the mean values for internal bonding (IB) ranged from 0.54 MPa (Paricá) to 1.54 MPa (Pinus). For IB, all treatments analyzed reached the minimum of 0.5 MPa value recommended for OSB type 4 [9].

Mean values of this study are consistent with those results reported in the literature for the OSB made with phenol formaldehyde resin and castor oil-based polyurethane resin. For example, Saldanha and Iwakiri [32], for the OSB with *Pinus taeda* L, obtained mean values of 0.39 MPa for IB; Nascimento et al. [25], for panels made with wood from Caatinga-Brazilian northeast regions such as Marmeleiro (*Croton sonderianus* Muell. Arg.) with the bulk density of 750 kg/m³ to 850 kg/m³, Jurema-branca (*Piptadenia stipulacea* (Benth.) Ducke) with the bulk density of 750 kg/m³ to

900 kg/m³, and Catanduva (*Piptadenia moniliformis* Benth.) with the bulk density between 800 kg/m³ and 940 kg/m³ obtained mean values of 0.45 MPa, 0.58 MPa, and 0.68 MPa, respectively.

For all properties, it was observed that wood species is a significant factor in panel performance, being the best one obtained for the panels manufactured with lower density woods. Panels manufactured with higher density wood species presented deficiency in adhesion, especially when considered that more dense woods present difficulties in the resin penetration.

However, Paricá and wood from the thinning operations are also lower density woods, and the OSB produced with these species presented lower mechanical performance when compared to Pinus species. The lower mechanical behavior may be associated with the strand's particle geometry, since during particle generation, a great amount of fines were obtained. The large amount of fines interfere in the mechanical behavior since the longitudinal properties of the wood contribute effectively to panel properties [36].

4. Conclusion

Results obtained in this study attest the viability of OSB production with some Brazilian species such as Cambará and Paricá and, besides that, the feasibility of using Pinus wood from first thinning and castor oil-based polyurethane resin. Panels produced could be introduced in Brazilian market, once they reach the code requirements. OSB is among the three main panels produced in Brazil with perspective of increase because it has more possibilities of use when compared to the other panels, such as MDF (medium-density fiberboards) and plywood, in civil construction and furniture industry.

For physical properties, wood species was an influence factor. It was observed that the lowest mean values, that is, better performance, were obtained for higher density woods such as Cambará, once those higher density woods tend to have lower porosity, and consequently, water absorption in relation to the lower density woods is smaller. Besides that, the OSB made with Pinus wood from first thinning presents better performance than the OSB made with Paricá and Pinus (all low-density species). This is the result of strand geometry, since that during particle generation from thinning wood were obtained a large amount of fines.

Regarding mechanical properties, it was observed that the wood species was a significant factor in panel performance. The better performance was obtained for OSBs made with low-density woods such as Pinus, due to the possible higher resin penetration in the wood. Even though they are manufactured with low-density woods, the OSB made with Paricá and wood from thinning operations presented worse properties than that made with Pinus. This is due to the strand geometry because the large amount of fines interfere in the mechanical property behavior, since the longitudinal properties of the wood contribute effectively to panel properties.

Conflicts of Interest

The authors declare that they have no conflicts of interest.

References

[1] A. Akrami, A. Fruehwald, and M. C. Barbu, "The effect of fine strands in core layer on physical and mechanical properties of oriented strand boards (OSB) made of beech (*Fagus sylvatica*) and poplar (*Populus tremula*)," *European Journal of Wood and Wood Products*, vol. 72, no. 4, pp. 521–525, 2014.

[2] F. S. Ferro, T. H. Almeida, D. H. Almeida, A. L. Christoforo, and F. A. Rocco Lahr, "Physical properties of OSB panels manufactured with CCA and CCB treated *Schizolobium amazonicum* and bonded with castor oil based polyurethane resin," *International Journal of Materials Engineering*, vol. 6, no. 5, pp. 151–154, 2016.

[3] P. G. Surdi, G. Bortoletto Filho, R. F. Mendes, and N. F. Almeida, "Use of hybrid *Pinus elliottii* var. *elliottii* x *Pinus caribaea* var. *hondurensis* and *Pinus taeda* L. in the production of OSB panels," *Scientia Forestalis*, vol. 43, no. 108, pp. 763–772, 2015.

[4] ABRAF-Brazilian Association of Forest Plantation Producers, *Yearbook Statistical ABRAF: Base Year 2013*, STCP Engenharia de Projetos, Brasília, Brazil, 2013.

[5] IBÁ-Brazilian Tree Industry, "Report 2016," 2017, http://iba.org/images/shared/Biblioteca/IBA_RelatorioAnual2017.pdf.

[6] C. Barbuta, P. Blanchet, A. Cloutier, V. Yadama, and E. Lowell, "OSB as substrate for engineered wood flooring," *European Journal of Wood and Wood Products*, vol. 70, no. 1–3, pp. 37–43, 2010.

[7] M. S. Bertolini, A. L. Christoforo, and F. A. Rocco Lahr, "Thermal insulation particleboards made with wastes from wood and tire rubber," *Key Engineering Materials*, vol. 668, pp. 263–269, 2015.

[8] L. B. de Macedo, M. R. da Silva, A. A. da Silva César, T. H. Panzera, A. L. Christoforo, and F. A. Rocco Lahr, "Painéis OSB de madeira *Pinus* sp. e adição de partículas de polipropileno biorientado (BOPP)," *Scientia Forestalis*, vol. 44, p. 120, 2016.

[9] European Committee for Standardization, *EN 300: Oriented Strand Boards (OSB): Definitions, Classification and Specification*, European Committee for Standardization, Brussels, Belgium, 2006.

[10] Grand View Research, "Oriented strand board (OSB) market worth $71 million by 2022," 2017, http://www.grandviewresearch.com/press-release/global-oriented-strand-board.

[11] K. Ozkaya, C. I. Abdullah, B. Erol, and A. Salih, "The effect of potassium carbonate, borax and wolmanit on the burning characteristics of oriented strand board (OSB)," *Construction and Building Materials*, vol. 21, no. 1, pp. 1457–1462, 2007.

[12] S. Veigel, J. Rathke, M. Weigl, and W. Gindl-Altmutter, "Particle board and oriented strand board prepared with nanocellulose-reinforced adhesive," *Journal of Nanomaterials*, vol. 2012, Article ID 158503, p. 8, 2012.

[13] M. F. Nascimento and E. A. M. Morales, "Fabricação de Painéis OSB Com Madeira Proveniente de Espécies de Madeira da Caatinga do Nordeste Brasileiro," F. A. R. Lahr, Ed., vol. 6, pp. 119–136, Produtos Derivados da Madeira, São Carlos, SP, Brazil, 2008.

[14] A. M. de Souza, L. D. Varanda, L. B. de Macedo et al., "Mechanical properties of OSB wood composites with resin derived from a renewable natural resource," *International Journal of Composite Materials*, vol. 4, no. 3, pp. 157–161, 2014.

[15] Canadian Forest Industries, "Wood-based panel products technology," 2017, http://s3.amazonaws.com/zanran_storage/www.ic.gc.ca/ContentPages/5685312.pdf.

[16] C. L. Eisfeld and R. Berger, "Análise das estruturas de mercado das indústrias de painéis de madeira (compensado, MDF e OSB) no estado do Paraná," *Floresta*, vol. 42, no. 1, pp. 21–34, 2012.

[17] J. F. Freitas, A. M. Souza, L. A. M. N. Branco, E. Chahud, A. L. Christoforo, and F. A. Rocco Lahr, "A preliminary study about the utilization of Cajueiro and Amescla for MDP panels production," *International Journal of Materials Engineering*, vol. 7, no. 2, pp. 21–24, 2017.

[18] E. A. M. Morales, M. S. Bertolini, M. F. Nascimento, F. A. R. Lahr, and A. W. Ballarin, "Study of Brazilian commercial oriented strand board panels using stress wave," *Wood Research*, vol. 58, no. 2, 2013.

[19] L. Gorski, A. B. Cunha, P. A. Rios et al., "Utilização da madeira de *Eucalyptus benthamii* na produção de painéis de partículas orientadas (OSB)," *Floresta*, vol. 45, no. 4, pp. 865–874, 2015.

[20] R. Trianoski, S. Iwakiri, and D. Chies, "Utilização da madeira de *Cryptomeria japonica* para produção de painéis de partículas orientadas (OSB)," *Scientia Forestalis, Piracicaba*, vol. 44, no. 110, pp. 487–496, 2016.

[21] R. T. Migita, C. I. de Campos, and C. A. O. de Matos, *Análise do Desempenho na Flexão Estática de Painéis de Partículas Orientadas (OSB–Oriented Strand Board)*, 21° CBECIMAT-Congresso Brasileiro de Engenharia e Ciência dos Materiais, Cuiabá, MT, Brazil, 2014, http://www.metallum.com.br/21cbecimat/CD/PDF/204-046.pdf.

[22] F. S. Ferro, "Painéis OSB com madeira schizolobium amazonicum e resina poliuretana à base de óleo de mamona: viabilidade técnica de produção," M.S. thesis, Escola de Engenharia de São Carlos, Universidade de São Paulo, São Carlos, SP, Brazil, 2013.

[23] A. Cloutier, "Oriented strand board (OSB): raw material, manufacturing process, properties of wood-base fiber and particle materials," in *Proceedings of the International Seminar on Solid Wood Products of High Technology*, 1., Anais, SIF, pp. 173–185, Belo Horinzonte, MG, Brazil, 1998.

[24] F. S. Ferro, F. H. Icimoto, A. M. D. Souza, D. H. D. Almeida, A. L. Christoforo, and F. A. R. Lahr, "Produção de painéis de partículas orientadas (OSB) com Schizolobium amazonicum e resina poliuretana à base de óleo de mamona," *Scientia Forestalis*, vol. 43, pp. 313–320, 2015.

[25] M. F. Nascimento, M. S. Bertolini, T. H. Panzera, A. L. Christoforo, and F. A. R. Lahr, "Painéis OSB fabricados com madeiras da caatinga do nordeste do Brasil," *Ambiente Construído*, vol. 15, no. 1, pp. 41–48, 2015.

[26] A. M. de Souza, L. D. Varanda, A. L. Christoforo et al., "Modulus of elasticity in static bending for oriented strand board (OSB)," *International Journal of Composite Materials*, vol. 4, no. 2, pp. 56–62, 2014.

[27] L. D. Varanda, A. L. Christofoto, D. H. Almeida, D. A. L. Silva, T. H. Panzera, and F. A. R. Lahr, "Evaluation of modulus of elasticity in static bending of particleboards manufactured with *Eucalyptus grandis* wood and oat hulls," *Acta Scientiarum Technology*, vol. 36, no. 3, pp. 405–411, 2014.

[28] European Committee for Standardization, *EN 312: Particle-boards Specifications*, European Committee for Standardization, Bruxelas, Belgium, 2006.

[29] Associação Brasileira de Normas Técnicas, *NBR 14810-2: Painéis de Partículas de Média Densidade. Parte 2: Requisitos e Métodos de Ensaio*, Associação Brasileira de Normas Técnicas, Rio de Janeiro, RJ, Brazil, 2013.

[30] R. F. Mendes, L. M. Mendes, J. B. Guimarães Júnior, R. C. Santos, and A. A. S. César, "Efeito da associação de bagaço de cana, do tipo e do teor de adesivo na produção de painéis aglomerados," *Ciência Florestal*, vol. 22, no. 1, 2012.

[31] L. K. Saldanha, "Alternativas tecnológicas para produção de chapas de partículas orientadas "OSB"," M.S. dissertation, Universidade Federal do Paraná, Curitiba, PR, Brazil, 2004.

[32] L. K. Saldanha and S. Iwakiri, "Influência da densidade e do tipo de resina nas propriedades tecnológicas de painéis OSB de *Pinus taeda* L," *Floresta*, vol. 39, no. 3, pp. 571–576, 2009.

[33] W. E. Hsu, "A process for stabilizing waferboard/OSB," in *Proceedings of Particleboard Symposium*, pp. 219–236, Washington State University, Washington, DC, USA, 1987.

[34] Instituto de Pesquisa Tecnológica-IPT, Relatótio Técnico, vol. 126, pp. 416–205, Instituto de Pesquisa Tecnológica, Butantã, SP, Brazil, 2002.

[35] T. Neimsuwan, "Effect of resin and wax ratio on OSB properties," M.S. dissertation, Universidade de Tennessee, Knoxville, TN, USA, 2004.

[36] S. Suzuki and K. Takeda, "Production and properties of Japanese oriented strand board I: effect of strand length and orientation on strength properties of sugi oriented strand board," *Journal of Wood Science*, vol. 46, pp. 289–295, 2000.

11

Formation and Physical Properties of *h*-BN Atomic Layers: A First-Principles Density-Functional Study

Yoshitaka Fujimoto

Department of Physics, Tokyo Institute of Technology, Tokyo, Japan

Correspondence should be addressed to Yoshitaka Fujimoto; fujimoto@stat.phys.titech.ac.jp

Academic Editor: Achim Trampert

Hexagonal boron nitride (*h*-BN) atomic layers have attracted much attention as a potential device material for future nanoelectronics, optoelectronics, and spintronics applications. This review aims to describe the recent works of the first-principles density-functional study on *h*-BN layers. We show physical properties induced by introduction of various kinds of defects in *h*-BN layers. We further discuss the relationship among the defect size, the strain, and the magnetic as well as the electronic properties.

1. Introduction

Since the discovery of a single atomic layer of graphite, that is, graphene, other atomic layers have received much attention [1–8]. Hexagonal boron nitride (*h*-BN) atomic layers, in which alternating boron and nitrogen atoms are arranged on the honeycomb lattices, have also gained a lot of interest because of the similar structural and mechanical properties to graphene [9–11]. On the other hand, electronic properties of *h*-BN layers are considerably different from those of graphene; graphene is a gapless material with metallic conduction [12], and the *h*-BN atomic layers possess a wide band gap [13–15]. Such wide band gap nature of *h*-BN layers is expected to give rise to new physical properties as well as relevant electronic/optoelectronic applications in nanotechnology.

For example, *h*-BN atomic layers exhibit high optical emission intensities in deep ultraviolet (UV) regions [13, 16–18], and the *h*-BN bilayers can transform from an indirect-gap semiconductor to a direct-gap one by applying strains [19]. Thus, *h*-BN layers are highly important optoelectronics device materials used in deep UV lasers and light-emitting diodes [13, 17, 20–23]. The atomic vacancies in *h*-BN layers can be introduced using electron beam irradiations and are shown to be useful as a source of magnetism [10, 24–27]. The introductions of C atoms and graphene-like flakes into *h*-BN

sheets have been also performed via electron beam irradiations due to analogous structural properties among B, C, and N elements [28–36], which can tune the electronic and the magnetic properties [37–39]. It is predicted theoretically that the energy band gaps are tunable depending on the size of the graphene flakes [37, 38]. The electronic transport properties of the *h*-BN atomic layers can be changed from insulating to conductive properties when C atoms are doped, suggesting a possibility to fabricate the novel opto/nanoelectronics applications [31, 32, 34–36, 40]. Furthermore, when the C atom is doped, the exotic conduction channel in *h*-BN monolayers would emerge, which might improve electronic transport properties [41–46]. Thus, *h*-BN atomic layers are promising materials to be used in nanoelectronics and optoelectronics applications.

The purpose of this review is to provide the recent progress of the first-principles density-functional calculations that clarify various physical properties of *h*-BN atomic layers with lattice defects including atomic vacancies and carbon defects. In Section 2, the magnetic properties of atomic vacancies in *h*-BN layer are discussed. Section 3 is devoted to discussions on the electronic properties of carbon-doped *h*-BN layers. In Section 4, the behaviors of the magnetic properties and the energy band gaps are shown with the variation of the graphene-like defect size. Finally, Section 5 summarizes this paper together with outlook.

$$V_2^N \rightarrow V_3^N \rightarrow V_4^N$$

○ N
○ B

(a)

(b)

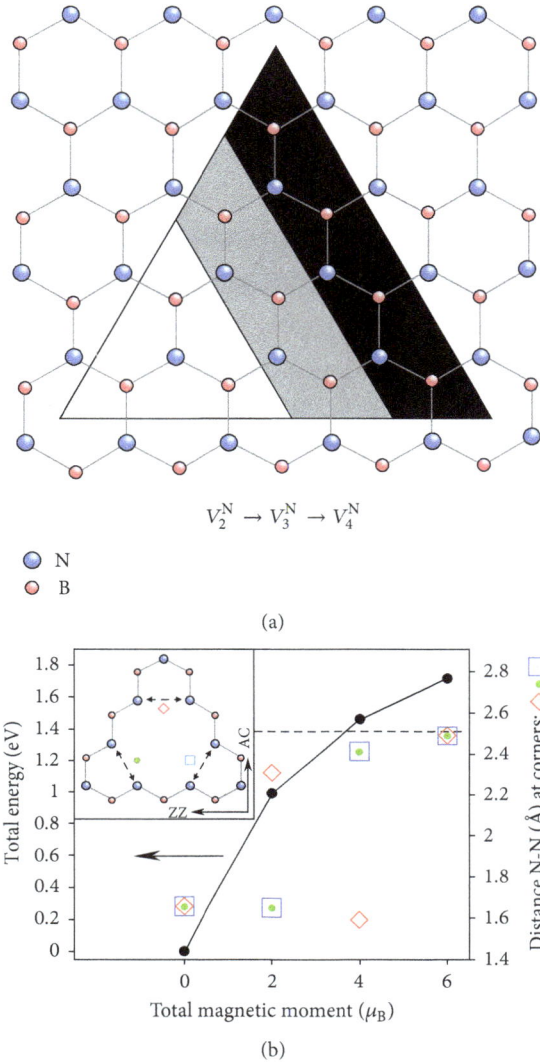

FIGURE 1: (a) Schematic view of h-BN monolayers for atomic vacancies. (b) Total energy and N-N distance near V_2^N defect as a function of imposed magnetic moment. Reproduced with permission from [24]; copyright 2012, the American Institute of Physics.

2. Atomic Vacancy

Atomic vacancies in h-BN layers can be created by electron beam irradiation [10, 25–27]. Triangular-shaped vacancies are observed experimentally by using high-resolution transmission electron microscopy and these vacancies in h-BN layers are arranged with various sizes and peculiar orientation [10]. It is therefore surmised that the B atoms are removed more easily than the N atoms from h-BN monolayers by electron beam irradiation [10].

The introductions of the atomic vacancies into the h-BN monolayers are reported to give rise to the magnetic moment [24, 47]. In Figure 1(a), the schematic view of the triangular atomic vacancies in the h-BN monolayer is shown, where the multivacancies V_T are defined with $T = \sqrt{n_B + n_N}$ and n_B and n_N are the numbers of removed B and N atoms,

respectively. The total magnetic moments of V_1^N, V_2^N, and V_3^N are calculated to be 1, 0, and $3\mu_B$, respectively. Unlike graphene [48], the behavior of vacancies V_T^N in the magnetic properties does not follow Lieb's theorem because of the structural reconstructions with the N-N bond formation [49]. When the N-N bond formation is broken, the total magnetic moment increases. The total magnetic moment of the V_2^N defect increases as the N-N distance around the vacancy increases, though the ground state of V_2^N exhibits zero magnetic moment (see Figure 1(b)).

3. Carbon Impurity

Substitutional carbon doping to h-BN layers and BN nanotubes has been performed using the electron beam irradiation technique [32, 33] and substitutionally doped C atoms in the h-BN monolayers have been observed via a transmission electron microscopy method [31]. It was found that the substitution with C atoms decomposed from hydrocarbon molecules takes place mostly at B atom sites, and it was surmised that the substitutional doping proceeds by repairing the B atom vacancies with C atoms broken by the knock-on electron beam [32, 34].

Recent first-principles calculations showed that the substitution of B atoms with C atoms needs less energy costs than that of N atoms under N-rich conditions. Moreover, it was shown that positive charging favors the substitution of B atom with C atom (C_B), whereas negative charging favors the substitution of N atom with C atom (C_N). In addition, it was shown that the substitution of B atom with C atoms may predominately take place even under B-rich conditions [34, 35].

It was reported that the electronic transport properties of h-BN layers can transform from insulating to conductive properties when C atom is doped [32]. The donor-like impurity states appear below the conduction-band minimum (CBM) when C atom is doped at B atom site, whereas the substitution of N atom with C atom gives rise to the acceptor-like impurity states above the valence-band maximum (VBM) [35, 41]. It was shown that ionization energies for acceptor-like as well as donor-like states can be controlled by applying strains (see Figure 2(a)) [41]. The strains can often produce not only the new structural properties but also the novel electronic properties [19, 50, 51]. Interestingly, in the case of the h-BN monolayers, the exotic transport channel will behave as an active state under more than ~1% compressive strains (Figure 2(b)). In addition, the band gaps are also tunable under strains [52–54].

Scanning tunneling microscopy (STM) measurements are one of the effective tools to observe the surface electronic structures at atomic level [55–61]. The B atoms and N atoms in graphene can be clearly identified experimentally and theoretically [62–66]. The STM image of the h-BN monolayer has a large bright triangular shape around the C defect surrounded by six bright spots above each N atom, since the C defect state consists of the triangular-shaped spatial distributions of local density of states (LDOS), and in addition to this defect state, the π states in the valence bands

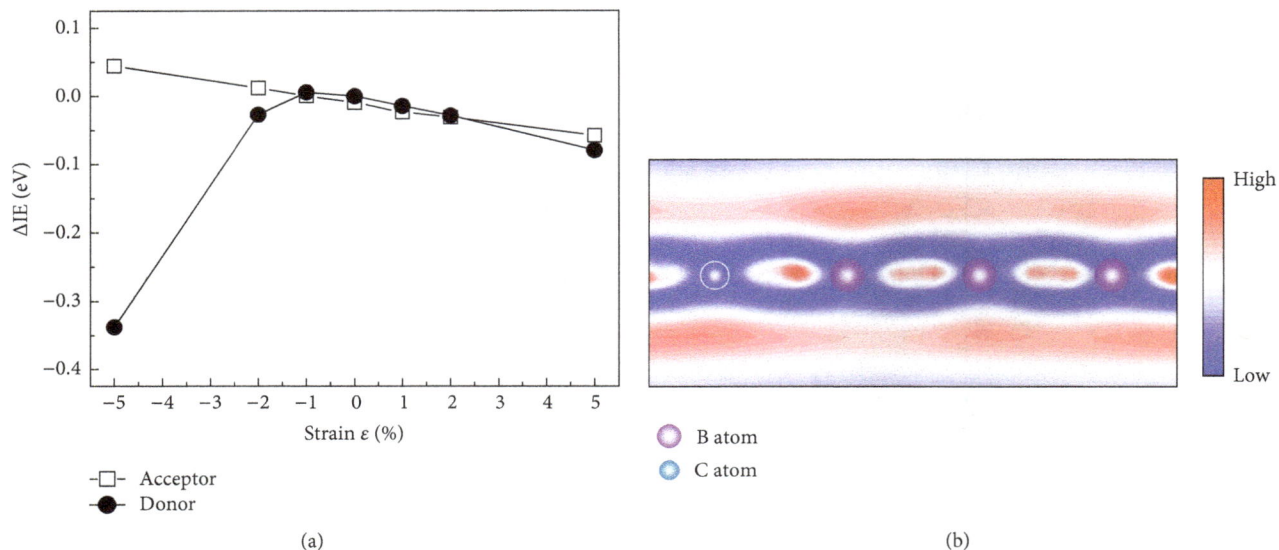

(a)   (b)

FIGURE 2: (a) Relative ionization energy (ΔIE) for acceptor and donor states in C-doped h-BN monolayers as a function of applied strain. (b) Contour plot of electron density of C-doped h-BN monolayer at the Γ point of the CBM, where B atom is replaced by C atom. Reproduced with permission from [41]; copyright 2016, the American Physical Society.

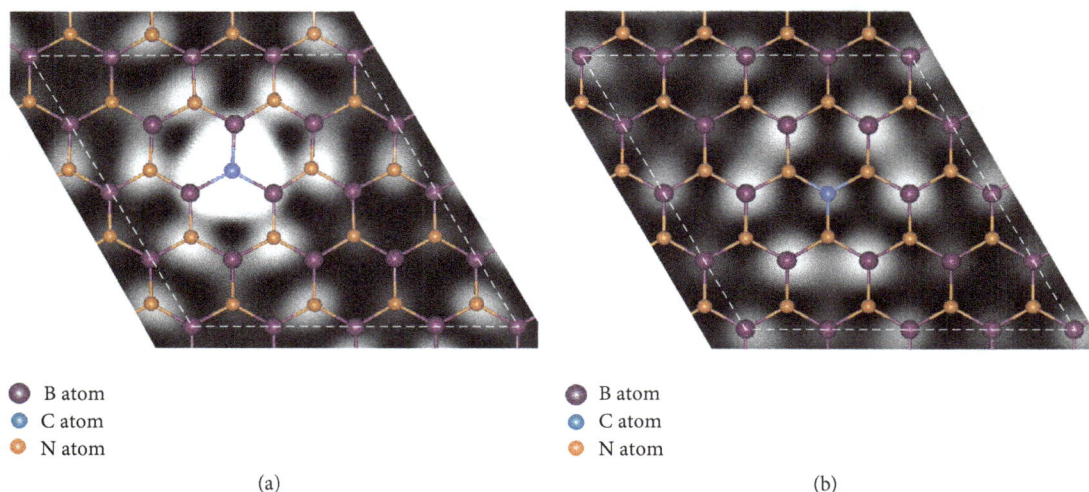

(a)   (b)

FIGURE 3: Calculated STM images of C-doped h-BN monolayers, where C atom is replaced at (a) the N site and (b) the B site. Reproduced with permission from [41]; copyright 2016, the American Physical Society.

contribute the STM image (Figure 3(a)). On the other hand, for the C substitution of the B atom, the STM image has a small dark spot above the C defect which is surrounded by six bright spots above each B atom (Figure 3(b)) [41]. The C defects in the h-BN monolayers might be identified by using STM methods.

4. Carbon Flake

The h-BN monolayers embedded with triangular graphene flakes can modify the electronic properties as well as the magnetic properties [34, 37]. The magnetic moments of the graphene-like flake embedded h-BN monolayer can be controlled depending on the number of the substituted C

atoms in the h-BN monolayers (Figure 4), where α and β are the numbers of the B atoms and the N atoms replaced with the C atoms and positive and negative ($\alpha - \beta$) values are used for the T1-$C_\alpha C_\beta$ and T2-$C_\alpha C_\beta$ structures, respectively. T1-$C_\alpha C_\beta$ and T2-$C_\alpha C_\beta$ structures give rise to $|\alpha - \beta|\mu_B$ magnetic moments per unit cell. By tuning the sizes of the triangle graphene flakes, the magnetic moments of T1- and T2-graphene-embedded BN sheet are controllable.

The energy band gap values of h-BN monolayers are shown to be tunable [37–39]. By replacing B and N atoms with graphene quantum dots (QD), the energy band gaps of h-BN monolayers decrease. As the size d of the graphene QD increases, it was shown that the band gaps diminish from ~3.6 eV to ~1.6 eV (Figure 5).

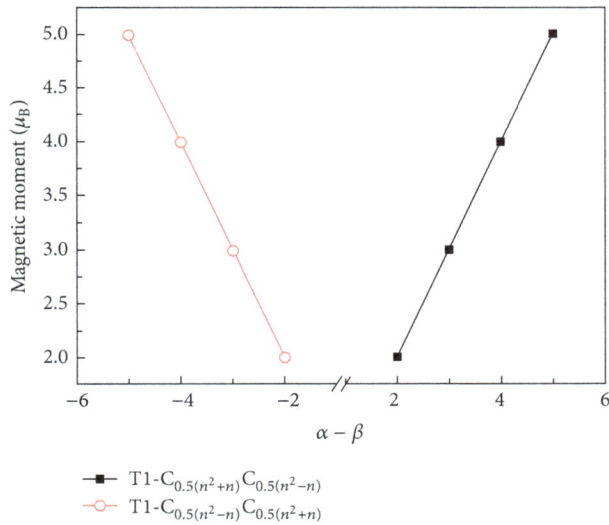

FIGURE 4: Variation of magnetic moments of T1-$C_\alpha C_\beta$ and T2-$C_\alpha C_\beta$ with $(\alpha - \beta)$. Reproduced with permission from [37]; copyright 2011, the American Physical Society.

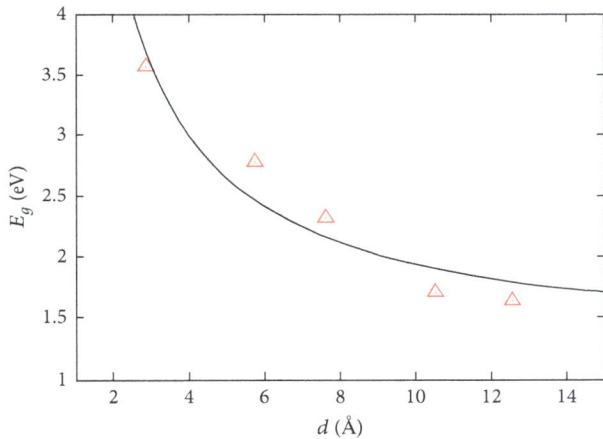

FIGURE 5: Energy band gap value E_g as a function of QD diameter d. Reproduced with permission from [38]; copyright 2011, the American Institute of Physics.

5. Concluding Remarks

We have reviewed the recent works of the first-principles density-functional study of h-BN atomic layers. The atomic vacancies can be created in the h-BN layers by electron beam irradiation techniques. The triangular-shaped atomic vacancies in h-BN layers can give rise to the magnetic moment depending on the vacancy sizes, and moreover the magnetic moment is shown to be tunable under strain.

The C atom can be doped to the h-BN layers by using electron beam irradiation combined with introducing the hydrocarbon molecules, and it is deduced that C atoms are doped more easily at B atom sites than at N atom sites. The first-principles density-functional calculations have revealed that the substitution of B atoms with C atoms becomes more favorable in energy than that of N atoms under N-rich conditions. The substitutions of the B atom and the N atom with the C atom induce the donor-like state below the CBM and the acceptor-like state above the VBM, respectively. The ionization energies are controllable for the donor and the acceptor states by applying strain, and furthermore the exotic electronic state can be opened as an active conduction channel under strain.

The graphene-like flake in the h-BN layers can modify the magnetic and the electronic properties. By substituting B atoms and N atoms at the two different sublattices with C atoms, the magnetic moment of the h-BN layers can be tuned. In addition, the variation of the size of the QD can tune the energy band gaps.

By tuning not only the sizes of the atomic vacancies and the C atoms flake but also strains, new magnetic and electronic properties of the h-BN layers would emerge, which might provide the novel nanoelectronics, optoelectronics, and spintronics devices.

Conflicts of Interest

The author declares that there are no conflicts of interest regarding the publication of this paper.

Acknowledgments

This work was partly supported by MEXT Elements Strategy Initiative to Form Core Research Center through Tokodai Institute for Element Strategy and JSPS KAKENHI (Grants nos. JP17K05053 and JP26390062). Computations were partly done at Institute for Solid State Physics, the University of Tokyo, and at Cybermedia Center of Osaka University.

References

[1] K. S. Novoselov, A. K. Geim, S. V. Morozov et al., "Electric field in atomically thin carbon films," *Science*, vol. 306, no. 5696, pp. 666–669, 2004.

[2] A. K. Geim and K. S. Novoselov, "The rise of graphene," *Nature Materials*, vol. 6, no. 3, pp. 183–191, 2007.

[3] C. Berger, Z. Song, X. Li et al., "Electronic confinement and coherence in patterned epitaxial graphene," *Science*, vol. 302, pp. 1191–1196, 2006.

[4] E. V. Castro, K. S. Novoselov, S. V. Morozov et al., "Biased bilayer graphene: semiconductor with a gap tunable by the electric field effect," *Physical Review Letters*, vol. 99, no. 21, Article ID 216802, 2007.

[5] Y. Zhang, Y.-W. Tan, H. L. Stormer, and P. Kim, "Experimental observation of the quantum Hall effect and Berry's phase in graphene," *Nature*, vol. 438, no. 7065, pp. 201–204, 2005.

[6] Y. Zhang, T.-T. Tang, C. Girit et al., "Direct observation of a widely tunable bandgap in bilayer graphene," *Nature*, vol. 459, no. 7248, pp. 820–823, 2009.

[7] J. R. Williams, L. DiCalro, and C. M. Marcus, "Quantum hall effect in a gate-controlled p-n junction of graphene," *Science*, vol. 317, no. 5838, pp. 638–641, 2007.

[8] A. F. Young and P. Kim, "Quantum interference and Klein tunnelling in graphene heterojunctions," *Nature Physics*, vol. 5, pp. 222–226, 2009.

[9] D. Pacile, J. C. Meyer, C. O. Girit, and A. Zettl, "In-line phase-contrast imaging of a biological specimen using a compact laser-Compton scattering-based x-ray source," *Applied Physics Letters*, vol. 92, no. 13, Article ID 133107, 2008.

[10] C. Jin, F. Lin, K. Suenaga, and S. Iijima, "Fabrication of a Freestanding Boron Nitride Single Layer and Its Defect Assignments," *Physical Review Letters*, vol. 102, no. 19, 2009.

[11] C. R. Dean, A. F. Young, I. Meric et al., "Boron nitride substrates for high-quality graphene electronics," *Nature Nanotechnology*, vol. 5, no. 10, pp. 722–726, 2010.

[12] N. H. Shon and T. Ando, "Quantum Transport in Two-Dimensional Graphite System," *Journal of the Physical Society of Japan*, vol. 67, pp. 2421–2429, 1998.

[13] K. Watanabe, T. Taniguchi, and H. Kanda, "Direct-bandgap properties and evidence for ultraviolet lasing of hexagonal boron nitride single crystal," *Nature Materials*, vol. 3, pp. 404–409, 2004.

[14] L. Song, L. Ci, H. Lu et al., "Large scale growth and characterization of atomic hexagonal boron nitride layers," *Nano Letters*, vol. 10, no. 8, pp. 3209–3215, 2010.

[15] X. Blase, A. Rubio, S. G. Louie, and M. L. Cohen, "Quasiparticle band structure of bulk hexagonal boron nitride and related systems," *Physical Review B*, vol. 51, no. 11, pp. 6868–6875, 1995.

[16] Y. Kubota, K. Watanabe, O. Tsuda, and T. Taniguchi, "Deep ultraviolet light-emitting hexagonal boron nitride synthesized at atmospheric pressure," *Science*, vol. 317, no. 5840, pp. 932–934, 2007.

[17] K. Watanabe, T. Taniguchi, T. Niiyama, K. Miya, and M. Taniguchi, "Far-ultraviolet plane-emission handheld device based on hexagonal boron nitride," *Nature Photonics*, vol. 3, no. 10, pp. 591–594, 2009.

[18] B. Huang, X. K. Cao, H. X. Jiang, J. Y. Lin, and S. H. Wei, "Origin of the significantly enhanced optical transitions in layered boron nitride," *Physical Review B*, vol. 86, Article ID 155202, 2012.

[19] Y. Fujimoto and S. Saito, "Band engineering and relative stabilities of hexagonal boron nitride bilayers under biaxial strain," *Physical Review B*, vol. 94, Article ID 245427, 2016.

[20] R. Dahal, J. Li, S. Majety et al., "Epitaxially grown semiconducting hexagonal boron nitride as a deep ultraviolet photonic material," *Applied Physics Letters*, vol. 98, no. 21, Article ID 211110, 2011.

[21] S. Majety, J. Li, X. K. Cao et al., "Epitaxial growth and demonstration of hexagonal BN/AlGaN *p-n* junctions for deep ultraviolet photonics," *Applied Physics Letters*, vol. 100, no. 6, Article ID 061121, 2012.

[22] T. Sugino, K. Tanioka, S. Kawasaki, and J. Shirafuji, "Characterization and field emission of sulfur-doped boron nitride synthesized by plasma-assisted chemical vapor deposition," *Japanese Journal of Applied Physics*, vol. 36, part 2, p. L463, 1997.

[23] J. Li, S. Majety, R. Dahal, W. P. Zhao, J. Y. Lin, and H. X. Jiang, "Dielectric strength, optical absorption, and deep ultraviolet detectors of hexagonal boron nitride epilayers," *Applied Physics Letters*, vol. 101, no. 17, Article ID 171112, 2012.

[24] E. Machado-Charry, P. Boulanger, L. Genovese, N. Mousseau, and P. Pochet, "Tunable magnetic states in hexagonal boron nitride sheets," *Applied Physics Letters*, vol. 101, no. 13, Article ID 132405, 2012.

[25] J. C. Meyer, A. Chuvilin, G. Algara-Siller, J. Biskupek, and U. Kaiser, "Selective sputtering and atomic resolution imaging of atomically thin boron nitride membranes," *Nano Letters*, vol. 9, no. 7, pp. 2683–2689, 2009.

[26] J. Kotakoski, C. H. Jin, O. Lehtinen, K. Suenaga, and A. V. Krasheninnikov, "Electron knock-on damage in hexagonal boron nitride monolayers," *Physical Review B—Condensed Matter and Materials Physics*, vol. 82, no. 11, Article ID 113404, 2010.

[27] N. Alem, R. Erni, C. Kisielowski, M. D. Rossell, W. Gannett, and A. Zettl, "Atomically thin hexagonal boron nitride probed by ultrahigh-resolution transmission electron microscopy," *Physical Review B—Condensed Matter and Materials Physics*, vol. 80, no. 15, Article ID 155425, 2009.

[28] Y. Fujimoto and S. Saito, "Energetics and electronic structures of pyridine-type defects in nitrogen-doped carbon nanotubes," *Physica E*, vol. 43, no. 3, pp. 677–680, 2011.

[29] Y. Fujimoto and S. Saito, "Structure and stability of hydrogen atom adsorbed on nitrogen-doped carbon nanotubes," *Journal of Physics: Conference Series*, vol. 302, no. 1, Article ID 012006, 2011.

[30] Y. Fujimoto and S. Saito, "Hydrogen adsorption and anomalous electronic properties of nitrogen-doped graphene," *Journal of Applied Physics*, vol. 115, no. 15, Article ID 153701, 2014.

[31] O. L. Krivanek, M. F. Chisholm, V. Nicolosi et al., "Atom-by-atom structural and chemical analysis by annular dark-field electron microscopy," *Nature*, vol. 464, no. 7288, pp. 571–574, 2010.

[32] X. Wei, M.-S. Wang, Y. Bando, and D. Golberg, "Electron-beam-induced substitutional carbon doping of boron nitride nanosheets, nanoribbons, and nanotubes," *ACS Nano*, vol. 5, no. 4, pp. 2916–2922, 2011.

[33] X. Wei, M.-S. Wang, Y. Bando, and D. Golberg, "Post-synthesis carbon doping of individual multiwalled boron nitride nanotubes via electron-beam irradiation," *Journal of the American Chemical Society*, vol. 132, no. 39, pp. 13592-13593, 2010.

[34] N. Berseneva, A. V. Krasheninnikov, and R. M. Nieminen, "Berseneva, Krasheninnikov, and Nieminen reply:," *Physical Review Letters*, vol. 107, no. 23, Article ID 239602, 2011.

[35] N. Berseneva, A. Gulans, A. V. Krasheninnikov, and R. M. Nieminen, "Electronic structure of boron nitride sheets doped with carbon from first-principles calculations," *Physical Review B—Condensed Matter and Materials Physics*, vol. 87, no. 3, Article ID 035404, 2013.

[36] J. Zhou, Q. Wang, Q. Sun, and P. Jena, "Electronic and magnetic properties of a BN sheet decorated with hydrogen and fluorine," *Physical Review B*, vol. 81, Article ID 085442, 2010.

[37] M. Kan, J. Zhou, Q. Wang, Q. Sun, and P. Jena, "Tuning the band gap and magnetic properties of BN sheets impregnated with graphene flakes," *Physical Review B*, vol. 84, no. 20, Article ID 205412, 2011.

[38] J. Li and V. B. Shenoy, "Graphene quantum dots embedded in hexagonal boron nitride sheets," *Applied Physics Letters*, vol. 98, no. 1, Article ID 013105, 2011.

[39] A. Ramasubramaniam and D. Naveh, "Carrier-induced antiferromagnet of graphene islands embedded in hexagonal boron nitride," *Physical Review B*, vol. 84, no. 7, Article ID 075405, 2011.

[40] L. Ci, L. Song, C. Jin et al., "Atomic layers of hybridized boron nitride and graphene domains," *Nature Materials*, vol. 9, no. 5, pp. 430–435, 2010.

[41] Y. Fujimoto and S. Saito, "Effects of strain on carbon donors and acceptors in hexagonal boron nitride monolayers," *Physical Review B*, vol. 93, Article ID 045402, 2016.

[42] Y. Fujimoto and K. Hirose, "First-principles treatments of electron transport properties for nanoscale junctions," *Physical Review B*, vol. 67, Article ID 195315, 2004.

[43] Y. Fujimoto, K. Hirose, and T. Ohno, "Calculations of surface electronic structures by the overbridging boundary-matching method," *Surface Science*, vol. 586, no. 1–3, pp. 74–82, 2005.

[44] Y. Fujimoto and K. Hirose, "First-principles calculation method of electron-transport properties of metallic nanowires," *Nanotechnology*, vol. 14, no. 2, pp. 147–151, 2003.

[45] Y. Fujimoto, Y. Asari, H. Kondo, J. Nara, and T. Ohno, "First-principles study of transport properties of Al wires: Comparison between crystalline and jellium electrodes," *Physical Review B*, vol. 72, no. 11, Article ID 113407, 2005.

[46] T. Ono, S. Tsukamoto, Y. Egami, and Y. Fujimoto, "Real-space calculations for electron transport properties of nanostructures," *Journal of Physics: Condensed Matter*, vol. 23, no. 39, Article ID 394203, 2011.

[47] A. J. Du, Y. Chen, Z. Zhu, R. Amal, G. Q. Lu, and S. C. Smith, "Dots versus antidots: computational exploration of structure, magnetism, and half-metallicity in boron-nitride nanostructures," *Journal of the American Chemical Society*, vol. 131, no. 47, pp. 17354–17359, 2009.

[48] H.-X. Yang, M. Chshiev, D. W. Boukhvalov, X. Waintal, and S. Roche, "Inducing and optimizing magnetism in graphene nanomeshes," *Physical Review B - Condensed Matter and Materials Physics*, vol. 84, no. 21, Article ID 214404, 2011.

[49] E. H. Lieb, "Two theorems on the Hubbard model," *Physical Review Letters*, vol. 62, no. 10, pp. 1201–1204, 1989.

[50] Y. Fujimoto, T. Koretsune, S. Saito, T. Miyake, and A. Oshiyama, "A new crystalline phase of four-fold coordinated silicon and germanium," *New Journal of Physics*, vol. 10, Article ID 083001, 2008.

[51] Y. Fujimoto and A. Oshiyama, "Formation and Stability of 90 Degree Dislocation Cores in Ge Films on Si(001)," *AIP Conference Proceedings*, vol. 1399, no. 1, p. 185, 2011.

[52] Y. Fujimoto and S. Saito, "Atomic geometries and electronic structures of hexagonal boron-nitride bilayers under strain," *Journal of the Ceramic Society of Japan*, vol. 123, no. 1439, pp. 576–578, 2015.

[53] Y. Fujimoto and S. Saito, "Interlayer distances and band-gap tuning of hexagonal boron-nitride bilayers," *Journal of the Ceramic Society of Japan*, vol. 124, no. 5, pp. 584–586, 2016.

[54] Y. Fujimoto, T. Koretsune, and S. Saito, "Electronic structures of hexagonal boron-nitride monolayer: strain-induced effects," *Journal of the Ceramic Society of Japan*, vol. 122, no. 1425, pp. 346–348, 2014.

[55] J. Tersoff and D. R. Hamann, "Theory and application for the scanning tunneling microscope," *Physical Review Letters*, vol. 50, no. 25, pp. 1998–2001, 1983.

[56] J. Tersoff and D. R. Hamann, "Theory of the scanning tunneling microscope," *Physical Review B*, vol. 31, Article ID 805, 1985.

[57] Y. Fujimoto, H. Okada, K. Endo, T. Ono, S. Tsukamoto, and K. Hirose, "Images of scanning tunneling microscopy on the Si(001)-p(2 × 2) reconstructed surface," *Materials Transactions*, vol. 42, no. 11, pp. 2247–2252, 2001.

[58] H. Okada, Y. Fujimoto, K. Endo, K. Hirose, and Y. Mori, "Detailed analysis of scanning tunneling microscopy images of the Si(001) reconstructed surface with buckled dimers," *Physical Review B*, vol. 63, no. 19, Article ID 195324, 2001.

[59] Y. Fujimoto and A. Oshiyama, "Structural stability and scanning tunneling microscopy images of strained Ge films on Si(001)," *Physical Review B*, vol. 87, Article ID 075323, 2013.

[60] Y. Fujimoto, H. Okada, K. Inagaki, H. Goto, K. Endo, and K. Hirose, "Theoretical study on the scanning tunneling microscopy image of Cl-Adsorbed Si(001)," *Japanese Journal of Applied Physics*, vol. 42, part 1, no. 8, p. 5267, 2003.

[61] Y. Fujimoto and A. Oshiyama, "Atomic structures and energetics of 90∘ dislocation cores in Ge films on Si(001)," *Physical Review B*, vol. 81, Article ID 205309, 2010.

[62] L. Zhao, M. Levendorf, S. Goncher et al., "Local atomic and electronic structure of boron chemical doping in monolayer graphene," *Nano Letters*, vol. 13, no. 10, pp. 4659–4665, 2013.

[63] Y. Fujimoto and S. Saito, "Energetics and scanning tunneling microscopy images of B and N defects in Graphene Bilayer," *Springer Proceedings in Physics*, vol. 186, pp. 107–112, 2017.

[64] Y. Fujimoto and S. Saito, "Formation, stabilities, and electronic properties of nitrogen defects in graphene," *Physical Review B*, vol. 84, Article ID 245446, 2011.

[65] Y. Fujimoto and S. Saito, "Gas adsorption, energetics and electronic properties of boron- and nitrogen-doped bilayer graphenes," *Chemical Physics*, vol. 478, pp. 55–61, 2016.

[66] Y. Fujimoto and S. Saito, "Electronic structures and stabilities of bilayer graphene doped with boron and nitrogen," *Surface Science*, vol. 634, pp. 57–61, 2015.

12

Physical and Compaction Properties of Granular Materials with Artificial Grading behind the Particle Size Distributions

Ming-liang Chen,[1] **Gao-jian Wu,**[2] **Bin-rui Gan,**[3] **Wan-hong Jiang,**[2] **and Jia-wen Zhou** ⓘ[1]

[1]*State Key Laboratory of Hydraulics and Mountain River Engineering, Sichuan University, Chengdu 610065, China*
[2]*Sinohydro Bureau 5 Co., Ltd., Power Construction Corporation of China, Chengdu 610066, China*
[3]*College of Water Resource and Hydropower, Sichuan University, Chengdu 610065, China*

Correspondence should be addressed to Jia-wen Zhou; jwzhou@scu.edu.cn

Academic Editor: Konstantinos Karamanos

Granular materials in geotechnical engineering is generally considered to be mixtures of clay, sand, and gravel that commonly appear in slopes, valleys, or river beds, and they are especially used for the construction of earth-rock-filled dams. The complexity of the constitution of granular materials leads to the complexity of their properties. Particle size distribution (PSD) has a great influence on the strength, permeability, and compaction behavior of granular materials, and some implicit correlation may exist between the PSD and the compaction properties of granular materials. Field testing and statistical analysis are used to study the physical and compaction properties of granular materials with artificial grading behind the particle size distributions. The statistical properties in PSD of dam granular materials and how the variation of PSD renders statistical constant are revealed. The statistical constants of three types of dam granular materials are 2.459, 2.475, and 2.499, respectively, on average. These statistical constants have a positive correlation with dry density and a negative correlation with moisture content. According to this characteristic and little deviation between two different calculation methods (from grading analysis and based on the Weibull distribution), the presentation of the statistical analysis ensures the validity of the Weibull function's description of the granular materials with artificial grading. After fitting the Weibull function to the PSD curve, the relationship between the Weibull parameters and the compaction degree in different soil samples is consistent with that in different types, providing guiding significance for evaluating and selecting dam granular materials.

1. Introduction

Granular material is widely distributed in nature and is especially relevant in the geotechnical engineering field; it includes landslide deposits, granular soil (fine-grained soil or coarse-grained soil), filling/construction material used for earth-rock dams, and other applications [1–3]. Various granular materials in geotechnical engineering exhibit different material composition, particle size distribution, moisture content, mass density, and other physical/mechanical properties [3–5]. Although the physical/mechanical properties of granular materials are very complicated and difficult to determine, there are potential implicit correlations between the particle size distribution (PSD) and the physical/mechanical properties of granular materials [6]. Discovering

these implicit correlations is very important for understanding the physical/mechanical behavior of granular materials.

The PSD of granular materials is an essential issue for the design of earth-rock-filled dams [7–9]. In practical engineering work, the preparation requires analyzing the range of particle sizes and determining the proportion of grains with various particle sizes. A granular material generally covers a range of particle sizes between 0.001 mm and 1,000 mm, and it is well known for containing wide-ranging grain composition [10]. Therefore, the scaling distribution for grain composition is a vital index for studying the physical and mechanical properties of soil systems. When the PSD of granular materials is changed, its permeability, deformation, strength, and other properties are altered [11]. During the compaction process of an earth-rock-filled dam,

TABLE 1: Grain compositions of tested dam granular materials (unit: %).

Type	Number	Particle size (mm)													
		<0.075	<0.25	<0.5	<1	<2	<5	<10	<20	<40	<60	<80	<100	<200	<300
A	S1	4.24	10.34	13.60	13.05	14.64	27.04	10.21	6.89						
	S2	4.40	11.20	12.63	14.42	14.47	24.60	10.62	7.66						
	S3	4.14	14.28	12.24	14.95	13.22	25.10	12.92	3.14						
	S4	4.76	10.27	15.34	15.60	19.31	23.10	9.02	2.60						
	S5	3.85	11.99	15.51	14.53	14.02	29.11	7.55	3.44						
	...	...	...	...	...	...	...	...	...						
B	S1	2.28	9.82	6.46	8.15	12.25	17.55	15.38	19.06	9.05					
	S2	2.89	11.87	7.06	6.84	10.36	19.81	14.03	22.00	5.14					
	S3	2.62	12.09	6.82	6.47	9.95	22.64	15.79	21.42	2.20					
	S4	2.46	9.51	5.98	7.22	9.52	20.41	19.62	18.63	6.63					
	S5	2.55	11.82	6.44	8.13	11.43	18.7	14.51	18.04	8.38					
	...	...	...	...	...	...	...	...	...	...					
C	S1	2.50	3.05	3.72	4.02	4.85	6.53	13.14	12.01	5.38	8.70	6.63	5.75	16.25	7.49
	S2	2.15	1.80	3.09	2.85	3.90	5.35	12.98	12.23	6.81	10.64	10.09	7.35	15.64	5.11
	S3	1.92	1.61	2.75	2.54	3.47	4.76	11.59	13.12	8.73	9.50	7.81	9.07	16.51	6.63
	S4	2.52	3.05	3.43	2.68	1.88	5.86	15.47	15.85	12.08	9.41	9.67	3.89	11.19	3.01
	S5	2.27	2.18	3.38	2.92	3.17	6.58	11.65	9.76	11.64	9.79	6.44	11.65	15.02	3.57
	...	...	...	...	...	...	...	...	...	...	...	...	...	...	...

the pore reduction between grains is caused by natural settlement and artificial rolling [12]. The compaction properties of granular materials are the key physical indices for the construction of an earth-rock-filled dam.

Previous studies indicated that the phenomenon of compaction dynamics is a slow relaxation influenced by the microscopic characteristics of the grains that may be grain shape, friction, or cohesion [13–15]. Through a large amount of field testing and observation, the macroscopic analysis can provide an opportunity to understand the relationship between PSD and physical/compaction properties [12, 16, 17]. To depict the PSD of soil systems for engineers to intuitively analyze, traditional methods usually use statistics analysis to get a series of characteristic parameters or a cumulative frequency curve [18]. The difficulty in studying a soil system with a complex and irregular clastic rock texture by quantitative characterization is due to the limitations and restrictions of traditional methods where one or two descriptive parameters are hardly suitable and adaptive to the complexity and irregularity of granular materials [19–21].

Since statistical distribution functions were used to describe the grain composition of soil systems [21–23], it has been thought of as a scientific and accurate tool for getting a greater knowledge of the relation between PSD and soil structure [24]. At present, the approach of analyzing the size distribution curve with the statistical analysis is popular for various kinds of soils [25–28]. This paper aims at exploring the physical and compaction properties of the granular materials behind the PSD curve analysis. Three different granular materials are taken as the samples for experiments and analyses, and 50 groups of field tests are carried out for each type of granular materials. The PSDs are analyzed by applying the field testing and statistical analysis methods. Combined with the grading analysis and the PSD functions, the statistical constants of granular materials are analyzed to verify the general applicability of the PSD function for dam filling material and to describe physical characteristics. Then, the parameters in the PSD function are also used to seek the implicit association with the compaction properties of granular materials.

2. Materials and Methods

2.1. Sampling and Measuring. The granular materials' samples used in this study are from the filling material for the construction of the Changheba Hydropower Station. The Changheba Hydropower Station is the 10th level hydropower station for cascade development of the Dadu River, and it is located in Kangding County, Sichuan Province, in the Southwest of China [29]. There are more than 10 types of granular materials used for the dam construction, each with different compositions and particle size distributions. Here, three typical types of dam granular materials are used for the field tests and the correlation studies. Table 1 provides some typical grain compositions of the tested dam granular materials. They are the filter dam material #1 (granular material A, Figure 1(a)), the filter dam material #3 (granular material B, Figure 1(b)), and the transition dam material (granular material C, Figure 1(c)). 50 groups of related field tests are carried out for each type of granular materials, with the total of 150 groups of field experiments.

The production method of the filter material is processing artificial aggregate by the sand and gravel processing system (contains a large number of fine particles with diameter less than 2.0 mm). However, the transition material is the rock material gained by blasting activity in the quarry (most are coarse particles with diameter larger than 2.0 mm). During the construction process of filling the dam, field tests are done before rolling compaction on each layer. The procedure takes two dam sections perpendicular to the dam axis as the fixed sections for examining the construction circumstances, and it sets a group every 5.0 m to measure the PSD of each layer. After compaction by vibration rolling

(a)

(b)

(c)

FIGURE 1: Photos for the three types of granular materials used in this study: (a) A (filter dam material #1); (b) B (filter dam material #3); (c) C (transition dam material).

machinery, the dry density, moisture content, and relative density are measured for each group.

2.2. Particle Size Distribution (PSD) Characteristics. The PSD of three granular materials are obtained by the artificial sieving method. Figure 2 shows the particle size distribution characteristics for the three types of tested granular materials. These three types of dam granular materials are produced in two different sites, where the filter dam material is produced at the soil material site and the transition dam material is produced at the rock-fill material site. As shown in Figure 2, there are large differences in particle size range and small differences in material composition (fine-grained soil and coarse rock blocks). The varied particle size distributions result in the different physical and construction properties of the dam granular materials.

2.3. Compaction Degree. The granular materials' compaction degree is a key index for controlling the construction of earth-rock-filled dams, and it has an important effect on the stability and seepage prevention of the dam. After rolling the granular materials to be dense, the grain size distribution reflects the pore size distribution to some extent. Judging the compaction degree of the soil system by porosity neglects the impact of grain shape and particle size distribution. Here, relative density D_r is used as an indicator of compaction. It measures compaction by comparing the porosity when the soil system is in the loosest situation and the densest. It is supposed that relative density is able to comprehensively reflect the impact of particle shape, particle size distribution, and other factors. The equation used to determine the relative density D_r is shown as follows:

$$D_r = \frac{e_{\max} - e}{e_{\max} - e_{\min}}, \tag{1}$$

where e is the porosity in the natural state of granular materials, $e_{\max}$ is the maximum value of porosity in the loosest state of granular materials, and $e_{\min}$ is the minimum value of porosity in the densest state of granular materials.

The above equation is the theoretical definition of relative density D_r. For engineering practice, the relative density is often determined from different dry density test results under different states as follows:

$$D_r = \frac{(\rho_d - \rho_{d\min})\rho_{d\max}}{(\rho_{d\max} - \rho_{d\min})\rho_d}, \tag{2}$$

where $\rho_{d\max}$ is the maximum dry density in the densest state of granular materials, $\rho_{d\min}$ is the minimum dry density in the loosest state of granular materials, and ρ_d is the dry density of granular materials after being rolling filled.

3. Statistical Properties of Granular Materials

The PSD potential can reflect the complexity and irregularity in structural properties of granular materials and can be used to study the statistical properties between the whole and the parts of granular materials [7]. By using mathematical representation and statistical analysis, one can describe the spatial structure of granular materials through a set of parameters and then try to discuss the correlation between the structural composition and physical properties of the granular materials [20]. Determination of the statistical constant is the core of studying particle size distribution combined with the statistical analysis, and the invariance in the spatial scale for the PSD curve of granular materials provides a scientific basis for depicting the structural composition [30].

3.1. PSD Analysis from Grading Analysis. The grading analysis is a direct method used for the PSD curve of the granular materials. The statistical constant is a mathematical expression of PSD of granular materials by a constant, in which the computational formula can be obtained with some mathematical processing. By using a logarithmic coordinate graph to explain the method and grading analysis, the statistical properties of granular materials can be

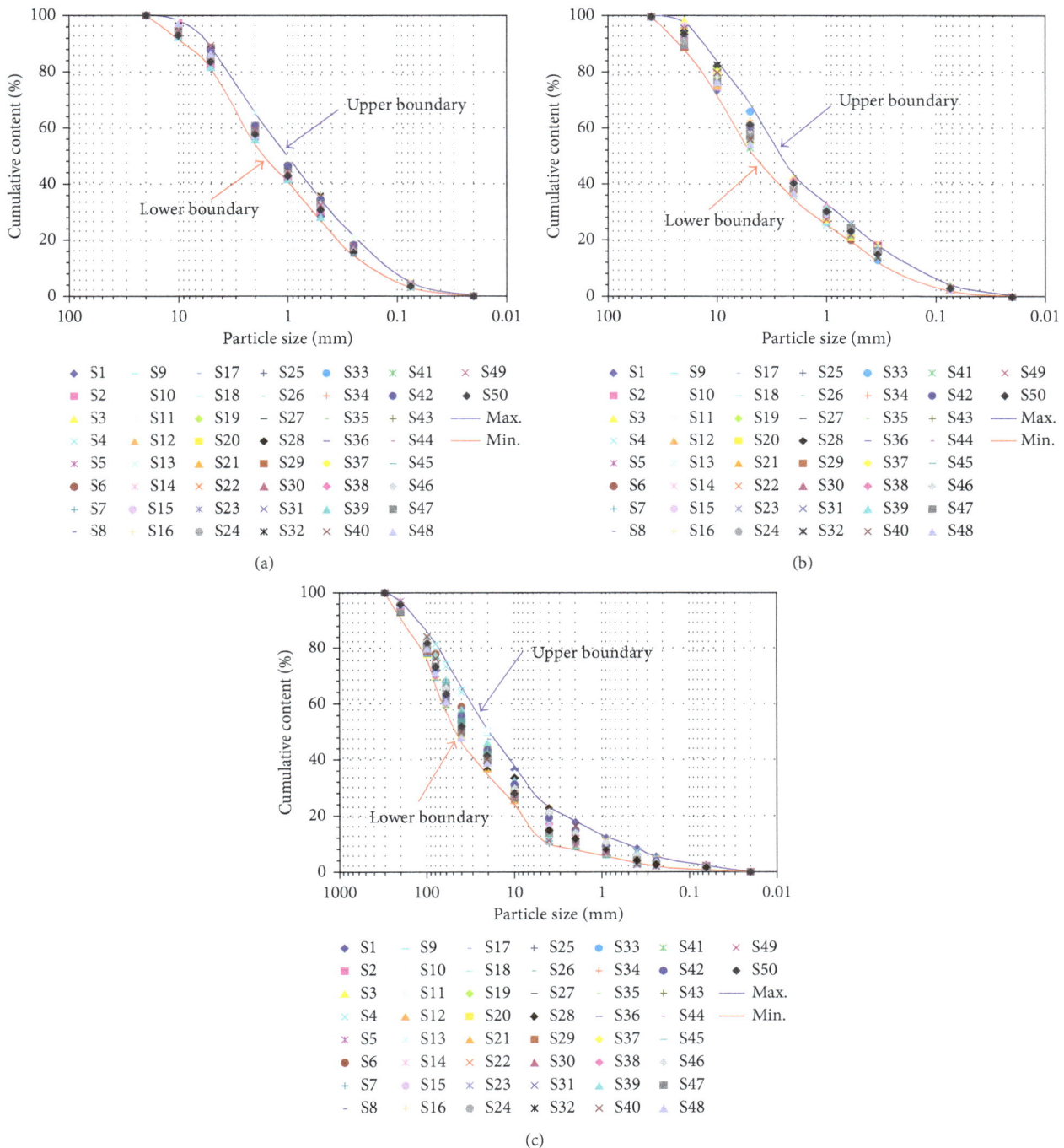

FIGURE 2: Particle size distribution characteristics of the tested granular materials: (a) A; (b) B; (c) C.

determined. As shown in Figures 3(a), 4(a), and 5(a), for granular materials with a good grading, the PSD of granular material can be almost viewed as a line. The computational result shows a possibility that PSD of a sample of granular material can be expressed by a constant, related with the gradient of the PSD curve under the log-log coordinate. Usually in engineering, uniformity coefficient C_U and curvature coefficient C_C are used as characteristic parameters for evaluating the grain composition of granular materials. As a control parameter can reflect whole performance, the

statistical constant has an advantage of simplifying calculation process. However, the validity of statistical constant need be verified for evaluating the grain composition of granular materials. Suppose that r represents the particle size of granular materials and $N(r)$ represents the corresponding cumulative content of the particles with size smaller than r. In general, a power law function is suitable for the particle size distribution. According to the definition of statistical constant D, the relationship between r and $N(r)$ can be drawn in this way [31]:

(a)

(b)

(c)

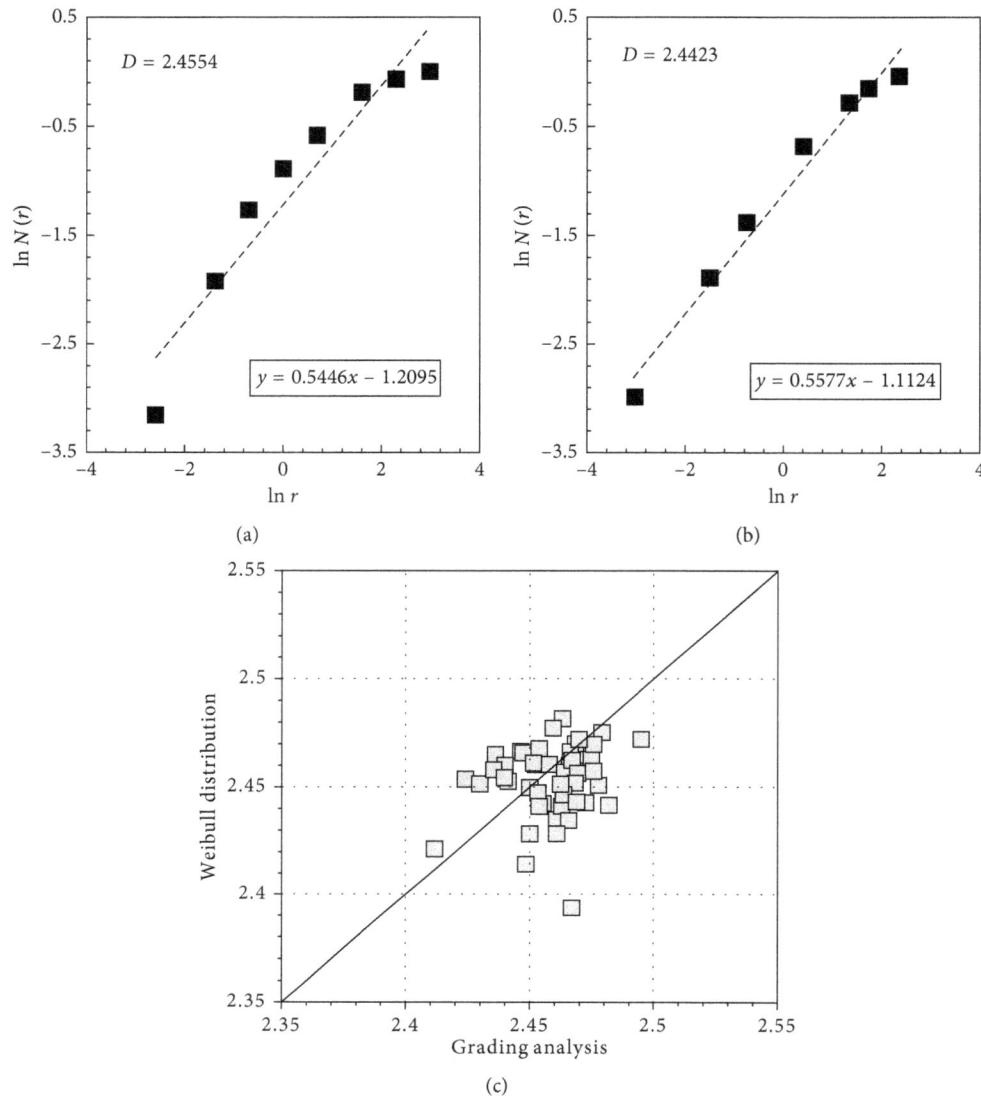

FIGURE 3: Statistical constant analysis results of dam granular material A and the comparison between different methods: (a) based on the grading analysis (S1); (b) based on the Weibull distribution (S1); (c) difference in the statistical constant results by using different methods (r represents particle size of granular materials; $N (<r)$ represents the corresponding cumulative content of the grains with the size smaller than r; N is the total content of all grains).

$$N(r) \propto r^{-D}, \tag{3}$$

where D is the statistical constant, which is a power law exponent. If one gives a differential operation for (3), it can be expressed as follows:

$$dN(r) \propto r^{-D-1}\, dr. \tag{4}$$

Suppose that N_0 is the total amount of particles in the granular materials, and the number of grains whose sizes are in $[r, r + dr]$ is

$$dN = N_0\, dN(r). \tag{5}$$

By simultaneously using (4) and (5), the following equation can be obtained:

$$dN = N_0 r^{-D-1}\, dr. \tag{6}$$

Assume that each particle in the granular materials can be expressed as an equivalent sphere and k_r represents the shape factor of the particles, or the proportion coefficient of particle volume and particle size. Then, the particle volume in granular materials can be calculated by the equation $V = k_r r^3$. If substituted into (6), it can be switched to the following form:

$$dV = N_0 k_r r^3 r^{-D-1}\, dr. \tag{7}$$

If V_0 represents the gross volume of the particles in the test sample, it can be obtained from the equation: $dV = V_0\, dN(r)$. Then, (7) can be expressed as follows:

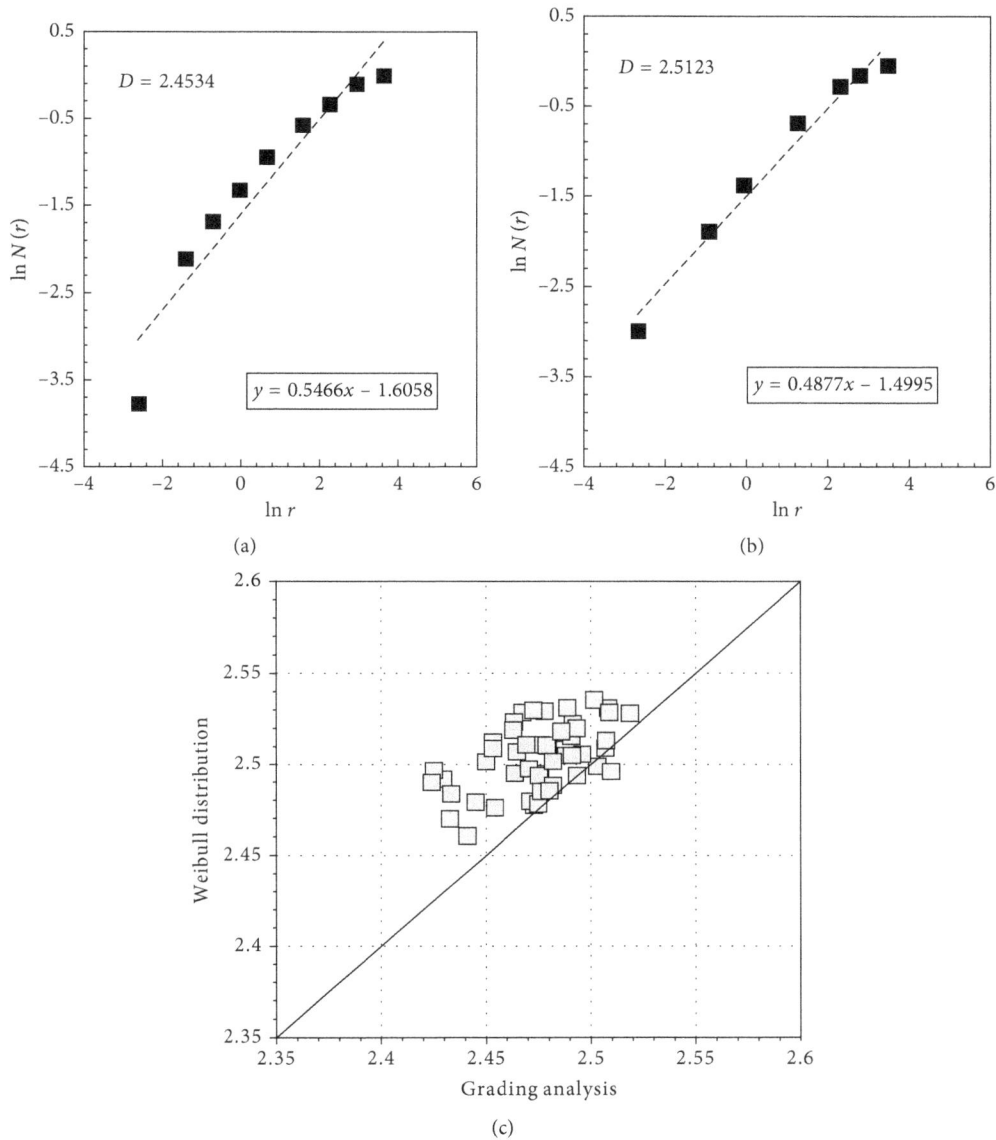

FIGURE 4: Statistical constant analysis results of dam granular material B and the comparison between different methods: (a) based on the grading analysis (S1); (b) based on the Weibull distribution (S1); (c) the difference in the statistical constant results by using different methods (r represents particle size of granular materials; $N\,(<r)$ represents the corresponding cumulative content of the grains with the size smaller than r; N is the total content of all grains).

$$dN\,(r) = \frac{N_0 k_r}{V_0} r^{2-D}\,dr. \tag{8}$$

Integrating (8), then

$$N\,(r) = \frac{N_0 k_r}{V_0\,(3-D)} r^{3-D}. \tag{9}$$

If assuming $N_0 k_r / [V_0\,(3-D)]$ in (9) is a constant term and b is defined as $b = 3 - D$, after the logarithmic changes, the relationship between the cumulative content of all the particles whose size is smaller than r can be described as follows:

$$\ln N\,(r) = b\ln r + C, \tag{10}$$

where $\ln(\cdot)$ represents the function of $\log_e(\cdot)$, C is a constant, and b is the gradient of the fitting curve between $\ln N\,(r)$ and $\ln r$. The relationship between the logarithm of the cumulative content of particles with size smaller than r in granular materials and the logarithm of particle size appears to be linear. Based on the above equation, the statistical constant of grains in granular materials can be calculated by the following equation:

$$D = 3 - b. \tag{11}$$

Equation (11) is the most general form for calculating the statistical constant in the PSD analysis of granular materials; here is a power law exponent.

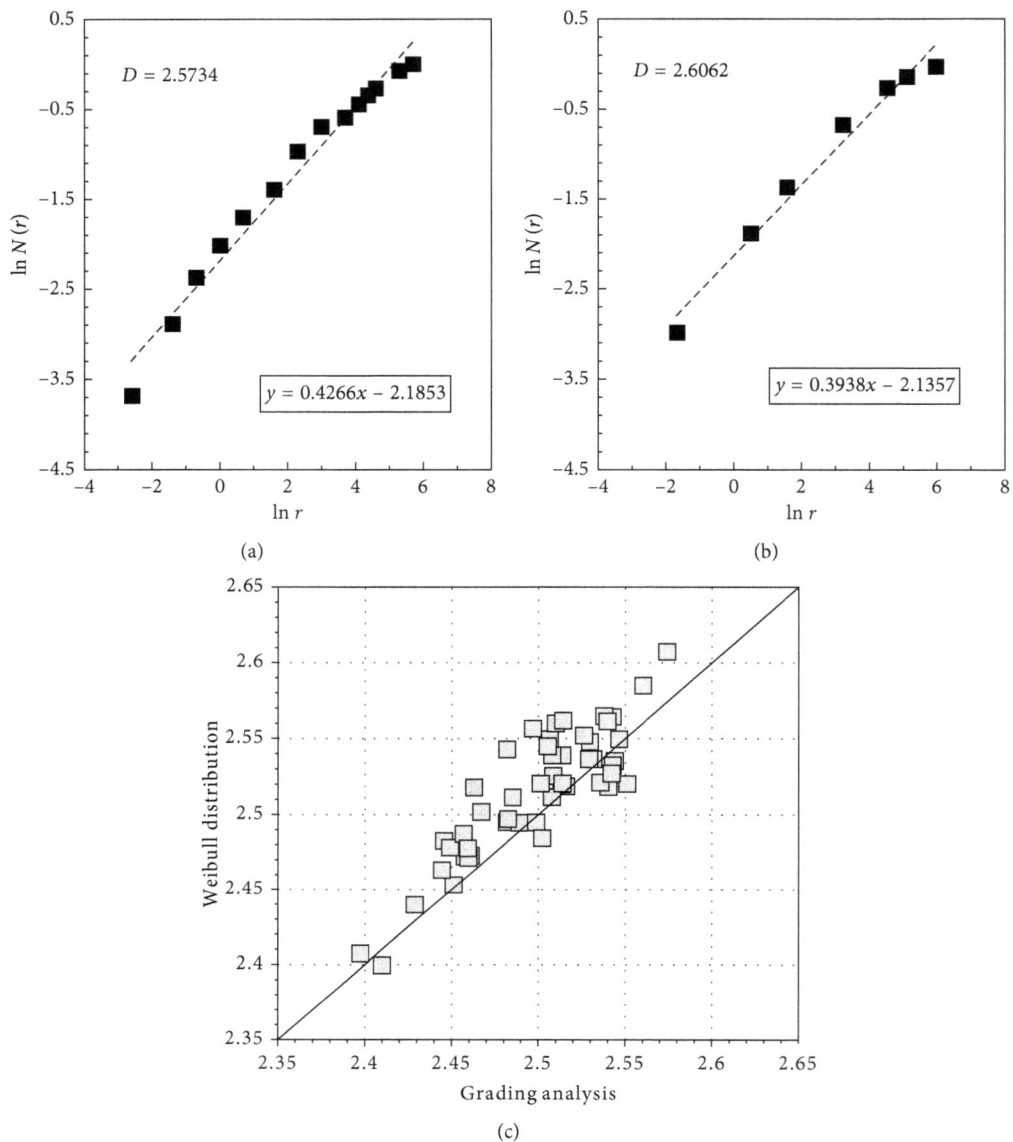

(a)

(b)

(c)

FIGURE 5: Statistical constant analysis results of dam granular material C and the comparison between different methods: (a) based on the grading analysis (S1); (b) based on the Weibull distribution (S1); (c) the difference in the statistical constant results by using different methods (r represents particle size of granular materials; N ($<r$) represents the corresponding cumulative content of the grains with the size smaller than r; N is the total content of all grains).

3.2. Statistical Analysis Based on the Weibull Distribution.
Frequently used mathematical representations for describing the PSD of granular materials are the Rosin–Rammler distribution function [32] and Weibull distribution function [33–35]. The particle groups of natural granular materials possibly exist with discontinuous distributions, due to complex geological formation histories, and may introduce some errors when using the mathematical distribution functions to analyze the PSD of natural granular materials. However, for the artificially prepared granular materials (which can be seen as having the same properties as natural granular materials), due to the requirements of stability and seepage prevention, the continuity of particle groups is often better than that for the natural granular materials. Therefore,

the mathematical distribution functions can better describe granular materials with artificial grading for the construction of earth-rock-filled dams.

Here, a two-parameter Weibull distribution with a size parameter r_0 and a shape parameter b is used to describe PSD of dam granular materials. According to the definition of two-parameter Weibull distribution, size parameter and shape parameter are strongly related with the shape of the PSD curve. The size parameter represents the average size of particles in a sample of granular materials, and the shape parameter reflects the distributed wide range of particle size. By using the Weibull distribution, the Weibull parameters have an advantage of directly reflecting statistical characteristics of PSD of granular materials than the uniformity coefficient and curvature coefficient. The formula of the Weibull distribution is shown as follows:

$$\frac{M\,(<r)}{M} = 1 - \exp\left[-\left(\frac{r}{r_0}\right)^b\right], \qquad (12)$$

where $M\,(<r)$ is the cumulative content of all particles whose size is smaller than r and M is the total volume of all particles.

By applying the Taylor progression for (12), a concise equation can be obtained as follows:

$$\frac{M\,(<r)}{M} = \left(\frac{r}{r_0}\right)^b. \qquad (13)$$

Then, by adopting a differential operation for (13), a simple power equation can be expressed as follows:

$$dM\,(<r) = r^{b-1}\,dr. \qquad (14)$$

The volume of the whole particles in granular materials can be calculated by $V = k_r r^3$, and ρ is a constant defined as the particle density. Then, the mass of the whole grains can be given by $M = \rho k_r r^3$. Assuming each particle in the granular materials can be represented as an equivalent sphere and applying $M = \rho k_r r^3$ to (7), the following equation can be obtained:

$$dM = \rho\,dV = N_0 \rho k_r r^3 r^{-D-1}\,dr. \qquad (15)$$

If M_0 represents the gross mass of the granular materials, it can be obtained from this equation: $dM = M_0\,dN\,(r)$. Then, (15) can be expressed as follows:

$$dN\,(r) = \frac{N_0 \rho k_r r^{2-D}}{M_0}\,dr. \qquad (16)$$

Integrating (16),

$$N\,(r) = \frac{N_0 \rho k_r}{M_0\,(3-D)} r^{3-D}. \qquad (17)$$

Equation (17) has the same form as (9), so the same treatment method for (17) can be used to determine the statistical constant of the PSD for granular materials with the Weibull distribution. Although there are some differences in the statistical constant results between using grading analyses or the Weibull distribution, the difference is very small if the continuity of gradation in the granular materials is high. The difference in the statistical constant between the grading analysis and the Weibull distribution can reflect some properties of the PSD, especially in the continuity of gradation for granular materials.

3.3. Relationship between PSD and Statistical Constant.

The physical properties of granular materials have a significantly positive correlation with the statistical constant (Figures 3–5), which further means that the statistical constant is qualified to become an evaluation indicator of the soil character. Based on the Weibull distribution and the gradation analysis, respectively, we calculate the statistical constant. Then, after comparing the results, the

TABLE 2: Computational results for the statistical constant of different dam granular materials (all of the field test samples).

Statistical constant, D		Granular material A	Granular material B	Granular material C
Grading analysis	Maximum	2.495	2.519	2.573
	Minimum	2.412	2.424	2.397
	Average	2.459	2.475	2.499
	Standard deviation	0.016	0.023	0.040
Weibull distribution	Maximum	2.482	2.535	2.606
	Minimum	2.394	2.460	2.399
	Average	2.453	2.504	2.515
	Standard deviation	0.018	0.018	0.042

difference value verifies the general applicability of the Weibull distribution to the grain composition of granular materials.

The diversity of lithologic compositions, physical properties, and grain size distributions results in statistical constant differences [7, 9, 16]. Table 2 summarizes the computational results for the statistical constant of different tested granular materials (all of the field test samples). As shown in Figures 3(c), 4(c), and 5(c), for all three types of granular materials, there is a good correlation for the statistical constant calculated based on the Weibull distribution and/or grading analysis, and the error is within the range from −3% to 3%. Although grain composition with artificial design violates the basic rule of particle aggregation in natural conditions to a certain degree, the calculation results indicate that the Weibull function has applicability in describing the PSD. The general validity of reflecting the PSD characteristics becomes the foundation for further exploring the question of whether PSD has a definite mathematical link with the physical and mechanical properties of dam granular materials.

Although the error of the statistical constant determined by the grading analysis and the Weibull distribution is small, there is an erratic fluctuation phenomenon in these two different methods (Figures 6(a) and 6(b)). The erratic fluctuation in statistical constants of dam granular material C is the largest and that of dam granular material A is the smallest. For example, the minimum statistical constant of granular material C is even lower than that of granular material A, although the average value of statistical constant for granular material C is larger than that of granular material A. For dam granular materials A and C, the fluctuation degree of the statistical constant from the grading analysis is higher than that from the Weibull distribution. However, the fluctuation degree of the statistical constant for dam granular material B from the grading analysis is lower than that from the Weibull distribution. If making an analysis according to the amplitude of fluctuations, the particle size range of granular materials influences the fluctuation degree of the statistical constant, as is proven to be true by the standard deviations of all the granular material samples of each type. Therefore, understanding how the behavior of PSD curves leads to the variation of the statistical

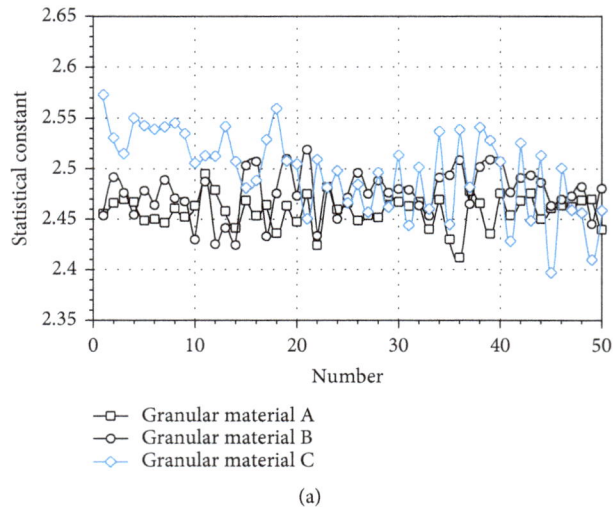

(a)

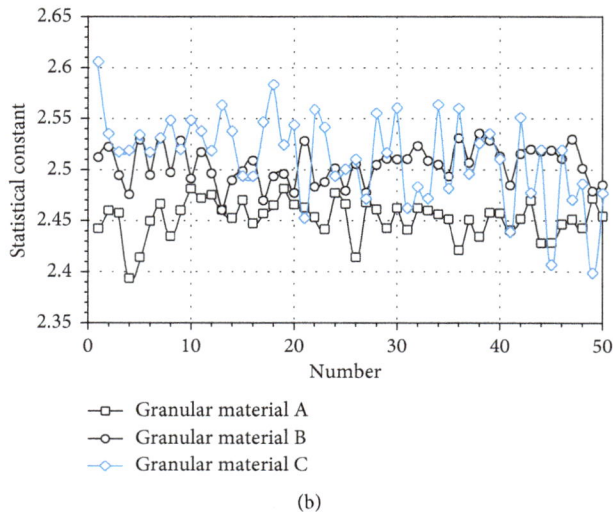

(b)

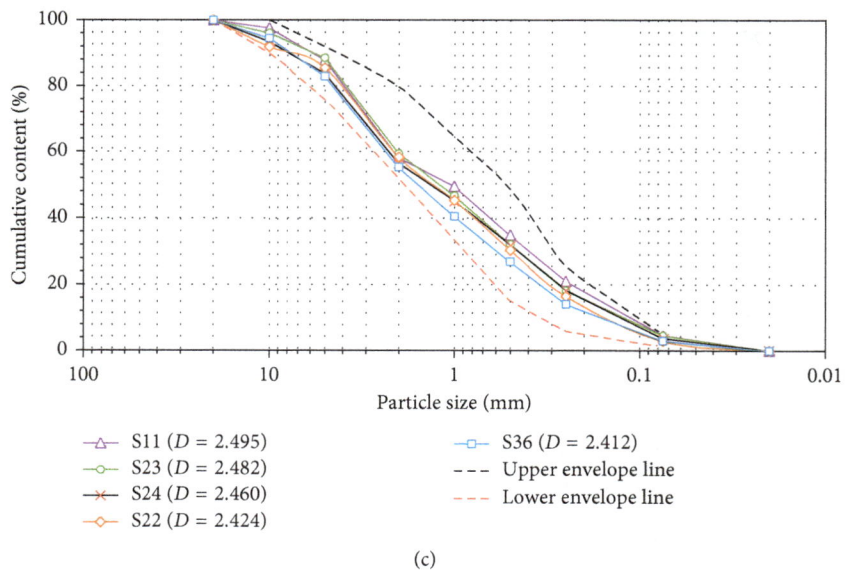

(c)

Figure 6: Continued.

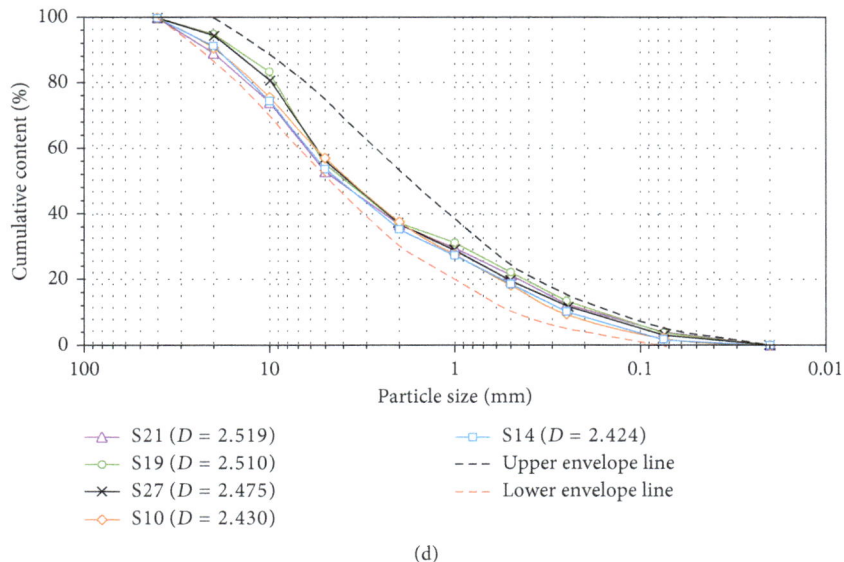

(d)

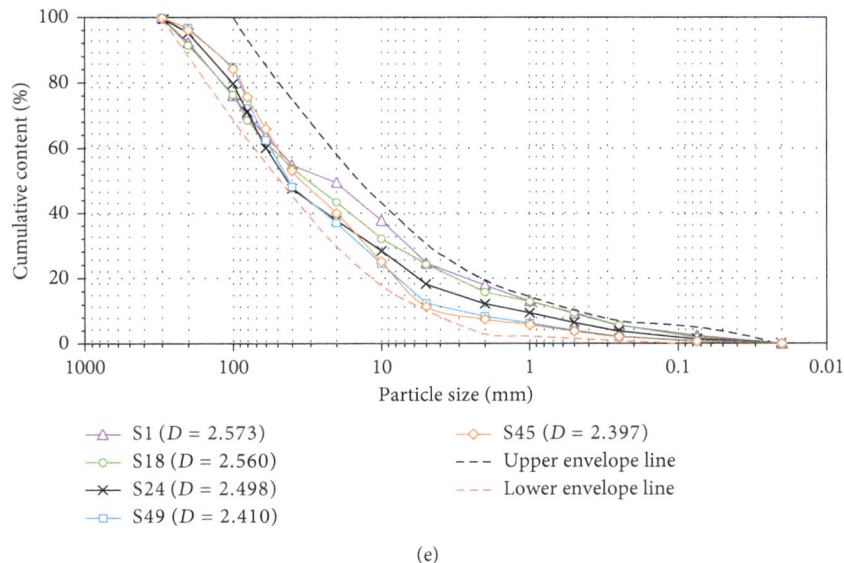

(e)

FIGURE 6: Statistical constant analysis results of different dam granular materials: (a) determined by the grading analysis; (b) determined by the Weibull distribution; typical PSD curve and the statistical constant for selected samples of granular materials (c) A, (d) B, and (e) C.

constant has vital importance for analyzing the physical characteristics of granular materials.

According to the design requirements for dam granular materials, the curves of PSD have been limited by the envelope lines (including the upper and lower envelope lines for different dam granular materials) for satisfying the requisite seepage and stability conditions (Figures 6(c)–6(e)). As shown in Figure 6 and Table 1, the samples of granular materials used in field tests are considered to be soils with continuous uniform grading and both have adequate compaction results for further construction. When the grain composition varies within the limited range restricted by the envelope lines, the statistical constant also varies within a certain range. For granular material A, the statistical constant is increasing where the particle size distribution curve is closer to the upper envelope line and decreasing where the distribution curve is closer to the lower envelope

line (Figure 6(c)). The analysis results indicate that the rising content of the particle group with size less than 5 mm and the dropping content of the particle group with size more than 5 mm result in the increase of the statistical constant. For granular material B, the PSD is found to have a similar developing principle and distribution regularity compared to granular material A (Figure 6(d)). During the preparation and production of granular materials with a relatively small particle size range, the content of the fine particle group is a vital matter owing to the fact that an inappropriate setting can cause the statistical constant to be far away from the average level. For granular material C, the PSD with higher statistical constant has a characteristic where its bottom part is closer to the upper envelope line and its top part is nearer to the lower envelop line (Figure 6(e)). Granular material C is supposed to have a wide range of particle size. The descent of the intermediate particle group whose size is between

5 mm and 40 mm can lead to the growth of statistical constant because the PSD has a good correlation with the statistical constant. Therefore, if the analysis is combined with Figure 2 and Table 2, it can be concluded that the particle size range and the uniformity of the grading distribution affect the physical properties and can further be represented by the statistical constant.

When studying PSD of granular materials, the practice normally derives characteristic coefficients to describe grain composition. The uniformity coefficient C_U reflects the uniformity of the grading distribution in granular materials. The curvature coefficient C_C describes the entire morphology about the gradation accumulation curve of grain composition. However, the statistical constant D can be used to predict the characteristic value of distribution curves as a comprehensive indicator reflecting the structure of soil samples. According to the different sections in the particle size distribution curve, the varied gradient b of the curve can be obtained as follows (here three typical characteristic values are used):

$$b_1 = \frac{\lg 30 - \lg 10}{\lg d_{30} - \lg d_{10}} = \frac{\lg 3}{\lg (d_{30}/d_{10})}, \tag{18a}$$

$$b_2 = \frac{\lg 60 - \lg 10}{\lg d_{60} - \lg d_{10}} = \frac{\lg 6}{\lg (d_{60}/d_{10})}, \tag{18b}$$

$$b_3 = \frac{\lg 60 - \lg 30}{\lg d_{60} - \lg d_{30}} = \frac{\lg 2}{\lg (d_{60}/d_{30})}. \tag{18c}$$

Assuming the PSD curve after the logarithmic operation is a linear function, which means that $b_1 = b_2 = b_3 = b$, then the following equations can be obtained:

$$\frac{d_{30}}{d_{10}} = 3^{1/b} = 3^{1/(3-D)}, \tag{19a}$$

$$\frac{d_{60}}{d_{10}} = 6^{1/b} = 6^{1/(3-D)}, \tag{19b}$$

$$\frac{d_{60}}{d_{30}} = 2^{1/b} = 2^{1/(3-D)}. \tag{19c}$$

On the basis of the definition of the grading index, the relationships between statistical constant and uniformity coefficient and curvature coefficient can be determined as follows:

$$C_U = \frac{d_{60}}{d_{10}} = 6^{1/(3-D)}, \tag{20}$$

$$C_C = \frac{d_{30}^2}{d_{10}d_{60}} = 1.5^{1/(3-D)}. \tag{21}$$

However, if we observe Figures 3–5, it will be found that the curve $\ln N(r) = b \ln r + C$ incompletely coincides with the grading distribution curve. The curve $C_C = C_U^{\ln 1.5/\ln 6} = C_U^{0.2263}$ can be gained by switching (20) and (21). Taking granular material C as an example, there is also a fitting curve $C_C = 0.2501 C_U^{0.4945}$ as a result of the characteristic coefficients (C_C and C_U) calculated by the linear interpolation

method (Figure 7(b)). The theoretical curve $C_C = C_U^{0.2263}$ calculated with the statistical constant is the same type function as the fitting curve $C_C = 0.2501 C_U^{0.4945}$. Although deviation exists in the two curves, it states that the method based on the statistical constant is feasible. Compared to the linear interpolation method, it is much more convenient and has a more stable range of calculated results. The characteristic coefficients are made from the analysis of the part distribution character. The definition of the characteristic coefficients on the basis of several characteristic particle sizes of a certain distribution curve is unable to present the integral distribution of grain composition. Because of its lack of integrality and adaptability, the accuracy of the definition is an issue in itself that needs to be investigated. However, the statistical constant is considered to be an indicator with integrality. Thinking of it in reverse, the statistical constant can be used as an engineering indicator to evaluate whether artificial aggregate is in good condition.

When comparing the results calculated from two different methods, the uniformity coefficient results are in basic agreement (Figure 7(a)). The assumption of the method based on the statistical constant is that the granular material is supposed to have a good degree of uniformity for grading distribution, which is the following condition: $b_1 = b_2 = b_3 = b$. It leads to keeping the curvature coefficient values between 1 and 3 (Figures 7(d) and 7(f)). The value of the uniformity coefficient is obtained from the ratio of two special particle sizes (d_{60} and d_{10}). The linear interpolation calculation method is, supposedly, commonly used with negligible error. As shown in Figures 3(a), 4(a), and 5(a), the uniformity of the particle group distribution of granular material C is better than granular materials A and B, which can be stated as the deviation of the calculated characteristic coefficient values for granular material C is lower than that for granular materials A and B. Therefore, for granular material C, the calculated value based on the statistical constant is more accurate. As shown in Figure 7(a), the uniformity coefficient based on the statistical constant is universally lower than the linear interpolation method for granular material C, and there are contrary results for granular materials A and B. When d_{10} is smaller than 1 mm and d_{60} is higher than 20 mm, just a little deviation may result in the calculated uniformity coefficient values getting much larger (Figures 7(a) and 7(e)). The calculation results indicate that the computational method based on the statistical constant can better control the results' range.

4. Physical and Compaction Properties behind the PSD

The compaction degree is a major construction control index for detecting the quality of core rock-fill dam. During the construction process, the factors affecting the compaction properties are complex and numerous, including grain composition, aggregate shape, moisture content, rolling technology, and compaction power. However, it is difficult to analyze each factor and understand its mechanism. If preparing the granular materials according to the traditional norms and requirements, the actual compaction degree will

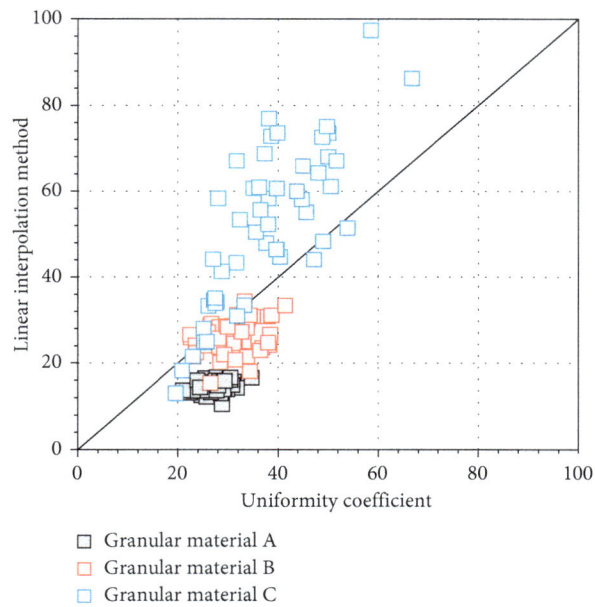

(a)

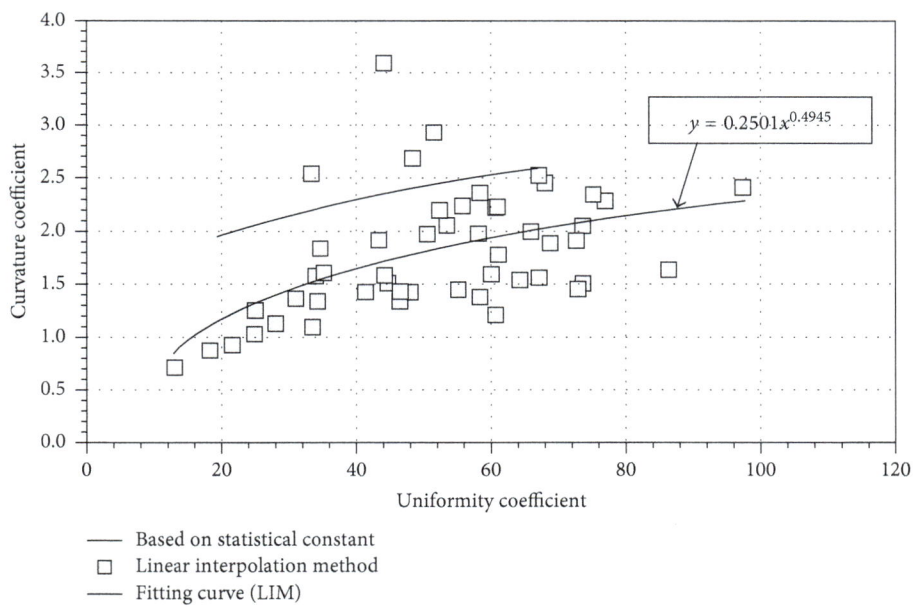

(b)

Figure 7: Continued.

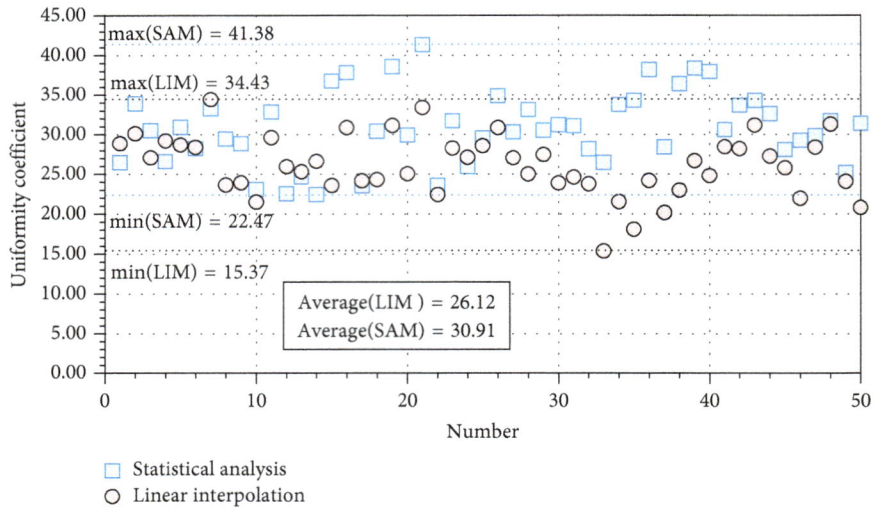

(c)

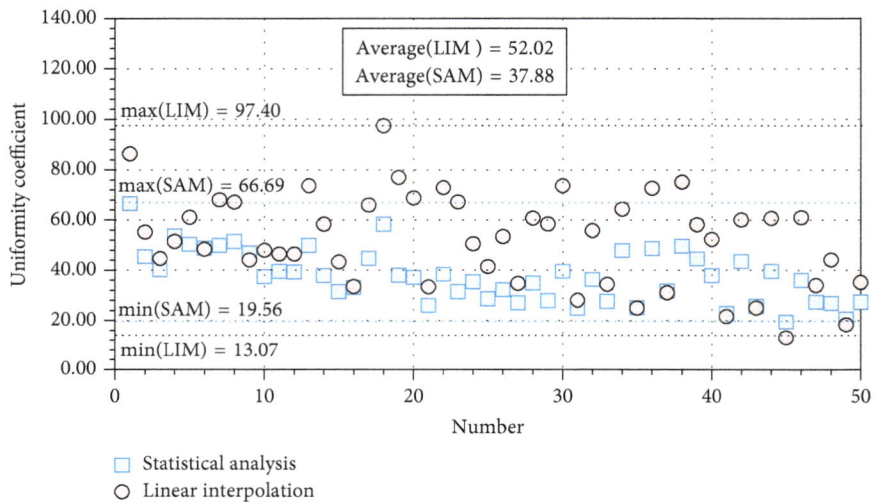

(d)

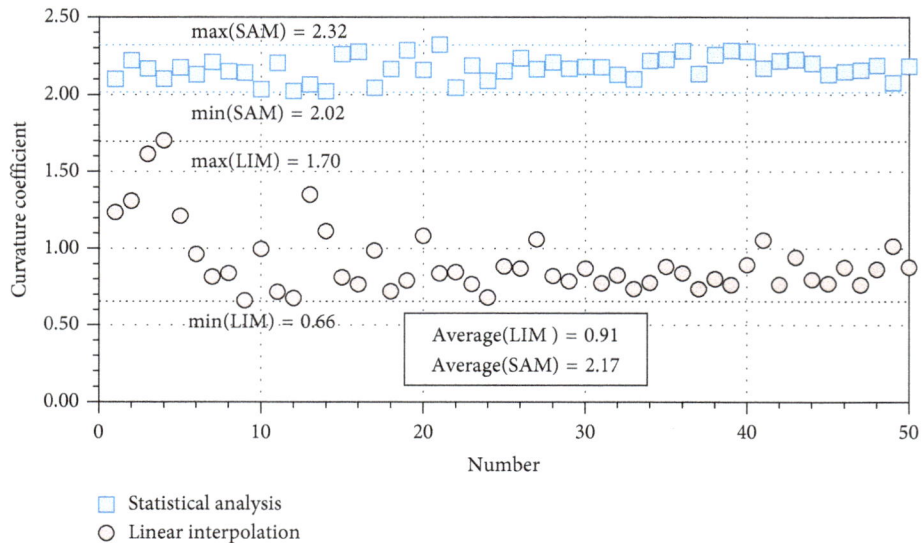

(e)

FIGURE 7: Continued.

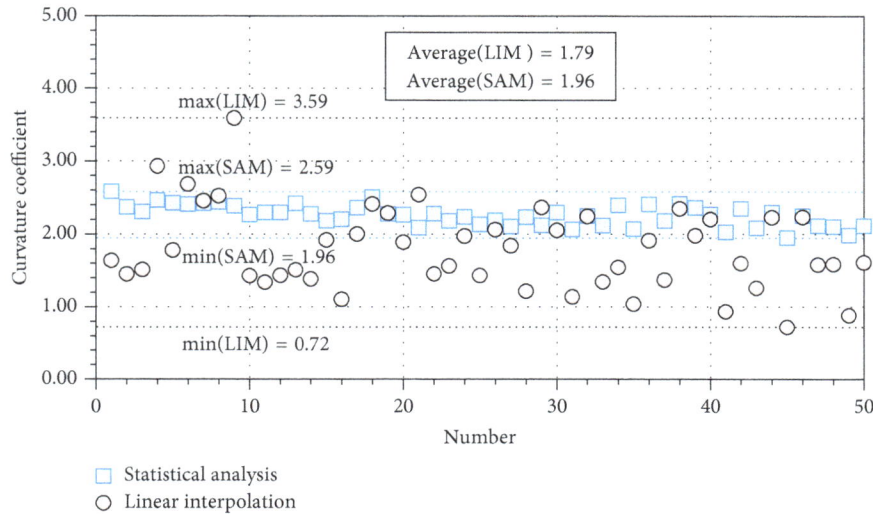

FIGURE 7: Characteristic values of the distribution curve of different granular materials: (a) correlation of the uniformity coefficient for PSD curves from the linear interpolation method and based on the statistical constant; (b) relationship between the uniformity coefficient and the curvature coefficient for PSD curves; (c) uniformity coefficient results for granular material B; (d) curvature coefficient results for granular material B; (e) uniformity coefficient results for granular material C; (f) curvature coefficient results for granular material C (LIM represents the linear interpolation method; min(LIM) and max(LIM) represent the minimum and the maximum uniformity coefficients of the PSD curve by using the linear interpolation method, resp.; min(SAM) and max(SAM) represent the minimum and the maximum uniformity coefficients of the PSD curve by using the statistical analysis method, resp.; average(LIM) and average(SAM) represent the average values for the uniformity coefficients of the PSD curve by using the linear interpolation method and statistical analysis method, resp.).

have difficulty in reaching the engineering standard. Studying the compaction properties only by particle size is short of preciseness and accuracy. However, based on the Weibull distribution, the size and shape parameters are beneficial for understanding how to impact the compaction process in practice. This work focuses on the potential mathematical relation between the Weibull parameters and compacting properties, which can testify as to whether the Weibull distribution has the ability to be taken as a supplement to construction specifications for evaluating and selecting dam granular materials.

4.1. The Relationship between the Statistical Constant and Physical Properties. Granular materials in a relatively loose state have more pores than relatively tight states making it possible to hold more water. Exploring the relationship between the statistical parameters and physical properties can validate the effectiveness of the statistical constant in describing the spatial structure of granular materials. As shown in Figure 8(c) and Table 3, the statistical constant has a positive correlation with dry density and a negative correlation with moisture content. The calculated result is in accord with the recent research for the link between the statistical parameters, water, and physical properties [36–38]. As a result, the good correlation between the statistical constant and physical properties indicates that the statistical constant can quantifiably integrate the structure of dam granular materials. However, the correlation between the statistical parameters and physical properties has obvious limitations. For certain granular materials, the variation of

the statistical constant has no possible effect on physical properties. As shown in Figures 8(a) and 8(b), moisture content and dry density have no obvious change associated with the statistical constant and stay within a reasonable range. However, the correlation can be established among different kinds of granular materials. These characteristics illustrate that the point is aimed specifically at the sorting of granular materials whose statistical constant is a character parameter for the spatial feature. In other words, the statistical constant is only suitable for representing the integral level of physical and compaction properties or PSD.

Therefore, we cannot choose the statistical constant as the character parameter for evaluating the compaction properties of granular materials owing to the lack of the validity specific to the same sort. However, every single dam granular sample has a specific set of Weibull parameters to match. Meanwhile, this set of Weibull parameters has the ability to show different performances compared to not only the granular material samples of a sort but those belonging to other sorts (Table 4). Based on the statistical analyses, the paper has proven that the Weibull distribution is generally applicable for describing granular materials. As above-mentioned, some latent correlation must exist between the variation of the Weibull parameters and the alteration of granular materials behavior and it can be explored.

4.2. The Relationship between the Weibull Parameters and Compaction Properties. The spatial distribution after rolling compaction of granular materials with a wide particle size range depends on a number of factors besides the size

Granular material A
Granular material B
Granular material C

Average line A
Average line B
Average line C

(a)

Granular material A
Granular material B
Granular material C

Average line A
Average line B
Average line C

(b)

Dry density
Moisture content

(c)

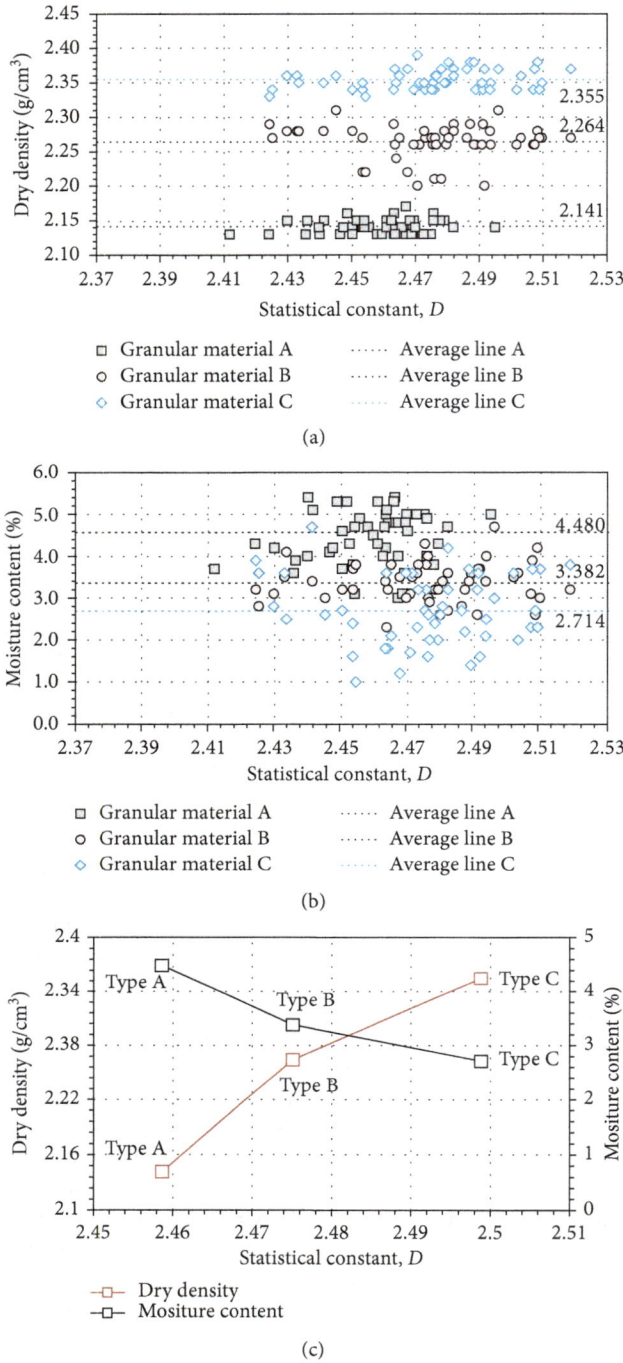

FIGURE 8: Relationship of the physical properties of granular materials and the statistical constant (all of the 50 groups for each type of granular materials are used): (a) the relationship behind the types; (b) the graph of moisture content and the statistical constant; (c) the graph of dry density and the statistical constant.

difference of the particle [14]. Therefore, adopting a single character parameter as a reference standard for evaluating the compaction degree is to be rejected owing to the lack of scientific rigor. This paper tries to apply the contour map to describe the variation rule of granular materials compaction degree using two Weibull parameters (the size and shape

TABLE 3: Computational results for the physical and compaction behaviors of different dam granular materials (all of the field test samples).

Properties		Granular material A	Granular material B	Granular material C
Moisture content (%)	Maximum	5.4	4.7	4.7
	Minimum	3.0	2.3	1.0
	Average	4.5	3.4	2.7
Dry density (g/cm³)	Maximum	2.17	2.31	2.39
	Minimum	2.13	2.20	2.33
	Average	2.14	2.26	2.36
Compaction degree	Maximum	0.93	0.97	0.99
	Minimum	0.85	0.85	0.90
	Average	0.87	0.88	0.94

parameters). As shown in Figure 9, the contours are distributed in an extremely scattered way and make great difficulty in concluding the dominant mathematical laws. However, when comparing the contour maps of three granular materials, the distribution is more messy and irregular when the particle size range is wider. Despite the influence of the particle itself, granular material samples with a wide range of particle sizes are much more susceptible to the effects of other factors in the process of rolling compaction. Above all, it is because the impacting factors are complex and difficult to control during the actual operation process that the approach based on quantitative experiment is unable to explore the explicit mathematical relationship between the Weibull parameters and relative density.

In general, various data indicators that can explain the difficulty in analyzing variables are provided with different data characteristics and different units. Data normalization is known to eliminate the impact of variable measurements, and it is a great assistance for discovering the recessive relationship between variables. When evaluating the structure status of granular material samples under the assumption that the particle is considered a sphere, the average level of particle size, scaling of particle groups' content, and the mutual arrangement of the particles are primary factors, and it is necessary to consider their influence. In this paper, these three evaluative dimensions of granular material structure are measured by the size parameter, the shape parameter, and the compaction degree, respectively. When a granular material system is compacted to a critical state and can no longer be tighter, it can be assumed that the impact of the three evaluative dimensions will remain in balance. Then, it can be supposed that the balance status is expressed by the following mathematical equation (granular materials A and B use (22) and granular material C uses (23)):

$$r_0^* + b^* + D_r^* = 1, \tag{22}$$

$$\text{or } r_0^* + b^* + D_r^* = 100. \tag{23}$$

Therefore, the data normalization referring to the ternary phase diagram can be processed by the following equations:

TABLE 4: Characteristic values behind the PSD curves of different dam granular materials (only shows the analysis results of first 10 groups).

Granular material type	Number	D			C_U		C_C		Weibull parameter	
		Weibull distribution	Grading analysis	Error (%)	Statistical analysis	Linear interpolation	Statistical analysis	Linear interpolation	Size parameter	Shape parameter
A	S1	2.442	2.455	−0.5	26.84	14.25	2.11	0.77	2.451	0.758
	S2	2.460	2.466	−0.3	28.67	14.47	2.14	0.83	2.437	0.734
	S3	2.458	2.470	−0.5	29.32	14.58	2.15	0.75	2.149	0.740
	S4	2.394	2.467	−3.0	28.84	10.51	2.14	0.86	1.871	0.826
	S5	2.414	2.449	−1.4	25.78	12.20	2.09	0.69	1.994	0.797
	S6	2.450	2.450	0.0	26.02	15.13	2.09	0.66	2.271	0.748
	S7	2.467	2.446	0.8	25.45	16.49	2.08	0.69	2.490	0.724
	S8	2.435	2.461	−1.1	27.76	16.70	2.12	0.77	2.276	0.768
	S9	2.460	2.452	0.3	26.36	15.13	2.10	0.62	2.206	0.732
	S10	2.482	2.463	0.7	28.18	15.63	2.13	0.67	2.268	0.703
B	S1	2.512	2.453	2.4	26.52	28.87	2.10	1.23	6.167	0.663
	S2	2.522	2.492	1.2	33.93	30.13	2.22	1.31	5.600	0.650
	S3	2.495	2.476	0.8	30.55	27.07	2.17	1.61	5.216	0.688
	S4	2.476	2.454	0.9	26.67	29.22	2.10	1.70	6.158	0.713
	S5	2.529	2.478	2.1	30.97	28.71	2.17	1.21	5.656	0.640
	S6	2.495	2.464	1.2	28.30	28.38	2.13	0.96	5.847	0.687
	S7	2.531	2.489	1.7	33.30	34.43	2.21	0.81	6.593	0.637
	S8	2.497	2.471	1.1	29.52	23.63	2.15	0.84	4.902	0.683
	S9	2.529	2.468	2.5	28.95	23.91	2.14	0.66	6.056	0.641
	S10	2.491	2.430	2.5	23.15	21.50	2.04	0.99	5.914	0.691
C	S1	2.606	2.573	1.3	66.69	86.33	2.59	1.64	47.852	0.536
	S2	2.536	2.531	0.2	45.66	55.16	2.37	1.45	52.615	0.632
	S3	2.518	2.515	0.1	40.34	44.70	2.31	1.51	58.655	0.655
	S4	2.520	2.551	−1.2	53.89	51.51	2.47	2.93	37.318	0.654
	S5	2.535	2.543	−0.3	50.57	61.12	2.43	1.78	51.232	0.633
	S6	2.518	2.540	−0.9	49.04	48.40	2.41	2.68	44.657	0.656
	S7	2.532	2.542	−0.4	50.05	68.01	2.42	2.45	56.590	0.637
	S8	2.549	2.546	0.1	51.67	67.11	2.44	2.52	43.539	0.614
	S9	2.520	2.535	−0.6	47.18	44.12	2.39	3.59	37.868	0.652
	S10	2.549	2.506	1.7	37.60	48.01	2.27	1.43	45.801	0.614

Note. D is the value of the statistical constant; C_U is the uniformity coefficient of the PSD curve; C_C is the curvature coefficient of the PSD curve.

$$r_0^* = \frac{r_0}{r_0 + b + D_r},$$

$$b^* = \frac{b}{r_0 + b + D_r}, \qquad (24)$$

$$D_r^* = \frac{D_r}{r_0 + b + D_r}.$$

Table 3 summarizes the statistical results for the physical and compaction behaviors of different dam granular materials including all of the field test samples. As shown in Table 3, the dry density of granular materials increases with the increasing content of coarse particles, where the granular material C is the maximum. The moisture content of granular materials is influenced by the particle size range and distribution (moisture content is an indirect reflection of the porosity of granular materials), and therefore, there is a vague relationship between the moisture content and the particle size distribution of granular materials. Here, the moisture content of granular materials decreases with the expanding range of particle size distribution, and it is also influenced by the continuity of particle gradation

(various size of particles are rationally distributed). Therefore, the statistical results indicate that the relative density of granular material C is the largest and shows the best compaction performance. The compaction performance of granular material A is close to granular material B, due to the small difference in the particle size distribution.

The regularity in granular material structure after data normalization (Figure 10) can be described as that the impact of compaction degree will decline, whereas the impact of the size parameter increases. At the same time, the impact of the compaction degree will add up when the impact of the shape parameter increases. The analysis indicates that it is much more difficult to have a tight rolling compaction when the average particle size shows a growth trend for a certain type of dam granular materials. In a similar way, the dam granular materials of a sort will be easier to compact at the point where the shape parameter is in an increasing trend. As shown in Figures 10(d) and 10(e), the above regularity also exists when comparing the average level of different types of granular materials. Figure 10(d) shows that the average level of granular material C, having a much wider range of particle sizes, is much larger than that of granular materials A and B. Figure 10(e) shows that

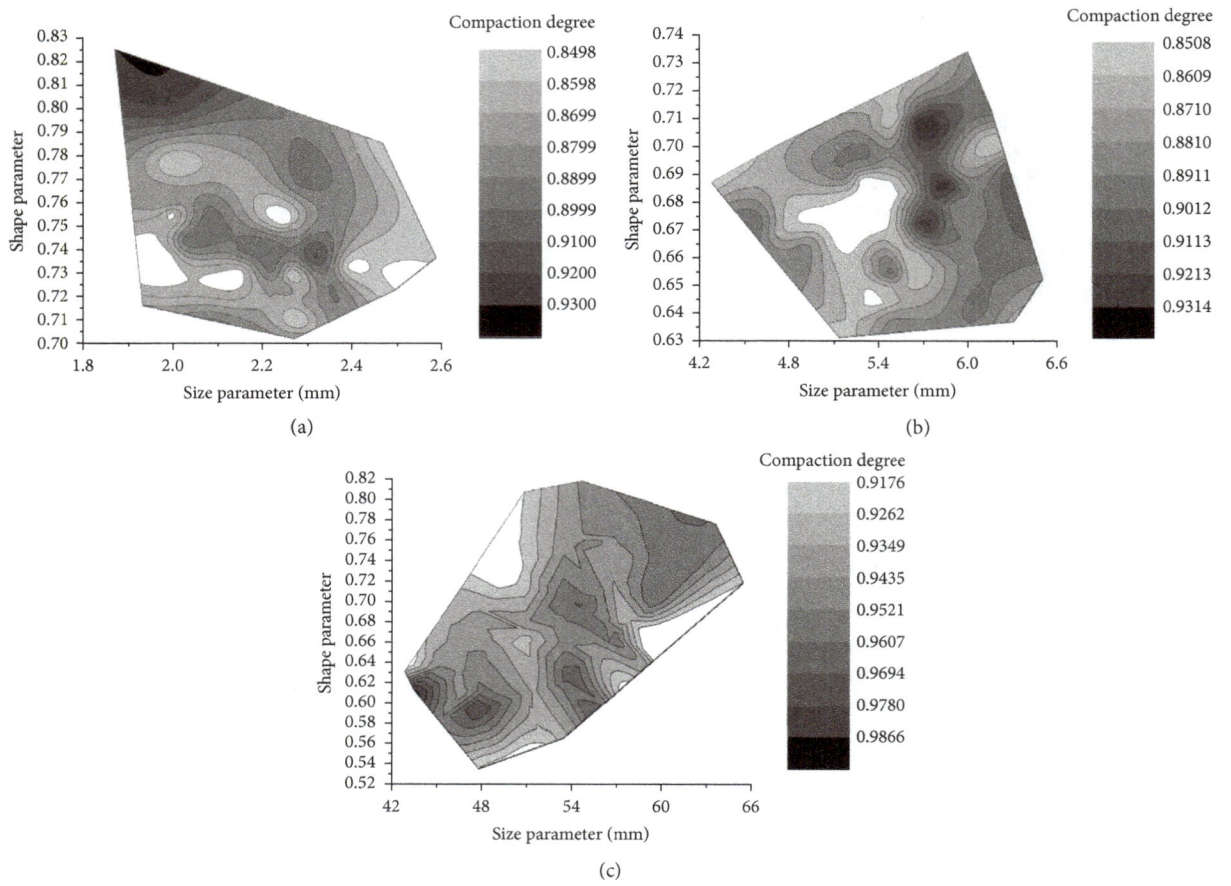

FIGURE 9: Relationship of the Weibull parameters and the compaction degree of granular materials: (a) A; (b) B; (c) C.

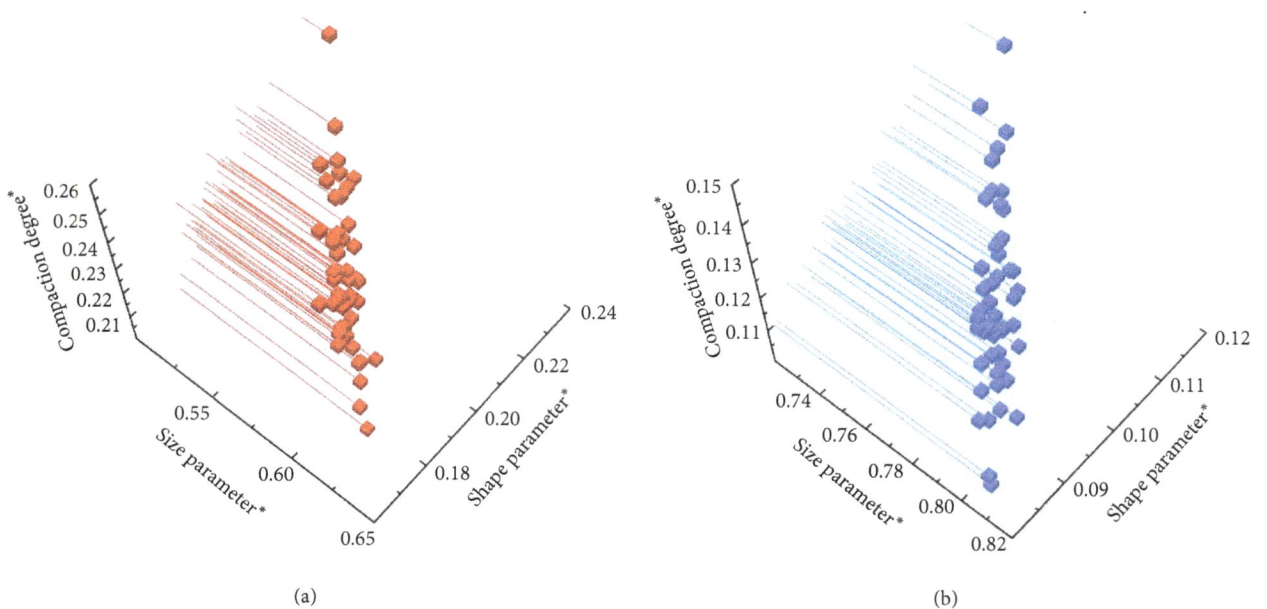

FIGURE 10: Continued.

(c)

(d)

(e)

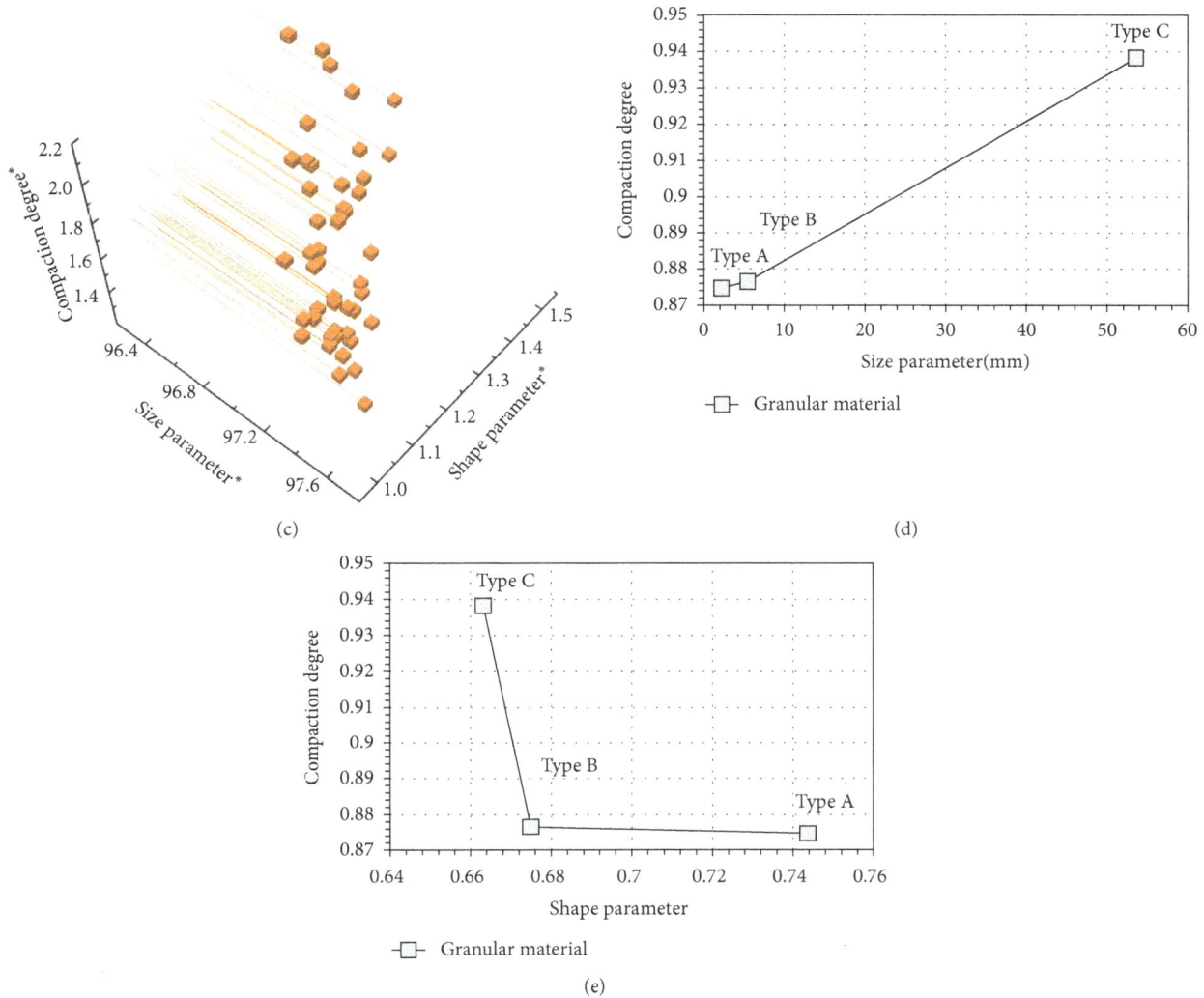

FIGURE 10: Relationship between the Weibull parameters and the compaction degree of granular materials after normalization: (a) granular material A; (b) granular material B; (c) granular material C; (d) correlation between the size parameter and the compaction degree of different granular materials; (e) correlation between the shape parameter and the compaction degree of different granular materials (size parameter* represents the value of the size parameter after normalization; shape parameter* represents the value of the shape parameter after normalization; compaction degree* represents the value of the compaction degree after normalization).

granular material C has a larger shape parameter separation between granular materials A and B. There is a closer contact between large and small particles when in a wider range of particle sizes. It further illustrates that the particle size range of granular materials can dramatically improve the compaction effect for artificial materials with a good grading. There are sufficient result foundations explaining that the size parameter has a positive correlation with the compaction degree and the shape parameter has a negative correlation with the compaction degree.

5. Discussion and Conclusions

The statistical analysis ensures the validity of the theoretical basis in quantifying the physical and compaction characteristics of granular materials on different spatial scales [36]. In regard to this possibility, this paper discusses the use of

the Weibull distribution function for describing the PSD of granular materials with artificial grading. It is well known that a two-parameter Weibull distribution function is capable of characterizing the PSD of soil mixtures [34]. However, the soil samples the paper uses are from the dam granular materials with artificial grading. To ensure the precision and scientific foundation, it is found that there is little error between the two different calculation methods of statistical constant (from analysis and based on the Weibull distribution), which states that the use of the Weibull distribution function is feasible here. A two-parameter Weibull distribution function has size parameter r_0 and shape parameter b. The size parameter is generally considered to be the average value of particle size. Meanwhile, from (19), there is a definite mathematical equation between the slope of any point on the PSD curves and the ratio of the cumulative contents of the particles' size less than two specific values. The shape parameter is characterizing the shape of the Weibull function curve. When using the Weibull function to fit the PSD curve, it is obvious that the shape parameter is connected with the particle size range, or more precisely, the scaling distribution of the particle group, and the larger the size range is, the less the shape parameter will be. However, this study is based on the assumption that all particles act as ideal spheres, though the relationship between particle shape and the critical state has been discussed [6]. Therefore, further study needs to take these factors into consideration, including the relationship between the shape parameter and the grading scaling distribution as well as the impact of particle shape on compaction.

This paper explores the physical and compaction properties of granular materials with four parameters (relative density D_r, statistical constant D, and Weibull parameters (size parameter r_0 and shape parameter b)). These parameters (D_r, r_0, and b) provide a uniform data standard for comparing measured values between granular material samples of different types. However, the statistical constant is not qualified to do this because the erratic fluctuation of statistical results appears as a stagger phenomenon on statistical constant ranges of different types. The stagger scope is connected with the particle size range, which leads to invalid comparisons of the statistical constants from different types of granular materials samples. The statistical constant can only reflect the integral level of certain granular materials, and it has the ability to show the type discrepancies (Figure 8(c)). Therefore, the paper uses these three parameters D_r, r_0 and b to get a comprehensive analysis of the structural status of three types of granular materials with the assistance of the statistical constant.

There is a direct and close link between PSD curve and statistical constant. The statistical constant can represent the variation of PSD curves, and the characteristic coefficients (C_U and C_C) of the PSD curve can be calculated by the statistical constant under postulated conditions. It can be said that the statistical constants are likely to be an indicative parameter for verifying the quality of artificial grading. The regularity of the Weibull parameters and compaction degree in different granular material samples of the same type and the regularity in different types are consistent, making a contribution to providing significant guidance on how to improve compaction during actual construction. However, the mechanism by which Weibull parameters influence compaction process needs to be further explored.

The final results are shown with some fluctuations. First, the granular material samples the paper uses are fill materials in construction and all are considered to have good grading; the analysis results of the compaction properties have limitations. The lack of a comparison between good and bad grading distribution calls for future research setting up a series of control groups. Second, measurement error, especially the significance figure caused in calculation, is existed in dry density and compactness of granular materials, and the measurement is achieved by engineering personnel, which causes that the paper cannot reach an accurate expression of how Weibull parameters influence compaction properties. Therefore, the future research needs

to make a better measurement of compaction degree (1) considering the use of a method of entropy approach for adding another evaluating indicator of the PSD curve and (2) viewing the particle of granular materials as a rectangular parallelepiped shape or a dodecahedron of which four surfaces are octagon and eight surface are triangle and making formula derivations for obtaining statistical constants under different assuming conditions.

Conflicts of Interest

The authors declare that there are no conflicts of interest regarding the publication of this paper.

Acknowledgments

The authors gratefully acknowledge the support of the Youth Science and Technology Fund of Sichuan Province (2016JQ0011) and the Key Project of the Power Construction Corporation of China (ZDZX-5).

References

[1] J. C. Davis and N. J. Hoboken, *Statistics and Data Analysis in Geology*, Wiley, New York, NY, USA, 3rd edition, 2012.

[2] G. Marketos and M. D. Bolton, "Quantifying the extent of crushing in granular materials: a probability-based predictive method," *Journal of the Mechanics and Physics of Solids*, vol. 55, no. 10, pp. 2142–2156, 2007.

[3] D. Hillel, *Fundamentals of Soil Physics*, Academic Press, New York, NY, USA, 1st edition, 1980.

[4] B. Caicedo, M. Ocampo, and L. E. Vallejo, "Modelling comminution of granular materials using a linear packing model and Markovian processes," *Computers and Geotechnics*, vol. 80, pp. 383–396, 2016.

[5] J. P. Hyslip and L. E. Vallejo, "Fractal analysis of the roughness and size distribution of granular materials," *Engineering Geology*, vol. 48, no. 3-4, pp. 231–244, 1997.

[6] J. Yang and X. D. Luo, "Exploring the relationship between critical state and particle shape for granular materials," *Journal of the Mechanics and Physics of Solids*, vol. 84, pp. 196–213, 2015.

[7] B. B. Mandelbrot, *The Fractal Geometry of Nature*, Macmillan Publishes Limited, London, UK, 1st edition, 1983.

[8] J. J. Wang, D. Zhao, Y. Liang, and H. B. Wen, "Angle of repose of landslide debris deposits induced by 2008 Sichuan Earthquake," *Engineering Geology*, vol. 156, pp. 103–110, 2013.

[9] M. Rieu and G. Sposito, "Fractal fragmentation, soil porosity, and soil water properties: I. Theory," *Soil Science Society of America Journal*, vol. 55, no. 5, pp. 1231–1238, 1991.

[10] Z. Y. Ma, F. N. Dang, and H. J. Liao, "Numerical study of the dynamic compaction of gravel soil ground using the discrete element method," *Granular Matter*, vol. 16, no. 6, pp. 881–889, 2014.

[11] G. Lumay and N. Vandewalle, "Compaction of anisotropic granular materials: experiments and simulations," *Physical Review E*, vol. 70, no. 1, 2004.

[12] H. S. Thilakasiri, M. Gunaratne, G. Mullins, P. Stinnette, and

B. Jory, "Investigation of impact stresses induced in laboratory dynamic compaction of soft soils," *International Journal for Numerical and Analytical Methods in Geomechanics*, vol. 20, no. 10, pp. 753–767, 1996.

[13] J. B. Knight, C. G. Fandrich, C. N. Lau, H. M. Jaeger, and S. R. Nagel, "Density relaxation in a vibrated granular material," *Physical Review E*, vol. 51, no. 5, pp. 3957–3963, 1995.

[14] N. Vandewalle, G. Lumay, O. Gerasimov, and F. Ludewig, "The influence of grain shape, friction and cohesion on granular compaction dynamics," *European Physical Journal E*, vol. 22, no. 3, pp. 241–248, 2007.

[15] P. Lu, I. F. Jefferson, M. S. Rosenbaum, and I. J. Smalley, "Fractal characteristics of loess formation: evidence from laboratory experiments," *Engineering Geology*, vol. 69, no. 3-4, pp. 287–293, 2003.

[16] M. A. Martín, M. Reyes, and F. J. Taguas, "Estimating soil bulk density with information metrics of soil texture," *Geoderma*, vol. 287, pp. 66–70, 2017.

[17] M. A. Martín, M. Reyes, and F. J. Taguas, "An entropy like parameter of particle size distributions as packing density index in complex granular media," *Granular Matter*, vol. 19, p. 9, 2017.

[18] M. W. Clark, "Some methods for statistical analysis of multimodal distributions and their application to grain-size data," *Journal of the International Association for Mathematical Geology*, vol. 8, no. 3, pp. 267–282, 1976.

[19] S. W. Tyler and S. W. Wheatcraft, "Application of fractal mathematics to soil water retention estimation," *Soil Science Society of America Journal*, vol. 53, no. 4, pp. 987–996, 1989.

[20] E. Perfect and B. D. Kay, "Fractal theory applied to soil aggregation," *Soil Science Society of America Journal*, vol. 55, no. 6, pp. 1552–1558, 1991.

[21] S. W. Tyler and S. W. Wheatcraft, "Fractal scaling of soil particle-size distributions: analysis and limitations," *Soil Science Society of America Journal*, vol. 56, no. 2, pp. 362–369, 1992.

[22] M. Rieu and G. Sposito, "Fractal fragmentation, soil porosity, and soil water properties: II. Applications," *Soil Science Society of America Journal*, vol. 55, no. 5, pp. 1239–1244, 1991.

[23] A. N. Anderson, A. B. McBratney, and J. W. Crawford, "Applications of fractals to soil studies," *Advances in Agronomy*, vol. 63, no. 8, pp. 1–76, 1997.

[24] E. Perfect and B. D. Kay, "Applications of fractals in soil and tillage research: a review," *Soil and Tillage Research*, vol. 36, no. 1-2, pp. 1–20, 1995.

[25] X. Wang, S. Liu, and G. Liu, "Fractal characteristics of soils under different land-use patterns in the arid and semi-arid regions of the Tibetan Plateau, China," *Wuhan University Journal of Natural Science*, vol. 134, no. 4, pp. 56–61, 2005.

[26] N. Prosperini and D. Perugini, "Particle size distributions of some soils from the Umbria Region (Italy): fractal analysis and numerical modeling," *Geoderma*, vol. 145, no. 3-4, pp. 185–195, 2008.

[27] S. Yu and C. T. Oguchi, "Role of pore size distribution in salt uptake, damage, and predicting salt susceptibility of eight types of Japanese building stones," *Engineering Geology*, vol. 115, no. 3-4, pp. 226–236, 2010.

[28] F. Yang, Z. Ning, and H. Liu, "Fractal characteristics of shales from a shale gas reservoir in the Sichuan Basin, China," *Fuel*, vol. 115, no. 1, pp. 378–384, 2014.

[29] S. Shao, X. G. Yang, and J. W. Zhou, "Numerical analysis of different ventilation schemes during the construction process of inclined tunnel groups at the Changheba Hydropower Station, China," *Tunnelling and Underground Space Technology*, vol. 59, pp. 157–169, 2016.

[30] D. L. Turcotte, "Fractals and fragmentation," *Journal of Geophysical Research-Solid Earth*, vol. 91, no. B2, pp. 1921–1926, 1986.

[31] H. P. Xie, *Fractals in Rock Mechanics*, CRC Press, Boca Raton, FL, USA, 1993.

[32] L. Kittleman, "Application of Rosin's distribution in size-frequency analysis of clastic rocks," *Journal of Sedimentary Research*, vol. 34, no. 3, pp. 483–502, 1964.

[33] W. Weibull, "A statistical theory of the strength of materials," in *Proceedings of the NR151 the Royal Swedish Institute for Engineering Research*, Stockholm, Sweden, 1939.

[34] Y. S. Cheong, G. K. Reynolds, A. D. Salman, and M. J. Hounslow, "Modelling fragment size distribution using two-parameter Weibull equation," *International Journal of Mineral Processing*, vol. 74, no. 50, pp. 227–237, 2004.

[35] F. Casini, G. M. Viggiani, and S. M. Springman, "Breakage of an artificial crushable material under loading," *Granular Matter*, vol. 15, no. 5, pp. 661–673, 2013.

[36] H. Millan, M. Gonzalez-Posada, M. Aguilar, J. Domínguez, and L. Cespedes, "On the fractal scaling of soil data. Particle-size distributions," *Geoderma*, vol. 117, no. 1-2, pp. 117–128, 2003.

[37] Y. X. Xiao, C. F. Lee, and S. J. Wang, "Particle-size distribution of interlayer shear zone material and its implications in geological processes—a case study in China," *Engineering Geology*, vol. 66, no. 3-4, pp. 221–232, 2002.

[38] Y. Z. Su, H. L. Zhao, W. Z. Zhao, and T. H. Zhang, "Fractal features of soil particle size distribution and the implication for indicating desertification," *Geoderma*, vol. 122, no. 1, pp. 43–49, 2004.

Thermodynamic Study of Tl_6SBr_4 Compound and Some Regularities in Thermodynamic Properties of Thallium Chalcohalides

Dunya Mahammad Babanly,[1] Qorkhmaz Mansur Huseynov,[2] Ziya Saxaveddin Aliev,[3] Dilgam Babir Tagiyev,[1] and Mahammad Baba Babanly[1]

[1]*Institute of Catalysis and Inorganic Chemistry, ANAS, 1143 Baku, Azerbaijan*
[2]*ANAS, Nakhchivan Branch, 7000 Nakhchivan, Azerbaijan*
[3]*Azerbaijani State Oil and Industrial University, Azadlıg Av. 16/21, 1010 Baku, Azerbaijan*

Correspondence should be addressed to Dunya Mahammad Babanly; dunyababanly2012@gmail.com

Academic Editor: Michael Aizenshtein

The solid-phase diagram of the Tl-TlBr-S system was clarified and the fundamental thermodynamic properties of Tl_6SBr_4 compound were studied on the basis of electromotive force (EMF) measurements of concentration cells relative to a thallium electrode. The EMF results were used to calculate the relative partial thermodynamic functions of thallium in alloys and the standard integral thermodynamic functions ($-\Delta_f G^0$, $-\Delta_f H^0$, and S^0_{298}) of Tl_6SBr_4 compound. All data regarding thermodynamic properties of thallium chalcogen-halides are generalized and comparatively analyzed. Consequently, certain regularities between thermodynamic functions of thallium chalcogen-halides and their binary constituents as well as degree of ionization (DI) of chemical bonding were revealed.

1. Introduction

Chalcohalides of p-elements are of considerable interest as promising functional materials of modern electronic engineering. Multitude of them exhibit semiconductor, thermoelectric, photoelectric, topological insulator, radiation detector, and magnetic properties [1–3]. $Tl_6S(Se)I_4$ ternary compounds are known as effective X-ray and γ-ray detectors, outperforming the current state-of-the-art material for room temperature operation, CdZnTe (CZT) [4, 5]. Given the promising performance of thallium-based chalcogenides and chalcogen-halides, these compounds continue to attract attention despite toxicity of thallium derivatives.

Knowledge of phase equilibria and thermodynamic properties of phases is of key importance in designing techniques and optimizing conditions for the fabrication, crystal growth of multicomponent inorganic materials.

The investigation of phase equilibria in Tl-X-Hal (X-S, Se, Te; Hal-Cl, Br, I) systems was started from the beginning of the 80th years of the last century. A special attention was paid to quasibinary systems Tl_2X-TlHal because of possibility of formation of ternary compounds [6–11]. Existence of 11 ternary compounds was established in the Tl_2X-TlHal systems: 6 ternary compounds of type Tl_5X_2Hal (all selenium and tellurium systems) and 4 ternary compounds of type Tl_6XHal_4 (all sulphurous systems and Tl-Se-I system). Tl_2TeHal_6 compounds were detected in the systems Tl-Te-Hal [12, 13].

Tl_6SHal_4 (Hal-Cl, Br, I), Tl_6SeI_4, and Tl_2TeBr_6 compounds crystallize in a tetragonal structure, with space group $P4/mnc$ [6, 7, 14, 15]; $Tl_5Se_2Br(I)$ and Tl_5Te_2Hal compounds have a tetragonal structure, with space group $I4/mcm$ [8, 11, 16–19]. Tl_2TeCl_6 compound crystallizes in a cubic system, Sp.gr. Fm-$3m$ and Tl_2TeCl_6 in a monoclinic system, Sp. gr. $P2_1/c$ [13, 15].

Physicochemical investigation of Tl-X-Hal (X-S, Se, Te; Hal-Cl, Br, I) systems was implemented in detail at a wide or full concentration range. A number of polythermal

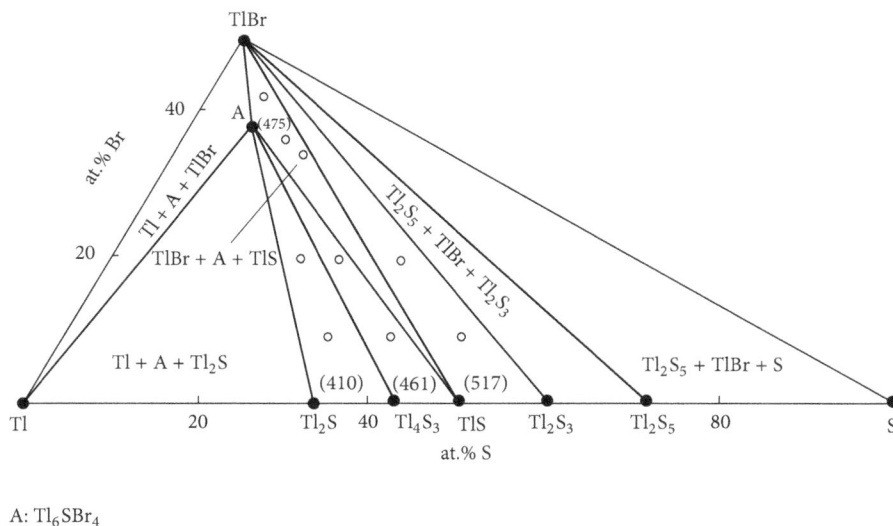

FIGURE 1: Solid-phase equilibrium diagram of the system Tl-TlBr-S at 300 K [24]. The compositions of the studied samples are given by hollow circles in the proper three-phase regions.

and isothermal sections as well as projections of liquidus surfaces were constructed. The primary crystallization and homogeneity areas of phases were fixed and the fundamental thermodynamic functions of ternary compounds and solid solutions were determined [20–23].

In this contribution, we present the thermodynamic study of Tl_6SBr_4 compound, summarize all data on thermodynamic properties of thallium chalcohalides, and carry out a comparative review between the latter. At the end of the paper, we present some regularities detected in the thermodynamic properties of thallium chalcohalides.

2. Experimental Details

2.1. Synthesis and Analysis. For planning experiments, we have used the solid-phase diagram of the Tl-TlBr-S system [24] which allowed us to determine the relevant compositions of samples for thermodynamic studies and to select conditions for their synthesis and thermal treatment (Figure 1). We have composed electrochemical cell of the type

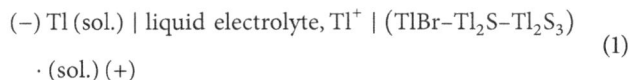

$$(-) \, Tl \, (sol.) \mid liquid \, electrolyte, Tl^+ \mid (TlBr–Tl_2S–Tl_2S_3) \quad (1)$$
$$\cdot \, (sol.) \, (+)$$

in which the left electrode was pure metallic thallium and the right electrodes were samples from the mentioned region. A saturated glycerin solution of KBr with the addition of 0.1 mass% TlBr was used as an electrolyte.

Initial compounds Tl_2S and TlBr were synthesized to prepare the right electrodes of the electrochemical cell of the type (1). Tl_2S was synthesized by alloying of stoichiometric amounts of high-purity elemental components (Tl, 99.999 mass%, Alfa Aesar; S, 99.999 mass%, Alfa Aesar) in an evacuated silica ampoules at temperatures 30–50 K above the melting point. Tl_2S, melting congruently at 728 K [17], readily crystallizes while slowly cooling the sample.

TlBr was prepared by an indirect method reported in [18]. At first, metallic thallium was dissolved in the dilute sulphuric acid (7–10 mol%) at 350 K to get the Tl_2SO_4 solution. Then diluted HBr was added to a hot 2% Tl_2SO_4 solution until complete precipitation of TlBr. Yellowish green TlBr was separated from the mother liquor and washed with icy distilled water. The product was dried over KOH in a desiccator at 380–400 K and stored in the dark to prevent its decomposition.

The synthesized compounds were identified by differential thermal analysis DTA (NETZSCH 404 F1 Pegasus system) and X-ray powder diffraction XRD (Bruker D8 ADVANCE diffractometer, $CuK\alpha_1$ radiation) methods.

Alloys of the subsystem $TlBr–Tl_2S–Tl_2S_3$ were prepared by blending and interacting TlBr and Tl_2S compounds and elemental sulphur in various ratios in evacuated quartz vessels. They were subjected to a long-term stepped homogenizing annealing under the conditions described in [24]. The total mass of samples was 1 g and, after determining the solidus temperature, samples were additionally held at 20–30 K below the solidus for 500 h.

2.2. EMF Measurements. To measure the EMF of the chains type (1), electrodes, electrolyte, and electrochemical cell were prepared. The left electrode was made by attaching metallic thallium to a molybdenum current collector. Taking into account oxidization of thallium even at room temperature, before assembling electrochemical cell, the left electrodes were kept in glycerin, since metallic thallium does not directly interact with it [19].

The right electrodes (see (1)) were prepared by pressing powdered equilibrium alloys of the system under study into current collectors in the form of cylindrical tablets with a weight of ~0.5 g and placed in a tubular furnace; the temperature was stabilized at 350 K for 40–50 h. The wires were sealed in glass jackets to protect them from contact

TABLE 1: Temperature dependences of EMF of the concentration chains type (1) in some phase regions of the subsystem $TlBr-Tl_2S-Tl_2S_3$.

Number	Phase region in Figure 1	$E, mV = a + bT \pm t \cdot S_E(T)$
1	$TlBr-TlS-Tl_2S_3$	$552{,}6 - 0{,}120T \pm 2\left[\left(\dfrac{1{,}8}{24}\right) + 8 \cdot 10^{-5}(T - 342{,}4)^2\right]^{1/2}$
2	$TlBr-TlS-Tl_6SBr_4$	$486{,}1 - 0{,}038T \pm 2\left[\left(\dfrac{0{,}9}{24}\right) + 5 \cdot 10^{-5}(T - 342{,}4)^2\right]^{1/2}$
3	$Tl_6SBr_4-Tl_4S_3-TlS$	$465{,}8 - 0{,}016T \pm 2\left[\left(\dfrac{1{,}5}{24}\right) + 7 \cdot 10^{-5}(T - 342{,}4)^2\right]^{1/2}$
4	$Tl_2S-Tl_6SBr_4-Tl_4S_3$	$383{,}4 + 0{,}086T \pm 2\left[\left(\dfrac{1{,}1}{24}\right) + 6 \cdot 10^{-5}(T - 342{,}4)^2\right]^{1/2}$

with the electrolyte. EMF was measured by the compensation method in the temperature range of 300–390 K with the accuracy of ±0.1 mV, using the high-resistance universal B7-34A digital voltmeter. In each experiment the first EMF reading was performed approximately 30 h after the start of the experiment and at least 2-3 h after reaching the desired temperature, which ensures the achievement of equilibrium. The EMF measurements were carried out 2–7 days or more in consideration of homogenization of the electrode alloys, after the cell had been assembled. Equilibrium values were considered the EMF readings that varied by no more than 0.5 mV irrespective of the direction of temperature change at repeated measurements at a given temperature. In order to eliminate the contribution of the thermopower, all contacts and leads were kept at the same temperature.

Techniques of assembling the electrochemical cell and EMF measurements are described in detail in [25, 26].

3. Results and Discussion

3.1. Thermodynamic Study of the Ternary Compound Tl_6SBr_4. The solid-phase equilibrium diagram of the Tl-TlBr-S subsystem that had been constructed in [24] is given in Figure 1. As can be seen from Figure 1, there are 4 three-phase regions in the $TlBr-Tl_2S-Tl_2S_3$ subsystem: $TlBr-TlS-Tl_2S_3$ (I), $TlBr-TlS-Tl_6SBr_4$ (II), $Tl_6SBr_4-Tl_4S_3-TlS$ (III), and $Tl_2S-Tl_6SBr_4-Tl_4S_3$ (IV). The EMF values measured in heterogeneous phase regions I–IV were processed by the least squares method [25, 26] and presented as the following linear equations:

$$E = a + bT \pm t\left[\frac{S_E^2}{n} + S_b^2\left(T - \overline{T}\right)^2\right]^{1/2}, \quad (2)$$

where n is the number of pairs of E and T values; S_E and S_b are the error variances of the EMF readings and b coefficient, respectively; $\overline{T}$ is the average of the absolute temperature; t is Student's test. At the confidence level of 95% and $n \geq 20$, Student's test is $t \leq 2$ [25, 26]. The composed equation of the mode (2) is presented in Table 1. The experimental data of T_i and E_i and steps of calculation for the phase region II (Table 1), which is of special interest in terms of calculation of thermodynamic functions of the Tl_6SBr_4 compound, are presented in Table 2. The values of $\overline{T}$, a, b, S_E^2, and S_b^2 quantities in (2) were calculated based on Table 2.

Our results of measuring the EMF of the cells type (1) in the $TlBr-Tl_2S-Tl_2S_3$ composition region were fully consistent with the solid-phase equilibrium diagram of the Tl-TlBr-S subsystem [24] (Figure 1). The EMF values at the given temperature, within I–IV three-phase regions, virtually coincide independently of the gross compositions of the right electrodes and vary discontinuously in transition from one of the regions to another. The EMF values at 300 K temperature for the phase areas showing in Table 1 are given in Figure 1.

It should also be noted that numerical values of EMF and equations of their temperature dependence in the phase regions I, III, and IV (Table 1) practically coincide with corresponding data [27] for the binary compounds TlS, Tl_4S_3, and Tl_2S, respectively. This testifies reversibility of the electrochemical cells type (1) and therefore indirectly points to the absence of appreciable regions of solid solutions based on the aforementioned sulfides in the $TlBr-Tl_2S-Tl_2S_3$ subsystem.

The EMF values measured in the phase region II (Table 1) can be assigned to the compound Tl_6SBr_4 and can be used in thermodynamic calculations. The calculation of the relative partial molar functions of thallium from the equation of the temperature dependence of the EMF in this phase region gave the quantities

$$\Delta\overline{G}_{Tl} = -45{,}80 \pm 0{,}07 \text{ kJ} \cdot \text{mol}^{-1};$$

$$\Delta\overline{H}_{Tl} = -46{,}90 \pm 0{,}48 \text{ kJ} \cdot \text{mol}^{-1}; \quad (3)$$

$$\Delta\overline{S}_{Tl} = -3{,}71 \pm 1{,}39 \text{ J} \cdot \text{mol}^{-1} \cdot \text{K}^{-1}$$

which are thermodynamic functions of the potential-forming reaction

$$Tl + 4TlBr + TlS = Tl_6SBr_4 \quad (4)$$

according to Figure 1.

Based on (4), we determined ΔZ^0, the standard thermodynamic functions of formation (ΔG^0, ΔH^0) of Tl_6SBr_4:

$$\Delta Z^0\left(Tl_6SBr_4\right) = \Delta\overline{Z}_{Tl} + 4\Delta Z^0(TlBr) + \Delta Z^0(TlS), \quad (5)$$

where ΔZ^0 is the standard Gibbs free energy $\Delta_f G^0{}_{298}$ and standard enthalpy of formation $\Delta_f H^0{}_{298}$ for the corresponding compound. The standard entropy of Tl_6SBr_4 was calculated as

$$S^o\left(Tl_6SBr_4\right) = \Delta\overline{S}_{Tl} + S^0(Tl) + 4S^0(TlBr) + S^0(TlS). \quad (6)$$

TABLE 2: The calculation steps of the EMF measurements for the $TlBr–TlS–Tl_6SBr_4$ phase region.

T_i, K	E_i, mV	$T_i - \overline{T}$	$E_i\left(T_i - \overline{T}\right)$	$\left(T_i - \overline{T}\right)^2$	E	$E_i - E$
300,5	476,2	−41,90	−19952,78	1755,61	474,57	1,63
301,2	476,1	−41,20	−19615,32	1697,44	474,54	1,56
305,1	473,5	−37,30	−17661,55	1391,29	474,39	−0,89
311,2	474,9	−31,20	−14816,88	973,44	474,16	0,74
313	473,2	−29,40	−13912,08	864,36	474,09	−0,89
317,7	473,4	−24,70	−11692,98	610,09	473,91	−0,51
319,3	473,2	−23,10	−10930,92	533,61	473,85	−0,65
324,1	472,8	−18,30	−8652,24	334,89	473,66	−0,86
328,7	474,1	−13,70	−6495,17	187,69	473,48	0,62
331,2	472,2	−11,20	−5288,64	125,44	473,39	−1,19
335,4	472,1	−7,00	−3304,70	49,00	473,23	−1,13
341,5	473,9	−0,90	−426,51	0,81	472,99	0,91
345,4	472,1	3,00	1416,30	9,00	472,84	−0,74
349,3	473,1	6,90	3264,39	47,61	472,69	0,41
351,6	472,1	9,20	4343,32	84,64	472,61	−0,51
356,1	472,8	13,70	6477,36	187,69	472,43	0,37
360,4	471,1	18,00	8479,80	324,00	472,27	−1,17
364,2	471,2	21,80	10272,16	475,24	472,12	−0,92
366,3	473,1	23,90	11307,09	571,21	472,04	1,06
369,3	472,9	26,90	12721,01	723,61	471,93	0,97
374,2	473,2	31,80	15047,76	1011,24	471,74	1,46
380,8	472,3	38,40	18136,32	1474,56	471,48	0,82
382,9	470,9	40,50	19071,45	1640,25	471,40	−0,50
388,2	470,6	45,80	21553,48	2097,64	471,20	−0,60

TABLE 3: Standard integral thermodynamic functions of ternary compounds in the Tl-X-Hal (X-S, Se, Te; Hal – I, Br, Cl) systems ($T = 298$ K).

Compound	$-\Delta_f G^0$ (298 K) kJ·mol^{-1}	$-\Delta_f H^0$ (298 K)	S^0_{298} J·K^{-1}·mol^{-1}
Tl_6SCl_4	833,5 ± 3,7	928,1 ± 14,0	599 ± 9
Tl_6SBr_4	768,2 ± 2,9	791,3 ± 5,2	644 ± 9
Tl_6SI_4	601,7 ± 2,5	595,1 ± 4,0	672 ± 10
Tl_6SeI_4	613,1 ± 1,5	609,7 ± 2,6	671 ± 5
Tl_5Se_2Cl	392,8 ± 1,1	421,6 ± 5,1	433,9 ± 7,2
Tl_5Se_2Br	374,3 ± 1,0	384,3 ± 2,7	447,6 ± 6,4
Tl_5Se_2I	341,7 ± 0,8	345,3 ± 2,5	449 ± 8
Tl_5Te_2Cl	355,9 ± 1,1	377,1 ± 5,0	474,1 ± 6,8
Tl_5Te_2Br	340,6 ± 1,6	344,5 ± 2,7	483,4 ± 6,2
Tl_5Te_2I	300,4 ± 1,3	301,1 ± 2,3	475,8 ± 6,6

For calculations besides our own $\Delta \overline{Z}_{Tl}$ data, we used the standard integral thermodynamic functions of TlS [27] and TlBr [28] and the standard entropy of thallium [28]:

TlS: $\Delta_f G^0 = -51,6 \pm 0,42$ kJ·mol^{-1}; $\Delta_f H^0 = -52,72 \pm 2,01$ kJ·mol^{-1}; $S^0_{298} = -3,6 \pm 5,8$ J·K^{-1}·mol^{-1}.

$TlBr$: $\Delta_f G^0 = -167,4 \pm 0,6$ kJ·mol^{-1}; $\Delta_f H^0 = -172,7 \pm 0,7$ kJ·mol^{-1}; $S^0_{298} = 122,6 \pm 0,2$ J·K^{-1}·mol^{-1}.

Tl: $S^0_{298} = 64,2 \pm 0,2$ J·mol^{-1}·K^{-1}.

The errors were found by the error accumulation method.

Calculated integral thermodynamic functions of Tl_6SBr_4 compound are given in Table 3. This table also summarizes the thermodynamic functions of other thallium chalcohalides.

3.2. Comparative Review of Thermodynamic Properties of Thallium Chalcohalides. Using obtained values of the standard integral thermodynamic functions of formation and standard entropies of thallium chalcohalides, other fundamental characteristics, standard thermodynamic functions of atomization of these compounds, were calculated. As is known, the atomization thermodynamic functions are quantities that characterize the change of relevant thermodynamic functions during decomposition of a compound to monatomic gas mixture. The atomization energy of ternary compounds was calculated using

$$\Delta_{at} H \left(comp\right) = \sum \Delta_{at} H \left(elem.\right) - \Delta_f H \left(comp\right), \quad (7)$$

where $\Delta_f H$ (comp) is the enthalpy of formation of a compound and $\sum \Delta_{at} H$ (elem.) is the sum of atomization

TABLE 4: Standard thermodynamic functions of atomization of thallium chalcogen-halides.

Compound	$\Delta_{at}G^0$ (298 K)	$\Delta_{at}H^0$ (298 K)	$\Delta_{at}S^0$ (298 K)
	kJ·mol^{-1}		J·K^{-1} mol^{-1}
Tl_5Se_2Cl	1564	1858	988
Tl_5Se_2Br	1531	1816	958
Tl_5Se_2I	1488	1775	962
Tl_5Te_2Cl	1480	1766	960
Tl_5Te_2Br	1442	1728	961
Tl_5Te_2I	1389	1669	939
Tl_6SCl_4	2142	2534	1314
Tl_6SBr_4	1987	2377	1308
Tl_6SI_4	1784	2172	1303
Tl_6SeI_4	1754	2145	1313

TABLE 5: Comparison of thermodynamic functions of atomization, formation from elemental components, and formation from binary compounds of thallium chalcogen-halides.

Compound	$\Delta_{at}G^0$	$-\Delta_fG^0$	$-\Delta_fG^0$ (b.c)	$\Delta_{at}H^0$	$-\Delta_fH^0$	$-\Delta_fH^0$ (b.c)
			kJ·mol^{-1}			
Tl_5Se_2Cl	1564	392,8	17,0	1858	421,6	28,1
Tl_5Se_2Br	1531	374,3	16,1	1816	384,3	26,4
Tl_5Se_2I	1488	334,0	17,9	1775	324,0	15,1
Tl_5Te_2Cl	1480	355,9	12,5	1766	377,1	7,8
Tl_5Te_2Br	1442	340,1	14,3	1728	343,9	10,2
Tl_5Te_2I	1389	296,5	12,8	1669	286,8	12,1
Tl_6SCl_4	2142	833,5	2,8	2534	928,1	3,6
Tl_6SBr_4	1987	787,2	6,6	2377	791,3	8,2
Tl_6SI_4	1784	601,7	9,7	2172	595,1	9,0
Tl_6SeI_4	1754	613,1	16,6	2145	609,7	22,3

enthalpies of elements in a compound. The atomization entropy of ternary compounds was calculated on the following equation:

$$\Delta_{at}S^0 \text{ (comp)} = \sum S^0 \text{ (at.gas.)} - S^0 \text{ (comp)}. \quad (8)$$

Here S^0 (comp) is the absolute standard entropy and $\sum S^0$ (at.gas.) is the sum of the absolute entropies of elemental constituents of the considered compound in a monoatomic gas state.

The Gibbs free energy of atomization of ternary compounds were calculated from Gibbs-Helmholtz equation using the data obtained from (7) and (8):

$$\Delta_{at}G \text{ (comp)} = \Delta_{at}H^0 \text{ (comp)} - T\Delta_{at}S^0 \text{ (comp)}. \quad (9)$$

Results are given in Table 4.

The standard thermodynamic functions of atomization, formation, and formation from appropriate binary compounds (Tl_2X and TlHal) of thallium chalcogen-halides are summarized in Table 5, for comparative analysis.

As can be seen from Table 5, the relationship between the standard thermodynamic functions of atomization, formation, and formation from appropriate binary compounds (Tl_2X and TlHal) for all ternary compounds is as follows:

$$\left|\Delta_{at}Z^0\right| \geq \left|\Delta_fZ^0\right| \geq \left|\Delta_fZ^0 \text{ (b.c.)}\right|, \quad (10)$$

where Z is Gibbs free energy (G) or enthalpy (H).

The atomization Gibbs free energies and enthalpies of all ternary compounds are very large positive quantities. The standard thermodynamic functions of formation of ternary compounds are 3–5 times smaller by absolute value than the proper atomization functions. This is associated with the fact that $\Delta_{at}H^0$ is the minimal energy required to split up the crystal lattice into separate atoms. However, $\Delta_{at}G^0$ is the driving force of the reverse process, combination of monatomic gases to form a crystal lattice.

The standard thermodynamic functions of formation of ternary compounds are much smaller than the Gibbs free energies and enthalpies of formation from binary constituents. The reason of such a sharp distinction can be explained by the fact that formation of a ternary compound from its binary constituents is not accompanied by a considerable energetic change in the system. Moreover, during

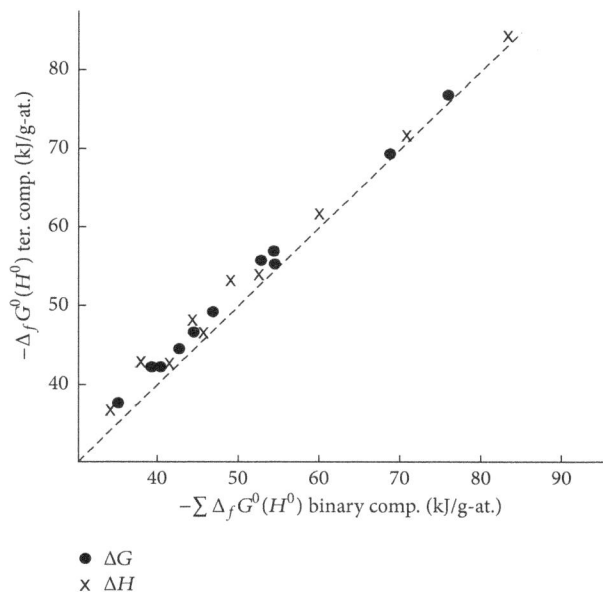

$\bullet$ ΔG
$\times$ ΔH

FIGURE 2: The correlation between $\Delta_f G^0$ and $\Delta_f H^0$ functions of thallium chalcohalides and the sum of the proper functions of binary compounds.

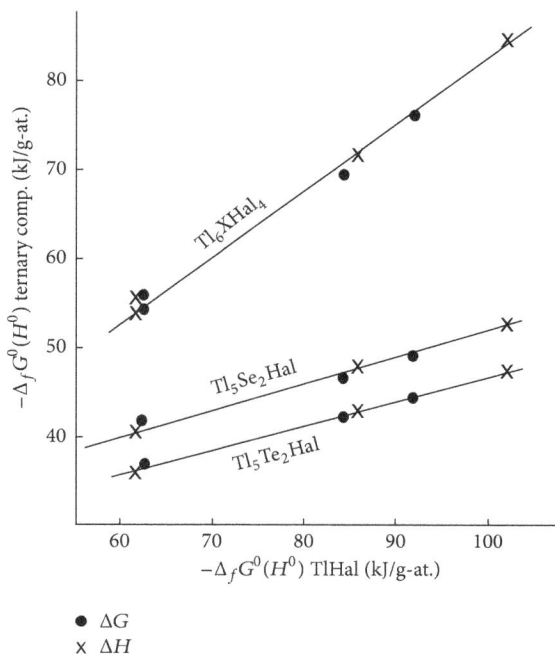

$\bullet$ ΔG
$\times$ ΔH

FIGURE 3: The correlation between $\Delta_f G^0$ and $\Delta_f H^0$ functions of ternary and binary TlHal compounds.

the latter process the oxidation state numbers of elements, consequently the type of chemical bonding do not alter significantly. However, during formation of ternary compound from elemental components, quantities change significantly and therefore total energy of the system decreases sharply.

The high numerical values of the atomization entropy of all ternary chalcogen-halides can be explained by the dramatically increase in disorder during the decomposition of their crystal lattice (Table 5).

3.3. Some Regularities in Thermodynamic Properties of Thallium Chalcogen-Halides.

The comparative analysis of different thermodynamic functions of thallium chalcohalides with degree of ionization of the chemical bonding in those compounds as well as with the proper thermodynamic functions of binary compounds have shown the availability of some regularities.

The conformity between the standard Gibbs free energies and enthalpies of formation of ternary (Tl$_6$XHal$_4$ and Tl$_5$X$_2$Hal) compounds and the sum of the proper functions of binary (TlHal and Tl$_2$X) compounds is demonstrated in Figure 2. As shown in Figure 2, the absolute values of $\Delta_f G^0$ and $\Delta_f H^0$ functions of all ternary compounds are higher (~3–7%) than the sum of the proper functions of binary compounds.

The correlations between $\Delta_f G^0_{298}$ and $\Delta_f H^0_{298}$ functions of ternary and binary TlHal compounds are represented in Figure 3. As can be seen, these dependencies are linear for all compounds of the type Tl$_6$XHal$_4$; however the linear dependencies for Tl$_5$Se$_2$Hal and Tl$_5$Te$_2$Hal compounds with the same halogen atoms differ from each other.

This is due to the fact that, TlHal compounds play a decisive role in the thermodynamic functions of the ternary compounds Tl$_6$XHal$_4$. Reversely, the main contribution in the thermodynamic functions of Tl$_5$X$_2$Hal compounds belongs to thallium chalcogenides Tl$_2$X (X–Se, Te). Since the thermodynamic functions of formation of Tl$_2$Se and Tl$_2$Te compounds are considerably distinctive from each other, the values of relevant functions for Tl$_5$Se$_2$Hal and Tl$_5$Te$_2$Hal ternary compounds also differ by magnitude (Figure 3).

The dependence graphs of the standard thermodynamic functions (Gibbs free energy and enthalpy) of formation and atomization of the ternary compounds Tl$_6$XHal$_4$ and Tl$_5$X$_2$Hal upon the ionization degree (ID) of chemical bonding are demonstrated in Figure 4.

The ID of a chemical bond in thallium chalcohalides was calculated by classical method [29]. For this aim, the chemical bond in CsF compound with the highest value of ID was considered pure ionic type and the difference of the relative electronegativities of elements (Δ_{REN}) in this compound was found to be $\Delta_{REN} = 3{,}2$. Taking into account the equalities $\Delta_{REN} = 0$ and ID = 0% for nonpolar covalent bonds, the following equation was obtained.

$$ID\ (\%) = 31{,}25 \cdot \Delta_{REN}. \qquad (11)$$

Calculation of the Δ_{REN} for thallium chalcohalides is demonstrated by an example of Tl$_6$SBr$_4$ compound.

All thallium atoms in Tl$_6$SBr$_4$ compound are in Tl^{1+} state. The values of relative electronegativities (REN) of Tl, S, and Br that were used in calculations are, respectively, 1,4; 2,6; and 2,9 [29].

(a)

(b)

(c)

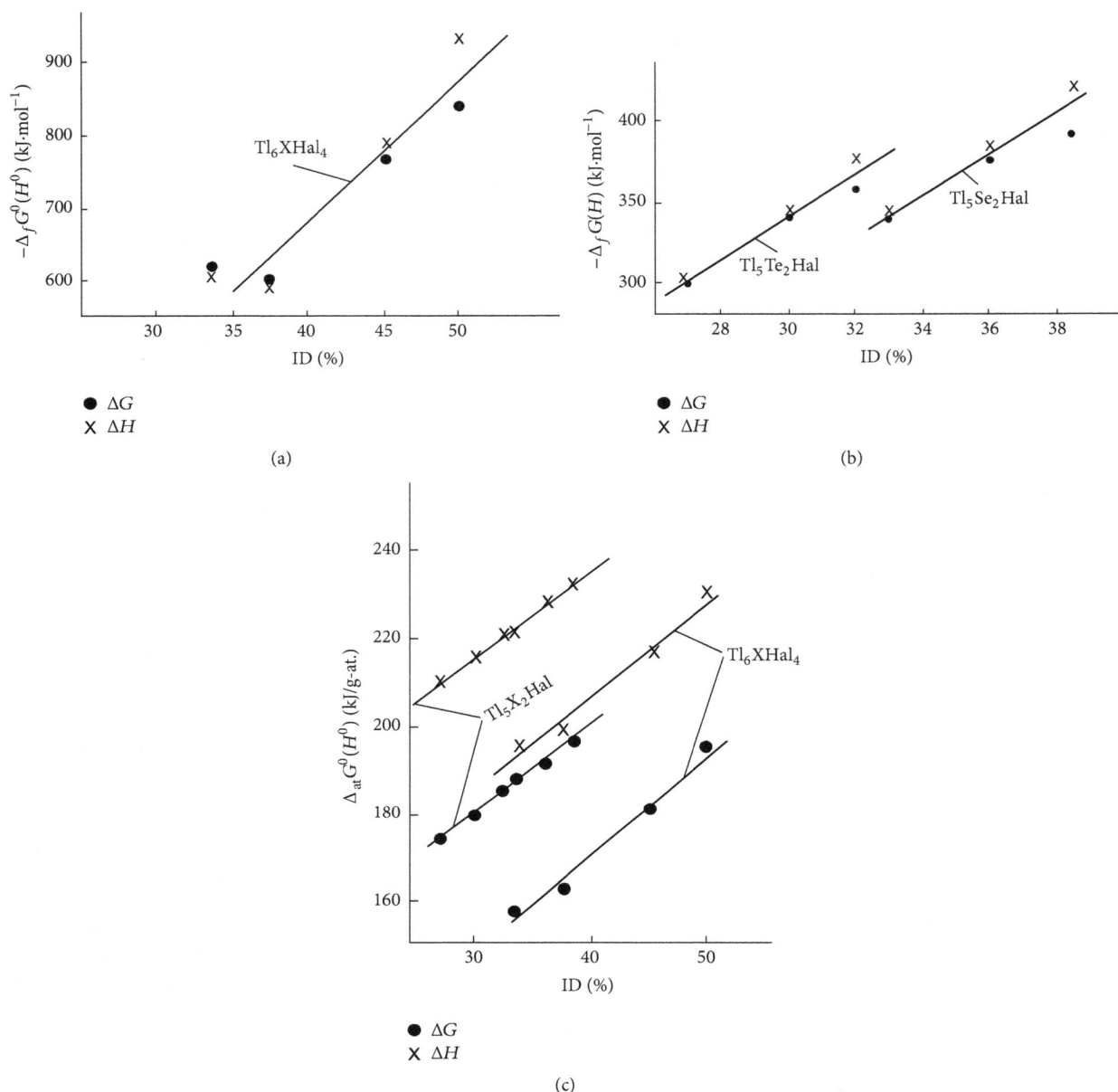

FIGURE 4: The dependence of the standard thermodynamic Gibbs free energy and enthalpies of formation (a, b) and atomization (c) of the ternary compounds on the degree of ionization (ID) of chemical bonding.

Since there are different anions (chalcogen and halogen) in the compound, firstly the average REN for the $[SBr_4]$ anion group and later Δ_{REN} for Tl_6SBr_4 compound were calculated:

$$REN\ [SBr_4] = (2{,}6 + 2{,}9 \cdot 4)/5 = 2{,}84;$$

$$\Delta_{REN}\ Tl_6SBr_4 = 2{,}84 - 1{,}4 = 1{,}44.$$

The averaged ID of the chemical bonds in thallium chalcogen-halides were calculated using (11): ID (%) = 31,25 · Δ_{REN} = 31,25·1,44 = 45%. The calculated Δ_{REN} and ID values for all thallium chalcohalides are summarized in Table 6.

Both thermodynamic properties of ternary compounds have a positive linear tendency with ID of bonding. Since

the lattice energy of substance rises with an increase in the ID of chemical bond, the extension of the above-mentioned thermodynamic functions is natural (Figure 4).

Unlike the energetical by nature thermodynamic functions (ΔG^0, ΔH^0) the atomization entropy of the ternary compounds Tl_6XHal_4 and Tl_5X_2Hal has virtually the same values within a certain error: $S^{at} = 120 \pm 3\ K \cdot mol^{-1} \cdot K^{-1}$ (Figure 5).

It can be explained by the fact that the entropy of atomization is an indicator of the rise of irregularity during decomposition of crystal lattice into monatomic gas mixture. The same numerical value of atomization entropies of all

TABLE 6: Δ_{REN} and ID values for thallium chalcohalides.

Compound	Δ_{REN}	ID, %
Tl_6SCl_4	1,6	50
Tl_6SBr_4	1,44	45
Tl_6SI_4	1,2	37,5
Tl_6SeI_4	1,04	33,5
Tl_5Se_2Cl	1,23	38,5
Tl_5Se_2Br	1,17	36
Tl_5Se_2I	1,07	33
Tl_5Te_2Cl	1,03	32
Tl_5Te_2Br	0,97	30
Tl_5Te_2I	0,87	27

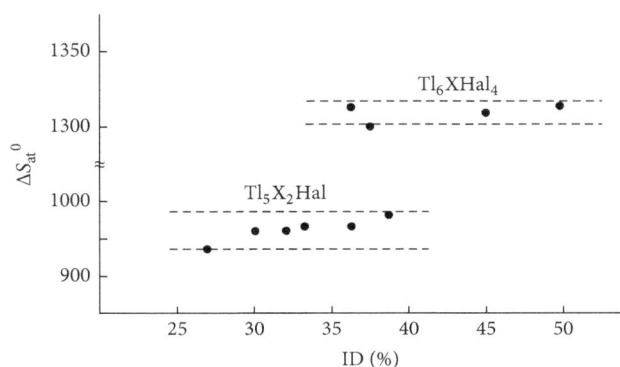

FIGURE 5: The correlation between the atomization entropy of ternary compounds and the ionization degree (ID) of chemical bonding.

considered ternary compounds is the result of the transformation of the system from regular crystallic state with the lowest value of entropy into the irregular atomic gas mixture during the atomization process. The difference between the entropy of a substance (with the same formula) in a crystal state and the entropy of a monoatomic gas mixture is nearly two orders smaller than the value of the atomization entropy and therefore does not affect the magnitude of the latter.

4. Conclusion

By using electromotive force (EMF) measurements a regulated complex of thermodynamic properties for the Tl_6SBr_4 compound was obtained. The data regarding the thermodynamic properties of thallium chalcogen-halides were systematized and comparatively analyzed. Some correlations between thermodynamic functions of thallium chalcogen-halides and their binary constituents as well as ionization degree of chemical bonding were revealed.

Conflicts of Interest

The authors declare that there are no conflicts of interest regarding the publishing of this paper.

References

[1] E. I. Gerzanich, V. A. Lyakhovitskaya, V. M. Fridkin, and B. A. Popovkin, "SbSI and other ferroelectric $A^V B^{VI} C^{VII}$ materials," *Current Topics in Materials Science*, vol. 10, pp. 55–190, 1982.

[2] G. Landolt, S. V. Eremeev, Y. M. Koroteev et al., "Disentanglement of surface and bulk rashba spin splittings in noncentrosymmetric BiTeI," *Physical Review Letters*, vol. 109, no. 11, Article ID 116403, 2012.

[3] K. Kurosaki, A. Kosuga, H. Muta, and S. Yamanaka, "Thermoelectric properties of thallium compounds with extremely low thermal conductivity," *Materials Transactions*, vol. 46, no. 7, pp. 1502–1505, 2005.

[4] S. L. Nguyen, C. D. Malliakas, J. A. Peters et al., "Photoconductivity in Tl_6SI_4: a novel semiconductor for hard radiation detection," *Chemistry of Materials*, vol. 25, no. 14, pp. 2868–2877, 2013.

[5] S. Johnsen, Z. Liu, J. A. Peters et al., "Thallium chalcohalides for X-ray and γ-ray detection," *Journal of the American Chemical Society*, vol. 133, no. 26, pp. 10030–10033, 2011.

[6] R. Blachnic and H. A. Dreisbach, "Tl_6X_4S-ein Neuer Chalkogen halogenid typ in thallium sulfid-thallium halogenid Systemen," Z. Naturforsch, 1981. B.36 (12), s. 1500–1503.

[7] R. Blachnik, H. A. Dreisbach, and B. Engelen, "The system thallous iodide - thallous selenide and the structure of the Tl_6X_4Y compounds," *Zeitschrift fur Naturforschung - Section B Journal of Chemical Sciences*, vol. 38, no. 2, pp. 139–142, 1983.

[8] R. Blachnik and H. A. Dreisbach, "Phase relations in the TlX-Tl_2Se systems (X = Cl, Br, I) and the crystal structure of Tl_5Se_2I," *Journal of Solid State Chemistry*, vol. 52, no. 1, pp. 53–60, 1984.

[9] D. M. Babanly, Y. A. Yusibov, and M. B. Babanly, "Phase equilibria and thermodynamic properties of the system Tl-TlCl-Se," *Russian Journal of Inorganic Chemistry*, vol. 52, no. 5, pp. 753–760, 2007.

[10] E. Yu. Peresh, V. B. Lazarev, V. V. Chigica, and O. I. Kornejchuk, "Homogeneity regions, preparation and properties of Tl_6SI_4, $Tl_5Se_2Br(I)$ monocrytals," *Russian Journal of Inorganic Chemistry*, vol. 27, no. 10, p. 2079, 1991.

[11] D. M. Babanly, A. A. Nadzhafova, M. I. Chiragov, and M. B. Babanly, New tellurium halogenides of thallium. *Kimya Problemleri – Chemical Problems*, 2005. V.2, pp.149-151. (In Azerbaijan).

[12] E. Y. Peresh, V. I. Sidei, and O. V. Zubaka, "Systems based on A_2TeC_6(A-Tl,K,Rb,Cs; C-Br,I) compounds with peritectic interactions," *Russian Journal of Inorganic Chemistry*, vol. 54, no. 2, pp. 315–318, 2009.

[13] M. G. Brik and I. V. Kityk, "Modeling of lattice constant and their relations with ionic radii and electronegativity of constituting ions of AXY cubic crystals (A = K, Cs, Rb, Tl; X = tetravalent cation, Y = F, Cl, Br, I)," *Journal of Physics and Chemistry of Solids*, vol. 72, no. 11, pp. 1256–1260, 2011.

[14] R. Blachnik, H. A. Dreisbach, and J. Pelzl, "The thallous chalcogenides Tl_6X_4Y (X=Cl, Br, I; Y = S,Se)," *Materials Research Bulletin*, vol. 19, no. 5, pp. 599–605, 1984.

[15] V. I. Sidey, O. V. Zubaka, A. M. Solomon, S. V. Kun, and E. Y. Peresh, "X-ray powder diffraction studies of Tl_2TeBr_6 and Tl_2TeI_6," *Journal of Alloys and Compounds*, vol. 367, no. 1-2, pp. 115–120, 2004.

[16] T. Doert, R. Asmuth, and P. Böttcher, "Syntheses and crystal structures of Tl_5Se_2Cl and Tl_5Se_2Br," *Journal of Alloys and Compounds*, vol. 209, no. 1-2, pp. 151–157, 1994.

[17] Binary alloy phase diagrams, Ed. Massalski T. B., second edition. ASM International, Materials park, Ohio. 2 (1990) 3589 p.

[18] Brauer G. (Ed.). Handbuch der Präparativen Anorganischen Chemie. Ferdinand Enke Verlag. Stuttgart, 1975.

[19] N. Y. Turova and A. V. Novoselova, "Alcohol derivatives of the alkali and alkaline earth metals, magnesium, and thallium (I)," *Russian Chemical Reviews*, vol. 34, no. 3, pp. 161–185, 1965.

[20] D. M. Babanly, Z. S. Aliev, F. Y. Dhafarli, and M. B. Babanly, "Phase equilibria in the Tl-TlCl-Te system and thermodynamic properties of the compound Tl_5Te_2Cl," *Russian Journal of Inorganic Chemistry*, vol. 56, no. 3, pp. 442–449, 2011.

[21] D. M. Babanly, "Composition range and thermodynamic properties of Tl_5Se_2Br – based solid solutions," *Inorganic Materials*, vol. 47, no. 6, pp. 583–587, 2011.

[22] D. M. Babanly, I. R. Amiraslanov, A. V. Shevelkov, and D. B. Tagiyev, "Phase equilibria in the Tl-TlI-Se system and thermodynamic properties of the ternary phases," *Journal of Alloys and Compounds*, vol. 644, pp. 106–112, 2015.

[23] D. M. Babanly, "Thermodynamic study of $Tl_5Te_{3-x}Br_x$ solid solutions by EMF measurements," *Inorganic Materials*, vol. 51, no. 4, pp. 326–330, 2015.

[24] M. B. Babanly, G. M. Guseinov, D. M. Babanly, and F. M. Sadygov, "Phase equilibria in the Tl-TlBr-S system," *Russian Journal of Inorganic Chemistry*, vol. 51, no. 5, pp. 810–813, 2006.

[25] M. Babanly, Y. Yusibov, and N. Babanly, "The EMF method with solid-state electrolyte in the thermodynamic investigation of ternary Copper and Silver Chalcogenides. Electromotive force and measurement in several systems. Ed. S. Kara. Intechweb.Org, 2011, pp.57-78. (ISBN 978-953-307-728-4)".

[26] A. G. Morachevskii, G. F. Voronin, and I. B. Kutsenok, Electrochemical Research Methods in Thermodynamics of Metallic Systems. 2003. ITSK "Akademkniga", ISBN 5-94628-064-3, Moscow: 335 p.

[27] V. P. Vasilev, A. V. Nikolskaya, V. V. Chernyshev, and Y. I. Gerasimov, "Thermodynamic properties of of thallium sulfides," Izv. Akad. Nauk SSSR, Neorg. Mater. 1973. V.9(6), pp.900–903.

[28] O. Kubaschewski, C. B. Alcock, and P. Spenser, *Materials Thermochemistry*, Pergamon Press, 1993.

[29] L. Pauling, *General Chemistry*, General Chemistry, San Francisco, Calif, USA, 1970.

14

Experimental Investigation on Embedding Strength Perpendicular to Grain of Parallel Strand Bamboo

Junwen Zhou [1,2] **Dongsheng Huang,** [2] **Yang Song,** [1] **and Chun Ni** [3]

[1] *School of Civil and Architecture Engineering, Changzhou Institute of Technology, Changzhou 213032, China*
[2] *School of Civil Engineering, Nanjing Forestry University, Nanjing 210037, China*
[3] *FPInnovations, Vancouver, BC, Canada V6T 1Z4*

Correspondence should be addressed to Junwen Zhou; zhoujw@czu.cn

Academic Editor: Nadezda Stevulova

Parallel strand bamboo (PSB) is a latest construction material; to know more about mechanical properties of PSB, 5 groups of specimens with difference only in bolt diameter were designed to study the impact of the fastener diameter on embedding strength perpendicular to grain of PSB. Based on the tested result, the feasibility for PSB of the theoretical equation in the American code and European code on embedding strength predication was assessed. A controlled displacement was used to load till specimen failure, the stress-displacement curve of all specimens was obtained in terms of the tested results, and the yielding tested strength based on 5% bolt diameter offset proposed by the American code was found. The tested results showed that the yielding strength perpendicular to grain of PSB was stable, the variable coefficient was between 5.88% and 13.34%, and the average yielding strength values were 80.84 MPa, 77.40 MPa, 76.52 MPa, 74.20 MPa, and 67.01 MPa, respectively, which decreased with the increase of bolt diameter, and the average yielding strength values are larger than the calculated results using theoretical formula. Therefore, the theoretical equation on embedding yielding strength of wood in the American code and European code applies to PSB.

1. Introduction

Nowadays, timber and bamboo buildings again have started gaining attention because low-carbon and ecological concepts are the new architecture tendency. Also, primary construction materials used in modern timber structures are renewable materials and are better than fossil materials such as steel and concrete in thermal performance, which is in conformity with the sustainable development principle [1, 2]; thus, timber buildings have become a new development trend in recent decades. In northern America, timber buildings are already preferred in low-carbon architecture, and almost 90% of low-rise buildings are timber structures. Nowadays, more and more midrise and tall timber buildings have been constructed in some countries, such as the most famous 18-storey Brock Commons Student Residence with a height of 53 meters in UBC [3]. The growth period of original structural wood from the time of planting to harvesting is longer, about decades, which is indirectly proportional to the existing demand, especially in Asian countries.

In addition, constructed wood is sparse due to cutting of trees in early stage; hence, bamboo is preferred for its short harvesting period and identical excellent mechanical performance compared with traditional wood.

Bamboo, widely planted in Asia, especially in Southeast Asia, is "the second forestry resource" because of the suitable temperature and humidity in lower geographic latitude.

Bamboo was already used as a construction material about thousand years ago [4]. Bamboo has a short harvesting period of about 3–5 years [5–8]. Bamboo is easily available and has higher mechanical strength compared to wood [9]; however, because of its thin wall and hollow and limited sections, large scale use of raw bamboo is restricted in building structures. As a result, raw bamboo is commonly used in rural houses, simple bridges, water channels, and some landscape architecture. With the progress in technology, some new engineered bamboo materials, such as

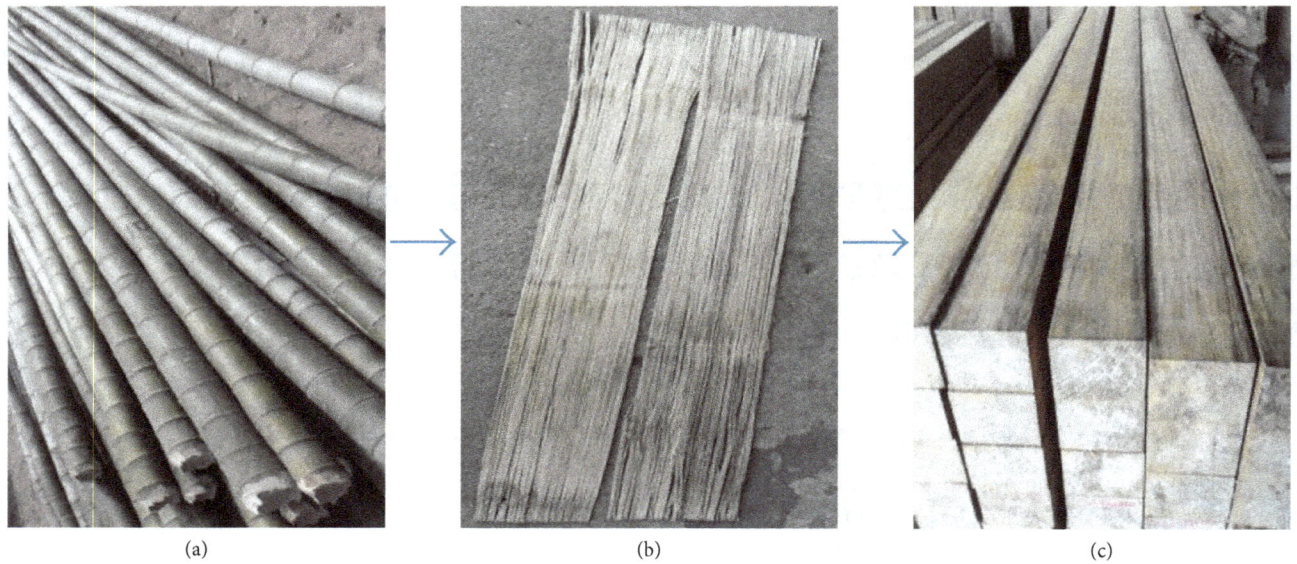

FIGURE 1: Manufacturing process of parallel strand bamboo. (a) Raw bamboo. (b) Bamboo fiber bundle. (c) Parallel strand bamboo.

laminated bamboo [10–13] and parallel strand bamboo (PSB) [14–16], have been developed and those materials have a pronounced breakthrough in dimensions and, therefore, have higher load-bearing capacity and posses a better application prospect in building structures.

PSB is a new engineering material made with raw bamboo. At plant, PSB is fabricated in the manufacturing process expressed in Figure 1; first, original bamboo is cut into certain length according to the demand parallel to grain and, then, is rolled into bamboo fiber bundle which is broken in the longitudinal direction and connected in the latitudinal direction, as shown in Figure 1(b). The bamboo fiber bundle is dried to reach a moisture content between 5% and 10%, immersed in glue well, and then put into billet under pressure and heat, and the fiber is aligned to be in parallel with the length of the member. Once the adhesive of phenolic resin is cured, the bamboo fiber is glued together to form a PSB member. Commonly, the shape of PSB is sheet or prism which is actually determined in terms of practical applications. PSB holds the continuity fiber of bamboo and overcomes some defect of section limit and concentration knot, and as a result, PSB owns the outstanding mechanical performances compared to raw bamboo [16]. A two-storey building has been constructed in China using PSB material [14]. The superiority mentioned above indicates that PSB has a good prospect to be an alternative construction material. At present, some studies on PSB mainly deal with the mechanical performance of small-scale specimens and basic components [15–17], showing that PSB has more superior qualities than engineered wood.

Like wooden buildings, bamboo buildings are also prefabricated structure buildings, and the mechanical performance of connection, such as bearing capacity, and stability are very important, concerning the safety of the overall structure. As a basic connection mode, bolted connection has been used widespread in modern wooden structures, and dowel bearing of wood in connection is the main behavior of bolted connection, impacting the mechanical performance of whole connection. Few studies were performed to determine the four important influence factors, fastener diameter, density of wood, moisture content, and load-to-grain angle, on embedding strength of wood.

Rammer [18] investigated the mechanical performance of embedding strength of hardwoods parallel to grain, and three diameters of steel nail and fastener were tested. The results showed that fastener diameter has affinity on embedding stiffness and no affinity on embedding strength; however, the nail diameter has obvious influence on embedding strength and stiffness. Rammer and Winistorfer [19] proposed the calculating formula, including the moisture content for embedding capacity of wood. Sawata and Yasumura [20] tested the dowel-bearing strength with four kinds of fastener diameter parallel to grain and perpendicular to grain, respectively; the embedding strength is decided based on the 5% bolt diameter offset method and 5 mm displacement maximum load, and the result shows that the embedding strength perpendicular to grain increases with the decrease in dowel diameter. Franke and Magnière [21] investigated the embedding behavior of European hardwoods through the testing analysis, and the influence of the load-to-grain angle and fastener diameter was studied. Schweigler et al. [22] studied the embedding performance of fastener with two kinds of diameter on different load-to-grain in the laminated veneer lumber, and the relative mechanical behavior was obtained. Seri et al. [23] concentrated on the embedding performance of glulam with and without glue line wood with two different fastener diameters, and the results show that the fastener diameter and manufacturing method have pronounced influence on the dowel-bearing strength of the specimen.

The dowel-bearing experiment with superiorities of convenience, practicability, and economy is employed to attain some important data; to the author's knowledge, only few studies are available on embedding strength of PSB perpendicular

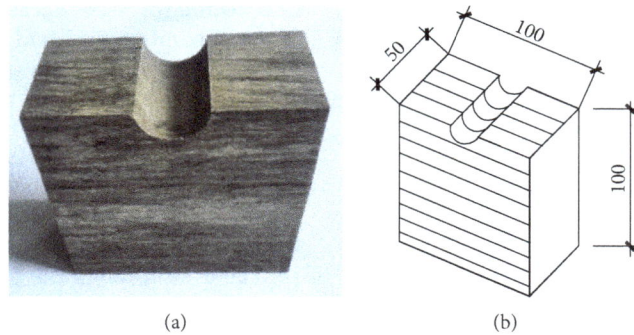

FIGURE 2: Drawing details of the specimen.

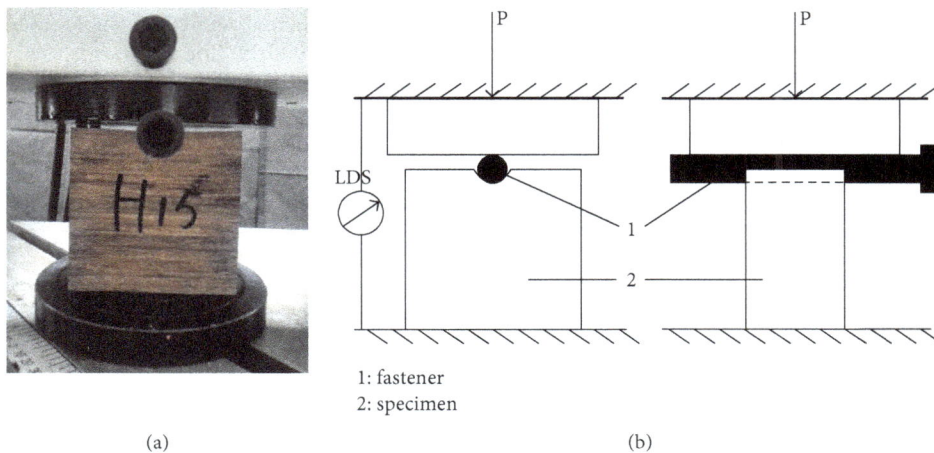

FIGURE 3: Experimental setup.

FIGURE 4: Typical failure mode of the specimen.

to grain, so more works are needed to be done to know about the mechanical behavior of PSB. To achieve the aim of this study, 40 specimens divided into 5 groups in terms of bolt diameter were tested to investigate the dowel-bearing property of PSB; based on the experiment and computational analysis, the formula for calculating the embedding strength will be proposed for PSB perpendicular to grain.

2. Test Procedure

PSB emerged recently as a new construction material; there are no tested criterion and method for PSB, concerning

about similar mechanical performance as engineering wood, and the test method for engineering wood was employed to test the PSB material.

According to ASTM D5764-97a [24], tests were conducted by a square PSB block with a half-circle-hole perpendicular to grain located in the top face and a hole of 1.0 mm larger than the fastener diameter. All the specimens were same in dimension with 100 mm length, 50 mm width, and 100 mm height, except for the magnitude of the circle hole, and a detailed drawing of the specimen is shown in Figure 2. PSB has bigger density because of tremendous pressure in a higher temperature condition during the

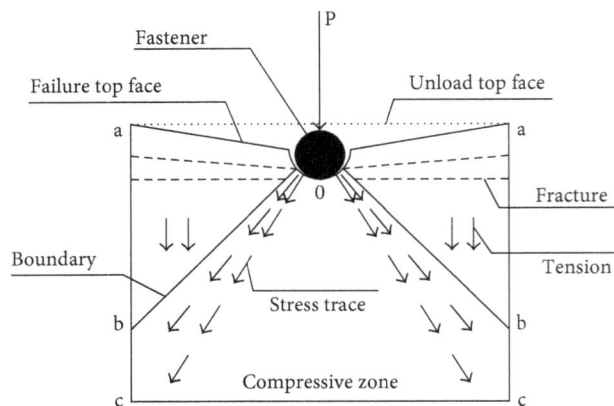

FIGURE 5: Stress state of the specimen.

manufacturing process, and the average tested value of gravity and moisture content are 1.08 g/cm^3 and 9.0%, respectively, in terms of ASTM D2395 [25], and the average tested value of compression strength perpendicular to grain is 62 N/mm^2 according to ASTM D143-09 [26].

According to the experimental method provided by ASTM D5764-97a [24] for evaluating the embedding strength of wood and wood-based material, a controlled displacement of 1 mm/min was adopted to apply the fastener till the specimen failure which can be defined that the side part PSB of hole meeting the loading plate and fastener embedded in PSB or load reaching 85% of the ultimate load. The laser displacement sensor (LDS) was employed to monitor the overall deformation of the specimen in the vertical direction, which is shown in Figure 3. Before the normal test, a small scale load was applied to check all devices at the start of the test.

3. Test Result

3.1. Failure Mode. When failure occurred, all fasteners were submerged into the specimen and the bearing zone under the fastener sank distinctly; unloaded PSB besides the fastener met the loading plate and cracked in the horizontal direction in all the specimens; PSB under the fastener bulged out of the plane; a vertical crack or a diagonal crack or both were observed in the broken specimen; the typical failure mode is shown in Figure 4.

The stress state of the specimen under loading is shown in Figure 5; the load was transferred though fastener from the loading plate to the specimen, and the bamboo fiber including under fastener and inside the stress dispersal boundary line was subjected to compressive force.

As shown in Figure 5, compression stress cannot reach the left and right zones oab, which is far away from the direct compressive zone o and is out of the stress dispersal boundary line; therefore, vertical deformation in zone oab is little. For zone obccb, especially contacting zone o, obvious deformation occurred on account of direct compression of the fastener; vertical deformation in zone bc brought out tension to bamboo fiber in zone obc; as a result, the horizontal fracture parallel to grain came out in zone ab, mainly

near to the top face. Because of concentrated stress under fastener, bamboo fiber had large deformation in the vertical direction out of the grain plane, which results in bulking of bamboo out of the plane. Furthermore, the larger horizontal deformation brought out one or more fractures in bamboo fiber near fastener; with the increase of load, the fracture developed in the direction of two lateral edges and going down or sloping down; at the end, the specimen broke due to fracture which is shown in Figure 4.

Table 1 shows all the tested results. The dowel-bearing strength perpendicular to grain was obtained by the 5% fastener diameter offset method proposed in ASTM D5764-97a [24]. In this way, the load-displacement curve was obtained first in terms of tested data. A straight line fit to the initial linear portion of the load-displacement curve was offset by a deformation of 5% fastener diameter, and the load at which the offset line intersects the load-displacement curve was used as the yield load; finally, the yield load was divided by the bolt diameter and specimen thickness to get the embedding strength, and the method of obtaining yielding strength is shown in Figure 6. Table 1 gives the average value of each group specimen.

3.2. Strength-Displacement Curve. The strength-displacement curve of the 5 group specimens are shown in Figures 7–11. These curves in Figures 7 and 8 are for the bolt diameter of 12 mm and 14 mm, respectively. It was obvious that the curve is similar in tendency and rises as the load increases, and at the end, the load stops due to the meeting between the loading plate and the top face of the specimen.

In two groups specimen of bigger diameter of 20 mm and 24 mm, especially the group of 24 mm diameter, when the curve reached the summit value point, the dowel-bearing capacity started decreasing and obvious discreteness in peak load was observed. The previous phenomenon may be due to the following reasons: for groups H1 and H2 with a small bolt diameter, the interstice in PSB under load started diminishing, before the bolt was immersed into PSB, and the specimen was extruded to be compact; therefore, the embedding capacity increased constantly. Moreover, the loading path is shorter because of small diameter, and the loading will stop when the loading plate meets the top face of the specimen before the specimen was compacted well. This makes that the defect in structure and construction of the specimen did not display during the test.

As for groups H4 and H5 with a bigger bolt diameter, due to the longer loading path, fracture perpendicular to grain because of large deformation of internal bamboo fiber under increasing load comes after the PSB was extruded fully to be dense.

It is also observed that bolts with bigger diameter have larger bearing stiffness, and the bearing stiffness of group H1 is minimum and group H5 is maximum. Because bolts with bigger diameter have larger bearing area, the deformation of the specimens under a larger bolt is less than that of the specimens under a smaller bolt in the same load.

According to the tested results in Table 1, it is shown that even the yielding strength of each group decreased with the

TABLE 1: Tested result.

Specimen group	Number	Diameter of bolt (mm)	①	②	③	④	⑤	⑥ ①/③	⑦ ①/④	⑧ ⑤/①
H1	8	12	80.84	5.88	72.16	68.42	133.26	1.12	1.18	1.65
H2	8	14	77.40	8.42	68.61	63.34	114.30	1.13	1.22	1.48
H3	8	16	76.52	10.63	65.25	59.26	105.35	1.17	1.29	1.38
H4	8	20	74.20	12.20	59.04	53.00	99.32	1.26	1.40	1.34
H5	8	24	67.01	13.34	53.42	48.38	84.54	1.25	1.39	1.26

Note. ① is the average value of yielding strength based on the 5% bolt diameter offset method (MPa). ② is the variable coefficient of tested yielding strength (%). ③ is the calculating value of formula in BS EN 1995 (MPa). ④ is the calculating value of formula in NDS-2015 (MPa). ⑤ is the average value of ultimate tested strength (MPa).

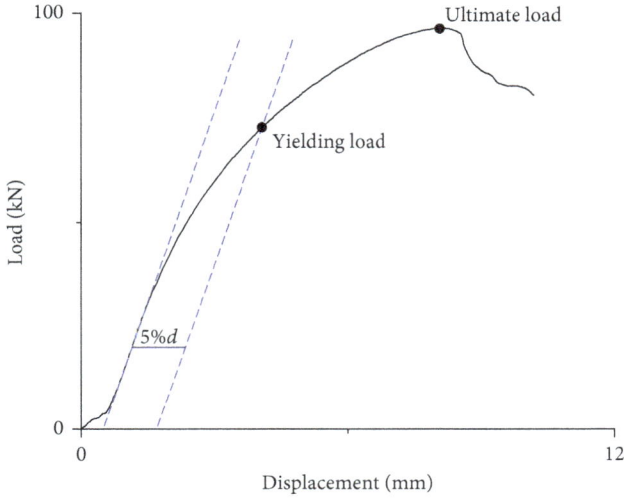

FIGURE 6: Method to evaluate the yielding load.

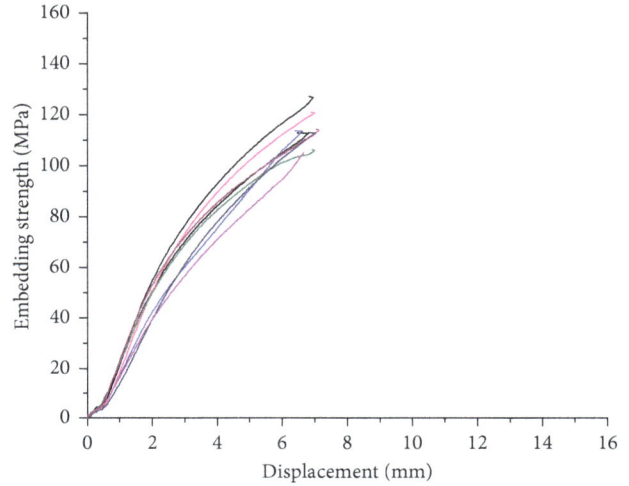

FIGURE 8: Strength-displacement curve of group H2.

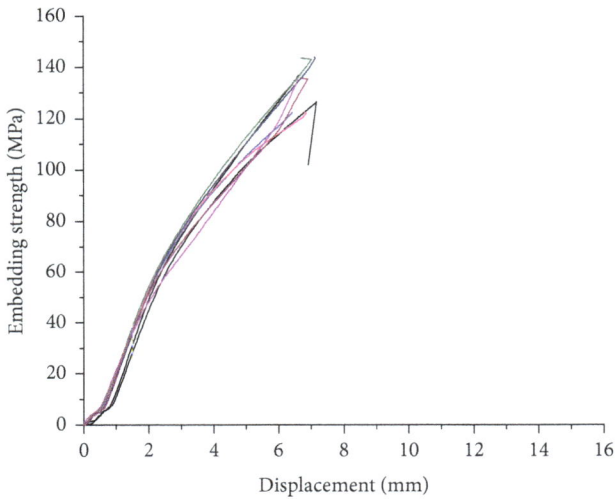

FIGURE 7: Strength-displacement curve of group H1.

increase in bolt diameter, which demonstrates that the magnitude of bolt has pronounced effect on dowel-bearing strength, which is in agreement with the tested results of Sawata and Yasumura [20]. Furthermore, the variable coefficient enlarges as the bolt diameter increases, which are in agreement with regularities of distribution of strength-displacement of the specimen.

In addition, from Table 1, it is also observed that the ratio between the maximum bearing strength and yielding strength of 5% diameter offset decreased with the increase in bolt diameter, which indicated that bolted connection in small bolt diameter that has larger bearing capacity was calculated with the yielding strength of 5% diameter offset.

3.3. Bearing Capacity Perpendicular to Grain. The strength-displacement curve in Figures 7–11 showed that it is not easy to find the obvious yielding point in the curve; therefore, other approaches were needed to seek the yielding strength for engineering design. At present, the yielding strength in dowel-bearing capacity was determined by the 5% diameter offset proposed in ASTM D5764-97a [24], which has already been accepted. The yielding strength value of each specimen is displayed in Figure 12, and the average value of yielding strength of each group specimen is displayed in Table 1. Due to higher cost and more time needed for on-site testing, BS EN 1995-1-1 [27] and NDS-2015 [28] both gave the calculating formula for the yielding value of dowel-bearing strength perpendicular to grain. In the formula proposed in BS EN 1995-1-1 [27], the embedding strength was based on the bolt diameter and material density and is shown as the following formula:

$$f_c = \frac{0.082\,(1 - 0.01d)\rho_k}{k_{90}}. \tag{1}$$

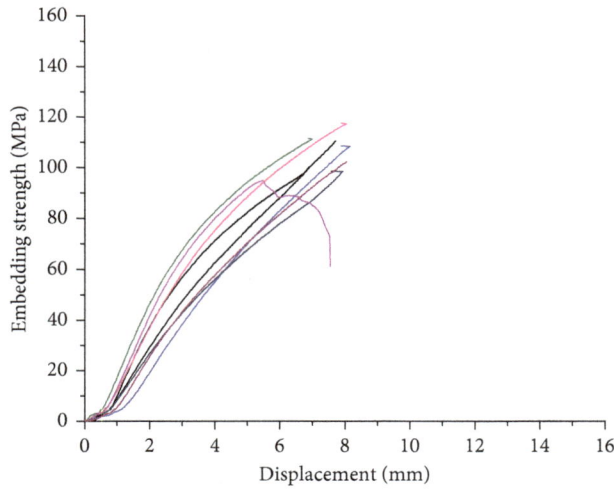

FIGURE 9: Strength-displacement curve of group H3.

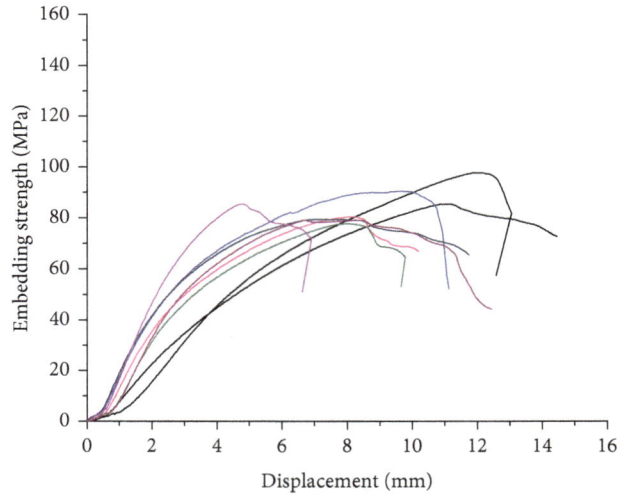

FIGURE 11: Strength-displacement curve of group H5.

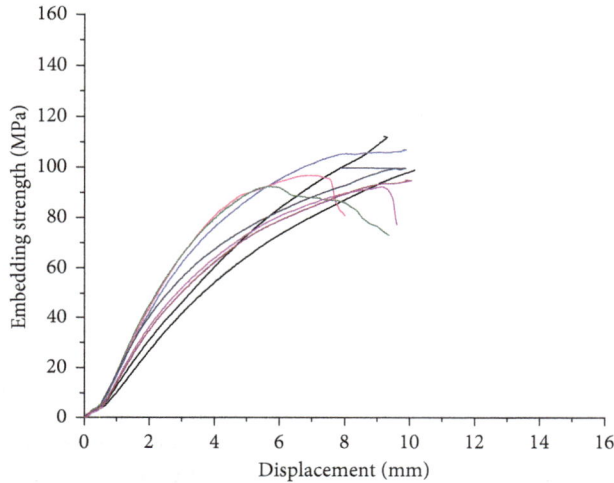

FIGURE 10: Strength-displacement curve of group H4.

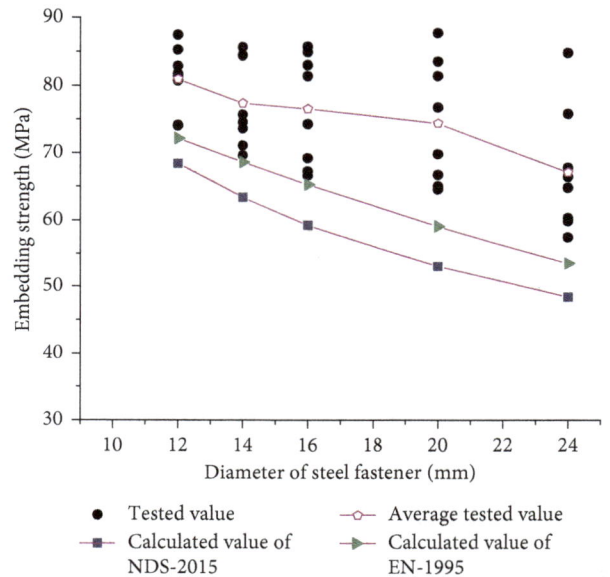

- ● Tested value
- —□— Calculated value of NDS-2015
- —○— Average tested value
- —▶— Calculated value of EN-1995

FIGURE 12: The comparison between the tested value and theoretical value.

For hardwoods,

$$k_{90} = 0.90 + 0.015d, \qquad (2)$$

where f_c is the dowel-bearing strength perpendicular to grain (MPa), d is the bolt diameter (mm), and ρ_k is the density of wood (kg/m^3).

In NDS-2015 [28], the dowel-bearing strength perpendicular to grain also concerning with bolt diameter and wood density is given by the following formula:

$$F_e = \frac{6100G^{1.45}}{\sqrt{D}}, \qquad (3)$$

where F_e is the dowel-bearing strength perpendicular to grain ((psi)^2), D is the bolt diameter (inch), and G is the relative density of wood.

For contrasting analysis of (1) easily, (3) was substituted to obtain the following formula:

$$f_e = \frac{212G^{1.45}}{\sqrt{D}}, \qquad (4)$$

where f_e is the dowel-bearing strength perpendicular to grain (MPa), D is the bolt diameter (mm), and G is the density of wood (g/cm^3).

The value of embedding strength in terms of (1) and (4) is shown in Table 1.

As shown in Table 1, the embedding strength value according to the formula is smaller when bigger bolt diameter was applied, which is in accord with the tendency of tested results. Moreover, the average bearing strength in terms of 5% bolt diameter offsetting is approximately the calculated value for small diameter bolt and is the obvious gap for big diameter bolt.

For PSB, the calculated value based on BS EN 1995-1-1 [27] is larger than that according to NDS-2015 [28] and approached the average tested result.

In Table 1 and Figure 12, the average tested values of 5 groups are all larger than the calculated values according to BS EN 1995-1-1 [27], which indicated that the dowel-bearing yielding strength of PSB perpendicular to grain can be obtained by using the formula proposed in BS EN 1995-1-1 [27].

4. Conclusions

Five group specimens in a total number of 40 were tested to investigate the mechanical behavior of embedding strength of PSB perpendicular to grain, and strength-displacement curves of each specimen were provided in this paper. According to the previous curve, the tested embedding yielding strength was obtained by the 5% bolt diameter offset; meanwhile, the calculated embedding strength in terms of the NDS-2015 15 and BS EN 1995-1-1 was provided. Some conclusions were obtained as following.

(1) The certain relationship between the embedding strength of PSB perpendicular to grain and bolt diameter was observed; the bigger the bolt diameter, the smaller the embedding strength.

(2) The mechanical properties of PSB is stable, and the variable coefficient of embedding strength adopted 5% bolt diameter offset is between 5.88% and 13.34%; moreover, the bigger the bolt diameter, the larger the variable coefficient.

(3) The average tested value of embedding yielding strength of PSB perpendicular to grain in accordance with the 5% bolt diameter is larger than the calculating value by using the formula proposed in NDS-2015 and BS EN 1995-1-1. So the method of using the calculating formula in BS EN 1995-1-1 or NDS-2015 to determine the embedding strength is feasible.

(4) The ratio between the tested ultimate strength and the yielding strength according to 5% bolt diameter offset decreased with the increase of bolt diameter.

Conflicts of Interest

The authors declare that they have no conflicts of interest.

Acknowledgments

The research was supported by the National Natural Science Foundation of China (nos. 51778299 and 51708049), the Project of Housing and Urban-Rural Development Ministry of China (no. 2014-K2-014), the Top-Notch Academic Programs Project of Jiangsu Higher Education Institutions (PPZY2015A041), the Science Foundation of Changzhou Institute of Technology (no. YN1615), and the Changzhou Sci & Tech Program (no. CJ20179043).

References

[1] R. H. Falk, "Wood as a sustainable building material," *Forest Products Journal*, vol. 59, no. 9, pp. 6–12, 2009.

[2] M. Mahdavi, P. L. Clouston, and S. R. Arwade, "Development of laminated bamboo lumber: review of processing, performance, and economical considerations," *Journal of Materials in Civil Engineering*, vol. 23, no. 7, pp. 1036–1042, 2011.

[3] M. Green and J. E. Karsh, *Tall Wood-the Case for Tall Wood Buildings*, Wood Enterprise Coalition, Vancouver, Canada, 2012.

[4] S. Rittironk, *Investigating Laminated Bamboo Lumbers Available Structural Material in Architectural Application*, Illinois Institute of Technology, Chicago, IL, USA, 2009.

[5] S. Amada, Y. Ichikawa, T. Munekata, Y. Nagase, and K. Shimizu, "Fiber texture and mechanical graded structure of bamboo," *Composites Part B: Engineering*, vol. 28, no. 1-2, pp. 13–20, 1997.

[6] A. Porras and A. Maranon, "Development and characterization of a laminate com-posite material from polylactic acid (PLA) and woven bamboo fabric," *Composites Part B: Engineering*, vol. 43, no. 7, pp. 2782–2788, 2012.

[7] W. Fuli, Z. Shao, and W. Yijun, "Mode II interlaminar fracture properties of Moso bamboo," *Composites Part B: Engineering*, vol. 44, no. 1, pp. 242–247, 2013.

[8] L. Tingju, J. Man, J. Zhongguo, H. David, W. Zeyong, and Z. Zuowan, "Effect of surface modification of bamboo cellulose fibers on mechanical properties of cellulose/epoxy composites," *Composites Part B: Engineering*, vol. 51, pp. 28–34, 2013.

[9] F. Correal Juan, "Mechanical properties of Colombian glued laminated bamboo," in *Proceedings of the 1st International Conference on Modern Bamboo Structures (ICBS'07)*, pp. 121–127, Changsha, China, October 2007.

[10] Y. Wei, S. X. Jiang, Q. F. Lv, Q. S. Zhang, L. B. Wang, and Z. T. Lv, "Flexural performance of glued laminated bamboo beams," *Advanced Materials Research*, vol. 168–170, pp. 1700–1703, 2010.

[11] H.-T. Li, Q.-S. Zhang, D.-S. Huang, and A. J. Deeks, "Compressive per-formance of laminated bamboo," *Composites Part B: Engineering*, vol. 54, pp. 319–328, 2013.

[12] S. Arijit, W. Daniel, and M. Skyler, "Structural performance of glued laminated bamboo beams," *Journal of Structural Engineering*, vol. 140, no. 1, p. 04013021, 2014.

[13] H.-T. Li, J.-W. Su, Q.-S. Zhang, A. J. Deeks, and D. Hui, "Mechanical performance of laminated bamboo column under axial compression," *Composites Part B: Engineering*, vol. 79, pp. 374–382, 2015.

[14] Z. Aiping, H. Dongsheng, L. Haitao, and S. Yi, "Hybrid approach to determine mechanical parameters of fibers and matrixes of bamboo," *Construction and Building Materials*, vol. 35, pp. 191–196, 2012.

[15] D. S. Huang, A. P. Zhou, and Y. L. Bian, "Experimental and analytical study on the nonlinear bending of parallel strand bamboo beams," *Construction and Building Materials*, vol. 44, pp. 585–592, 2013.

[16] D. S. Huang, Y. L. Bian, A. P. Zhou et al., "Experimental study on stress-strain relationships and failure mechanisms of parallel strand bamboo made from phyllostachys," *Construction and Building Materials*, vol. 77, pp. 130–138, 2015.

[17] D. S. Huang, Y. L. Bian, D. M. Huang et al., "An ultimate-state-based-model for inelastic analysis of intermediate slenderness PSB columns under eccentrically compressive load," *Construction and Building Materials*, vol. 94, pp. 306–314, 2015.

[18] D. R. Rammer, "Parallel-to-grain dowel-bearing strength of two Guatemalan hardwoods," *Forest Products Journal*, vol. 49, no. 6, pp. 77–87, 2014.

[19] D. R. Rammer and S. G. Winistorfer, "Effect of moisture content on dowel bearing strength," *Wood and Fiber Science*, vol. 33, no. 1, pp. 126–139, 2001.

[20] K. Sawata and M. Yasumura, "Determination of embedding strength of wood for dowel-type fasteners," *Journal of Wood Science*, vol. 48, no. 2, pp. 138–146, 2002.

[21] S. Franke and N. Magnière, "The embedment failure of european beech compared to Spruce wood and standards," *Materials and Joints in Timber Structures*, vol. 9, no. 1, pp. 221–229, 2014.

[22] M. Schweigler, T. K. Bader, G. Hochreiner, G. Unger, and J. Eberhardsteiner, "Load-to-grain angle dependence of the embedment behavior of dowel-type fasteners in laminated veneer lumber," *Construction and Building Materials*, vol. 126, pp. 1020–1033, 2016.

[23] N. A. B. Seri, M. F. B. Nurddin, and R. B. Hassan, "Dowel-bearing strength properties of glulam with and without glue line made of Mengkulang species," in *InCIEC 2015*, pp. 725–734, Springer, Berlin, Germany, 2016.

[24] American Society for Testing and Materials, *Standard Test Method for Evaluating Dowel–Bearing Strength of Wood and Wood-Based Products (ASTM D5764-97a)*, ASTM, West Conshohocken, PA, USA, 2013.

[25] American Society for Testing and Materials, "ASTM D2395 Standard test methods for specific gravity of wood and wood-based materials," in *Annual Book of ASTM Standard*, ASTM, West Conshohocken, PA, USA, 2002.

[26] American Society for Testing and Materials, *ASTM D143-09 Standard Test Methods for Small Clear Specimens of Timber*, ASTM, West Conshohocken, PA, USA, 2009.

[27] British Standards Institution, BS EN 1995-1-1, *Eurocode 5: Design of Timber Structures-Part 1-1: General-Common Rules and Rules for Buildings*, BSI, London, UK, 2004.

[28] American National Standards Institute, *NDS-2015 National Design Specification for Wood Construction*, American Forest and Paper Association, Washington, DC, USA, 2012.

15

Superposed Incremental Deformations of an Elastic Solid Reinforced with Fibers Resistant to Extension and Flexure

Chun IL Kim (iD)

Department of Mechanical Engineering, University of Alberta, Edmonton, Alberta T6G 1H9, Canada

Correspondence should be addressed to Chun IL Kim; cikim@ualberta.ca

Guest Editor: Syed W. Haider

A comprehensive linear model for an elastic solid reinforced with fibers resistant to extension and flexure is presented. This includes the analysis of both unidirectional and bidirectional fiber-reinforced composites subjected to in-plane deformations. Within the prescription of the superposed incremental deformations, the fiber kinematics are approximated and used to determine the Euler equilibrium equations. The constraints of bulk incompressibility and admissible boundary conditions are also discussed for completeness. In particular, the complete systems of differential equations are obtained for the cases of Neo-Hookean and Mooney–Rivlin types of materials from which analytical solutions can be obtained.

1. Introduction

The mechanics of microstructured solids have consistently been the subject of intense research [1–5] for their practical importance in materials science and engineering. In particular, a considerable amount of attention is committed to the development of continuum models and analyses in an effort to predict the mechanical responses of fiber-matrix systems subjected to external forces and/or induced deformations (see, for example, [6, 7] and references therein). Continuum-based approaches postulate continuous distribution of fibers within the matrix materials so as to establish the idealized description of homogenized fiber-matrix composites. This is framed in the setting of anisotropic elasticity where the response function depends on the first gradient of deformations, typically augmented by the constraints of bulk incompressibility and/or fiber inextensibility. The latter condition often results highly constrained prediction models so that the corresponding deformation fields are essentially kinematically determinate, especially that arises in fibers [6, 7]. The approach has clear advantages in the prediction of deformation profiles of the system via the deformation mapping of an individual fiber, yet rather insufficient in the estimation of overall material properties of the fiber-matrix system. Nonetheless, the aforementioned models have been widely adopted in the analysis of composite materials for their merit in the continuum description and the associated mathematical framework [5–7]. For the estimation of resultant properties, one may also consider multiscale modeling method which integrates the material properties of continua assigned in different length scales by means of the Cauchy–Born rule. Examples of such practice can be found in the works of Shahabodini et al. [8, 9].

A considerable advance in the continuum theory of fiber-reinforced solids was made in recent years. This includes the incorporation of the bending resistance of fibers into the models of deformations where elastic resistance is assigned to changes in curvature of the fibers [10–12]. More precisely, the fibers are regarded as convected curves so that the bending deformation of fibers can be formulated via the second gradient of deformations explicitly [13]. The concept has been successfully adopted in a wide range of problems arising in materials science [14–17], and the mathematical perspective of the subject is discussed in [18–20]. The authors in [21] proposed a general theory for the mechanics of an elastic solids with fibers resistant to flexure, stretch, and twist within the simplified setting of the Cosserat theory of nonlinear elasticity [10, 22].

Further, the second gradient theory of elasticity for the mechanics of meshed structures is presented in [23–25]. To this end, the authors in [26–29] developed continuum-based models in the analysis of fiber-reinforced composites, where the extension and bending resistance of fibers are incorporated via the computations of the first and second gradient of deformations. The majority of the aforementioned studies address nonlinear continuum theory of fiber composites so that little has been devoted to the development of compatible linear models describing the mechanics of an elastic medium reinforced with fibers.

In the present work, we present comprehensive linear theory of the strain gradient elasticity for the mechanics of fiber-reinforced composites with fibers resistant to bending and extension. The kinematics of fibers is approximated with the prescription of the superposed incremental deformations. We formulate, in cases of Neo-Hookean types of materials, complete expressions of the linearized Piola stresses from which the Euler equilibrium equations and the admissible boundary conditions are obtained. Bidirectional fiber composites are accommodated via the decomposition of the deformation gradient tensor along the directions of fibers. In addition, the reduction of simultaneous differential equations into a single differential equation is demonstrated by utilizing the compatibility conditions of the response function. More importantly, we show that, even with the introduction of the second gradient of deformations, two different bases (referential and current coordinate) do indeed merge for "small" deformations superposed in large. Lastly, Mooney–Rivlin types of materials are elaborated where we show that the resulting Euler equilibrium equations are of the same form as those obtained from the Neo-Hookean model. The corresponding Piola stresses, on the other hand, are distinguished in that they demonstrate clear dependency on both the first and second invariant of the deformations gradient tensor. The obtained model can be easily adopted in the study of composite structures subjected to small in-plane deformations. For example, in the design stage of crystalline nanocelluloses (CNCs) composite, the overall responses and deformation profiles can be predetermined by utilizing the proposed model. In addition, the underlined theory can also be extended to the deformation analysis of lipid membranes (e.g., budding and thickness distension) [30–32], where phospholipids (microstructure) are aligned to the normal direction of the membrane, and therefore, their deformations can be mapped and computed using the proposed model.

Throughout the paper, we make use of a number of well-established symbols and conventions such as A^T, A^{-1}, A^*, and $\mathrm{tr}(A)$. These are the transpose, the inverse, the cofactor, and the trance of a tensor A, respectively. The tensor product of vectors is indicated by interposing the symbol $\otimes$, and the Euclidian inner product of tensors $\mathbf{A}$ and $\mathbf{B}$ is defined by $\mathbf{A} \cdot \mathbf{B} = \mathrm{tr}(\mathbf{AB}^\mathrm{T})$; the associated norm is $|\mathbf{A}| = \sqrt{\mathbf{A} \cdot \mathbf{A}}$. The symbol $|*|$ is also used to denote the usual Euclidian norm of three vectors. Latin and Greek indices take values in $\{1, 2\}$ and, when repeated, are summed over their ranges. Lastly, the notation $\mathbf{F}_\mathbf{A}$ stands for the tensor-valued derivatives of a scalar-valued function $\mathbf{F}(\mathbf{A})$.

2. Incremental Elastic Deformations and Equilibrium Equations

The incremental deformation is defined by (see, for example, [33] and the references therein) the following equation:

$$\chi = \chi_\mathrm{o} + \varepsilon\dot\chi, \quad |\varepsilon| \ll 1, \tag{1}$$

where $(*)_\mathrm{o}$ denotes configurations of $*$ evaluated at $\varepsilon = 0$ and $(\dot *) = \partial(*)/\partial\varepsilon$. In the forthcoming derivations, we define $\dot\chi = (\partial\chi/\partial\varepsilon) = \mathbf{u}$ for the sake of convenience and clarity. From equation (1), the gradient of the deformation function $\chi(\mathbf{X})$ can be approximated up to the leading order as shown in the following equation:

$$\mathbf{F} = \frac{\partial(\chi_\mathrm{o} + \varepsilon\dot\chi)}{\partial\mathbf{X}} = \mathbf{F}_\mathrm{o} + \varepsilon\nabla\mathbf{u}, \tag{2}$$

where $\dot{\mathbf{F}} = (\partial\dot\chi/\partial\mathbf{X}) = \nabla\mathbf{u}$.

In the above equation, we assume that the body is initially undeformed and stress free at $\varepsilon = 0$, that is, $\mathbf{F}_\mathrm{o} = \mathbf{I}$ and $\mathbf{P}_\mathrm{o} = 0$, where $\mathbf{P}$ is the Piola stress. Thus, equation (2) becomes the following equation:

$$\mathbf{F} = \frac{\partial(\chi_\mathrm{o} + \varepsilon\dot\chi)}{\partial\mathbf{X}} = \mathbf{I} + \varepsilon\nabla\mathbf{u}, \tag{3}$$

and successively yields

$$\mathbf{F}^{-1} = \mathbf{I} - \varepsilon\nabla\mathbf{u} + o(\varepsilon), \tag{4}$$

which can be found by using the identity $\mathbf{FF}^{-1} = \mathbf{I}$. The determinant of $\mathbf{F}$ can be approximated similarly as

$$J = J_o + \varepsilon\dot{J} + o(\varepsilon), \tag{5}$$

where we evaluate $\dot{J} = (J_\mathrm{F})_\mathrm{o} \cdot \dot{\mathbf{F}} = J_\mathrm{o}\mathrm{tr}(\mathbf{F}_\mathrm{o}^{-1}\dot{\mathbf{F}})$ and $\dot{\mathbf{F}} = (\mathrm{grad}\mathbf{u})\mathbf{F}_\mathrm{o}$. Since $\mathrm{tr}(\mathbf{F}_\mathrm{o}^{-1}\dot{\mathbf{F}}) = \mathrm{tr}(\mathrm{grad}\mathbf{u}) = \mathrm{div}\mathbf{u}$, we obtain from equation (5) that

$$J = 1 + \varepsilon\mathrm{div}\mathbf{u} + o(\varepsilon). \tag{6}$$

In addition, the Euler equilibrium equation can be expanded as the following equation:

$$\mathrm{Div}(\mathbf{P}) = \mathrm{Div}(\mathbf{P}_\mathrm{o})1 + \varepsilon\mathrm{Div}\dot{\mathbf{P}} + o(\varepsilon) = 0. \tag{7}$$

Dividing the above equation by ε and letting $\varepsilon \longrightarrow 0$, we obtain

$$\mathrm{Div}\dot{\mathbf{P}} = 0, \tag{8}$$

which serves as the linearized Euler equilibrium equation. The expression of Piola stresses in the case of fiber composites reinforced with fibers resistant to flexure is given by [27]:

$$\mathbf{P} = W_\mathrm{F} - p\mathbf{F}^* - C\mathrm{Div}(\mathbf{g} \otimes \mathbf{D} \otimes \mathbf{D}), \tag{9}$$

where W, p, and C are the energy density function, Lagrange multiplier, and material constant of fibers (C = constant), respectively. Also, $\mathbf{g}$ is the geodesic curvature of fibers' trajectory (i.e., $\mathbf{g} = \mathbf{G}(\mathbf{D} \otimes \mathbf{D})$) and $\mathbf{D}$ is the initial director filed of fibers' where $\mathbf{D} = (\partial\mathbf{X}(S)/\partial S$. Here, S is the arc length parameter on the reference configuration. In general, most of fibers are straight prior to deformations. Even slightly curved fibers can be regarded as "fairly straight" fibers, considering their length scales with respect to that of matrix materials. Thus, from here and forthcoming derivations, it is assumed that

$$\frac{\partial \mathbf{D}}{\partial \mathbf{X}} = 0, \tag{10}$$

$$\dot{\mathbf{D}} = \frac{\partial \mathbf{D}}{\partial \varepsilon} = 0.$$

Accordingly, we find $\mathrm{Div}(\mathbf{D}) = 0$ and thereby reduce equation (10) to the following equation:

$$\mathbf{P} = W_{\mathrm{F}} - p\mathbf{F}^* - C\nabla\mathbf{g}(\mathbf{D}\otimes\mathbf{D}), \tag{11}$$

where the last term of the above equation is obtained by using the identity $g_{i,B}D_A D_B(\mathbf{e}_i\otimes\mathbf{E}_A) = \nabla\mathbf{g}(\mathbf{D}\otimes\mathbf{D})$. Thus, from equation (11), $\dot{\mathbf{P}}$ can be evaluated as the following equation:

$$\begin{aligned}
\dot{\mathbf{P}} &= (W_{\mathrm{F}})^{\cdot} - (p\mathbf{F}^*)^{\cdot} - [C\nabla\mathbf{g}(\mathbf{D}\otimes\mathbf{D})]^{\cdot} \\
&= W_{\mathrm{FF}}\cdot\dot{\mathbf{F}} - \dot{p}\mathbf{F}_{\mathrm{o}}^* - p_{\mathrm{o}}\dot{\mathbf{F}}^* - C\nabla\dot{\mathbf{g}}(\mathbf{D}\otimes\mathbf{D}).
\end{aligned} \tag{12}$$

In the case of an incompressible Neo-Hookean material, the energy density function is given by the following equation:

$$W(\mathbf{F}) = \frac{\mu}{2}(\mathbf{F}\cdot\mathbf{F} - 3), \tag{13}$$

and we find

$$\dot{\mathbf{P}} = \mu\dot{\mathbf{F}} - \dot{p}\mathbf{F}_{\mathrm{o}}^* - p_{\mathrm{o}}\dot{\mathbf{F}}^* - C\nabla\dot{\mathbf{g}}(\mathbf{D}\otimes\mathbf{D}), \quad \because W_{\mathrm{FF}} = \mu. \tag{14}$$

In view of equations (8) and (14), the linearized Euler equilibrium equation then satisfies

$$\mathrm{Div}(\mu\dot{\mathbf{F}}) - \mathrm{Div}(\dot{p}\mathbf{F}_{\mathrm{o}}^*) - \mathrm{Div}(p_{\mathrm{o}}\dot{\mathbf{F}}^*) - C\mathrm{Div}(\nabla\dot{\mathbf{g}}(\mathbf{D}\otimes\mathbf{D})) = 0. \tag{15}$$

The evaluation of equation (15) is essential to extract boundary value problems (BVPs); nonetheless, the details are often heavily omitted in the literature (see, for example, [26–29]). To see this, we first compute the following equation:

$$\mathrm{Div}(\mu\dot{\mathbf{F}}) = \mathrm{Div}(\mu\nabla\mathbf{u}) = \mu u_{i,AA}\mathbf{e}_i. \tag{16}$$

In the above equation, caution needs to be taken, in which $\nabla(*)$ and $\mathrm{div}(*)$ are the operators in the reference frame. Although, there are no clear distinction between the reference and current configurations for "*small*" incremental deformations, the mathematical procedure should reasonably address their connections especially when dealing with tensors with mixed bases (e.g., $\nabla\mathbf{u}$, $\mathbf{F}$). The collapse of bases for the present problem will be discussed in the later sections. The second term in equation (15) becomes

$$\begin{aligned}
\mathrm{Div}(\dot{p}\mathbf{F}_{\mathrm{o}}^*) &= \mathrm{Div}[\dot{p}(\mathbf{F}_{\mathrm{o}}^*)_{iA}\mathbf{e}_i\otimes\mathbf{E}_A] = [\dot{p}(\mathbf{F}_{\mathrm{o}}^*)_{iA}]_{,A}\mathbf{e}_i \\
&= \dot{p}_{,A}(\mathbf{F}_{\mathrm{o}}^*)_{iA}\mathbf{e}_i = \dot{p}_i\mathbf{e}_i,
\end{aligned} \tag{17}$$

where we use the Piola identity (i.e., $\mathbf{F}_{iA,A}^* = 0$) and $(\mathbf{F}_{\mathrm{o}}^*)_{iA} = \delta_{iA}$. Similarly, it can be easily shown that

$$\mathrm{Div}(p_{\mathrm{o}}\dot{\mathbf{F}}^*) = \left[(p_{\mathrm{o}})_A\dot{\mathbf{F}}_{iA}^* + p_{\mathrm{o}}(\dot{\mathbf{F}}^*)_{iA,A}\right]\mathbf{e}_i = (p_{\mathrm{o}})_A\dot{\mathbf{F}}_{iA}^*\mathbf{e}_i. \tag{18}$$

However, in order to recover the initial stress free state at $\varepsilon = 0$, we require from equations (11) and (13) that

$$\mathbf{P}_{\mathrm{o}} = \mu\mathbf{F}_{\mathrm{o}} - p_{\mathrm{o}}\mathbf{F}_{\mathrm{o}}^* - C\nabla\mathbf{g}_{\mathrm{o}}(\mathbf{D}\otimes\mathbf{D}) = \mu\mathbf{I} - p_{\mathrm{o}}\mathbf{I} = 0, \tag{19}$$

and thus find $p_o = \mu = $ constant. Therefore, equation (18) becomes

$$\begin{aligned}
\mathrm{Div}(p_{\mathrm{o}}\dot{\mathbf{F}}^*) &= \mu_A\dot{\mathbf{F}}_{iA}^*\mathbf{e}_i + \mu\left(\mathrm{Div}(\mathbf{F}^*)\right)^{\cdot} = 0, \\
&\quad \because (p_{\mathrm{o}})_A = (\mu)_A = 0,
\end{aligned} \tag{20}$$

and $\mathrm{Div}(\mathbf{F}^*) = 0$ (Piola's identity). Also, $\mathrm{Div}(\nabla\dot{\mathbf{g}}(\mathbf{D}\otimes\mathbf{D}))$ can be evaluated as

$$\begin{aligned}
\mathrm{Div}\left[\dot{g}_{i,B}D_A D_B(\mathbf{e}_i\otimes\mathbf{E}_A)\right] &= (\dot{g}_{i,B}D_A D_B)_A\mathbf{e}_i \\
&= \dot{G}_{iAB,CD}D_A D_B D_C D_C\mathbf{e}_i,
\end{aligned} \tag{21}$$

where $g_i = G_{iAB}D_A D_B$ (see [27]) and $\mathbf{G} = G_{iAB}(\mathbf{e}_i\otimes\mathbf{E}_A\otimes\mathbf{E}_B)$ is the second gradient of deformations (i.e., $\nabla\mathbf{F} = \mathbf{G}$). Consequently, by substituting equations (16), (17), (20), and (21) into equation (15), the linearized Euler equation can be obtained as follows:

$$\begin{aligned}
\left(\mu u_{i,AA} - \dot{p}_i - Cu_{i,ABCD}D_A D_B D_C D_D\right)\mathbf{e}_i &= 0\mathbf{e}_i, \\
\because \dot{G}_{iAB} = \dot{\mathbf{F}}_{i,AB} &= u_{i,AB}.
\end{aligned} \tag{22}$$

For a single family of fibers (i.e., $\mathbf{D} = \mathbf{E}_1$, $D_1 = 1$, $D_2 = 0$), equation (22) reduces to

$$\left(\mu(u_{i,11} + u_{i,22}) - \dot{p}_i - Cu_{i,1111}\right)\mathbf{e}_i = 0\mathbf{e}_i. \tag{23}$$

3. Boundary Conditions and Solution to the Linearized System

The corresponding boundary conditions are given by [27]

$$\mathbf{t} = \mathbf{PN} - \frac{d}{ds}[C\mathbf{g}(\mathbf{D}\cdot\mathbf{T})(\mathbf{D}\cdot\mathbf{N})],$$

$$\mathbf{m} = C\mathbf{g}(\mathbf{D}\cdot\mathbf{N})^2, \tag{24}$$

$$\mathbf{f} = C\mathbf{g}(\mathbf{D}\cdot\mathbf{T})(\mathbf{D}\cdot\mathbf{N}),$$

where $\mathbf{t}$, $\mathbf{m}$, and $\mathbf{f}$ are, respectively, the expressions of edge tractions, edge moments, and the corner forces. Further, $\mathbf{N}$ and $\mathbf{T}$ are unit normal and tangent to the boundary. The "*small*" increment of boundary forces are then computed as follows:

$$\dot{\mathbf{t}} = \dot{\mathbf{P}}\mathbf{N} - \frac{d}{ds}[C\dot{\mathbf{g}}(\mathbf{D}\cdot\mathbf{T})(\mathbf{D}\cdot\mathbf{N})],$$

$$\dot{\mathbf{m}} = C\dot{\mathbf{g}}(\mathbf{D}\cdot\mathbf{N})^2, \tag{25}$$

$$\dot{\mathbf{f}} = C\dot{\mathbf{g}}(\mathbf{D}\cdot\mathbf{T})(\mathbf{D}\cdot\mathbf{N}).$$

The expression of $\dot{\mathbf{P}}$ can be obtained from equation (14) that

$$\dot{\mathbf{P}} = \dot{\mathbf{P}}_{iA}\left(\mathbf{e}_i \otimes \mathbf{E}_A\right)$$
$$= \left[\mu u_{i,A} - \dot{p}\left(\mathbf{F}_{iA}^*\right)_o - p_o \dot{\mathbf{F}}_{iA}^* - C\dot{g}_{i,B}D_AD_B\right]\left(\mathbf{e}_i \otimes \mathbf{E}_A\right), \tag{26}$$

$$\dot{\mathbf{g}}_i\mathbf{e}_i = \dot{G}_{iAB}D_AD_B\mathbf{e}_i = \dot{F}_{iA,B}D_AD_B\mathbf{e}_i = u_{i,AB}D_AD_B\mathbf{e}_i. \tag{27}$$

In order to apply boundary tractions (e.g., $\dot{\mathbf{P}}_{11}$), it is necessary to compute equation (26) as a function of $\mathbf{u} = \mathbf{u}(X_1, X_2)$. For the purpose, we first find the following equation:

$$\left(\mathbf{F}_{iA}^*\right)_o = J_o\left(\mathbf{F}_{iA}^T\right)_o^{-1} = \left(\delta_{iA}\right)^{-1} = \delta_{iA}, \tag{28}$$

where $\left(\mathbf{F}_{iA}\right)_o = \delta_{iA}$ at $\varepsilon = 0$. Also, to compute $\dot{\mathbf{F}}_{iA}^*\left(\mathbf{e}_i \otimes \mathbf{E}_A\right)$, we use the chain rule in the form of the following equation:

$$\dot{\mathbf{F}}_{iA}^*\left(\mathbf{e}_i \otimes \mathbf{E}_A\right) = \left(\mathbf{F}_{\mathbf{F}}^*\right)_o\left[\dot{\mathbf{F}}\right] = \left(\frac{\partial \mathbf{F}_{iA}^*}{\partial \mathbf{F}_{jB}}\right)_o \dot{\mathbf{F}}_{kC}\left(\mathbf{e}_i \otimes \mathbf{E}_A \otimes \mathbf{e}_j \otimes \mathbf{E}_B\right)\left(\mathbf{e}_k \otimes \mathbf{E}_C\right)$$
$$= \left(\frac{\partial \mathbf{F}_{iA}^*}{\partial \mathbf{F}_{jB}}\right)_o u_{j,B}\left(\mathbf{e}_i \otimes \mathbf{E}_A\right). \tag{29}$$

Here, the expression of $\left(\partial \mathbf{F}_{iA}^*/\partial \mathbf{F}_{jB}\right)$ can be found via the connection [34]:

$$J\left(\frac{\partial \mathbf{F}_{iA}^*}{\partial \mathbf{F}_{jB}}\right) = \mathbf{F}_{iA}^*\mathbf{F}_{jB}^* - \mathbf{F}_{jA}^*\mathbf{F}_{iB}^*. \tag{30}$$

Thus, at $\varepsilon = 0$, we have from the above equation that

$$\left(\frac{\partial \mathbf{F}_{iA}^*}{\partial \mathbf{F}_{jB}}\right)_o = \delta_{iA}\delta_{jB} - \delta_{jA}\delta_{iB}, \quad (\delta_{iA}: \text{ Kronecker delta}), \tag{31}$$

and thereby obtain

$$\left(\frac{\partial \mathbf{F}_{iA}^*}{\partial \mathbf{F}_{jB}}\right)_o u_{j,B} = \left(\delta_{iA}\delta_{jB} - \delta_{jA}\delta_{iB}\right)u_{j,B} = \delta_{iA}\left(\text{Div}\mathbf{u}\right) - u_{A,i}. \tag{32}$$

Consequently, substituting equations (27), (28), and (32) into equation (26) furnishes

$$\dot{\mathbf{P}}_{iA}\left(\mathbf{e}_i \otimes \mathbf{E}_A\right) = \left[\mu u_{i,A} - \dot{p}\delta_{iA} - p_o\left(\delta_{iA}\left(\text{Div}\mathbf{u}\right) - u_{A,i}\right)\right.$$
$$\left. - Cu_{i,BCD}D_AD_BD_CD_D\right]\left(\mathbf{e}_i \otimes \mathbf{E}_A\right), \tag{33}$$

from which the expression of boundary tractions is completely determined in terms of $\mathbf{u}$.

Remark 1. In the above equation, $\delta_{iA}\delta_{jB}u_{j,B}$ is interpreted as $\delta_{iA}u_{B,B}$ resulting $\delta_{iA}\delta_{jB}u_{j,B} = \delta_{iA}\text{Div}\mathbf{u}$ in the reference configuration (Eulerian). However, one may also find $\delta_{iA}\delta_{jB}u_{j,B} = u_{j,j}$ and thus obtain $\delta_{iA}\delta_{jB}u_{j,B} = \delta_{iA}\text{div}\mathbf{u}$ in the current configuration (Lagrangian). This confirms the well-known result from the linear elasticity theory that the bases collapse in the event of small deformations superposed on large (i.e., $\text{Div}\mathbf{u} = \text{div}\mathbf{u}$ and $\mathbf{E}_A = \mathbf{e}_i$). In the present problem, the result allows one to formulate linearized Euler equations without conflicting bases mismatch especially

when operating linear transform of mixed basis tensors. Although there are no clear distinctions between the current and deformed configurations, caution needs to be taken that the Euler equation ($\text{Div}\dot{\mathbf{P}}$) is, in principal, defined in the reference frame together with the boundary conditions.

Now, in view of equations (5) and (6), the constraints of bulk incompressibility reduce to the following equation:

$$(J-1)^\cdot = \dot{J} = \mathbf{F}_o^* \cdot \dot{\mathbf{F}} = \text{tr}\left(\mathbf{F}_o^{-1}\dot{\mathbf{F}}\right) = \text{tr}\left(\text{grad}\mathbf{u}\right) = \text{div}\mathbf{u} = 0. \tag{34}$$

Equation (34) serves as the linearized bulk incompressibility condition (i.e., $u_{i,i} = 0$), which needs to be solved together with equation (23). In addition, since $\text{Div}\mathbf{u} = \text{Div}\mathbf{u}$ for small deformations (see Remark 1), we find $\text{Div}\mathbf{u} = \text{Div}\mathbf{u} = 0$ and thereby reduce equation (33) to the following equation:

$$\dot{\mathbf{P}}_{iA}\left(\mathbf{e}_i \otimes \mathbf{E}_A\right) = \left[\mu u_{i,A} - \dot{p}\delta_{iA} + p_o u_{A,i}\right.$$
$$\left. - Cu_{i,BCD}D_AD_BD_CD_D\right]\left(\mathbf{e}_i \otimes \mathbf{E}_A\right), \tag{35}$$

which serves as an explicit form of the linearized Piola stress for elastic solids reinforced with single family of fibers. In particular, if the fibers' directions are either normal or tangential to the boundary (i.e., $(\mathbf{D} \cdot \mathbf{T})(\mathbf{D} \cdot \mathbf{N}) = 0$), equation (25) becomes

$$\dot{\mathbf{t}} = \dot{\mathbf{P}}\mathbf{N},$$
$$\dot{\mathbf{m}} = C\dot{\mathbf{g}}(\mathbf{D} \cdot \mathbf{N})^2, \tag{36}$$
$$\dot{\mathbf{f}} = 0,$$

and therefore, we compute, for example,

$$\dot{\mathbf{t}}_i\mathbf{e}_i = \left(\mu u_{i,A} - \dot{p}\delta_{iA} + p_o u_{A,1} - Cu_{i,111}\delta_{1A}D_1D_1D_1\right)\left(\mathbf{e}_i \otimes \mathbf{E}_A\right)\mathbf{E}_1$$
$$= \left(\mu u_{i,1} - \dot{p}\delta_{i1} + p_o u_{1,i} - Cu_{i,111}\right)\mathbf{e}_i, \tag{37}$$

where $\mathbf{N}$ is assumed to be parallel to the fiber's director field (i.e., $\mathbf{N} = \mathbf{D} = \mathbf{E}_1$). Lastly, from equations (27) and (36), the expression of boundary moments are obtained by

$$\dot{m}_i\mathbf{e}_i = Cu_{i,AB}D_AD_B\mathbf{e}_i. \tag{38}$$

3.1. Solutions to the Linearized System. In the case of single family of fibers (i.e., $\mathbf{D} = \mathbf{E}_1$), the linearized system of partial differential equations (PDEs) is given by

$$\left(\mu\left(u_{i,11} + u_{i,22}\right) - \dot{p}_i - Cu_{i,1111}\right)\mathbf{e}_i = 0, \tag{39}$$
$$u_{i,i} = 0.$$

The second one in the above equation can be automatically satisfied by introducing the following scalar field φ:

$$\mathbf{u} = \mathbf{k} \times \nabla\varphi, \quad \mathbf{k}\,(\text{unit normal}),$$
$$u_i = \varepsilon_{\lambda i}\varphi_\lambda, \tag{40}$$

so that $u_{1,1} + u_{2,2} = \varphi_{12} - \varphi_{21} = 0$. Now, we recast the first one of equation (39) and thereby find the following equation:

$$\dot{p}_i \mathbf{e}_i = \left[\mu \varepsilon_{\lambda i} \left(\varphi_{\lambda 11} + \varphi_{\lambda 22} \right) - C \varepsilon_{\lambda i} \varphi_{\lambda 1111} \right] \mathbf{e}_i. \qquad (41)$$

In addition, using the compatibility conditions of p_i (i.e., $\dot{p}_{ij} = \dot{p}_{ji}$), the first one of equation (41) becomes

$$\dot{p}_{21} - \dot{p}_{12} = \mu \left(\varphi_{1111} + 2\varphi_{1122} + \varphi_{2222} \right) - C \left(\varphi_{11} + \varphi_{22} \right)_{1111} = 0. \qquad (42)$$

Consequently, equation (42) further reduces to

$$\nabla H - \alpha H_{1111} = 0, \quad \text{where } H = \Delta \varphi,$$

$$\alpha = \frac{C}{\mu} > 0. \qquad (43)$$

An analytical solution of the above exists (see [27]) and is completely determined by imposing admissible boundary conditions as discussed in equation (25). For example, the symmetric bending can be imposed as

$$\dot{\mathbf{m}} = \dot{m}_1 \mathbf{e}_1 + \dot{m}_2 \mathbf{e}_2 = C u_{1,11} \mathbf{e}_1 + 0 \mathbf{e}_2 = -C \varphi_{112} \mathbf{e}_1, \qquad (44)$$

which serves as the boundary conditions for the equation (42).

4. Extensible Fibers

The Piola stress in the case of initially straight and extensible fibers is given by Zeidi and Kim [28]:

$$\mathbf{P} = W_{\mathbf{F}} + \frac{E}{2} \left(\mathbf{FD} \cdot \mathbf{FD} - 1 \right) \left(\mathbf{FD} \otimes \mathbf{D} \right) - p\mathbf{F}^* - C\nabla \mathbf{g} \left(\mathbf{D} \otimes \mathbf{D} \right), \qquad (45)$$

where E is the elastic modulus of fibers (extension). Further, the fibers' stretch λ can be computed as

$$\lambda^2 = \mathbf{FD} \cdot \mathbf{FD}. \qquad (46)$$

In the case of Neo-Hookean type materials (see equation (13)), equation (45) becomes

$$\mathbf{P} = \mu \mathbf{F} + \frac{E}{2} \left(\mathbf{FD} \cdot \mathbf{FD} - 1 \right) \left(\mathbf{FD} \otimes \mathbf{D} \right) - p\mathbf{F}^* - C\nabla \mathbf{g} \left(\mathbf{D} \otimes \mathbf{D} \right). \qquad (47)$$

Thus, equations (1) and (47) can be approximated as

$$\dot{\mathbf{P}} = \mu \dot{\mathbf{F}} + E \left(\dot{\mathbf{F}} \mathbf{D} \cdot \mathbf{F}_o \mathbf{D} \right) \left(\mathbf{F}_o \mathbf{D} \otimes \mathbf{D} \right)$$

$$+ \frac{1}{2} E \left(\mathbf{F}_o \mathbf{D} \cdot \mathbf{F}_o \mathbf{D} - 1 \right) \left(\dot{\mathbf{F}} \mathbf{D} \otimes \mathbf{D} \right) - p_o \dot{\mathbf{F}}^* \qquad (48)$$

$$- \dot{p} \mathbf{F}_o^* - C\nabla \dot{\mathbf{g}} \left(\mathbf{D} \otimes \mathbf{D} \right).$$

Since $\mathbf{F}_o = \mathbf{F}_o^* = \mathbf{I}$, the above equation further reduces to the following equation:

$$\dot{\mathbf{P}} = \mu \dot{\mathbf{F}} + E \left(\dot{\mathbf{F}} \mathbf{D} \cdot \mathbf{ID} \right) \left(\mathbf{ID} \otimes \mathbf{D} \right) - p_o \dot{\mathbf{F}}^* - \dot{p} \mathbf{I} - C\nabla \dot{\mathbf{g}} \left(\mathbf{D} \otimes \mathbf{D} \right), \qquad (49)$$

where $\mathbf{F}_o \mathbf{D} \cdot \mathbf{F}_o \mathbf{D} = \mathbf{D} \cdot \mathbf{D} = 1$. To obtain the desired expression, it is required to compute the following equation:

$$\mathbf{ID} = \delta_{iA} \left(\mathbf{e}_i \otimes \mathbf{E}_A \right) D_B \mathbf{E}_B = \delta_{iA} \delta_{AB} D_B \mathbf{e}_i = D_i \mathbf{e}_i. \qquad (50)$$

In the above equation, the initial director field ($\mathbf{D}$) is represented by the current frame (i.e., $\mathbf{e}_i$). However, $\mathbf{D}$ is, in principal, a vector in the reference coordinate (i.e., $\mathbf{D} = D_A \mathbf{E}_A$). This is due to the collapse of bases as discussed in Remark 1. Caution needs to be taken when applying the Einstein summation. Thus, we obtain from equations (32), (34), (49), and (50) that

$$\dot{\mathbf{P}}_{iA} \left(\mathbf{e}_i \otimes \mathbf{E}_A \right) = \left[\mu u_{i,A} + E u_{j,B} D_i D_j D_A D_B + p_o u_{A,i} \right.$$

$$\left. - \dot{p} \delta_{iA} - C u_{i,BCD} D_A D_B D_C D_D \right] \left(\mathbf{e}_i \otimes \mathbf{E}_A \right), \qquad (51)$$

where $\dot{\mathbf{F}} \mathbf{D} \cdot \mathbf{ID} = \dot{\mathbf{F}}_{jB} D_j D_B = u_{j,B} D_j D_B$. Equation (51) can be used as the expression of boundary tractions in the case of extensible fibers. For example, if $\mathbf{D} = \mathbf{E}_1$ (single family of fibers), the above equation yields the following equation:

$$\dot{\mathbf{P}}_{iA} \left(\mathbf{e}_i \otimes \mathbf{E}_A \right) = \left[\mu u_{i,A} + E u_{1,1} \delta_{i1} \delta_{A1} + p_o u_{A,i} \right.$$

$$\left. - \dot{p} \delta_{iA} - C u_{i,111} \delta_{A1} \right] \left(\mathbf{e}_i \otimes \mathbf{E}_A \right). \qquad (52)$$

Further, from equations (8) and (49), the corresponding equilibrium equation then satisfies the following equation:

$$\text{Div} \left(\dot{\mathbf{P}} \right) = \text{Div} \left(\mu \dot{\mathbf{F}} \right) - \text{Div} \left(\dot{p} \mathbf{F}_o^* \right) + \text{Div} \left(E \left(\dot{\mathbf{F}} \cdot \mathbf{ID} \right) \left(\mathbf{ID} \otimes \mathbf{D} \right) \right)$$

$$- \text{Div} \left(p_o \dot{\mathbf{F}}^* \right) - C\text{Div} \left(\nabla \dot{\mathbf{g}} \left(\mathbf{D} \otimes \mathbf{D} \right) \right) = 0. \qquad (53)$$

The evaluation of the above equation is well discussed through equations (16)–(21) except the stretch term which is now equated as

$$\text{Div} \left(E \left(\dot{\mathbf{F}} \cdot \mathbf{ID} \right) \left(\mathbf{ID} \otimes \mathbf{D} \right) \right) = \text{Div} \left(E u_{j,B} D_i D_j D_A D_B \left(\mathbf{e}_i \otimes \mathbf{E}_A \right) \right)$$

$$= E u_{j,AB} D_i D_j D_A D_B \mathbf{e}_i. \qquad (54)$$

Therefore, we obtain the following equation:

$$\mu u_{i,AA} - \dot{p}_i + E u_{j,AB} D_i D_j D_A D_B$$

$$- C u_{i,ABCD} D_A D_B D_C D_D \mathbf{e}_i = 0. \qquad (55)$$

The linearized boundary conditions in the case of extensible fibers remain intact (see [28]), except the expression of $\dot{\mathbf{P}}$ where the explicit expression is obtained in equation (52). Thus, for example, in the case of unidirectional fiber composited where a boundary vector $\mathbf{N}$ is parallel to the fiber's director field (i.e., $\mathbf{N} = \mathbf{D} = \mathbf{E}_1$), the complete set of equations can be found as

$$\left(\mu u_{i,11} - \dot{p}_i + E u_{1,1} \delta_{i1} - C u_{i,1111} \right) \mathbf{e}_i = 0,$$

$$u_{i,i} = 0. \qquad (56)$$

Also, the corresponding boundary conditions are given by

$$\dot{t}_i \mathbf{e}_i = \dot{\mathbf{P}} \mathbf{E}_1 = \dot{P}_{i1} \mathbf{e}_i,$$

$$\dot{m}_i \mathbf{e}_i = C\dot{g} (\mathbf{E}_1 \cdot \mathbf{E}_1)^2 = C u_{i,11} \mathbf{e}_i, \qquad (57)$$

$$\dot{f}_i \mathbf{e}_i = 0 \mathbf{e}_i,$$

where from equation (52), $\dot{P}_{iA} = [\mu u_{i,A} + E u_{1,1} D_i D_A - p_o (\delta_{iA} u_{1,1} - u_{A,i}) - \dot{p}\delta_{iA} - C u_{i,111} D_A]\,(\mathbf{e}_i \otimes \mathbf{E}_A)$. Lastly, we note here that the value of p_o, in the case of extensible fibers, can be obtained by evaluating the corresponding stresses (see equation (47)) at $\varepsilon = 0$ which is again found to be $p_o = \mu$. By applying the similar scheme as applied in Section 3.1, we rearrange equation (56) and thereby obtain

$$\dot{p}_{21} - \dot{p}_{12} = \Delta \left[\Delta\varphi - \frac{C}{\mu}\varphi_{1111} \right] + \frac{E}{\mu}\varphi_{1122} = 0. \qquad (58)$$

The solution of equation (58) is not accommodated by conventional methods such as the separation of variables method, Fourier transform, and polynomial solutions. Instead, a particular form of solution ($\varphi = X(x)\sin(my)$) can be proposed, inspired by the modified Helmholtz equation.

5. Extensible Bidirectional Fibers

In the forthcoming derivations, we confine our attention to the case of initially orthogonal fiber families. The bidirectional fibers can be accommodated by using the following decompositions of the deformation gradient:

$$\mathbf{F} = \lambda l \otimes \mathbf{L} + \gamma m \otimes \mathbf{M},$$

$$\mathbf{M} \cdot \mathbf{L} = 0, \qquad (59)$$

where $\mathbf{L}$ and $\mathbf{M}$ are the director fields of each fiber family in the reference configuration and l and m are their counterparts in the current configuration. The fibers' stretches are computed similarly as in equation (46) that

$$\lambda^2 = \mathbf{FL} \cdot \mathbf{FM},$$

$$\gamma^2 = \mathbf{FM} \cdot \mathbf{FM}. \qquad (60)$$

Equations (59) and (60) complete the deformation mapping, for example, $\mathbf{L} = L_A \mathbf{E}_A$ and $\mathbf{l} = l_i \mathbf{e}_i$ to yield

$$\lambda l_i \mathbf{e}_i = \mathbf{FL} = F_{iA} L_A \mathbf{e}_i,$$

$$\because \mathbf{M} \cdot \mathbf{L} = 0 \text{ for orthogonal fiber families.} \qquad (61)$$

Further, by employing the variational principles, the Piola stress for the bidirectional and extensible fiber composites can be found as

$$\mathbf{P} = W_{\mathbf{F}} + \frac{E_1}{2} ((\mathbf{FL} \cdot \mathbf{FL} - 1)(\mathbf{FL} \otimes \mathbf{L}))$$

$$+ \frac{E_2}{2} ((\mathbf{FM} \cdot \mathbf{FM} - 1)(\mathbf{FM} \otimes \mathbf{M})) - C_1 \nabla \mathbf{g}_1 (\mathbf{L} \otimes \mathbf{L})$$

$$- C_2 \nabla \mathbf{g}_2 (\mathbf{M} \otimes \mathbf{M}) - p\mathbf{F}^*, \qquad (62)$$

where E_i and C_i are the material constants of fiber families, respectively, to extension and flexure. Applying the similar

scheme as in Section 4, we approximate equation (62) and successively obtain

$$\dot{P}_{iA}(\mathbf{e}_i \otimes \mathbf{E}_A) = \big[\mu u_{i,A} + E_1 u_{j,B} L_i L_j L_A L_B + E_2 u_{j,B} M_i M_j M_A M_B$$

$$- C_1 u_{i,BCD} L_A L_B L_C L_D - C_2 u_{i,BCD} M_A M_B M_C M_D$$

$$+ p_o u_{A,i} - \dot{p}\delta_{iA} \big] (\mathbf{e}_i \otimes \mathbf{E}_A). \qquad (63)$$

Therefore, equations (8) and (63) furnish

$$0\mathbf{e}_i = \dot{P}_{iA,A}\mathbf{e}_i = \mu u_{i,AA} - \dot{p}_i + E_1 u_{j,AB} L_i L_j L_A L_B$$

$$+ E_2 u_{j,AB} M_i M_j M_A M_B - C_1 u_{i,ABCD} L_A L_B L_C L_D$$

$$- C_2 u_{i,ABCD} M_A M_B M_C M_D \mathbf{e}_i, \qquad (64)$$

where we evaluate

$$\mathrm{Div}\big[p_o u_{A,i}(\mathbf{e}_i \otimes \mathbf{E}_A) \big] = p_o u_{A,iA}\mathbf{e}_i = p_o (u_{A,A})_i \mathbf{e}_i = 0,$$

$$\because u_{A,A} = 0 \,(\text{bulk incompressibility}), \qquad (65)$$

$$\mathrm{Div}\big[\dot{p}\delta_{iA}(\mathbf{e}_i \otimes \mathbf{E}_A) \big] = \dot{p}_A \delta_{iA}\mathbf{e}_i = \dot{p}_i\mathbf{e}_i.$$

Compatible results can be also found directly from equations (17) and (20) (i.e., $\mathrm{Div}(\dot{p}\mathbf{F}_o^*) = \dot{p}_i\mathbf{e}_i$ and $\mathrm{Div}(p_o\dot{\mathbf{F}}^*) = 0$). In the case of bidirectional and extensible fibers, the corresponding boundary conditions are given by (see [26]) the following equation:

$$\mathbf{t} = \mathbf{PN} - \frac{d}{ds}\big[C_1 \mathbf{g}_1 (\mathbf{L} \cdot \mathbf{T})(\mathbf{L} \cdot \mathbf{N}) + C_2 \mathbf{g}_2 (\mathbf{M} \cdot \mathbf{T})(\mathbf{M} \cdot \mathbf{N}) \big],$$

$$\mathbf{m} = C_1 \mathbf{g}_1 (\mathbf{L} \cdot \mathbf{N})^{2+} + C_2 \mathbf{g}_2 (\mathbf{M} \cdot \mathbf{N})^2,$$

$$\mathbf{f} = C_1 \mathbf{g}_1 (\mathbf{L} \cdot \mathbf{T})(\mathbf{L} \cdot \mathbf{N}) + C_2 \mathbf{g}_2 (\mathbf{M} \cdot \mathbf{T})(\mathbf{M} \cdot \mathbf{N}). \qquad (66)$$

Now, we approximate the above boundary conditions and thus obtain

$$\dot{\mathbf{t}} = \dot{\mathbf{P}}\mathbf{N} - \frac{d}{ds}\big[C_1 \dot{\mathbf{g}}_1 (\mathbf{L} \cdot \mathbf{T})(\mathbf{L} \cdot \mathbf{N}) + C_2 \dot{\mathbf{g}}_2 (\mathbf{M} \cdot \mathbf{T})(\mathbf{M} \cdot \mathbf{N}) \big],$$

$$\dot{\mathbf{m}} = C_1 \dot{\mathbf{g}}_1 (\mathbf{L} \cdot \mathbf{N})^{2+} + C_2 \dot{\mathbf{g}}_2 (\mathbf{M} \cdot \mathbf{N})^2,$$

$$\dot{\mathbf{f}} = C_1 \dot{\mathbf{g}}_1 (\mathbf{L} \cdot \mathbf{T})(\mathbf{L} \cdot \mathbf{N}) + C_2 \dot{\mathbf{g}}_2 (\mathbf{M} \cdot \mathbf{T})(\mathbf{M} \cdot \mathbf{N}). \qquad (67)$$

If the fibers' directions are aligned with the axes of Cartesian coordinates (i.e., $\mathbf{L} = \mathbf{E}_1$ and $\mathbf{M} = \mathbf{E}_2$) and are either normal or tangential to the boundary (i.e., $(\mathbf{L} \cdot \mathbf{T})(\mathbf{L} \cdot \mathbf{N}) = (\mathbf{M} \cdot \mathbf{T})(\mathbf{M} \cdot \mathbf{N}) = 0$), equations (64) and (67) further reduces to the following equation:

$$[\mu u_{i,AA} - \dot{p}_i + E_1 u_{1,11}\delta_{i1} + E_2 u_{2,22}\delta_{i2}$$

$$- C_1 u_{i,1111} - C_2 u_{i,1111}]\mathbf{e}_i = 0,$$

$$\dot{\mathbf{t}} = \dot{\mathbf{P}}\mathbf{N}, \qquad (68)$$

$$\dot{\mathbf{m}} = C_1 \dot{\mathbf{g}}_1 (\mathbf{L} \cdot \mathbf{N})^2 + C_2 \dot{\mathbf{g}}_2 (\mathbf{M} \cdot \mathbf{N})^2,$$

$$\dot{\mathbf{f}} = 0,$$

which together with the incompressibility condition ($u_{i,i} = 0$) constitutes the complete set of the PDEs system to solve the final deformation profiles. Similarly, equation (63) now becomes

$$\dot{\mathbf{P}}_{iA}\left(\mathbf{e}_i \otimes \mathbf{E}_A\right) = \left[\mu u_{i,A} + E_1 \overrightarrow{u}_{1,1}\delta_{i1}\delta_{A1} + E_2 u_{2,2}\delta_{i2}\delta_{A2}\right.$$
$$- C_1 u_{i,111}\delta_{A1} - C_2 u_{i,222}\delta_{A2}$$
$$\left. + p_0 u_{A,i} - \dot{p}\delta_{iA}\right]\left(\mathbf{e}_i \otimes \mathbf{E}_A\right).$$

(69)

For example, if the unit normal of a boundary is $\mathbf{N} = \mathbf{E}_1$, the boundary traction can be computed as follows:

$$\dot{\mathbf{t}}_i \mathbf{e}_i = \dot{\mathbf{P}}\mathbf{E}_1 = \left[\mu u_{i,A} + E_1 u_{1,1}\delta_{i1}\delta_{A1} + E_2 \overrightarrow{u}_{2,2}\delta_{i2}\delta_{A2}\right.$$
$$\left. - C_1 u_{i,111}\delta_{A1} - C_2 u_{i,222}\delta_{A2} + p_0 u_{A,i} - \dot{p}\delta_{iA}\right]\left(\mathbf{e}_i \otimes \mathbf{E}_A\right)\mathbf{E}_1$$
$$= \left(\mu u_{i,1} + E_1 u_{1,1}\delta_{i1} - C_1 u_{i,111} + p_0 u_{1,i} - \dot{p}\delta_{i1}\right)\mathbf{e}_i.$$

(70)

Again, by introducing the scalar field φ (see equation (40)) and employing the compatibility conditions (i.e., $\dot{p}_{ij} = \dot{p}_{ji}$), we reduce the above equation to the following equation:

$$\dot{p}_{21} - \dot{p}_{12} = \Delta\left[\Delta\varphi - \frac{C_1}{\mu}\varphi_{1111} - \frac{C_2}{\mu}\varphi_{2222}\right]$$
$$+ \left(\frac{E_1}{\mu} - \frac{E_2}{\mu}\right)\varphi_{1122} = 0.$$

(71)

A complete analytical solution for equation (71) is available via the methods of iterative reduction and the principle of eigenfunction expansion (see, more details, [35–37]). Figures 1–4 illustrate the deformation profiles for the cases discussed through Sections 3–5. As the equations become mathematically compact, the corresponding deformation fields experience less oscillatory behaviors. This is clearly demonstrated by the results in Figures 3-4, bidirectional cases, which show a close agreement with the predictions from nonlinear models. We also mention here that equation (71) reduces to equations (42) and (58) in the limit of vanishing C_2 and E_2, respectively. This, in turn, suggests that unidirectional cases can be assimilated within the systems of equation (71) by setting $C_2 = E_2 = 0$ (see Figure 5). In fact, the latter model is recommended to use, since the bidirectional and extension fibers case is the most general and compact form (with minimal singular behaviors) and therefore produces more accurate prediction results.

To elaborate the proposed model, we present comparisons with the experimental data obtained from the 3-point bending test of CNC fiber composite ($C = 150$ GPa, $\mu = 1$ GPa). This is a particular case of the proposed model, when $C/\mu = 150$ and vanishing E (i.e., equation (20)). Using the method presented in Section 3.1, we find the solution of equation (20) as

$$\varphi(x, y) = \sum_{m=1}^{\infty}\left[\left\{e^{a_m x}\left(-C_m \cos b_m x + D_m \sin b_m x\right)\right.\right.$$
$$\left.\left. + e^{-a_m x}\left(C_m \cos b_m x + D_m \sin b_m x\right) \times \sin\left(\frac{\pi}{2d}\right)y\right]\right\},$$

(72)

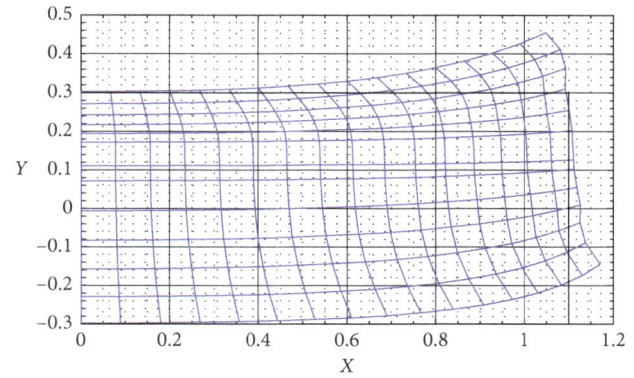

FIGURE 1: Unidirectional fibers.

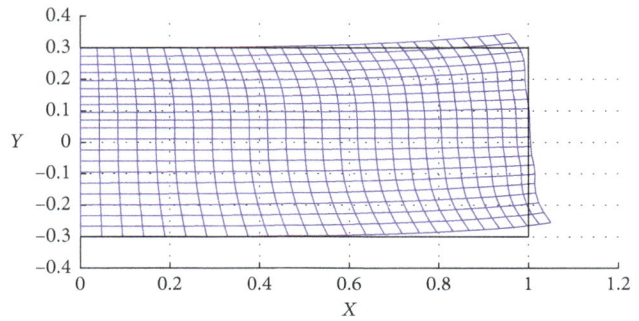

FIGURE 2: Unidirectional extensible fibers.

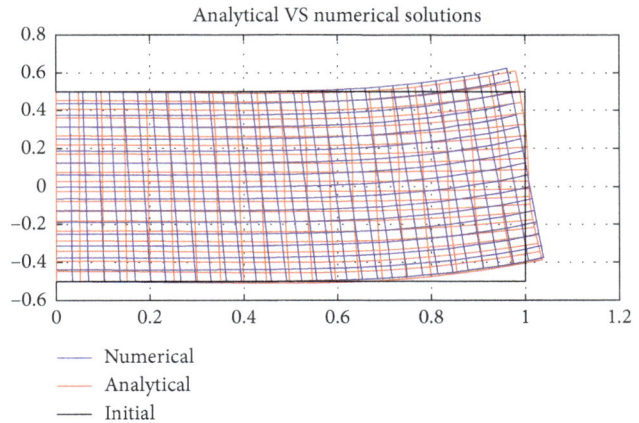

FIGURE 3: Bidirectional fibers.

where

$$m = \frac{\pi n}{2d},$$
$$\alpha = \frac{C}{\mu},$$
$$a_m = \frac{\sqrt{2m\sqrt{\alpha} + 1}}{2\sqrt{\alpha}},$$
$$b_m = \frac{\sqrt{2m\sqrt{\alpha} - 1}}{2\sqrt{\alpha}},$$

(73)

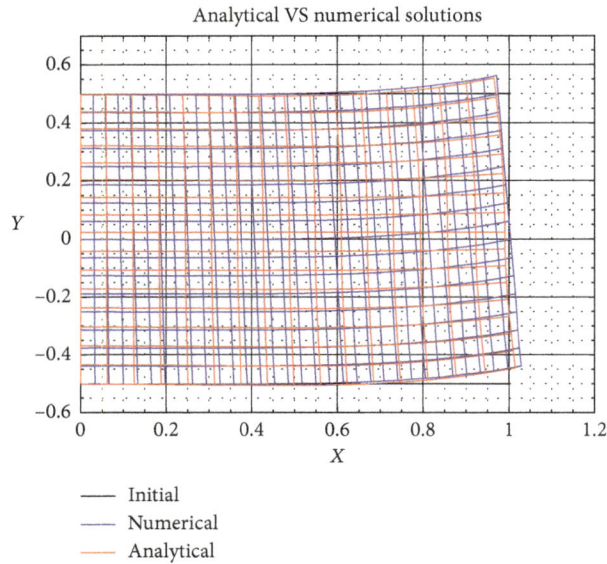

FIGURE 4: Bidirectional extensible fibers.

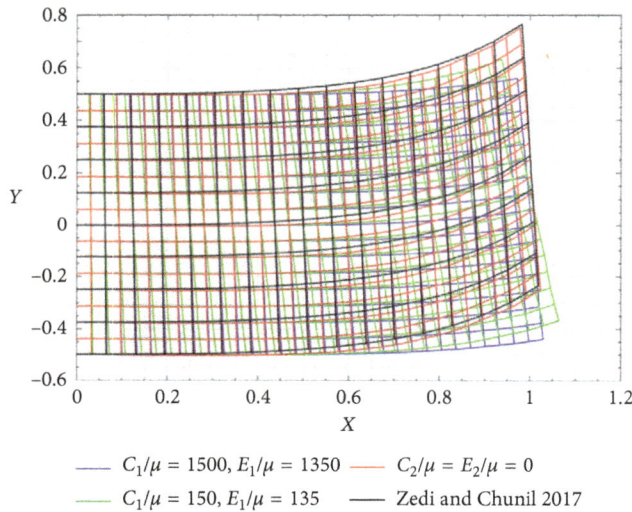

FIGURE 5: Reduction of solutions in the limit of $C_2 = E_2 = 0$.

and the unknowns C_m and D_m can be determined via the applied moment in the form of Fourier expansion; i.e.,

$$\dot{m}_1 = Cu_{1,11} = -\varphi_{211} = \sum_{n=1}^{30} \frac{20}{\pi n} (-1)^{n-1/2} \cos\left(\frac{\pi n}{2d}\right) y. \quad (74)$$

Using equation (74), we compute the maximum deflections of the CNC composite and illustrate the results in Figure 6. It is shown in Figure 6 that the proposed model successfully predicts the normal deflections of the CNC composite strip. Since the slope of the graph in Figure 6 indicates the moduli of the composite (when divided by the corresponding length scale), it can also be used in the estimation of the overall material properties of the composite. However, due to the paucity of available data, we are not able to provide the quantitative analysis other than

those presented in Figure 6 at this time. The study is currently underway, and our intention is to present elsewhere when ready. We also mention here that the overall responses of composite materials can also be estimated using multiscale modeling methods (see, for example, Shahabodini et al. [8, 9]).

Lastly, we assimilate the case of single family of fiber composites subjected to the axial tension. The linear solution obtained from equation (56) demonstrates good agreement with the compatible nonlinear model [28] for small deformations superposed on large, while it shows discrepancies in the prediction of large deformations (See Figure 7). It is also noted here that we reserve the details in solving the corresponding differential equations (equation (56)) for the sake of conciseness. However, the procedures for the particular case of the present example can be found in [28].

6. Further Considerations

For the analysis of soft materials-based composites, such as carbon rubber-fiber composites and polymer composites, a different type of energy potential may be suggested instead of the Neo-Hookean model discussed in the previous sections. The Mooney–Rivlin model is one of the most commonly used energy potentials for the aforementioned cases (see, for example, [33]). In the forgoing development, we present a compatible linear model for the Mooney–Rivlin types of materials for the desired applications. The expression of the Mooney–Rivlin potential is given by the following equation:

$$W(\mathbf{F}) = \frac{\mu}{2}(\mathbf{F} \cdot \mathbf{F} - 3) + \frac{\lambda}{2}\left[\frac{1}{2}(\text{tr}\mathbf{F})^2 - \text{tr}(\mathbf{F})^2\right], \quad \mu, \lambda > 0. \quad (75)$$

The derivative of equation (75) with respect to the deformation gradient tensor then yields the following equation:

$$W_{\mathbf{F}} = \mu\mathbf{F} + \lambda\mathbf{F}\left[(\mathbf{F} \cdot \mathbf{F})\mathbf{I} - \mathbf{F}^T\mathbf{F}\right]. \quad (76)$$

Substituting equation (76) into equation (9) furnishes the following expression of the Piola stresses:

$$\mathbf{P} = \mu\mathbf{F} + \lambda\mathbf{F}\left[(\mathbf{F} \cdot \mathbf{F})\mathbf{I} - \mathbf{F}^T\mathbf{F}\right] - p\mathbf{F}^* - C\text{Div}(\mathbf{g} \otimes \mathbf{D} \otimes \mathbf{D}). \quad (77)$$

Now, by applying the same schemes as adopted in the previous sections, we find the following equation:

$$\dot{\mathbf{P}} = \mu\dot{\mathbf{F}} - \left[\lambda\mathbf{F}\{(\mathbf{F} \cdot \mathbf{F})\mathbf{I} - \mathbf{F}^T\mathbf{F}\}\right]^{\cdot} - \dot{p}\mathbf{F}_o^* - p_o\dot{\mathbf{F}}^* - C\nabla\dot{\mathbf{g}}(\mathbf{D} \otimes \mathbf{D}), \quad (78)$$

where the second term of the right side of equation (78) can be evaluated at $\varepsilon = 0$ as

$$\left[\lambda\mathbf{F}\{(\mathbf{F} \cdot \mathbf{F})\mathbf{I} - \mathbf{F}^T\mathbf{F}\}\right]^{\cdot} = \lambda\dot{\mathbf{F}}\{(\mathbf{F}_o \cdot \mathbf{F}_o)\mathbf{I} - \mathbf{F}_o^T\mathbf{F}_o\}$$
$$+ \lambda\mathbf{F}_o\left\{2(\dot{\mathbf{F}} \cdot \mathbf{F}_o)\mathbf{I}_o - \dot{\mathbf{F}}^T\mathbf{F}_o - \mathbf{F}_o^T\dot{\mathbf{F}}\right\}. \quad (79)$$

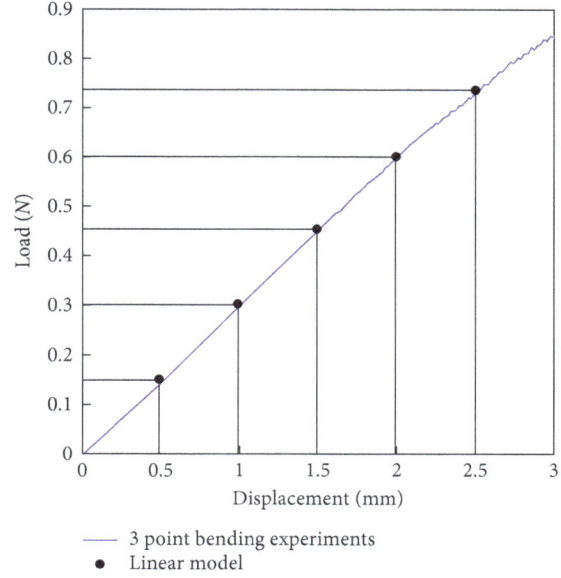

(a) (b)

FIGURE 6: (a) Experimental setting. (b) Experiment vs theoretical prediction.

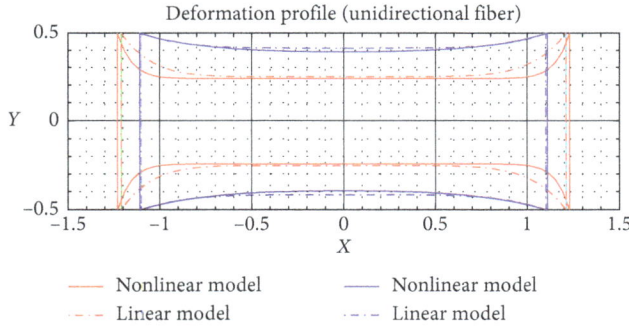

FIGURE 7: Deformation profiles of unidirectional fiber composites under axial tension.

Since $\mathbf{F}_o^T = \mathbf{F}_o = \mathbf{I}$, the above equation becomes the following equation:

$$\lambda \dot{\mathbf{F}}(\delta_{AA} - 1) + \lambda \left\{ 2(\dot{\mathbf{F}} \cdot \mathbf{I})\mathbf{I} - \dot{\mathbf{F}}^T - \dot{\mathbf{F}} \right\}$$
$$= \left(\lambda \dot{\mathbf{F}}_{iA} + 2\lambda \dot{\mathbf{F}}_{BB}\delta_{iA} - \lambda \dot{\mathbf{F}}_{Ai} - \lambda \dot{\mathbf{F}}_{iA} \right)(\mathbf{e}_i \otimes \mathbf{E}_A), \qquad \delta_{AA} = 2. \tag{80}$$

Thus, from the results in equations (2), (27), (28), (33), (78), and (80), we find the following stress expression for the cases of unidirectional fiber composites as

$$\dot{\mathbf{P}}_{iA}(\mathbf{e}_i \otimes \mathbf{E}_A) = \left[\mu u_{i,A} - \dot{p}\delta_{iA} + p_o u_{A,i} \right.$$
$$\left. - C u_{i,BCD}D_A D_B D_C D_D \right](\mathbf{e}_i \otimes \mathbf{E}_A), \tag{81}$$

where $\dot{\mathbf{F}}_{BB} = u_{B,B} = 0$ from the conditions of bulk incompressibility (see equation (34)). Consequently, the corresponding Euler equilibrium equation satisfies the following equation:

$$\text{Div}(\dot{\mathbf{P}}) = \lambda u_{A,i} \left[\mu u_{i,AA} - \lambda u_{A,iA} - p_i + p_o u_{A,iA} \right.$$
$$\left. - C u_{i,ABCD}D_A D_B D_C D_D \right]\mathbf{e}_i = 0\mathbf{e}_i. \tag{82}$$

Using the compatibility condition of $u_{A,iA}$ (i.e. $u_{A,iA} = u_{A,Ai}$) and the constraints of bulk incompressibility, we find that $u_{A,Ai} = (u_{A,A})_i = 0$ and thereby reduce equation (83) to the following equation:

$$\left(\mu u_{i,AA} - p_i - C u_{i,ABCD}D_A D_B D_C D_D \right)\mathbf{e}_i = 0\mathbf{e}_i. \tag{83}$$

The resulting equilibrium equation (equation (83)) implies that there is no clear distinction between the Neo-Hookean model and Mooney–Rivlin model as far as the superposed incremental deformations with the augmented bulk incompressibility conditions are considered. This is due to the fact that the divergence of high-strain terms identically vanishes upon the imposition of linearized bulk incompressibility conditions (i.e., $\text{Div}\mathbf{u} = \text{div}\mathbf{u} = 0$). However, the expressions of Piola stresses are affected by the introduction of Mooney–Rivlin type materials (see equations (33) and (81)). Thus, for example, if $\mathbf{N} = \mathbf{D} = \mathbf{E}_1$, we compute from equation (25) that

$$\dot{t}_i \mathbf{e}_i = \dot{\mathbf{P}}\mathbf{E}_1 = \mu u_{i,A} - \lambda u_{A,i} - \dot{p}\delta_{iA} + p_o u_{A,i}$$
$$- C u_{i,111}\delta_{1A}D_1 D_1 D_1 (\mathbf{e}_i \otimes \mathbf{E}_A)\mathbf{E}_1 \tag{84}$$
$$= \left(\mu u_{i,1} - \lambda u_{1,i} - \dot{p}\delta_{i1} + p_o u_{1,i} - C u_{i,111} \right)\mathbf{e}_i,$$

which is the boundary tractions for the case of the Mooney–Rivlin energy potential. Lastly, from equation (77), we find at $\varepsilon = 0$ that

$$\mathbf{P}_o = \mu\mathbf{I} + \lambda\mathbf{I}[\mathbf{I} - \mathbf{I}] - p_o\mathbf{I} - C\text{Div}(\mathbf{I} \otimes \mathbf{D} \otimes \mathbf{D}) = 0, \tag{85}$$

and thereby obtain $\mu = p_o$. Equation (83) together with the imposed boundary conditions (i.e., equations (25) and (84)) constitutes a complete set of PDEs system which can be solved using the similar schemes as presented in Section 3.1. Further, because the resulting equilibrium equations are of the same form (see equations (22) and (84)), equations (81) and (84) may be used in the determination of the parameters associated with high-strain terms (λ) by comparing stresses in equations (33) and (81). Investigations in this respect (including the implementations of the developed linear theory) are currently underway. Our intention is to report elsewhere.

7. Conclusions

We present complete linear models for the mechanics of an elastic solid reinforced with fibers resistant to flexure and extension. Within the prescription of the superposed incremental deformations, the first and second gradient of deformations is approximated through which the kinematics of fibers is explicitly determined. The linearized Euler equilibrium equations and the Piola stress are then formulated for the Neo-Hookean types of materials. We also derive the corresponding boundary conditions and the conditions of bulk incompressibility for the sake of completeness. In addition, the cases of the bidirectional fiber composites are considered via the fiber decomposition of the deformation gradient tensor. It is found that the systems of PDEs for the bidirectional and extensible fibers reduce to those from single family of fibers in the limit of vanishing material parameters of fibers.

In particular, we demonstrate the well-known result from the linear theory of elasticity that the merging of the bases remains valid even with the incorporation of fibers' bending and extension into the models of deformations. A Mooney–Rivlin material is also elaborated, where we show that there is no clear distinctions in the resulting Euler equations obtained from the two different strain energy models (i.e., Neo-Hookean and Mooney–Rivlin) as long as the small deformations are concerned. However, the resulting Piola stresses are of different forms in that the one from the Mooney–Rivlin model demonstrates clear dependency on both the first and second invariant of deformation gradient tensors, whereas the Piola stress, in the case of the Neo-Hookean model, depends only on the first invariant. Lastly, we mention here that the present model can be used in the approximation of the nonlinear theory of strain gradient elasticity arising in finite plane elastostatics.

Conflicts of Interest

The authors declare that they have no conflicts of interest.

Acknowledgments

This work was supported by the Natural Sciences and Engineering Research Council of Canada via Grant #RGPIN 04742 and the University of Alberta through a start-up grant. The author would like to thank Dr. David Steigmann for many useful discussions.

References

[1] J. E. Adkins, "Finite plane deformation of thin elastic sheets reinforced with inextensible cords," *Philosophical Transactions of the Royal Society A: Mathematical, Physical and Engineering Sciences*, vol. 249, no. 961, pp. 125–150, 1956.

[2] J. L. Ericksen and R. S. Rivlin, "Large elastic deformations of homogeneous anisotropic materials," *Indiana University Mathematics Journal*, vol. 3, no. 3, pp. 281–301, 1954.

[3] A. J. M. Spencer, *Deformations of Fibre-Reinforced Materials*, Oxford University Press, Oxford, UK, 1972.

[4] A. C. Pipkin, "Stress analysis for fiber-reinforced materials," *Advances in Applied Mechanics*, vol. 19, pp. 1–51, 1979.

[5] J. F. Mulhern, T. G. Rogers, and A. J. M. Spencer, "A continuum model for fibre-reinforced plastic materials," *Proceedings of the Royal Society A: Mathematical, Physical and Engineering Sciences*, vol. 301, no. 1467, pp. 473–492, 1967.

[6] J. F. Mulhern, T. G. Rogers, and A. J. M. Spencer, "A continuum theory of a plastic-elastic fibre-reinforced material," *International Journal of Engineering Science*, vol. 7, no. 2, pp. 129–152, 1969.

[7] A. C. Pipkin and T. G. Rogers, "Plane deformations of incompressible fiber-reinforced materials," *Journal of Applied Mechanics*, vol. 38, no. 3, 1971.

[8] A. Shahabodini, R. Ansari, and M. Darvizeh, "Atomistic-continuum modeling of vibrational behavior of carbon nanotubes using the variational differential quadrature method," *Composite Structures*, vol. 185, pp. 728–747, 2018.

[9] A. Shahabodini, R. Ansari, and M. Darvizeh, "Multiscale modeling of embedded graphene sheets based on the higher-order Cauchy-Born rule: nonlinear static analysis," *Composite Structures*, vol. 165, pp. 25–43, 2017.

[10] R. A. Toupin, "Theories of elasticity with couple-stress," *Archive for Rational Mechanics and Analysis*, vol. 17, no. 2, pp. 85–112, 1964.

[11] R. D. Mindlin and H. F. Tiersten, "Effects of couple-stresses in linear elasticity," *Archive for Rational Mechanics and Analysis*, vol. 11, no. 1, pp. 415–448, 1962.

[12] W. T. Koiter, "Couple-stresses in the theory of elasticity," *Proceedings of the Knononklijke Nederlandse Akademie van Wetenschappen B*, vol. 67, pp. 17–44, 1964.

[13] A. J. M. Spencer and K. P. Soldatos, "Finite deformations of fibre-reinforced elastic solids with fibre bending stiffness," *International Journal of Non-Linear Mechanics*, vol. 42, no. 2, pp. 355–368, 2007.

[14] H. C. Park and R. S. Lakes, "Torsion of a micropolar elastic prism of square cross-section," *International Journal of Solids and Structures*, vol. 23, no. 4, pp. 485–503, 1987.

[15] G. A. Maugin and A. V. Metrikine, *Mechanics of Generalized Continua: One Hundred Years after the Cosserats*, Springer, New York, NY, USA, 2010.

[16] P. Neff, "A finite-strain elastic.plastic Cosserat theory for polycrystals with grain rotations," *International Journal of Engineering Science*, vol. 44, no. 8-9, pp. 574–594, 2006.

[17] I. Munch, P. Neff, and W. Wagner, "Transversely isotropic material: nonlinear Cosserat vs. classical approach," *Continuum Mechanics and Thermodynamics*, vol. 23, no. 1, pp. 27–34, 2011.

[18] P. Neff, "Existence of minimizers for a finite-strain micromorphic elastic solid," *Proceedings of the Royal Society of*

Edinburgh: Section A Mathematics, vol. 136, no. 5, pp. 997–1012, 2006.

[19] S. K. Park and X. L. Gao, "Variational formulation of a modified couple-stress theory and its application to a simple shear problem," *Zeitschrift für angewandte Mathematik und Physik*, vol. 59, no. 5, pp. 904–917, 2008.

[20] E. Fried and M. E. Gurtin, "Gradient nanoscale polycrystalline elasticity: intergrain interactions and triple-junction conditions," *Journal of the Mechanics and Physics of Solids*, vol. 57, no. 10, pp. 1749–1779, 2009.

[21] D. J. Steigmann, "Theory of elastic solids reinforced with fibers resistant to extension, flexure and twist," *International Journal of Non-Linear Mechanics*, vol. 47, no. 7, pp. 734–742, 2012.

[22] C. Truesdell and W. Noll, "The non-linear field theories of mechanics," in *Handbuch der Physik, III/3*, S. Flugge, Ed., Springer, Berlin, Germany, 1965.

[23] F. D. Isola, M. Cuomo, L. Greco, and A. Della Corte, "Bias extension test for pantographic sheets: numerical simulations based on second gradient shear energies," *Journal of Engineering Mathematics*, vol. 103, no. 1, pp. 127–157, 2017.

[24] F. D. Isola, A. Della Corte, L. Greco, and A. Luongo, "Plane bias extension test for a continuum with two inextensible families of fibers: a variational treatment with Lagrange multipliers and a perturbation solution," *International Journal of Solids and Structures*, vol. 81, pp. 1–12, 2016.

[25] E. Turco, "Non-standard coupled extensional and bending bias tests for planar pantographic lattices. Part I: numerical simulations," *Zeitschrift für angewandte Mathematik und Physik*, vol. 67, no. 5, p. 122, 2016.

[26] C. Kim and M. Zeidi, "Gradient elasticity theory for fiber composites with fibers resistant to extension and flexure," *International Journal of Engineering Science*, vol. 131, pp. 80–99, 2018.

[27] M. Zeidi and C. Kim, "Finite plane deformations of elastic solids reinforced with fibers resistant to flexure: complete solution," *Archive of Applied Mechanics*, vol. 88, no. 5, pp. 819–835, 2017.

[28] M. Zeidi and C. Kim, "Mechanics of fiber composites with fibers resistant to extension and flexure," *Mathematics and Mechanics of Solids*, 2017.

[29] M. Zeidi and C. Kim, "Mechanics of an elastic solid reinforced with bidirectional fiber in finite plane elastostatics: complete analysis," *Continuum Mechanics and Thermodynamics*, vol. 30, no. 3, pp. 573–592, 2018.

[30] M. Zeidi and C. Kim, "The effects of intra-membrane viscosity on lipid membrane morphology: complete analytical solution," *Scientific Reprots. Nature*, vol. 8, no. 1, 2018.

[31] T. Belay, C. I. Kim, and P. Schiavone, "Analytical solution of lipid membrane morphology subjected to boundary forces on the edges of rectangular membrane," *Continuum Mechanics and Thermodynamics*, vol. 28, no. 1-2, pp. 305–315, 2015.

[32] C. Kim and D. J. Steigmann, "Distension-induced gradient capillarity in lipid membranes," *Continuum Mechanics and Thermodynamics*, vol. 27, no. 4-5, pp. 609–315, 2015.

[33] R. W. Ogden, *Non-Linear Elastic Deformations*, Ellis Horwood Ltd., Chichester, England, 1984.

[34] D. J. Steigmann, "Invariants of the stretch tensors and their application to finite elasticity theory," *Mathematics and Mechanics of Solids*, vol. 7, no. 4, pp. 393–404, 2002.

[35] W. W. Read, "Analytical solutions for a Helmholtz equation with dirichlet boundary conditions and arbitrary boundaries," *Mathematical and Computer Modelling*, vol. 24, no. 2, pp. 23–34, 1996.

[36] W. W. Read, "Series solutions for Laplace's equation with nonhomogeneous mixed boundary conditions and irregular boundaries," *Mathematical and Computer Modelling*, vol. 17, no. 12, pp. 9–19, 1993.

[37] Y. Huang and X. Zhang, "General analytical solution of transverse vibration for orthotropic rectangular thin plates," *Journal of Marine Science and Application*, vol. 1, no. 2, pp. 78–82, 2002.

16

Shape Stability of Polyethylene Glycol/Acetylene Black Phase Change Composites for Latent Heat Storage

Jingjing Zhang [ID],[1] **Hairong Li** [ID],[2] **Junyang Tu** [ID],[1] **Ruonan Shi** [ID],[1] **Zhiping Luo** [ID],[1,3] **Chuanxi Xiong,**[1] **and Ming Jiang** [ID][1]

[1]*School of Materials Science and Engineering, State Key Lab for New Textile Materials and Advanced Processing Technology, Wuhan Textile University, Wuhan 430200, China*
[2]*Mechanical Metrology Division, Hubei Institute of Measurement and Testing Technology, Wuhan 430223, China*
[3]*Department of Chemistry and Physics, Fayetteville State University, Fayetteville, NC 28301, USA*

Correspondence should be addressed to Hairong Li; ronghailee@163.com and Ming Jiang; mjiang@wtu.edu.cn

Academic Editor: Veronica Calado

Sufficient shape stability is essential for a high-performance phase change material (PCM). Although significant advances have been made to develop form-stable composites, technical development in the field of polymer-based PCMs is currently limited by an incomplete understanding of the shape stability. Form-stable polyethylene glycol/acetylene black (PEG/AB) PCMs containing PEGs with different average molecular weights have been obtained by melt mixing to investigate the shape stability of the PEG/AB composites. It was found that the phase change behaviors of the PEG/AB composites were not only attributed to the interactions between the AB and PEG, but also to the intermolecular interactions of the PEG chains, depending on the varying molecular weights of the PEGs. Physically crosslinked structure with temporary junctions was formed through hydrogen bonding, capillary, surface tension forces, intermolecular friction, and macromolecular entanglement, which contributed to the constrained chain motion and thus the solid-solid phase change behavior of the PEG/AB composites. The physically crosslinked structure was more stable with longer length of the PEG molecular chains, resulting in higher critical impregnated contents of the PEG into the AB and thus improved latent heat.

1. Introduction

Energy storage can optimize the energy flows to overcome the mismatch between the demand for energy and the supply of energy, especially for energy generated by variable and renewable resources, such as water, wind, and solar energy [1]. Among the principal energy-storage technologies, phase change materials (PCMs) have drawn tremendous attention because they can store and release large amounts of latent heat as they change from one physical state to another for effective use of thermal energy [2–5]. To increase the energy-storage ability, PCMs with large latent heat are essential for future thermal energy storage and management [6, 7].

Among the PCMs investigated, polyethylene glycol (PEG) is one of the prospective candidates in this regard as it exhibits relatively high latent heat-storage capacity, tunable

solid-liquid phase change temperature, and non-corrosiveness, which leads to compact storage devices with high energy-storage density. Moreover, PEG has much better cycling stability during services and higher security than small-molecule organic PCMs due to nondetectable vapor pressure when melts [8–11]. Nevertheless, as a solid-liquid PCM, the applications of PEG are largely limited, as the liquid phase leaks above the phase change point. The leakage is usually circumvented by introducing form-stable support. Considerable efforts have been devoted to develop form-stable composites with high energy-storage densities [12–23]. The form stabilization strategies of PCMs include encapsulation [12, 13], grafting [14, 15], and physical adsorptions [16–23]. Clay, diatomite, halloysite, active carbon, silica, expanded graphite, and graphene have been employed as fillers to prevent melted PCMs from leaking during the phase transition via physical adsorptions, relying on their predominantly polar surface and/or high porous structures [16–23]. However, although significant advances have been made, technical development in the field of polymer-based PCMs is currently limited by an incomplete understanding of the shape stability. The affinity of support fillers to PCMs depends on their structures, so the structure of the fillers usually is a primary consideration for the shape stability. Thus far, few attentions have been paid on the difference in shape stabilities between polymeric PCMs and low molecular weight organic PCMs. The effect of some critical structure factors for polymeric PCMs, such as molecular weight, on the shape stability was largely neglected. However, the shape stability of polymer-based composite PCMs could not be entirely determined by the affinity of support fillers to PCMs or the interactions between them and might be partly attributed to the structure of the polymeric PCMs. Therefore, a further understanding of the form stabilization mechanisms would ultimately produce a more rational and general approach to the thermal property optimisation of PEG-based composite PCMs.

The present work is an attempt to develop a better understanding of the shape stability of polymeric PCM-based composites by comparing the effects of PEGs with different molecular weights (M_w) on the thermal properties of PCM composites filled by commercially available acetylene black (AB). The phase change behaviors of the form-stable PEG/AB PCMs were verified by polarized optical microscopy (POM) and filter paper test. Furthermore, the structures and thermophysical properties of the composite PCMs were analyzed by scanning electron microscopy (SEM), X-ray diffractometer (XRD), thermogravimetric analysis (TGA), and differential scanning calorimetry (DSC). Finally, the existence of an intermolecular interaction mechanism of solid-solid phase change besides simple physical adsorptions in the PEG/AB composite PCMs was found.

2. Materials and Methods

2.1. Materials. Chemically pure PEGs were purchased from Sinopharm Chemical Reagent Beijing Co., Ltd. PEG800, PEG4000, PEG6000, and PEG10000 were utilized in this work. AB (purity > 99.99%, particle size = 150–200 nm based on the supplier's information) was purchased from Shanghai Chemical Reagent Co. Ltd. and used after drying.

2.2. Preparation of PEG/AB Composites. To fabricate the PEG/AB composites, the PEGs with various average molecular weights were first melt at 50–70°C, and then the AB was slowly added under stirring and vacuum. Finally, the mixture was dried in a vacuum oven at 70°C for 48 h.

2.3. Characterization. The Fourier-transform infrared (FTIR) spectrum was observed at room temperature (Thermo Nicolet Nexus) in the range of 4,000 to 400 cm^{-1} using KBr pellets. Nitrogen adsorption/desorption isotherms were measured on a Micromeritics Model TriStar II 3020

volumetric analyzer (USA). POM observation was conducted on an Olympus BX51-P polarizing optical microscope equipped with a LTS 350 hot stage. TGA was carried out on a TG/DTA 220U (Seiko Instrument Co. Ltd, Tokyo Japan) with the Exstar 6000 Station. The samples were scanned from 20 to 800°C with the heating rate of 10°C/min and nitrogen gas purging. The X-ray measurements were carried out on a Rigaku D/Max-IIIA X-ray diffractometer with Cu-Kα radiation at a wavelength of 1.54 Å. Thermal analysis was performed using a DSC (PE Co., USA). The thermal conductivities of the various samples were measured at 28°C by the guarded heat flow test method (TA Instruments DTC 300, USA). Thermograms were obtained at a heating and cooling rate of 5°C·min^{-1} at a temperature range from 20°C to 80°C under nitrogen atmosphere. To confirm the reproducibility of the results, at least two samples were measured for each composition. For each sample, at least three DSC traces were recorded. SEM measurements were performed with JSM-IT300A (JEOL Ltd., Japan).

3. Results and Discussion

3.1. Chemical and Physical Characterizations of the AB Particles. Chemical characterizations of the AB particles and the PEG/AB composites are carried out using FTIR spectroscopy, as shown in Figure 1. In the spectrum of the AB (Figure 1(a)), the peaks at 2,888 cm^{-1}, 1,146 cm^{-1}, and 1,114 cm^{-1} represent the stretching vibration of C-O groups. The peak at 1,637 cm^{-1} is attributed to C=O stretching vibration. O-H stretching vibrations are characterized by the broadband in the region of 3,300–3,650 cm^{-1}. The spectrum of pristine PEG (Figure 1(b)) is characterized by O-H stretching vibration at 3433 cm^{-1} and the strong absorption peaks of C-H bonds at 2888 cm^{-1}, 953 cm^{-1}, and 839 cm^{-1}. The C-O stretching vibration was at 1106 cm^{-1}. The FTIR results indicate that no significant new peaks were observed other than characteristic peaks of PEGs and AB at FTIR spectrums of the PEG/AB composites (Figure 1(b)), which proves that the interactions between the functional groups of PEG and AB are physical in nature. Moreover, the existence of weak hydrogen bonding between the AB and PEG, which could be able to increase PEG-absorbing capacity of the AB to some extent.

To further examine the pore adsorption capacity of the AB particles, nitrogen adsorption/desorption isotherms are measured. The pore-size distribution of the AB is shown in Figure 2. The AB particles possess a specific Brunauer–Emmett–Teller (BET) surface area of 53.79 m^2/g, with a most probable pore size of about 2.6 nm. The specific surface area is much lower than those of porous carbon materials, such as expanded graphite or active carbon. Moreover, the average pore size of the AB particles is smaller than the equivalent sphere diameter of PEG4000 or PEG with higher M_w. The equivalent radius of PEG random coils can be calculated according to previous reports [24]. The above analysis shows a relatively low pore adsorption capacity of the AB particles.

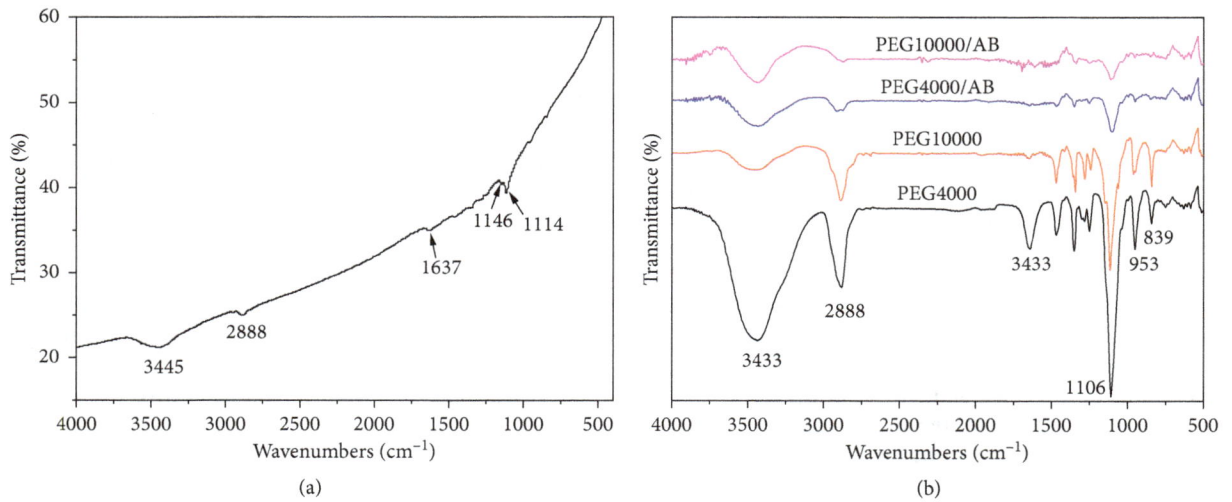

FIGURE 1: FTIR spectrums of the AB particles (a) and PEG/AB composites (b).

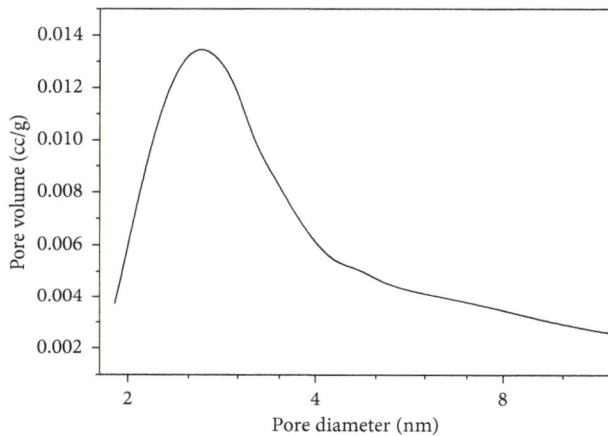

FIGURE 2: Pore-size distribution of the AB particles.

3.2. Leakage Behavior of the PEG/AB Composites.

The leakage behavior of the PEG/AB composites is investigated by POM and filter paper test. The POM images are taken below the crystallization temperature of the PEG10,000/AB composites, as shown in Figure 3. The darker regions correspond to the saturated AB, whereas the lighter regions marked by arrows indicate the excess PEG, as shown in Figure 3(a). When the mass fraction of the AB is increased to 11.6 wt.%, the excess PEG1,000 cannot be observed in the POM image in Figure 3(b), indicating the complete absorption of the PEG in the AB particles.

The leakage behavior of the PEG/AB composites at elevated temperature above the phase change point is further studied. Typically, the PEG/AB samples are kept in filter paper envelopes above the phase change temperatures for 1.5 h and the change in the weight of the filter paper envelopes before and after the test are measured, and the results were presented in Table 1. No significant change in weight is observed under 80°C, indicating that no obvious PEG leakage occurred during the test. Consequently, the optimum absorption ratio of

PEG800, PEG4,000, PEG6,000, and PEG10,000 in AB should be around 0.76, 0.84, 0.87, and 0.88 (responding to the results obtained by TGA below), respectively, and these mass fractions are adequate to obtain form-stable PEG/AB composite PCMs.

3.3. Thermal Reliability of the PEG/AB Composites.

The TG curves of the pure PEG and PEG/AB composite PCMs are shown in Figure 4. As can be seen from the curves, all samples have a single thermal degradation stage. The degradation of the PEGs with relatively large molecular weights and their composite PCMs starts around 350°C and ends at 430°C, while the PEG800 starts below 300°C and ends at 430°C. It indicates that the AB does not significantly affect the thermal stability of the PEGs with relatively large M_w but improves the thermal stability of PEG with relatively small M_w (PEG800). This is attributed to the stronger interaction between AB and PEG800 induced by pore adsorption due to the smaller equivalent sphere diameter of PEG800 than the average pore size of AB. That is, the PEG chains are trapped inside the AB pores. In addition, these results also demonstrate that no chemical reactions take place during simple melt blending and impregnation process. Moreover, the TG curves of the form-stable composite PCMs also confirm the optimum impregnation ratios of PEG800, PEG4,000, PEG6,000, and PEG10,000 as 76.4 wt%, 83.9 wt%, 86.8 wt%, and 88.4 wt%, respectively.

3.4. Crystalline Properties of the PEG/AB Composites.

The XRD patterns of the PEG and PEG/AB composites at room temperature are displayed in Figure 5. It is observed that the AB only has a weak and broad peak at 25°, indicating an amorphous structure. The XRD patterns of the PEG4,000, PEG6,000, and PEG10,000 are characterized by two strong 2θ peaks at 19.1° and 23.6°, which are assigned to the typical plane of (120) and (112) of the PEGs. The XRD curves of the PEG/AB composites are similar to those pure PEGs, and the position of the diffraction peaks remains unchanged after the

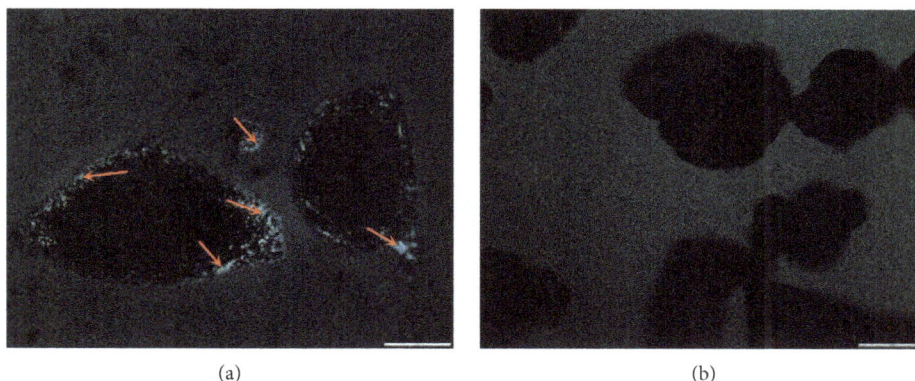

(a)   (b)

FIGURE 3: POM images taken below the crystallization temperature of PEG10000/9 wt.%AB (a) and PEG10000/11.6 wt.%AB (b) Scale bar: 100 μm.

TABLE 1: Relative weight changes of the PEG/AB composites before and after the leakage tests at 80°C for 1.5 h.

Samples	Average weight $_{before}$ (mg)	Average weight $_{after}$ (mg)	Δ (%)	Standard deviation (mg)
PEG800/AB	313.2	317.2	1.27	0.88
PEG4000/AB	318.5	321.2	0.85	0.31
PEG6000/AB	319.4	322.7	1.03	0.76
PEG10000/AB	312.3	315.3	0.96	0.56

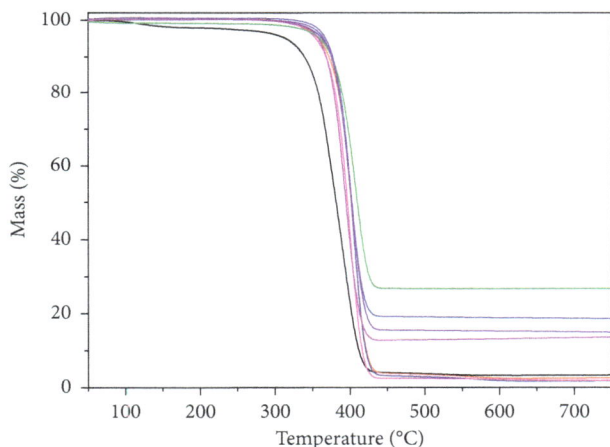

FIGURE 4: TG plots of the PEGs and PEG/AB composites.

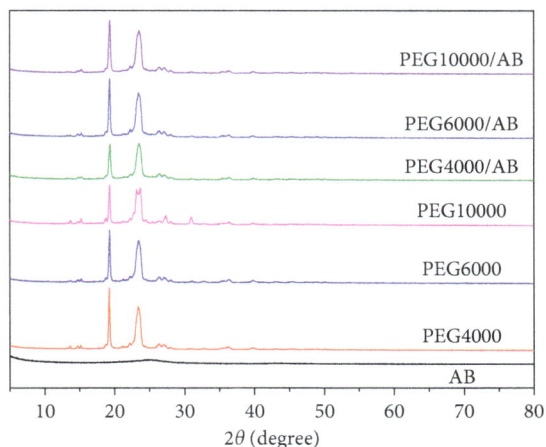

FIGURE 5: XRD spectra of the AB, pristine PEGs, and PEG/AB composites.

melt blending process, which demonstrates that the pristine PEG and the PEG/AB have the similar crystal structure and crystal cell type.

3.5. *Thermal Conductivities of the PEG/AB Composites.* The thermal conductivities of different mass ratios of AB are presented in Figure 6. It can be seen that almost linear correlation exists between the thermal conductivity and content of the AB for each PEG system. A slow increase in thermal conductivity from ~0.15 W/mK to ~0.26 W/mK with the increase of the mass ratio of the AB is observed due to the intrinsically low thermal conductivity of the AB induced by its amorphous structure. AB is chosen as the supporting material to verify the effect of molecular weight of PEG on the shape stability so as to exclude the strong interactions between PEGs and supporting material, but actually it is not a promising thermally conductive filler for PCMs.

3.6. *Thermal Performances of the PEG/AB Composites.* The latent heat and phase change temperature of the samples are measured using DSC. Figure 7 shows the melting and freezing DSC curves of the pristine PEGs and form-stable PEG/AB

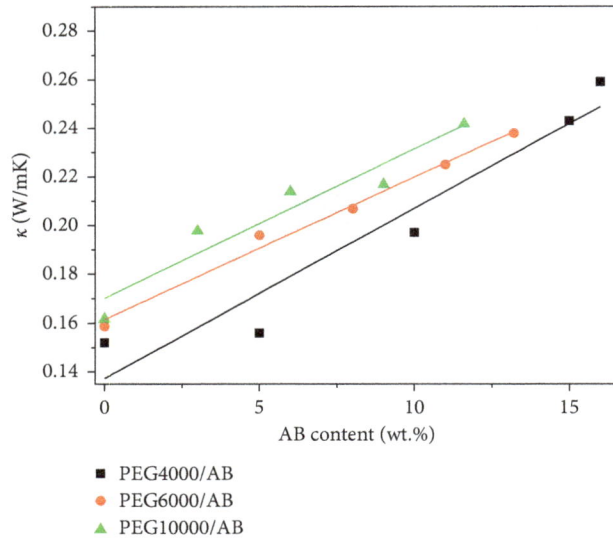

FIGURE 6: Thermal conductivities with various mass ratios of AB.

composite PCMs. The detailed results of DSC thermal analysis of the peak melting/crystallization temperature (T_m/T_c) ratio, the measured melting/crystallization enthalpy ($\Delta H_m/\Delta H_c$), and the theoretical melting/crystallization enthalpy ($\Delta H_m{}^*/\Delta H_c{}^*$) obtained are presented in Table 1. The theoretical melting/crystallization enthalpies of the PEG/AB composites are calculated based on the impregnated contents of the PEG into the AB and the measured melting/crystallization enthalpies of pristine PEGs, assuming that the AB cannot contribute to the phase change latent heat.

The latent heat capacities of the composite PCMs containing the PEGs with different M_w are obviously lower than those of the pristine PEGs due to the incorporation of the AB that does not undergo phase change. Consequently, in most cases in this study, the melting enthalpies of the PEG/AB composites almost show agreement with the relevant mixture rules. The measured latent heats of the PEG/AB composites with varying PEG M_w from ~4,000 to ~10,000 are comparable with the theoretical latent heats. By adding a proper loading of the AB, relatively high latent heats are retained, especially for the PEG/AB composites containing the PEGs with relatively large M_w. The incorporation of the AB does not induce a significant change of the crystal structure of the PEG with relatively large M_w. Notably, in the case of the PEG800/AB composite, the melting enthalpy is significantly lower by 55.7% than the theoretical value at the optimum loading ratio. This is partially expected given that some of the PCM weight (23.6%) is replaced by the AB particles that are not subject to phase change. In addition, it is quite possible that other mechanisms also affect the phase change process of the PEG800 in the PEG800/AB composites. The presumed reason may be that some of the PEG chains are trapped inside the AB pores when M_w of the PEG is low (PEG800) due to the small equivalent sphere diameter of PEG800 compared to the most

probable pore size of the AB. The phase transition of the PEG from amorphous phase into stable crystal phase can be inhibited to a large extent due to the confined chain motion of the PEG induced by the strong adsorbability of the AB pores, and thus the obviously reduced latent heat of phase change is detected by DSC. As a result, the phase change of the composite is prone to take place at lower temperature (Table 2). After AB incorporation, the changes in crystallization temperature identify their tendency to supercool when cooling. However, the incorporation of the AB does not induce a significant supercooling of the PEGs with relatively large M_w because of decreased restriction from the AB pores. By contrast, when the M_w of the PEG increased to ~4,000, most of the PEG chains are completely outside of the AB pores. These chains, different from both the free PEG chains and the chains trapped into the AB pores, are confined by the surface of the AB particles. The longer PEG chains may offer more interacting site for the surface absorption of the PEG on the AB particles. Furthermore, it should be noted that the optimum impregnation ratios of the PEG into the AB for the PEG/AB composites increase with higher M_w of the PEG. These imply the existence of some other mechanisms leading to solid-solid phase change in the PEG/AB composite PCMs besides physical adsorptions. The phase change behavior of the PEG as a polymeric PCM is significantly different from small-molecule PCM, namely, much stronger intermolecular interactions than them. It thus explains in terms of the stronger intermolecular interactions, such as intermolecular friction and macro-molecular entanglement, between the PEG4000 chains than those of the PEG800. Therefore, both the critical impregnated content of the PEG into the AB and enthalpy of the PEG4000/AB composite increase remarkably. When M_w increases to ~6,000, the confinement effect from intermolecular interactions is strengthened further with the increase of chain length, and thereby the declining amplitude in T_m of the PEG/AB composite is reduced; the critical impregnated content of the PEG into the AB also shows an increasing trend. With the increase of M_w to 10,000, the critical impregnated content of the PEG into the AB increases further and thus the enthalpy of the PEG10,000/AB composite is improved as well. In other words, the phase change behaviors for the PEG/AB composites are not only attributed to the interactions between the AB and PEG, but also to the intermolecular interactions of the PEG chains depending on the varying M_w of the PEG. The optimum absorption ratio of the PEG in the form-stable PEG800/AB, PEG4,000/AB, PEG6,000/AB, and PEG10,000/AB composite PCMs reaches 0.76, 0.84, 0.87, and 0.88, respectively.

3.7. Shape-Stability Model of the PEG/AB Composites. A model is proposed to explain the shape stability of the PEG/AB composites, based on a physically crosslinked structure with temporary junctions (Figure 8). A part of PEG

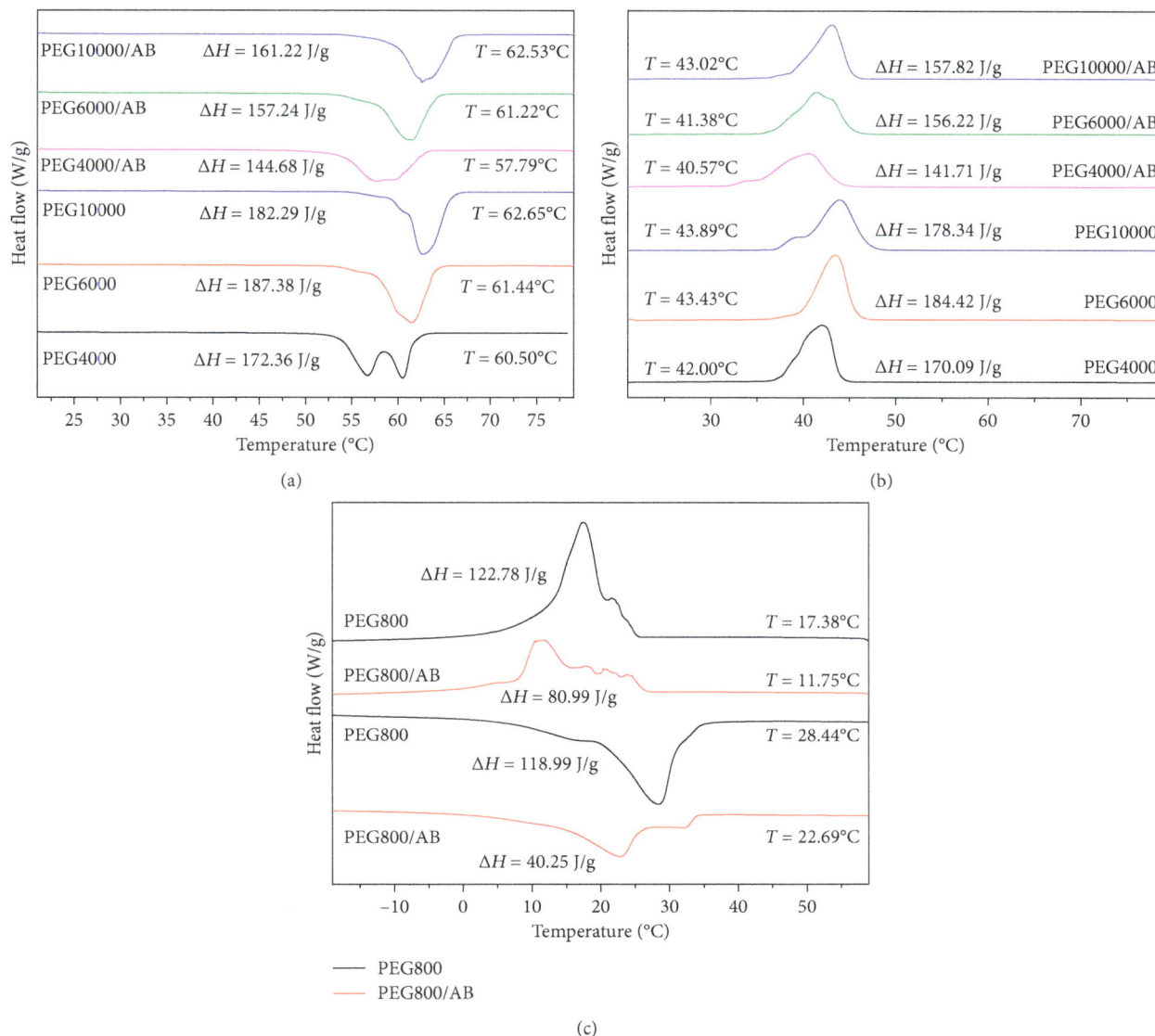

FIGURE 7: Cyclic DSC heating (a) and cooling (b) curves of pristine PEG4000, PEG6000, PEG10000, and their shape-stabilized PEG/AB composites. (c) DSC curves of PEG800 and PEG800/AB composites.

TABLE 2: Thermal analysis of the PEG/AB composites and pure PEGs with varying molecular weights.

Samples	Impregnation ratio (wt.%)	T_m (°C)	ΔH_m (J g^{-1})	ΔH_m^* (J·g^{-1})	T_c (°C)	ΔH_c (J·g^{-1})	ΔH_c^* (J·g^{-1})
PEG800	—	28.44	118.99	—	17.38	122.78	—
PEG4000	—	60.50	172.36	—	42.00	170.09	—
PEG6000	—	61.44	187.38	—	43.43	184.42	—
PEG10000	—	62.65	182.29	—	43.89	178.34	—
PEG800/AB	76.40	22.69	40.25	90.91	11.75	80.99	93.80
PEG4000/AB	83.90	57.79	144.68	144.68	40.57	141.71	142.71
PEG6000/AB	86.80	61.22	157.24	162.65	41.38	156.22	160.08
PEG10000/AB	88.40	62.53	161.22	161.14	43.02	157.82	157.65

random coils with small equivalent radius and segments are trapped into the pores of the AB particles by physical interactions, such as capillary and surface tension forces (abbreviated as trapped PEG in Figure 8). In addition, some of the PEG chains that are closely adjacent to the AB are adsorbed on the surface of the AB particles by hydrogen

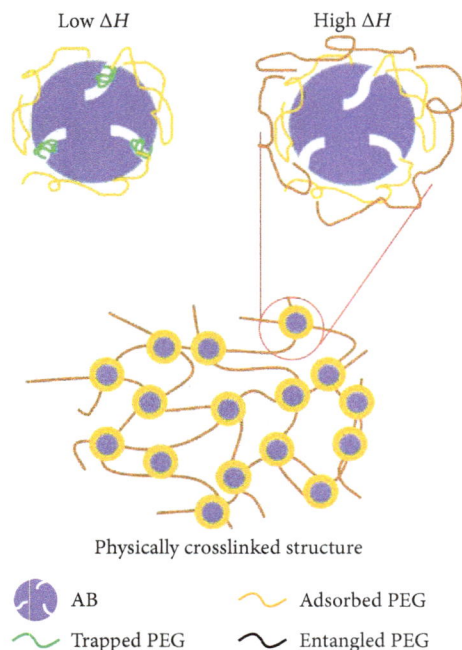

FIGURE 8: Schematic representation of shape stability.

bonding (abbreviated as adsorbed PEG in Figure 8); and some of the PEG chains with relatively large M_w are stabilized by the macromolecular entanglement and intermolecular friction (abbreviated as entangled PEG in Figure 8). The melting process of the PEG in the composite is confined by a physically crosslinked structure with temporary junctions above T_m of the PEG, which inhibits the phase change of form-stabilized solid phase into liquid phase. The structure becomes more stable with longer length of the PEG molecular chains, resulting in higher critical impregnated content of the PEG into the AB. When the PEG content is increased further above the critical impregnated content, some of the PEG chains would be unbound above the phase change point of the composite, eliminating the shape-stabilized composites.

3.8. Structural Analysis of the PEG/AB Composites. Comparison between the AB and the PEG/AB composites is made by SEM. Figure 9 shows the SEM images of the AB before PCM impregnation and the form-stable PEG/AB composites containing the critical impregnated contents of the PEGs. The morphology of the PEG800/AB composite is similar to that of the AB particles. The AB particles as the supporting material are covered with PEG layers, and the structure of AB framework can be seen because the PEG layer is thin. A part of the PEG800 diffuses into the pores of the AB particles, which is consistent with the DSC results. As discussed before, the PEG molecules with relatively small M_w can be easily trapped by AB, despite the fact that the AB can only provide limited adsorption capability due to the relatively low specific surface area and small pore size. In the

case of the composite containing the PEG with relatively large M_w, they appear to follow the different trend. Through SEM analysis, we confirm that most of the PEGs are not dispersed into the pores of the AB. With an increase of M_w to ~4,000 (Figure 9(c)), the AB framework is not obvious since the PEG layer outside of the AB is thicker. When M_w increases to ~6,000 (Figure 9(d)) or ~10,000 (Figure 9(e)), the AB is covered with PEG completely and an irregular morphology can be found since the PEG layer is much thicker due to the longer molecular chain and higher mass fraction of the PEG. Based on the physically crosslinked structure model as discussed above, a part of the PEG segments and random coils with small equivalent radius are trapped into pores of the AB particles by capillary or surface tension forces. The hydrophilic character of the AB particles and cavities in them help us to improve the interfacial compatibility and wetting between the PEG and AB particles; thus some of the PEG chains adjacent to the AB are adsorbed on AB particle surfaces by hydrogen bonding, capillary, or surface tension forces. Other PEG chains are anchored by intermolecular friction and entanglement because there are more chances of intermolecular interactions between the longer PEG chains compared to the PEG with relatively low M_w. Physically crosslinked structure comprising the PEG chains and AB particles is formed through hydrogen bonding, capillary, surface tension forces, intermolecular friction, and macromolecular entanglement, which contributes to constrained chain motion and thus the solid-solid phase change behavior of the PEG/AB composites. Therefore, the optimum impregnation ratio of the PEG into the AB for the PEG/AB composites without seepage of the melted PEG increases with the increasing M_w of the PEG.

4. Conclusions

The PEG/AB composite PCMs exhibited desirable properties in term of shape stability and large latent heats. The AB particles kept the composite form stable without inducing a large reduction in their phase change enthalpies. The fusion enthalpies of the form-stable PEG4,000/AB, PEG6,000/AB, and PEG10,000/AB composite PCMs reached 144.68, 157.24, and 161.22 J/g, respectively. Our results suggest that the phase change behaviors for the PEG/AB composites are attributed not only to the interactions between the AB and PEGs but also to the intermolecular interactions of the PEG chains depending on the varying M_w of the PEGs. The shape stability of the PEG/AB composites were explained based on a physically crosslinked structure model. Physically crosslinked structure with temporary junctions was formed through hydrogen bonding, capillary, surface tension forces, intermolecular friction, and macromolecular entanglement. The structure became stabler with increasing length of the PEG molecular chains, resulting in the increase of the critical impregnated content of the PEG into the AB.

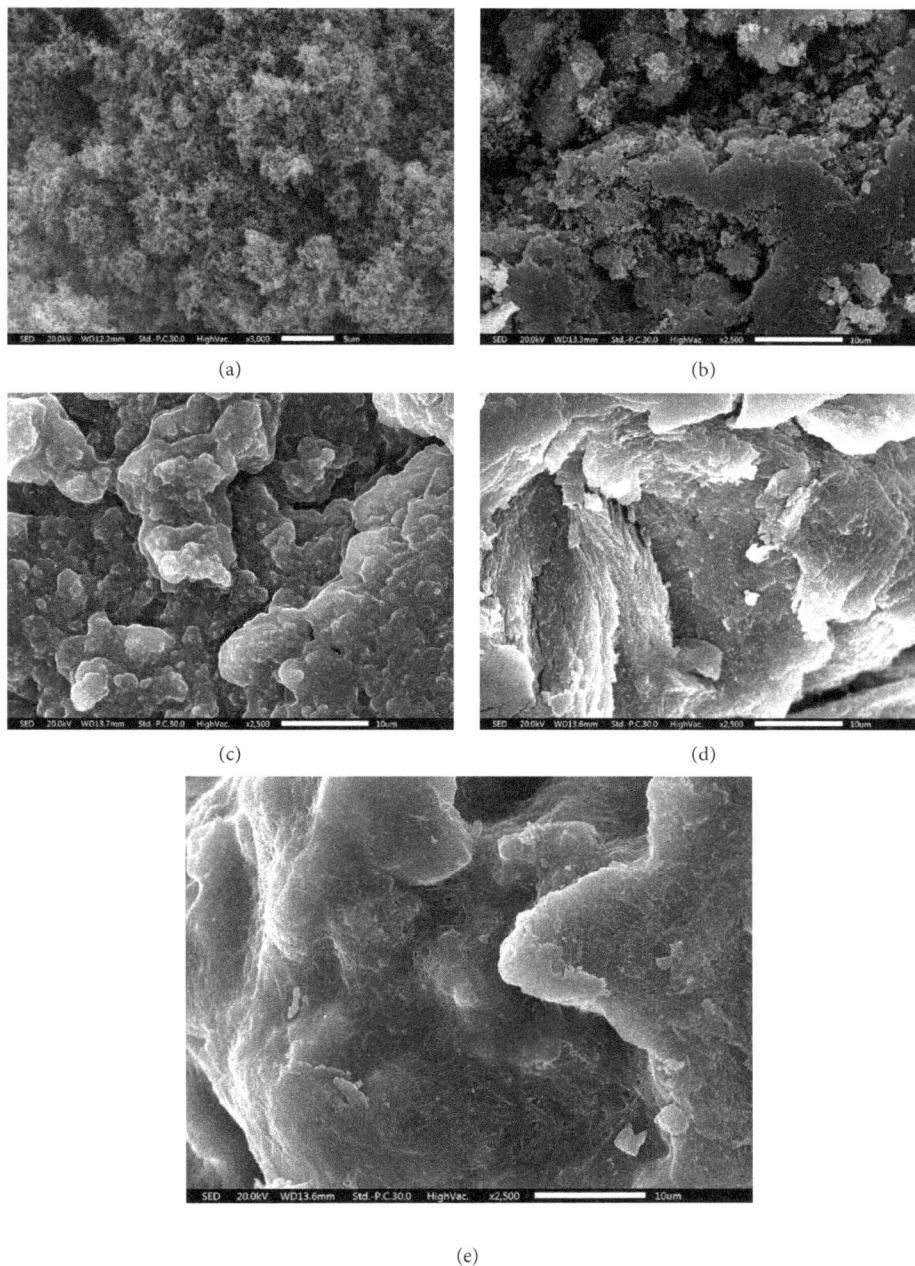

FIGURE 9: SEM images of the shape-stabilized composite PCMs. (a) AB. (b) PEG800/AB. (c) PEG4000/AB. (d) PEG6000/AB. (e) PEG10000/AB.

Conflicts of Interest

The authors declare that there are no conflicts of interest regarding the publication of this paper.

Acknowledgments

We gratefully acknowledge the support by the National Natural Science Foundation of China (No. 51503158).

References

[1] Q. Li, L. Chen, M. R. Gadinski et al., "Flexible high-temperature dielectric materials from polymer nano-composites," *Nature*, vol. 523, no. 7562, pp. 576–579, 2015.

[2] B. Zalba, J. M. Marín, L. F. Cabeza, and H. Mehling, "Review on thermal energy storage with phase change: materials, heat transfer analysis and applications," *Applied Thermal Engineering*, vol. 23, no. 3, pp. 251–283, 2003.

[3] M. Mehrali, S. T. Latibari, M. Mehrali, H. Metselaar, and M. Silakhori, "Shape-stabilized phase change materials with high thermal conductivity based on paraffin/graphene oxide composite," *Energy Conversion and Management*, vol. 67, pp. 275–282, 2013.

[4] A. Abhat, "Low temperature latent heat thermal energy storage: heat storage materials," *Solar Energy*, vol. 30, no. 4, pp. 313–332, 1983.

[5] S. D. Sharma and K. Sagara, "Latent heat storage materials and systems: a review," *International Journal of Green Energy*, vol. 2, no. 1, pp. 1–56, 2005.

[6] Y. Tian and C. Y. Zhao, "A Review of solar collectors and thermal energy storage in solar thermal applications," *Applied Energy*, vol. 104, pp. 538–553, 2013.

[7] A. Váz Sá, R. Almeida, H. Sousa, and J. Delgado, "Numerical analysis of the energy improvement of plastering mortars with phase change materials," *Advances in Materials Science and Engineering*, vol. 2014, Article ID 245473, 13 pages, 2014.

[8] C. Alkan, A. Sari, and O. Uzun, "Poly (ethylene glycol)/acrylic polymer blends for latent heat thermal energy storage," *AIChE Journal*, vol. 52, no. 9, pp. 3310–3314, 2006.

[9] K. Pielichowski and K. Flejtuch, "Differential scanning calorimetric studies on poly(ethylene glycol) with different molecular weights for thermal energy storage materials," *Polymers for Advanced Technologies*, vol. 13, no. 10-12, pp. 690–696, 2002.

[10] S. Han, C. Kim, and D. Kwon, "Thermal degradation of poly (ethylene glycol)," *Polymer Degradation and Stability*, vol. 47, no. 2, pp. 203–208, 1995.

[11] H. Li, M. Jiang, Q. Li et al., "Aqueous preparation of polyethylene glycol/sulfonated graphene phase change composite with enhanced thermal performance," *Energy Conversion and Management*, vol. 75, pp. 482–487, 2013.

[12] X. Jiao, D. Zhao, Y. Zhang et al., "Microencapsulated dimethyl terephthalate phase change material for heat transfer fluid performance enhancement," *Colloid and Polymer Science*, vol. 294, no. 4, pp. 639–646, 2016.

[13] H. Li, M. Jiang, Q. Li et al., "Facile preparation and thermal performances of hexadecanol/crosslinked polystyrene core/shell nanocapsules as phase change material," *Polymer Composites*, vol. 35, no. 11, pp. 2154–2158, 2014.

[14] A. Sarı, C. Alkan, and A. Biçer, "Synthesis and thermal properties of polystyrene-graft-PEG copolymers as new kinds of solid-solid phase change materials for thermal energy storage," *Materials Chemistry and Physics*, vol. 133, no. 1, pp. 87–94, 2012.

[15] R. Cao, H. Liu, S. Chen, D. Pei, J. Miao, and X. Zhang, "Fabrication and properties of graphene oxide-grafted-poly (hexadecyl acrylate) as a solid-solid phase change material," *Composites Science and Technology*, vol. 149, pp. 262–268, 2017.

[16] N. Sarier, E. Onder, S. Ozay, and Y. Ozkilic, "Preparation of phase change material-montmorillonite composites suitable for thermal energy storage," *Thermochimica Acta*, vol. 524, no. 1-2, pp. 39–46, 2011.

[17] X. Fu, Z. Liu, B. Wu, J. Wang, and J. Lei, "Preparation and thermal properties of stearic acid/diatomite composites as form-stable phase change materials for thermal energy storage via direct impregnation method," *Journal of Thermal Analysis and Calorimetry*, vol. 123, no. 2, pp. 1173–1181, 2016.

[18] D. Mei, B. Zhang, R. Liu, Y. Zhang, and J. Liu, "Preparation of capric acid/halloysite nanotube composite as form-stable phase change material for thermal energy storage," *Solar Energy Materials and Solar Cells*, vol. 95, no. 10, pp. 2772–2777, 2011.

[19] L. Feng, J. Zheng, H. Yang, Y. Guo, W. Li, and X. Li, "Preparation and characterization of polyethylene glycol/active carbon composites as shape-stabilized phase change materials," *Solar Energy Materials and Solar Cells*, vol. 95, no. 2, pp. 644–650, 2011.

[20] Y. Wang, T. D. Xia, H. Zheng, and H. X. Feng, "Stearic acid/silica fume composite as form-stable phase change material for thermal energy storage," *Energy and Buildings*, vol. 43, no. 9, pp. 2365–2370, 2011.

[21] Z. Zhang, N. Zhang, J. Peng, X. Fang, X. Gao, and Y. Fang, "Preparation and thermal energy storage properties of paraffin/expanded graphite composite phase change material," *Applied Energy*, vol. 91, no. 1, pp. 426–431, 2012.

[22] J. Yang, G. Qi, Y. Liu et al., "Hybrid graphene aerogels/phase change material composites: thermal conductivity, shape-stabilization and light-to-thermal energy storage," *Carbon*, vol. 100, pp. 693–702, 2016.

[23] K. Yuan, Y. Zhou, W. Sun, X. Fang, and Z. Zhang, "A polymer-coated calcium chloride hexahydrate/expanded graphite composite phase change material with enhanced thermal reliability and good applicability," *Composites Science and Technology*, vol. 156, pp. 78–86, 2018.

[24] D. H. Atha and K. C. Ingham, "Mechanism of precipitation of proteins by polyethylene glycols. Analysis in terms of excluded volume," *Journal of Biological Chemistry*, vol. 256, no. 23, pp. 12108–12117, 1981.

Representative Stress-Strain Curve by Spherical Indentation on Elastic-Plastic Materials

Chao Chang [ID],[1] **M. A. Garrido,**[2] **J. Ruiz-Hervias,**[3] **Zhu Zhang,**[1] **and Le-le Zhang**[4]

[1]*School of Applied Science, Taiyuan University of Science and Technology, Taiyuan 030024, China*
[2]*Departamento de Tecnología Química y Ambiental, Tecnología Química y Energética y Tecnología Mecánica, Escuela Superior de Ciencias Experimentales y Tecnología, Universidad Rey Juan Carlos, c/Tulipán s/n Móstoles, Madrid, Spain*
[3]*Departamento de Ciencia de Materiales, UPM, E.T.S.I. Caminos, Canales y Puertos, c/ Professor Aranguren s/n, 28040 Madrid, Spain*
[4]*School of Mechanical, Electronic and Control Engineering, Beijing Jiaotong University, Beijing 100044, China*

Correspondence should be addressed to Chao Chang; chao_chang_tyust@163.com

Academic Editor: Baozhong Sun

Tensile stress-strain curve of metallic materials can be determined by the representative stress-strain curve from the spherical indentation. Tabor empirically determined the stress constraint factor (stress CF), ψ, and strain constraint factor (strain CF), β, but the choice of value for ψ and β is still under discussion. In this study, a new insight into the relationship between constraint factors of stress and strain is analytically described based on the formation of Tabor's equation. Experiment tests were performed to evaluate these constraint factors. From the results, representative stress-strain curves using a proposed strain constraint factor can fit better with nominal stress-strain curve than those using Tabor's constraint factors.

1. Introduction

The instrumented indentation technique consists of applying load to the sample by means of an indenter of known geometry, while the applied load and the penetration depth of the indenter are recorded simultaneously during a loading and unloading cycle. The load-penetration data can be used to determine the mechanical properties of the material without having to image the residual impression left on the material's surface. Consequently, this technique can be applied at different scales, from macro to nano. The main mechanical properties measured by this technique are Young's modulus, E, and hardness, H, by sharp indenter [1, 2]. This methodology could also measure material properties like yield stress, creep property, and fracture [3–6]. However, these properties are not sufficient to characterize a material. The present work focuses on the methodology to determine the stress-strain curve of metallic materials by the depth-sensing indentation technique using a spherical indenter. Sharp indenters like Berkovich or Vicker are characterized by inducing a constant strain on the indented materials. This deformation depends

on the indenter angle. Consequently, if the complete stress-strain curve is needed, the sharp indenters are not the best option because they would give only a point of such curve. This is the reason why the interest for the spherical indenter has been recently grown. The experimental curve from spherical indenter provides more information, as this type of indenter has a smooth transition from elastic to elastic-plastic contact [7].

There were many methodologies for the extraction of flow stress with the spherical indentation technique. One category belongs to mathematical and numerical methods (e.g., dimensionless method and artificial neural network (ANN)). After running many finite element models within certain range of material properties, the indentation curves, contact stiffness, or the work of loading-unloading is obtained, which are verified by corresponding experimental indentation and establish a relation between the indentation characteristics and material properties [8–15]. However, these methods consume great computational effort. The unique solution of these methods is still under discussion [16, 17]. Moreover, the contamination in the experimental process like deficient indenter tip, roughness of specimen, or intrinsic noise of instrument machine is not considered in these methods.

Another category is to determine the material parameters from spherical indentation by defining the representative strain and stress. Regarding the experimental measurements, these methods are more convenient and can be used directly. Following the work of Meyer [18], Tabor defined the representative indentation strain, ε_r, at contact edge of the spherical indenter as $\varepsilon_r = 0.2\,(a/R)$, where the values of ψ and β empirically were determined as 2.8 and 0.2, respectively, based on the quantity of tensile tests on common metals [19]. Most of the researchers agreed with Tabor's indentation strain but emerge with the controversy on the choice of values of ψ and β [5, 20–25]. Richmond et al. [21] predicted that the mean pressure was approximately equal to 3 times the yield stress and that the representative strain was approximately equal to 0.32 times the impression to ball diameter ratio. Herbert et al. [23] had used higher stress constraint factor of 3.7 to obtain the representative stress-strain curve for Al 6061-T6 by spherical indentation with a radius of 385 nm. Based on the work of Matthews [22], Tirupataiah and Sundararajan [26] derived an expression for the indentation stress factor, ψ, which was related with hardening exponent, n. Similarly, Yetna N'Jock et al. [27] determined the tensile property by spherical indentation by simply using the expression of mean pressure to stress ratio as a function of hardening exponent which was derived from the FEM works of Taljat et al. [28]. Using the representative strain under spherical indentation, Ahn and Kwon [29] developed a shear strain definition by differentiating the displacement in the depth direction, in which the ratio of mean pressure to representative stress was equal to 3 in the fully plastic period by conducting instrumented spherical indentation on the steel specimen. By carrying out extensive forward analyses in FEM, Xu and Chen [24] using the indentation strain by Ahn and Kwon [29] found the indentation stress constraint factors, ψ, depended almost linearly on hardening exponent, n. Additionally, the corresponding indentation strain constraint factor β depended on both n and the ratio of Young's modulus to yield stress. Milman et al. [30] assumed that the fully plastic zone beneath indenter was incompressible and proposed a new representative strain which was related with contact radius and contact depth. Fu et al. [31] used the Milman representative strain [30] and proposed a novel iterative process to determine the tensile stress-strain curve. Recently, Kalidindi et al. [32, 33] argued that the definition of indentation strain as a/R lacks any physical interpretation as a measure of strain and proposed a new definition of the indentation strain consistent with the Hertz theory, which was evaluated from several FEM simulations as well as from the analysis of experimental measurements.

From the literature, there is not uniform agreement among investigations concerning the representative stress and strain equations and also the values for the factors involved in the expressions. The main purpose of the present investigation is to systematically study Tabor's indentation strain and propose the possibility to develop an analytical procedure to extract the stress-strain curve using experimental data from spherical indentations, which would be comparable to that obtained from a uniaxial test (i.e., tensile test).

2. Theoretical Background

2.1. The Analytical Relationship between Stress and Strain Factors. In 1908, Meyer had found that for many materials, the mean pressure increased with a/R according to the simple power law [18]

$$p_m = k\left(\frac{a}{R}\right)^m, \tag{1}$$

in which p_m is the mean pressure, a/R is the ratio of indentation radius to the ball radius, and m is the Meyer index.

In (1), k and m are constants. The Meyer equation was verified by further experimental studies [19, 34], and they suggested that the Meyer index, m, was related with the hardening exponent, n. Following Meyer's work, Tabor had proposed the concept of representative strain or stress, by which the mean pressure in Meyer's equation and the a/R ratio can be converted into the true stress-strain curve. He assumed that the mean pressure, p_m, at the fully plastic regime was proportional to the representative stress, σ_r, and the impression radius was proportional to the corresponding representative strain, ε_r. Consequently, the representative stress and strain can be expressed as [19]

$$\varepsilon_r = \beta \frac{a}{R},$$
$$\sigma_r = \frac{F}{\psi \pi a^2}, \tag{2}$$

where β is the indentation strain constraint factor, a is the contact radius, R is the radius of the indenter, ψ is the indentation stress constraint factor, and F is the indentation load.

Tabor determined the parameters β and ψ from empirically experimental data from spherical indentations under the fully plasticity regime. Generally, the indentation strain constraint factor β is considered to be equal to 0.2, and the indentation stress constraint factor ψ ranges from 2.8 to 3.2 [35]. It should be emphasized that the representative strain and stress defined by Tabor are an average value of the stress and strain states induced inside the material [36, 37]. The stress and strain constraint factors allow us to establish equivalence between the stress-strain indentation curve and the corresponding one obtained from a uniaxial test. The true stress-strain curve from uniaxial tensile or compression test can be expressed as

$$\sigma = E\varepsilon \quad \text{for } \sigma \le \sigma_y \text{ elastic regime}, \tag{3}$$

$$\sigma = K\varepsilon^n \quad \text{for } \sigma \ge \sigma_y \text{ Hollomon's equation for plastic regime}, \tag{4}$$

where E is the elastic modulus, σ_y is the yield' stress, K is the strength coefficient, and n is the strain-hardening exponent.

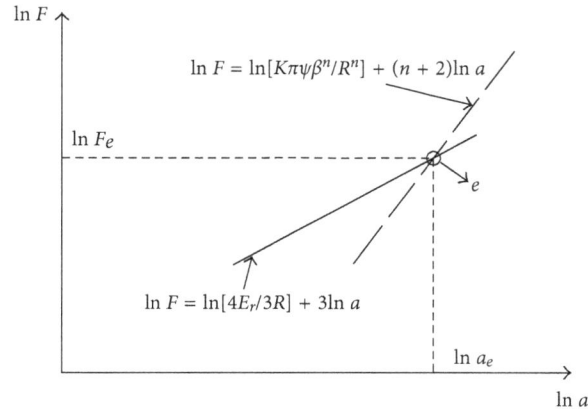

FIGURE 1: Equations (6) and (8) are in the same $\ln a$-$\ln F$ coordinate, and "e" is the intersectional point.

TABLE 1: Mechanical properties of steel F114, brass, and Al alloy.

Materials	E (GPa)	Yield stress (MPa)	Hardening exponent n	Hardening exponent n determined by indentation	Grain size (μm)
Steel F114	210	820	0.26	0.20	15
Brass alloy	100	320	0.15	0.19	20
Al alloy	70	220	0.1	0.15	10

Substituting (2) into the Hollomon equation:

$$\frac{F}{\psi \pi a^2} = K\left[\beta\left(\frac{a}{R}\right)\right]^n, \tag{5}$$

which represents the relationship between load, F, and the ratio of contact radius to indenter radius according to the power law at the fully plastic regime. Transforming (5) into the natural logarithm as

$$\ln F = \ln\left[\frac{K\pi\psi\beta^n}{R^n}\right] + (n+2)\ln a. \tag{6}$$

According to Hertz equation under the elastic regime, the loading and contact radius can be expressed as

$$F = \frac{4E_r}{3R}a^3, \tag{7}$$

where $1/E_r = (1-v^2/E) + (1-v_i^2/E_i)$, E, v and E_i and v_i are Young's modulus and Poisson's ratio for the bulk material to be measured and for the indenter, respectively.

Transforming (7) into the natural logarithm

$$\ln F = \ln\left[\frac{4E_r}{3R}\right] + 3\ln a. \tag{8}$$

Most metals have an n value between 0.10 and 0.50, plotting (6) and (8) in the $\ln a$ versus $\ln F$ coordinate, as shown in Figure 1. Inversely extending the two lines, there should be an intersectional point "e" for different linear slopes.

Therefore, the representative strain and stress at the point "e" can be expressed as

$$\varepsilon_r = E\beta\left(\frac{a_e}{R}\right), \tag{9}$$

$$\sigma_r = \frac{F_e}{\psi \pi a_e^e}. \tag{10}$$

Combining (9) and (10), we can obtain

$$\psi * \beta = \frac{R}{\pi E} \cdot \frac{F_e}{a_e^3}. \tag{11}$$

Comparing (10) with Hertz equation in (7), we can obtain

$$\phi = \psi * \beta = \frac{4}{3\pi} \cdot \frac{E_r}{E}. \tag{12}$$

In (12), the product, ϕ, of stress constraint factor ψ and strain constraint factor β is constant, which is dependent on the ratio of reduced Young's modulus E_r to Young's modulus E. According to (12), the constraint factors can be determined if one of them is known.

It should be noted that Tabor's representative method is restricted within the fully plastic regime. In this study, the procedure assumes that elastic-plastic transition is negligible, and the relationship between stress and strain constraint factors is affirmed from Hertz's theory. In Kalidindi et al.'s [32, 33] work, they utilized indentation strain derived from the Hertz theory for extracting the full stress-strain curve of metallic materials. Similarly, we utilize the Hertz equation to study the relationship between the stress and strain constraint factors to extract the stress-strain curve based on Tabor's representative method. For the soft metallic materials with high E/σ_y which has very short elastic-plastic transition, the intersection point "e" can be considered as a simplistic interpretation of the elastic-plastic transition regime. Especially, in experimental practice, when performing the nanoindentation test on the soft metallic

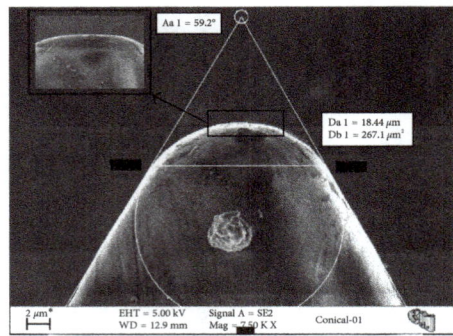

FIGURE 2: Scanning electron microscope image of a worn diamond conical-spherical indenter.

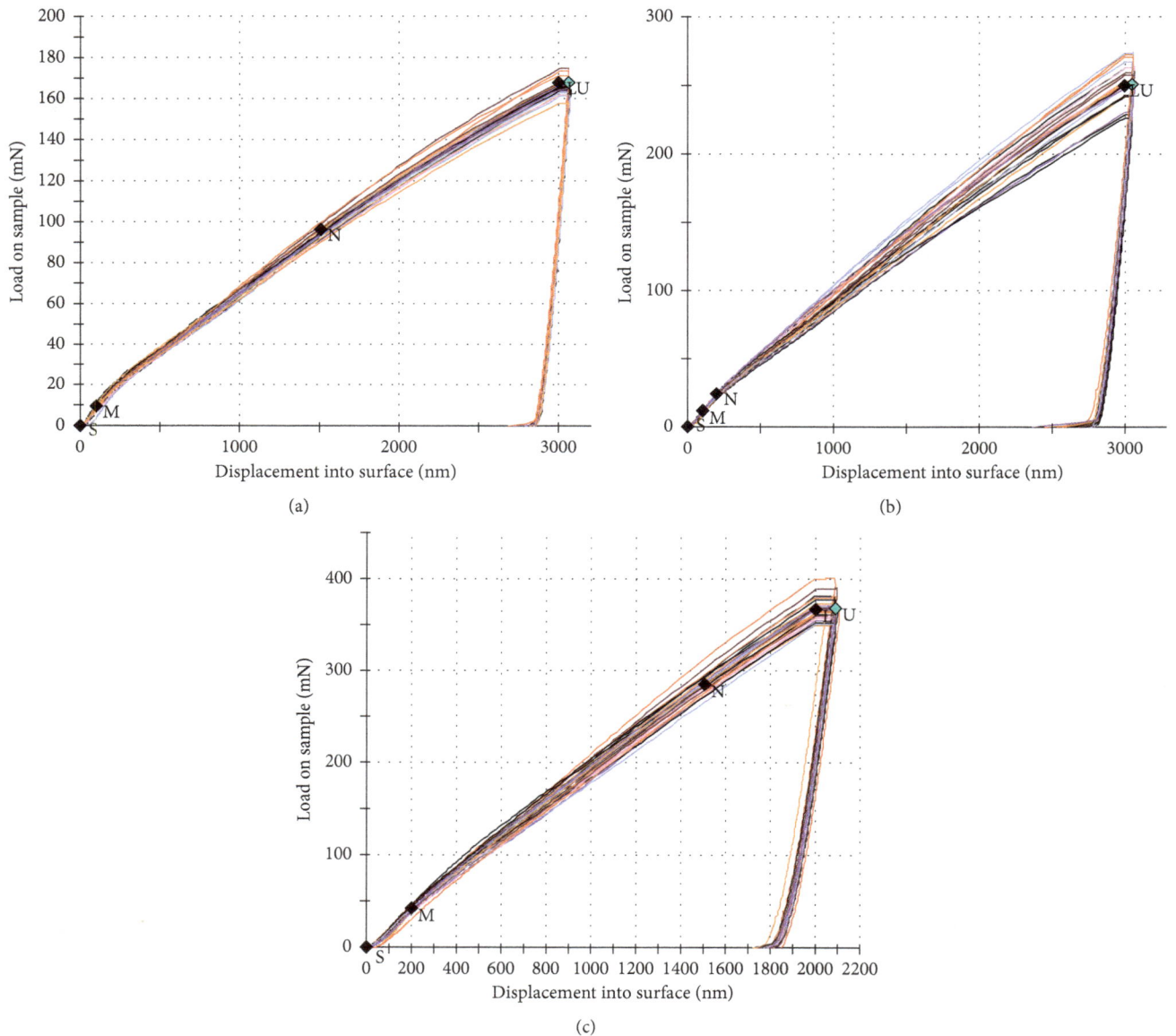

FIGURE 3: The load versus penetration depth curves for the three indented materials performed with spherical tip. (a) Al-1050, (b) brass, and (c) steel.

materials, the penetration depth is very shallow to turn into the fully plastic regime. Consequently, this assumption is reasonable only for certain soft metallic materials. The experimental verification will be discussed later.

In Tabor's equation, the value of stress constraint factor is selected as 3 for common engineering metals in literature [5, 31, 38]. However, there are few studies on the analytical solution for the stress constraint factor. The indentation pressure is found to be proportional to the yield strength of the material by the constraint factor, C. The magnitude of the constraint factor has been found to be depended on the material properties, particularly on the ratio of the modulus to the yield stress. For most metals, $E/\sigma_y > 100$, and the constraint factor is approximately equal to 3 ($C \approx 3$) [19]. For polymers, the ratio of $E/\sigma_y < 10$ and the constraint factor is less than 3 [39]. An approach to large strain plasticity problems in which the material is considered to behave in a plastic-elastic fashion, instead of as a plastic-rigid body, is applied to the axisymmetric blunt indenter. The ratio of the mean stress on the punch face to the uniaxial flow stress of the material (constraint factor C) is found to be 2.82 for an extensive specimen. However, it is shown that a small part of the punch face is elastically loaded, and if the loaded punch area is assumed to be equal to the size of the plastic impression, then the constraint factor to be used is 3 instead of 2.82. This is the value to be used in interpreting the ordinary Brinell test[40]. Additionally, Shield found the exact solution for the axisymmetric flat punch. He found a maximum pressure under the punch of $2.8\sigma_y$ at the centre [41]. Von Mises yield conditions are more representative of engineering materials than Tresca. Although the slip line field theory can be applied to both types of yield conditions, Tresca is generally selected as it leads to a simpler equation that can be solved analytically. However, Von Mises yield can give up to 15.5% higher limit load value than Tresca, which would lead to $C = 3.2$ [42]. Unlike the punch indenter, where the contact area is constant, the spherical one offers a gradual transition of the contact area. It could be assumed that the spherical indenter is corresponding to a punch indenter at a given penetration depth. As a result, the value of stress constraint factor is taken as 3.2, which is consistent with the analytical solution by Shield [41]. As the product of stress and strain constraint factors is constant, according to (12), the strain constraint factor can be determined.

2.2. Estimation of Contact Radius. In order to calculate the representative stress and strain, an accurate measurement of the contact radius at defined indentation load would be obtained. In the case of a flat punch indenter, Sneddon's analysis proposed an estimation expression of contact radius, a:

$$a = \frac{S_u}{2E_r}, \qquad (13)$$

where S_u is the contact stiffness.

From Eq. (13), the contanct radius and Young's modulus can be extracted from the initial unloading slope [43]. According to (13), the contact radius could be determined

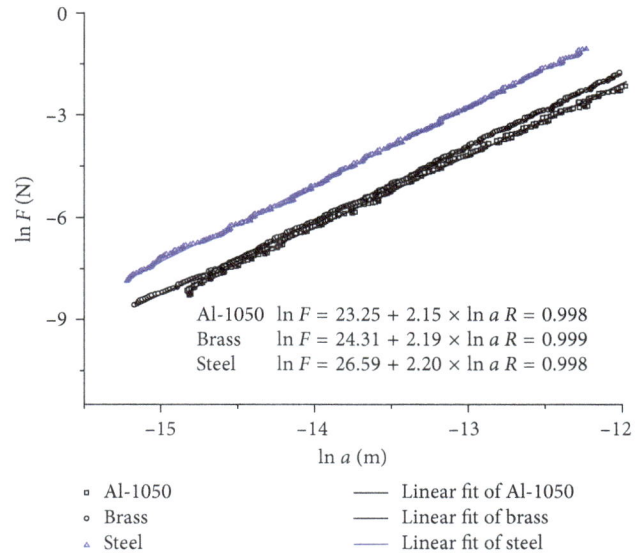

FIGURE 4: Linear regression of $\ln(F)$ versus $\ln(a)$ for the experimental materials.

when reduced Young's modulus and initial unloading slope of the unloading are known. The procedure to estimate the contact radius is consistent with Kalidindi et al.'s [32, 33] work.

3. Experimental Procedure

Depth-sensing indentation (DSI) tests were performed on a Nanoindenter XP (Nanoinstruments Innovation Center, MTS systems, TN, USA) by using the continuous stiffness measurement (CSM) methodology. A conical (angel 60°) spherical diamond indenter with a tip radius of 10 μm was selected. To verify the radius of the indenter, scanning electron microscope was used to measure the radius of the tip before the test. Continuous loading and unloading cycles were conducted during the loading branch by imposing a small dynamic oscillation of 2 nm and 45 Hz on the displacement signal and measuring the amplitude and phase of the corresponding force [44, 45]. Consequently, the contact stiffness was continuously measured as a function of the penetration depth during the experiment. A tip calibration procedure was carried out using the fused silica in accordance with the CSM methodology [46]. Prior to the indentation tests, the samples were ground and polished in two steps, using a mechanical polisher (Labopol-5, Struers, Copenhagen, Denmark) and finished with colloidal suspension of silica of 0.05 μm. A total of 25 indentations were performed in displacement control on three samples of commercial metallic alloys: a carbon steel (F-114), a brass alloy, and an aluminum alloy (Al-1050). The penetration depth was selected according to the grain size of indented materials: 3000 nm for Al-1050 and brass and 2000 nm for steel. The average of these curves is selected to determine the representative stress-strain curve by spherical indentation which can be comparable with the true stress-strain curve by tensile tests [3]. Tensile tests were also conducted

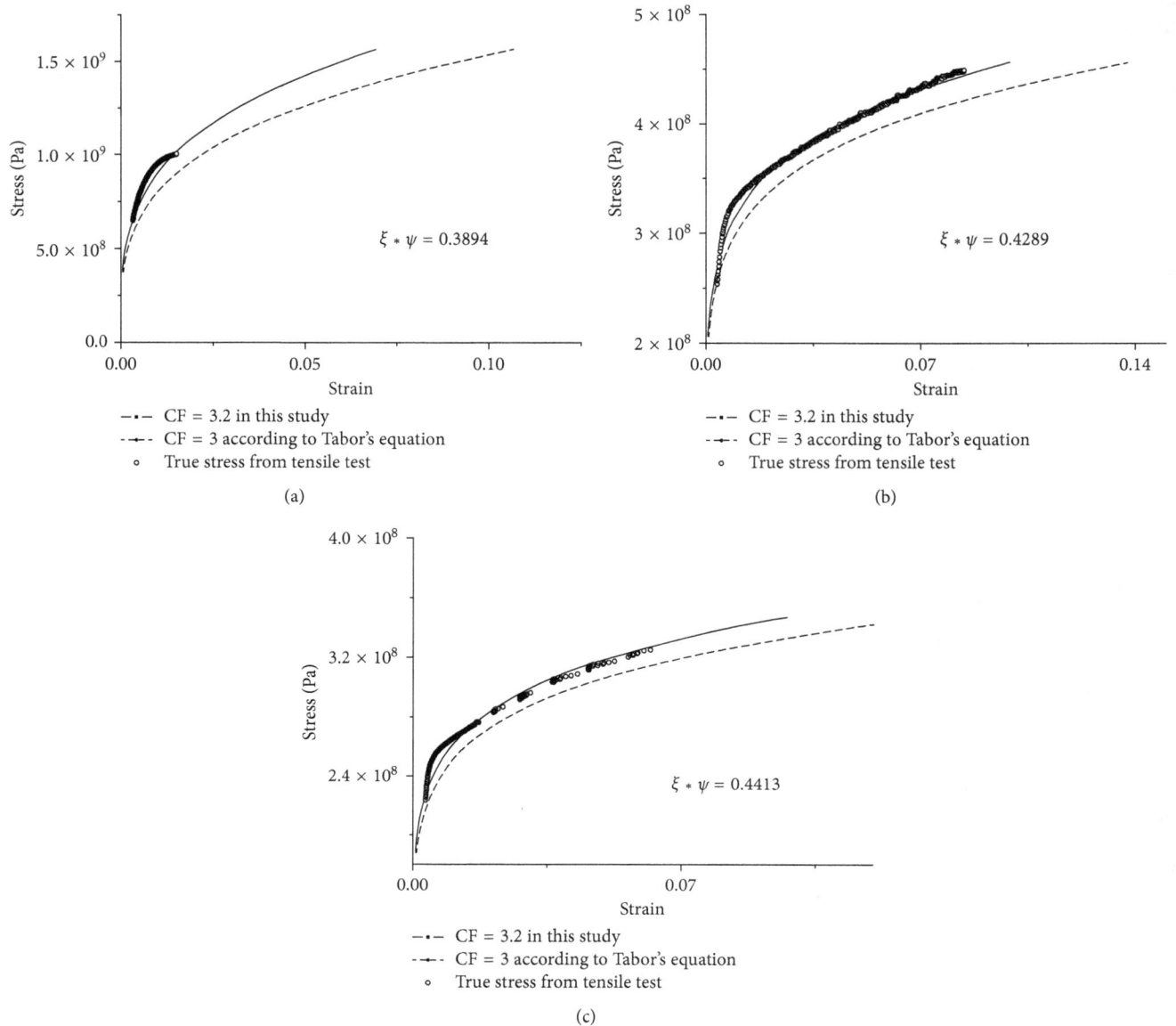

$\xi * \psi = 0.3894$

(a)

$\xi * \psi = 0.4289$

(b)

$\xi * \psi = 0.4413$

(c)

FIGURE 5: Curve plots for the representative stress-strain curve, Tabor's stress-strain curve, and the true stress-strain curve extracted from the tensile test. (a) Steel, (b) brass, and (c) aluminum.

on carbon steel (F-114), aluminum (Al-1050), and brass alloy samples. They were performed on a universal testing machine INSTRON 8501 with a 10 kN load cell at room temperature under displacement control by using a displacement rate of 0.5 mm/min. Cylindrical specimens were machined according to the ASTM E8 standard [47]. The gage length was 12.5 mm for the samples.

4. Results and Discussion

In the tensile test, according to Hollomon's equation, the mechanical properties can be obtained from the stress-strain curve. Young's modulus, yield stress which is determined by means of the 0.2% strain rule, and the hardening exponent are shown in Table 1.

In order to check the radius of the spherical indenter, as shown in Figure 2, the best fitting spherical radius is found to be about 9.2 μm from SEM measurement which is smaller than the nominal value. From the zoomed-in figure, the spherical indenter is worn at shallow penetration depth. In addition, inherent noise of continuous stiffness measurement and the roughness of specimen could scatter the representative stress-strain curve. Consequently, penetration depths less than 10 nm will not be utilized for the extraction of representative stress-strain curve.

Figure 3 shows the load versus penetration depth curves for the three indented materials performed with spherical tip. The indenter curves of brass exhibit considerable scatter, as the diameter of the indenter is comparative to the grain size of brass listed in Table 1. On the contrary, the indenter curves of steel and Al-1050 are much more reproducible for the multiple grain indents. The average of these curves is selected to determine the representative stress-strain curve. The procedure is consistent with the previous work [3].

When Young's modulus was confirmed by the tensile test and the contact stiffness can be measured by CSM, the contact radius can be estimated by Sneddon's equation [48]. It should be noted that, when using (13) to estimate the contact radius, Young's modulus is assumed as a constant during the indentation process [32, 49]. Tabor had concluded that the slope of the linear regression is Meyer's index m which is equal to $n + 2$, where n is the hardening exponent in the Hollomon equation. Figure 4 shows the linear regression of $Log(F)$ versus $Log(a)$ for the experimental materials. The values of hardening exponent from the slope are in good agreement with those obtained from the tensile experiments in Table 1.

In order to check the reasonable determination of constraint factors in Tabor's equation, the tensile stress-strain curves are compared with representative stress-strain curves from a different constraint factor (CF): stress CF = 3.2 in this study and stress CF = 3 according to Tabor's equation. In the experimental analysis, the value of stress constraint factor is taken as 3.2, and the strain constraint factor is confirmed from (12). In order to verify the reasonability of the confirmation of constraint factors from experimental data, the representative stress-strain curve extracted from this study will be compared with the curve extracted from Tabor's method.

Figure 5 shows the plots for the indentation stress-strain curve using the constraint factors in the study, the indentation stress-strain curve using the constraint factors in Tabor's equation, and the true stress-strain curve extracted from the tensile test. The values of E/σ_y for studied materials range from 250 to 320. Although the initial part of the true stress-strain curve and the representative stress-strain curves shows little derivation between them, the representative stress-strain curve using the constraint factors in the study shows well agreement with the true stress-strain curve from the tensile test in the fully plastic regime. In this study, the product ϕ is attained from the intersection between the elastic regime and the fully plastic regime, in which the effect of the transition from the elastic to plastic regime is ignored. This assumption is only suitable for the soft metal which has short transition of the elastic-plastic regime, and the fully plastic regime is developed at shallow penetration depth. For metals with longer transition of the elastic-plastic regime, it can be seen that there is significant derivation in the initial part of the representative stress-strain curve regarding the true stress strain curve obtained from tensile test. As noted in literature [32, 33, 50, 51], it is very difficult to make a comparable stress-strain curve with the tensile test in the elastic-plastic transition by Tabor's representative method for the metallic materials with low E/σ_y. The FEM simulations also showed that representative stress-strain from Kalidindi's equation led to large deviation for some materials with large transition of the elastic-plastic regime [52]. However, in the study, utilizing the proposed procedure to determine the constraint factors could make a comparable stress-strain curve in the fully plastic regime.

5. Conclusion

The main conclusion in the paper can be summarized as follows:

(1) In this study, a new insight into the constraint factors in the formation of Tabor's equation is analytically described. The product ϕ of strain constraint factor β and stress constraint factor ψ is constant at the transition position between the elastic regime and fully plastic regime, which is related to Young's modulus of contact materials.

(2) An experimental procedure is performed to evaluate the analytical analysis for extracting the representative stress-strain curve from the spherical indentation. The representative stress-strain curve using the proposed constraint factors in the study shows well agreement with the nominal true stress-strain curve.

(3) Due to the drawback of the formation of Tabor's equation, the representative stress-strain curve derived from the Tabor's formation could have large deviation at the initial strain period for the materials with longer elastic-plastic transition. In future, formation of new representative strain for spherical indentation needs more study.

Nomenclature

p_m: Mean pressure
a: Contact radius
R: Indenter radius
m: Meyer's index
k: Coefficient in Meyer's equation
ε_r: Representative strain
σ_r: Representative stress
β: Indentation strain constraint factor
ψ: Indentation stress constraint factor
E: Elastic modulus
σ_y: Yield stress
K: Strength coefficient in Hollomon's equation
n: Strain-hardening exponent
ν: Poisson's ratio
ϕ: Product of stress constraint factor ψ and strain constraint factor β
e: Elastic component.

Conflicts of Interest

The authors declare that they have no conflicts of interest regarding the publication of this paper.

Acknowledgments

This work was supported by the National Key R&D Program of China (2016YFB1200403-A-02), the Natural Science Foundation of Shanxi Province (no. 201701D221008), the

PhD Research Startup Foundation of Taiyuan University of Science and Technology (no. 201704), the Shanxi Province Science Foundation for Youths (no. 2015021021), and the National Science Foundation for Young Scientists of China (no. 11702182).

References

[1] M. F. Doerner and W. D. Nix, "A method for interpreting the data from depth-sensing indentation instruments," *Journal of Materials Research*, vol. 1, no. 4, pp. 601–609, 1986.

[2] W. C. Oliver and G. M. Pharr, "An improved technique for determining hardness and elastic modulus using load and displacement sensing indentation experiments," *Journal of Materials Research*, vol. 7, no. 6, pp. 1564–1583, 1992.

[3] J. Dean, J. Wheeler, and T. Clyne, "Use of quasi-static nanoindentation data to obtain stress–strain characteristics for metallic materials," *Acta Materialia*, vol. 58, no. 10, pp. 3613–3623, 2010.

[4] M. Á. Garrido, I. Giráldez, L. Ceballos, and J. Rodríguez, "On the possibility of estimating the fracture toughness of enamel," *Dental Materials*, vol. 30, no. 11, pp. 1224–1233, 2014.

[5] P. Haušild, A. Materna, and J. Nohava, "On the identification of stress–strain relation by instrumented indentation with spherical indenter," *Materials & Design*, vol. 37, pp. 373–378, 2012.

[6] J. S. Field and M. V. Swain, "Determining the mechanical properties of small volumes of material from submicrometer spherical indentations," *Journal of Materials Research*, vol. 10, no. 1, pp. 101–112, 1995.

[7] A. C. Fischer-Cripps, *Factors Affecting Nanoindentation Test Data*, Springer, Berlin, Germany, 2000.

[8] D. Ma, C. W. Ong, J. Lu, and J. He, "Methodology for the evaluation of yield strength and hardening behavior of metallic materials by indentation with spherical tip," *Journal of Applied Physics*, vol. 94, no. 1, pp. 288–294, 2003.

[9] Y. P. Cao and J. Lu, "A new method to extract the plastic properties of metal materials from an instrumented spherical indentation loading curve," *Acta Materialia*, vol. 52, no. 13, pp. 4023–4032, 2004.

[10] H. Lee, J. H. Lee, and G. M. Pharr, "A numerical approach to spherical indentation techniques for material property evaluation," *Journal of the Mechanics and Physics of Solids*, vol. 53, no. 9, pp. 2037–2069, 2005.

[11] M. Zhao, N. Ogasawara, N. Chiba, and X. Chen, "A new approach to measure the elastic–plastic properties of bulk materials using spherical indentation," *Acta Materialia*, vol. 54, no. 1, pp. 23–32, 2006.

[12] P. Jiang, T. Zhang, Y. Feng, R. Yang, and N. Liang, "Determination of plastic properties by instrumented spherical indentation: expanding cavity model and similarity solution approach," *Journal of Materials Research*, vol. 24, no. 3, pp. 1045–1053, 2009.

[13] M.-Q. Le, "Material characterization by instrumented spherical indentation," *Mechanics of Materials*, vol. 46, pp. 42–56, 2012.

[14] A. Muliana, R. M. Haj-Ali, R. Steward, and A. Saxena, "Artificial neural network and finite element modeling of nanoindentation tests," *Metallurgical and Materials Transactions A*, vol. 33, pp. 1939–1947, 2002.

[15] N. Huber, W. Nix, and H. Gao, "Identification of elastic-plastic material parameters from pyramidal indentation of thin films," *Proceedings of the Royal Society of London A: Mathematical, Physical and Engineering Sciences: The Royal Society*, vol. 458, no. 2023, pp. 1593–1620, 2002.

[16] N. A. Branch, G. Subhash, N. K. Arakere, and M. A. Klecka, "Material-dependent representative plastic strain for the prediction of indentation hardness," *Acta Materialia*, vol. 58, no. 19, pp. 6487–6494, 2010.

[17] B. Guelorget, M. François, C. Liu, and J. Lu, "Extracting the plastic properties of metal materials from micro-indentation tests: experimental comparison of recently published methods," *Journal of Materials Research*, vol. 22, no. 6, pp. 1512–1519, 2007.

[18] E. Meyer, "Investigations of hardness testing and hardness," *Phys Z*, vol. 9, p. 66, 1908.

[19] D. Tabor, *The Hardness of Metals*, Clarendon, Oxford, UK, 1951.

[20] F. O. Sonmez and A. Demir, "Analytical relations between hardness and strain for cold formed parts," *Journal of Materials Processing Technology*, vol. 186, no. 1–3, pp. 163–173, 2007.

[21] O. Richmond, H. Morrison, and M. Devenpeck, "Sphere indentation with application to the Brinell hardness test," *International Journal of Mechanical Sciences*, vol. 16, no. 1, pp. 75–82, 1974.

[22] J. R. Matthews, "Indentation hardness and hot pressing," *Acta Metallurgica*, vol. 28, no. 3, pp. 311–318, 1980.

[23] E. G. Herbert, W. Oliver, and G. Pharr, "On the measurement of yield strength by spherical indentation," *Philosophical Magazine*, vol. 86, no. 33–35, pp. 5521–5539, 2006.

[24] B. Xu and X. Chen, "Determining engineering stress–strain curve directly from the load–depth curve of spherical indentation test," *Journal of Materials Research*, vol. 25, no. 12, pp. 2297–2307, 2010.

[25] H. Habbab, B. Mellor, and S. Syngellakis, "Post-yield characterisation of metals with significant pile-up through spherical indentations," *Acta Materialia*, vol. 54, no. 7, pp. 1965–1973, 2006.

[26] Y. Tirupataiah and G. Sundararajan, "On the constraint factor associated with the indentation of work-hardening materials with a spherical ball," *Metallurgical Transactions A*, vol. 22, no. 10, pp. 2375–2384, 1991.

[27] M. Yetna N'Jock, D. Chicot, X. Decoopman, J. Lesage, J. M. Ndjaka, and A. Pertuz, "Mechanical tensile properties by spherical macroindentation using an indentation strain-hardening exponent," *International Journal of Mechanical Sciences*, vol. 75, pp. 257–264, 2013.

[28] B. Taljat, T. Zacharia, and F. Kosel, "New analytical procedure to determine stress-strain curve from spherical indentation data," *International Journal of Solids and Structures*, vol. 35, no. 33, pp. 4411–4426, 1998.

[29] J.-H. Ahn and D. Kwon, "Derivation of plastic stress–strain relationship from ball indentations: examination of strain definition and pileup effect," *Journal of Materials Research*, vol. 16, no. 11, pp. 3170–3178, 2001.

[30] Y. V. Milman, B. Galanov, and S. Chugunova, "Plasticity characteristic obtained through hardness measurement," *Acta Metallurgica et Materialia*, vol. 41, no. 9, pp. 2523–2532, 1993.

[31] K. Fu, L. Chang, B. Zheng, Y. Tang, and H. Wang, "On the determination of representative stress–strain relation of metallic materials using instrumented indentation," *Materials & Design*, vol. 65, pp. 989–994, 2015.

[32] S. R. Kalidindi and S. Pathak, "Determination of the effective zero-point and the extraction of spherical nanoindentation stress–strain curves," *Acta Materialia*, vol. 56, no. 14, pp. 3523–3532, 2008.

[33] D. K. Patel and S. R. Kalidindi, "Correlation of spherical nanoindentation stress-strain curves to simple compression stress-strain curves for elastic-plastic isotropic materials using finite element models," *Acta Materialia*, vol. 112, pp. 295–302, 2016.

[34] M. M. Chaudhri, "Subsurface plastic strain distribution around spherical indentations in metals," *Philosophical Magazine A*, vol. 74, no. 5, pp. 1213–1224, 1996.

[35] F. Yang and J. C.-M. Li, *Micro And Nano Mechanical Testing of Materials And Devices*, Springer, Berlin, Germany, 2008.

[36] K. L. Johnson, "The correlation of indentation experiments," *Journal of the Mechanics and Physics of Solids*, vol. 18, no. 2, pp. 115–126, 1970.

[37] J.-L. Bucaille, S. Stauss, E. Felder, and J. Michler, "Determination of plastic properties of metals by instrumented indentation using different sharp indenters," *Acta Materialia*, vol. 51, no. 6, pp. 1663–1678, 2003.

[38] E. G. Herbert, G. M. Pharr, W. C. Oliver, B. N. Lucas, and J. L. Hay, "On the measurement of stress–strain curves by spherical indentation," *Thin Solid Films*, vol. 398-399, pp. 331–335, 2001.

[39] H. O'Neill, *Hardness Measurement of Metals and Alloys*, Chapman & Hall, London, UK, 1967.

[40] M. C. Shaw and G. J. DeSalvo, "On the plastic flow beneath a blunt axisymmetric indenter," *Journal of Engineering for Industry*, vol. 92, no. 2, pp. 480–492, 1970.

[41] R. T. Shield, "On the plastic flow of metals under conditions of axial symmetry," *Proceedings of the royal society of London A: Mathematical, Physical and Engineering Sciences: The Royal Society*, vol. 233, no. 1193, pp. 267–287, 1955.

[42] M. Scibetta, E. Lucon, R. Chaouadi, and E. van Walle, "Instrumented hardness testing using a flat punch," *International Journal of Pressure Vessels and Piping*, vol. 80, no. 6, pp. 345–349, 2003.

[43] G. Pharr, W. Oliver, and F. Brotzen, "On the generality of the relationship among contact stiffness, contact area, and elastic modulus during indentation," *Journal of Materials Research*, vol. 7, no. 3, pp. 613–617, 1992.

[44] X. Li and B. Bhushan, "A review of nanoindentation continuous stiffness measurement technique and its applications," *Materials Characterization*, vol. 48, no. 1, pp. 11–36, 2002.

[45] S. Vachhani, R. Doherty, and S. Kalidindi, "Effect of the continuous stiffness measurement on the mechanical properties extracted using spherical nanoindentation," *Acta Materialia*, vol. 61, no. 10, pp. 3744–3751, 2013.

[46] W. C. Oliver and G. M. Pharr, "Measurement of hardness and elastic modulus by instrumented indentation: advances in understanding and refinements to methodology," *Journal of Materials Research*, vol. 19, no. 1, pp. 3–20, 2004.

[47] Standard A. E8, "Standard test methods for tension testing of metallic materials," *Annual Book of ASTM Standards*, vol. 3, pp. 57–72, 2004.

[48] I. N. Sneddon, "The relation between load and penetration in the axisymmetric Boussinesq problem for a punch of arbitrary profile," *International Journal of Engineering Science*, vol. 3, no. 1, pp. 47–57, 1965.

[49] S. R. Kalidindi, C. A. Bronkhorst, and L. Anand, "Crystallographic texture evolution in bulk deformation processing of FCC metals," *Journal of the Mechanics and Physics of Solids*, vol. 40, no. 3, pp. 537–569, 1992.

[50] S. Pathak and S. R. Kalidindi, "Spherical nanoindentation stress–strain curves," *Materials Science and Engineering: R: Reports*, vol. 91, pp. 1–36, 2015.

[51] K.-H. Kim, Y.-C. Kim, E.-c. Jeon, and D. Kwon, "Evaluation of indentation tensile properties of Ti alloys by considering plastic constraint effect," *Materials Science and Engineering: A*, vol. 528, pp. 5259–5263, 2011.

[52] C. Chang, *Determining the Elastic-Plastic Properties of Metallic Materials Through Instrumented Indentation*, Doctoral dissertation, Caminos, Universidad Politécnica de Madrid, Madrid, Spain, 2016.

18

Study on the Thermal Properties of Hollow Shale Blocks as Self-Insulating Wall Materials

Guo-liang Bai, Ning-jun Du, Ya-zhou Xu, and Chao-gang Qin

School of Civil Engineering, Xi'an University of Architecture and Technology, Shaanxi 710055, China

Correspondence should be addressed to Ning-jun Du; duningjun@live.xauat.edu.cn

Academic Editor: Wei Zhou

To reduce energy consumption and protect the environment, a type of hollow shale block with 29 rows of holes was designed and produced. This paper investigated the thermal properties of hollow shale blocks and walls. First, the guarding heat-box method was used to obtain the heat transfer coefficient of the hollow shale block walls. The experimental heat transfer coefficient is $0.726\,\text{W/m}^2\cdot\text{K}$, which would save energy compared to traditional wall materials. Then, the theoretical value of the heat transfer coefficient was calculated to be $0.546\,\text{W/m}^2\cdot\text{K}$. Furthermore, the one-dimensional steady heat conduction process for the block and walls was simulated using the finite element analysis software ANSYS. The predicted heat transfer coefficient for the walls was $0.671\,\text{W/m}^2\cdot\text{K}$, which was in good agreement with the test results. With the outstanding self-insulation properties, this type of hollow shale block could be used as a wall material without any additional insulation measures in masonry structures.

1. Introduction

Worldwide, economic development has been increasingly restricted by a shortage of natural resources [1]. Furthermore, economic growth results in problems such as destruction of the environment and resource waste. To improve this situation and to promote building energy efficiency, the traditional solid clay bricks have been officially forbidden in building construction, promoting the study and application of new wall materials [2].

Currently, there are many types of new wall materials, such as small concrete hollow block, aerated concrete block, and small hollow fly ash block. However, none of these wall materials are self-insulating, and certain external wall thermal insulation measures are required. External insulation measures for exterior walls are widely used in construction despite some obvious shortcomings such as easily falling off, short service life, and low safety. Furthermore, in traditional brick masonry, the thickness of mortar joints varies from 8 mm to 12 mm, easily forming obvious thermal bridges and resulting in significant energy loss.

Over the past 40 years, a variety of insulation sintered hollow blocks have been developed, such as those proposed by Porothem, Klimation, Poroton, Thermopor, Unipor, Monomur, and Thermoarcilla [3]. All of these blocks have the merits of low density, high hole rate, high surface smoothness, and good thermal performance. Zhu et al. [4] investigated the thermal properties of recycled aggregate concrete (RAC) and recycled concrete blocks. Sodupe-Ortega et al. [5] manufactured a type of rubberized long hollow block and studied the technical and economic feasibility of producing these blocks using automated brick machines. Zhang et al. [6] studied the thermal performance of concrete hollow blocks using FEM simulations. Fan et al. [7] described a new building material named expanded polystyrene recycled concrete and conducted corresponding numerical simulations for hollow EPSRC blocks and thermally insulating walls based on thermodynamic principles. In recent works, numerical simulation methods have been proposed by Del Coz Díaz et al. [8–11] for studying various types of walls made of different light concrete hollow bricks. Li et al. [12] presented the development of a simplified heat transfer model of hollow blocks for simple and efficient heat flow calculation.

Hollow shale block consists of shale as the main raw material, sawdust as the pore-forming agent, and industrial wastes such as fly ash, steel slag, and wastepaper chips as auxiliary materials. All of these raw materials are fired following a certain production process to make a new energy-saving

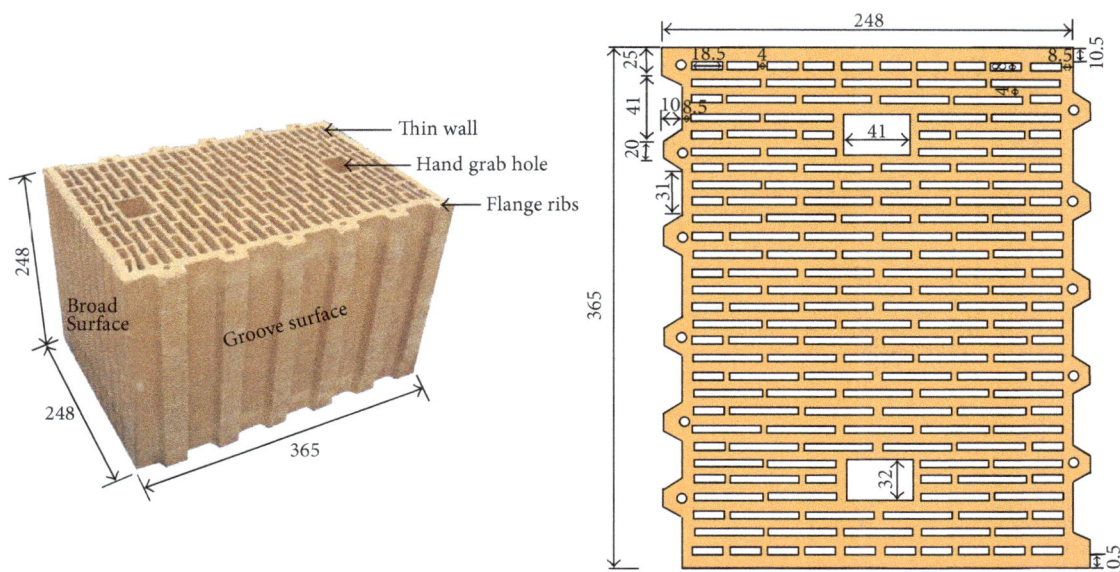

FIGURE 1: Detailed dimensions of a hollow shale block (in mm).

and environmentally friendly wall material that has the advantages of light weight, large size, high hole rate, and high smoothness. Meanwhile, hollow shale blocks make full use of the abundant shale resources to save farmland. During the process of building walls with hollow shale blocks, a mortar joint construction technology with a thickness of 1~ 2 mm is developed to significantly reduce the heat loss caused by structural thermal bridges. Excellent thermal insulation properties and energy efficiency of residential buildings in severe cold and cold areas are expected to be achieved in external walls without external insulation measures. Wu et al. [13] investigated the mechanical and thermal properties of fired hollow block walls. Bai et al. [14, 15] studied the seismic behavior of fired heat-insulated shale block walls with ultrathin mortar joints.

The heat transfer coefficient is one of the most important parameters for evaluating the thermal performance of walls. For a specified ambient temperature, the lower the heat transfer coefficient is, the less the heat dissipated through the wall is. Currently, the heat transfer coefficients of walls are mainly determined by on-site measurements or lab testing [16]. In this study, the heat transfer coefficients of hollow shale block walls were obtained from lab testing and compared with the theoretical calculation and finite element simulation results. Section 2 presents detailed dimensions, production procedures, chemical components, and mineral composition of the hollow shale block.

2. Hollow Shale Block

2.1. Details of the Hollow Shale Block. The dimensions of the blocks are 365 mm × 248 mm × 248 mm with 29 rows of holes; the density is 850 kg/m^3, which could significantly

TABLE 1: Chemical composition of shale.

Chemical constituents	Content (wt%)
SiO$_2$	62.91
Al$_2$O$_3$	17.01
Fe$_2$O$_3$	6.83
CaO	6.13
MgO	2.78
K$_2$O	1.88
Na$_2$O	1.04
SO$_3$	0.65
TiO$_2$	0.77

reduce the building weight and improve the heat insulation efficiency of the blocks. The detailed dimensions are shown in Figure 1.

2.2. Raw Materials

2.2.1. Shale. Shale is an ancient sedimentary rock that was formed through long-term geological processes. Ancient rocks are broken into clay minerals and a small amount of clastic minerals through weathering and are then transported to the sedimentary location in a state of suspension. All of these minerals are deposited mechanically and became clay rocks with a lamellation structure under low temperature and low pressure due to external forces and the diagenesis effect. In China, more than 75% of the land surface is covered by sedimentary rocks, of which shale represents 77.5% [17].

The chemical composition of shale is shown in Table 1; the main mineral components of shale are quartz, calcite,

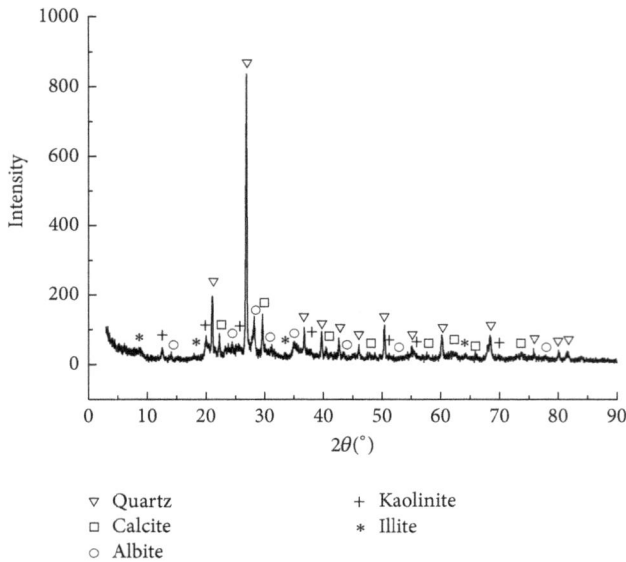

FIGURE 2: XRD spectrum of shale.

Legend:
▽ Quartz   + Kaolinite
□ Calcite   * Illite
○ Albite

3. Experimental Details

To verify the applicability of the hollow shale blocks, the thermal performance testing of the masonry walls was conducted in accordance with Chinese codes [18].

3.1. Specimens. Test walls with dimensions of 1650 mm × 1650 mm × 365 mm (length × height × width) were built using the hollow shale blocks (see Figure 4).

The void ratio of a hollow shale block is as high as 54%, and its compression strength grade reaches 10 MPa. In addition, its honeycomb mesh structure can provide excellent thermal insulation performance. Three specimens were built, and the horizontal mortar joint thickness ranged from 1 mm to 2 mm. Because there was no vertical mortar joint in the test walls, tongue-and-groove connections were used to interlock and strengthen the hollow shale block walls. After the specimens had been fully dried with maintenance for 20 days, the thermal performance was tested.

3.2. Test Apparatus. The schematic of the steady-state heat transfer performance test device is shown in Figure 5, which was designed according to Chinese code GB/T13475-2008 [18] and the guarding heat-box method, as illustrated in Figure 6.

Because the protection box in the guarding heat-box method encircles the metering box, the heat flux through the metering box wall (Q_3) and the heat flux of the flanking loss (Q_2) can be reduced to negligible levels if the internal air temperatures of the protection box and metering box are equal. Theoretically, if a homogeneous specimen is installed in the device whose inside and outside temperatures are uniform, the specimen's surface temperature will be at steady state. In other words, the heat flux through the metering box walls would be equal to the heat flux of the flanking loss ($Q_2 = Q_3 = 0$). However, the heat transfer coefficient of a real homogeneous specimen is always uneven, especially for the parts near the edges of the metering box. Therefore, the surface temperatures of the specimens and close to the metering box are uneven, and the heat flux through the metering box wall (Q_3) and the heat flux of the flanking loss (Q_2) cannot actually be reduced to zero. In the present work, Q_2 and Q_3 can be obtained using the standard calibration test. Furthermore, the heat transfer coefficient K can be calculated using

$$Q_1 = Q_P - Q_3 - Q_2$$
$$R = \frac{A\left(T_{si} - T_{se}\right)}{Q_1}$$
$$K = \frac{Q_1}{A\left(T_{ni} - T_{ne}\right)}. \tag{1}$$

sodium feldspar, kaolinite, and illite. The corresponding XRD spectrum is shown in Figure 2. After mining, crushing, and fine grinding, shale is one of the most promising new wall materials to replace sintered clay brick due to its abundant storage quantities and easy mining.

2.2.2. Pore-Forming Agent. The function of the pore-forming agent is to produce a large number of pores during the sintering process to take advantage of the lower thermal conductivity coefficient of air. Therefore, the pore-forming agent can effectively enhance the insulation performance of hollow shale blocks and reduce their weight, which improves the seismic performance. Considering energy conservation, resource recycling, and environmental protection, sawdust was selected as the pore-forming agent for the hollow shale blocks. As the scrap material of wood processing, sawdust has many advantages when used as the pore-forming agent. Sawdust is mainly made of plant fiber that is stable, and loss on ignition can be as high as 98.49%. The pore-forming can generate many pores inside blocks and enhance the thermal insulation property. In addition, sawdust is also plentiful, cheap, and easy to obtain.

2.2.3. Industrial Wastes. Fly ash, steel slag, and waste paper chips were added during the sintering process as auxiliary materials.

2.3. Production Procedure. As a new type of energy-saving wall material, the production process of hollow shale blocks includes grinding, aging, stirring, extrusion, incision, drying, setting, and high temperature sintering. Most of the processes are automated. The production process of hollow shale blocks is illustrated in Figure 3.

Eq. (1) includes the following variables: thermal power inlet Q_P, heat flux through the specimen Q_1, surface temperature on the warm side T_{si}, surface temperature on the cold side T_{se}, air temperature on the warm side T_{ni}, air temperature on the cold side T_{ne}, surface area of the specimen A, and thermal resistance R.

TABLE 2: Thermal parameters of the hollow shale block walls.

Specimens	Heat transfer coefficient K (W/m^2·K)	Thermal resistance R (m^2·K/W)	Total thermal resistance R_u (m^2·K/W)
A	0.751	1.275	1.332
B	0.726	1.080	1.377
C	0.703	1.342	1.422
Mean	0.726	1.232	1.377

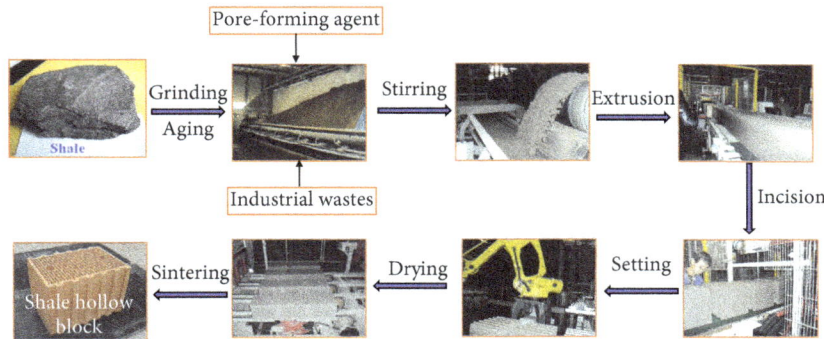

FIGURE 3: Production procedure for hollow shale blocks.

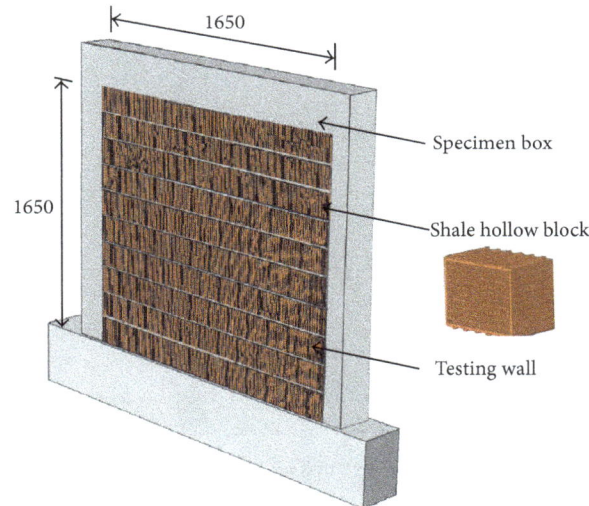

FIGURE 4: Details of the test wall.

3.3. Testing Procedure

(1) After 20 days of natural air drying, the specimens were installed in the test machine. The seam crossing parts between the specimen and the specimen box were filled with a foaming insulation agent for sealing, as shown in Figure 7(a).

(2) The length of the locating rods connected to the temperature sensors inside the cold box and heat metering box were checked and adjusted, as shown in Figure 7(b).

(3) After the test machine had run for more than 20 hours for each specimen and the range of heating power values was between 0.5 W and 3 W, the whole system could be regarded to be at a thermal steady state. Then, the measured data were collected every half hour, and the average value of the test results was calculated.

3.4. Experimental Results and Discussion

Based on the test results of the three hollow shale block walls, thermal parameters such as the heat transfer coefficient, thermal resistance, and total thermal resistance were calculated and are listed in Table 2.

The results indicate that the heat transfer coefficient of hollow shale block walls is 0.726 W/(m^2·K), which meets the

TABLE 3: Comparison with other wall materials.

Wall material	Heat transfer coefficient K (W/m²·K)	Thermal resistance R (m²·K/W)	Dimensions
Hollow shale block	0.726	1.232	365 mm × 248 mm × 248 mm with 29 rows of holes
Clay brick	2.240	0.296	240 mm × 115 mm × 53 mm
Concrete block	2.220	0.300	390 mm × 190 mm × 190 mm with three rows of holes
Recycled concrete blocks	1.620	0.457	390 mm × 240 mm × 190 mm with three rows of holes

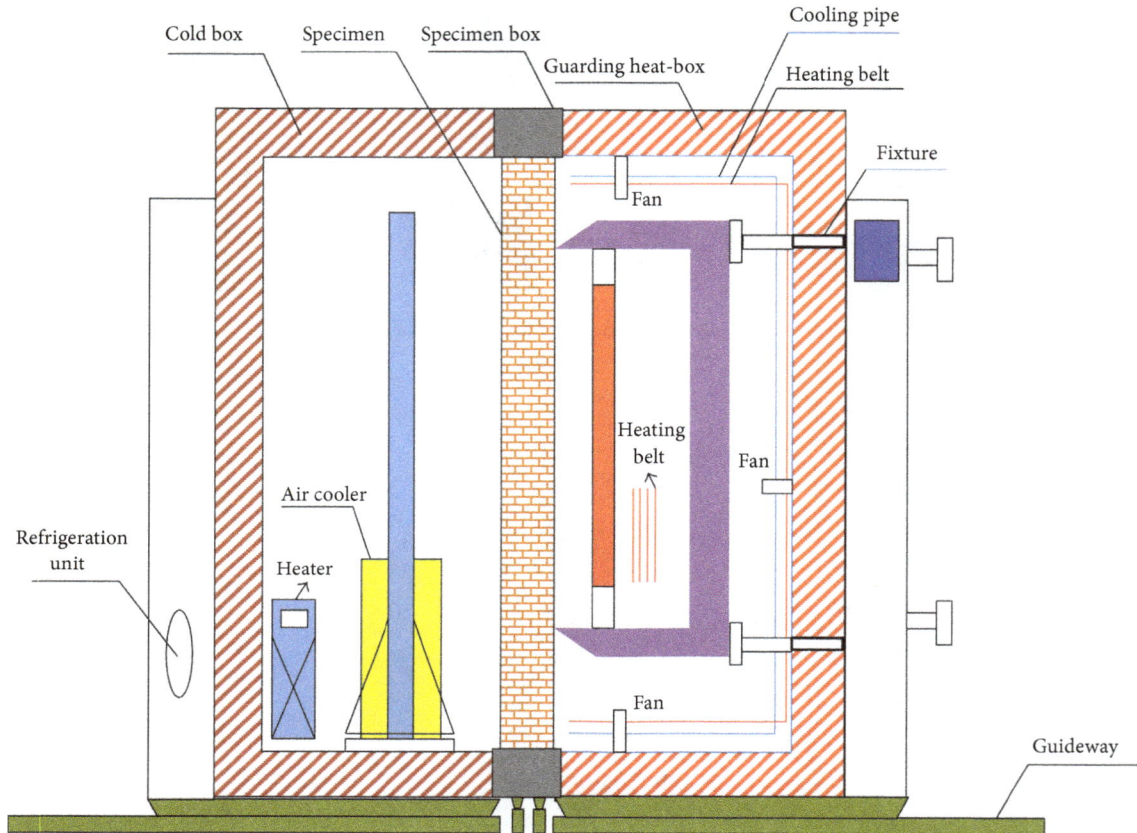

FIGURE 5: Schematic of the test machine for steady-state heat transfer.

design standard for the energy efficiency of public buildings in GB50189-2005 [19].

Heat transfer coefficient K and thermal resistance R of different wall materials which are measured by the same equipment and same testing methods are shown in Table 3 according to the research of Yang et al. [20] and Wu et al. [13] and technical specification for concrete small-sized hollow block masonry buildings of China JGJ/T2011 [21]. The heat preservation effect of the hollow shale block walls is 3.16 times higher than that of traditional clay brick walls, 3.11 times higher than that of concrete block walls, and 1.69 times higher than that of recycled concrete blocks walls. As a building envelope material, hollow shale blocks can not only improve the heat preservation and heat insulation performance of buildings but also make the indoor thermal environment more comfortable, especially in cold regions.

4. Theoretical Calculation of the Heat Transfer Coefficient of Hollow Shale Block Walls

Building envelopes can be categorized as single-layer walls, multilayer walls, and combination walls according to their composition. A multilayer wall, such as a double-sided plastered brick wall, is composed of several layers of different wall materials along the direction of heat flow. The total thermal resistance of a multilayer wall is the sum of the thermal resistance of each single-layer wall. Assuming that the heat transfer is a one-dimensional steady heat transfer

TABLE 4: Thermal parameters of hollow shale block walls.

Area number	1, 21	2, 4, 6, 8, 14, 16, 18, 20	3, 7, 15, 19	5, 17	9, 13	10, 12	11
F_i (mm)	14×248	18.5×248	4×248	4×248	4×248	18.5×248	4×248
R_0	0.938	3.317	2.976	2.074	1.568	3.082	1.767

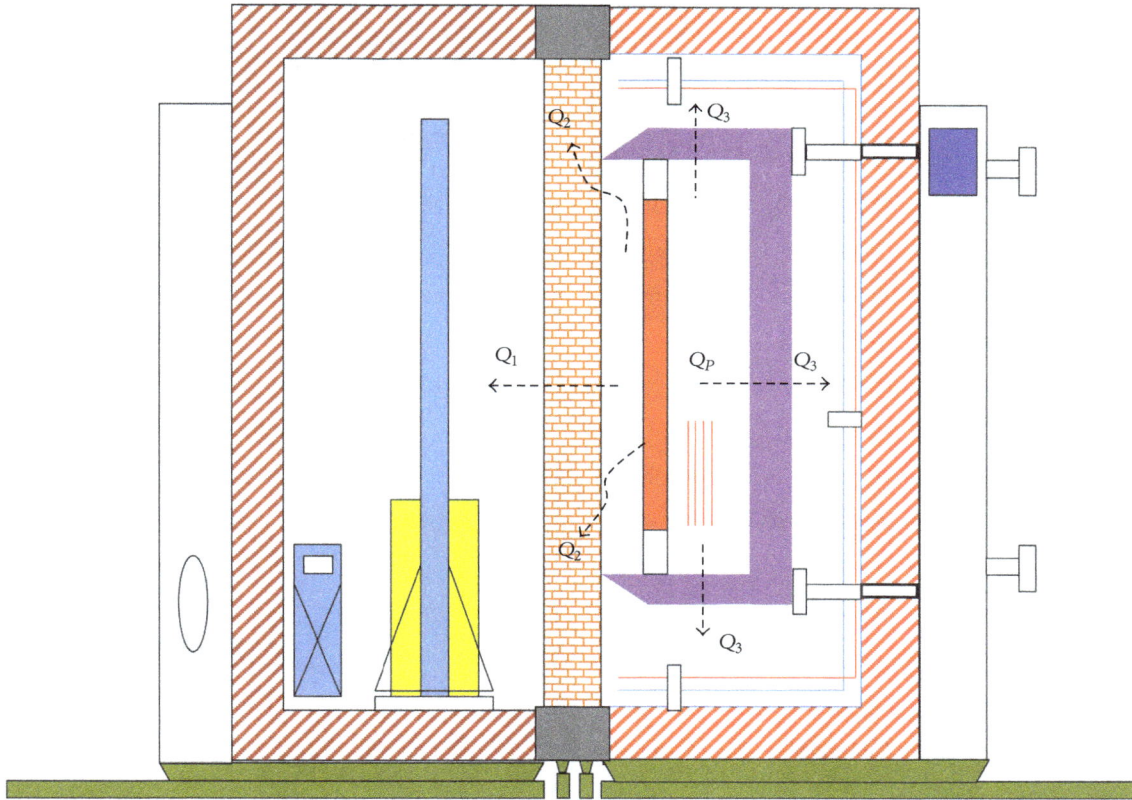

FIGURE 6: Illustration of the test theory.

process, the multilayer wall parallel to the direction of heat flow can be divided into several areas whose interfaces are determined according to the material layer composition [22]. The average thermal resistance of a multilayer wall can be calculated as follows [18]:

$$\overline{R} = \left[\frac{F_0}{F_1/R_{0,1} + F_2/R_{0,2} + \cdots + F_n/R_{0,n}} \left(R_i + R_e \right) \right] \varphi, \quad (2)$$

where $\overline{R}$ is the average thermal resistance, F_0 is the total heat transfer area perpendicular to the direction of heat flow, φ is a correction factor, which is 0.86 for hollow shale block, $F_1, F_2, \ldots, F_n$ are the divided areas parallel to the direction of heat flow, $R_{0,1}, R_{0,2}, \ldots, R_{0,n}$ are the thermal resistances of heat transfer surfaces, R_i is the thermal resistance of the inner surface, which is $0.11 \, \mathrm{m^2 \cdot K/W}$, and R_e is the thermal resistance of the outer surface, which is $0.04 \, \mathrm{m^2 \cdot K/W}$ [18].

Hollow shale blocks with 29 rows of holes are multilayer walls. Their average thermal resistance can be calculated using the method mentioned above. For convenience, the tongue-and-grooves on the side surfaces are neglected. The detailed area division is illustrated in Figure 8.

The total heat transfer surface of a hollow shale block perpendicular to the direction of heat flow is divided into 21 areas. All of these heat transfer areas are multilayer except for areas 1 and 2. The thermal conductivity of sintered shale material is $0.463 \, \mathrm{W/(m \cdot K)}$, the thermal resistance of an 8 mm air layer is $0.12 \, \mathrm{m^2 \cdot K/W}$, and the thermal resistance of a 32 mm air layer is $0.17 \, \mathrm{m^2 \cdot K/W}$. The results of the thermal resistance calculation are listed in Table 4.

The average thermal resistance of hollow shale blocks can be obtained using (2): $\overline{R} = 1.688 \, \mathrm{m^2 \cdot K/W}$. The average heat transfer coefficient can be obtained as follows:

$$K_0 = \frac{1}{R_i + \overline{R} + R_e} = 0.544 \left(\mathrm{W/m^2 \cdot K} \right). \quad (3)$$

Assuming that the thickness of the horizontal mortar is 2 mm and taking a block and a horizontal mortar joint as the typical unit, the heat transfer coefficients is

$$K = \frac{K_0 A_0 + K_1 A_1}{A_0 + A_1} = 0.546 \left(\mathrm{W/m^2 \cdot K} \right), \quad (4)$$

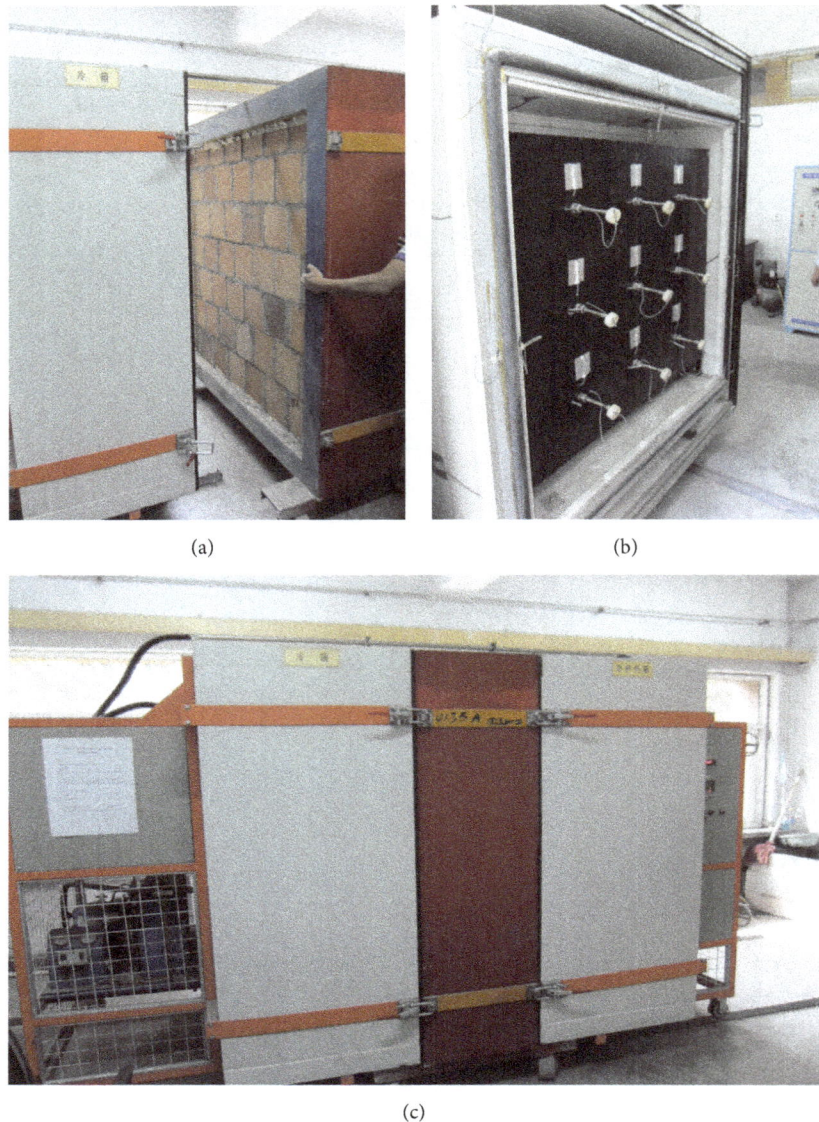

(a)

(b)

(c)

FIGURE 7: Testing procedure.

where A_0 and A_1 are the lateral areas of the hollow shale block and mortar joint, respectively, and K_0 and K_1 are the heat transfer coefficients of the hollow shale blocks and mortar joint, respectively. Compared with the experimental test results, the theoretical calculated values of K_0 and K of the hollow shale blocks are smaller due to the simplification on both sides of the hollow shale block.

5. Numerical Simulation Using the Finite Element Method

5.1. FEM Model. To provide an alternative thermal analysis and design of the hollow shale block, an FEM model using the three-dimensional thermal element SOLID70 was developed using the ANSYS package, as shown in Figure 9.

Considering the thermal resistance effect between air layers, the holes in the blocks were treated as solid elements

with the parameters of the air interlayer property. The heat flow between different materials was regarded as a continuous process. According to the temperatures of the hot chamber and the cold chamber, the heat transfer coefficient and temperature loads were defined on the surfaces of the blocks. The internal surface temperature is 30°C, and the external surface temperature is −10°C.

In fact, parameters for FEM simulation are crucial to reasonable calculation results. In present FEM models, values of the parameters needed to be given were set based on the thermal design code for civil building of China [23]. The convective heat transfer coefficients of the inner surface (guarding heat-box) and outer surface (cold box) of the hollow shale block wall are 8.7 W/(m²·K) and 23.0 W/(m²·K), respectively. The thermal conductivity of sintered shale material is 0.463 W/(m·K), the thermal conductivity of an 8 mm air layer is 0.067 W/(m·K), and the thermal conductivity of a

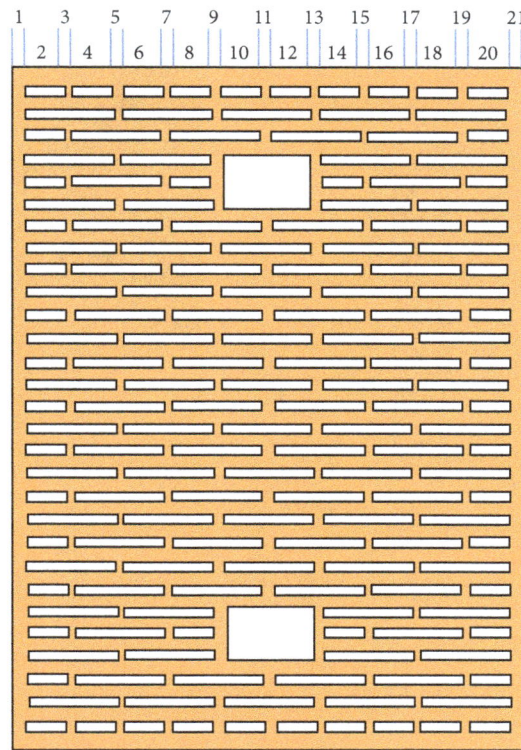

FIGURE 8: Area division of a hollow shale block.

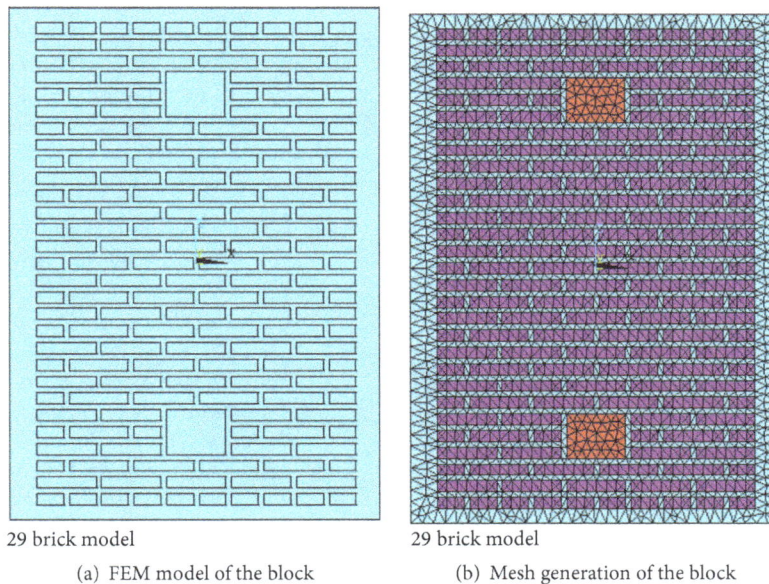

29 brick model

(a) FEM model of the block

29 brick model

(b) Mesh generation of the block

FIGURE 9: FEM model of the hollow shale block.

32 mm air layer is 0.188 W/(m·K). The thermal conductivity of the mortar is 0.339 W/(m·K).

Because there is no vertical mortar joint, the influence of the vertical connections can be neglected in the FEM model. The vertical joint between shale blocks was symmetrical, and the symmetry plane was considered to be an adiabatic boundary, meaning that there was no heat exchange on either

side of the symmetry plane. The corresponding FEM meshes and loading process of the walls are shown in Figure 10, in which the boundary conditions and temperature simulation are the same as those of the shale block.

5.2. Simulation Results. The simulated temperature field and heat flow density for the hollow shale block are shown in

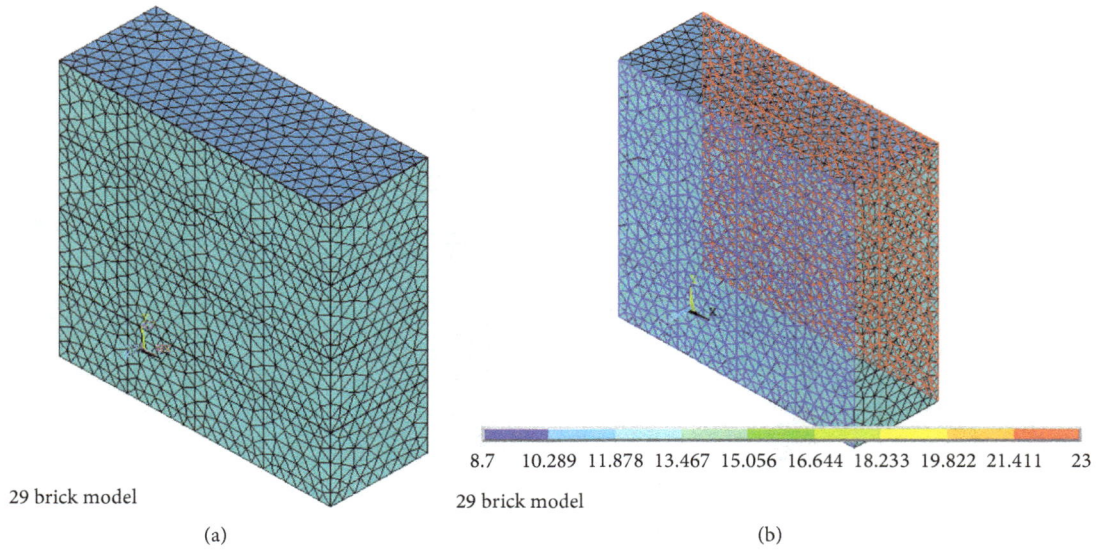

29 brick model   29 brick model

(a)   (b)

FIGURE 10: FEM meshes (a) and loading process (b) for the hollow shale block wall.

Figure 11. It is observed that the temperature distribution in the block varies linearly along the direction of heat flow and is distributed uniformly. The heat flux density and temperature gradient of the hollow shale block gradually increase from outside to inside. The heat flux density and temperature gradient are small for the air interlayer inside the block but larger on the rib between air interlayers along the direction of heat flow. In addition, the most heat dissipation per unit area occurs in the ribs of the hollow shale block. It is easy to determine that the internal air layer is beneficial for prevent the heat loss.

Figure 12 displays the simulation results of the hollow shale block wall. In the vertical junction of two blocks, there is no air interlayer along the direction of heat flow, especially around the edges of the blocks, where the heat flow is strong and the temperature gradient significantly changes. Conversely, the heat flow is small and the change in the temperature gradient is not as large on the horizontal mortar joints. The heat flux density vector also indicates that there is less heat loss through the horizontal mortar joints. The heat transfer effect of the hollow shale blocks depends on the masonry mortar, the quality of the wall masonry, and the mortar joint thickness. The 2 mm mortar joints of the hollow shale block wall are sufficiently thin that their influence on the thermal properties can reasonably be neglected.

Although the heat transfer coefficient cannot be directly obtained from the FEM simulation results, it can be calculated based on the following formula:

$$\lambda = \frac{q\delta}{\Delta t}$$

$$\overline{R} = \frac{\delta}{\lambda} \qquad (5)$$

$$K_0 = \frac{1}{R_i + \overline{R} + R_e},$$

where q is the average value of heat flux, which can be taken from the heat flow density distribution map, δ is the thickness of the wall, and Δt is the temperature difference between the internal and external surfaces of the wall. The heat transfer coefficient of the hollow shale block walls obtained using this method is $0.671 \, \text{W/m}^2 \cdot \text{K}$, which is less than the experimental value but greater than the theoretical result in Section 4.

Compared with the experimental results, the theoretical values and FEM simulation results of the heat transfer coefficients of the hollow shale blocks are smaller. The possible reasons for the difference are as follows:

(1) There are cracks on the surface or internal damage formed during transportation of the blocks, which will affect the thermal performance of the masonry wall.

(2) In the process of masonry, when two blocks interlock each other tightly, several closed air layers will emerge between two blocks theoretically. However, due to the deflections of the blocks in the production process, the air layers between two blocks may be interlinked inside and outside the wall which will cause heat loss through this channel and affect the thermal performance of the wall.

Besides the experimental and numerical methods, analytical methods, for example, homogenization method, are alternative ways to investigate the equivalent thermal properties. Homogenization is a quite general strategy which predicts the macrobehavior of a medium based on its microstructures and properties. Masonry structure may be approximately considered as a periodic composite continuum; it is made up of two different materials (brick or block and mortar) arranged in a periodic way. The homogenization theory for periodic media allows the global behavior of masonry to be derived from the behavior of the constitutive materials. So far, the homogenization approach has been used to study the mechanical

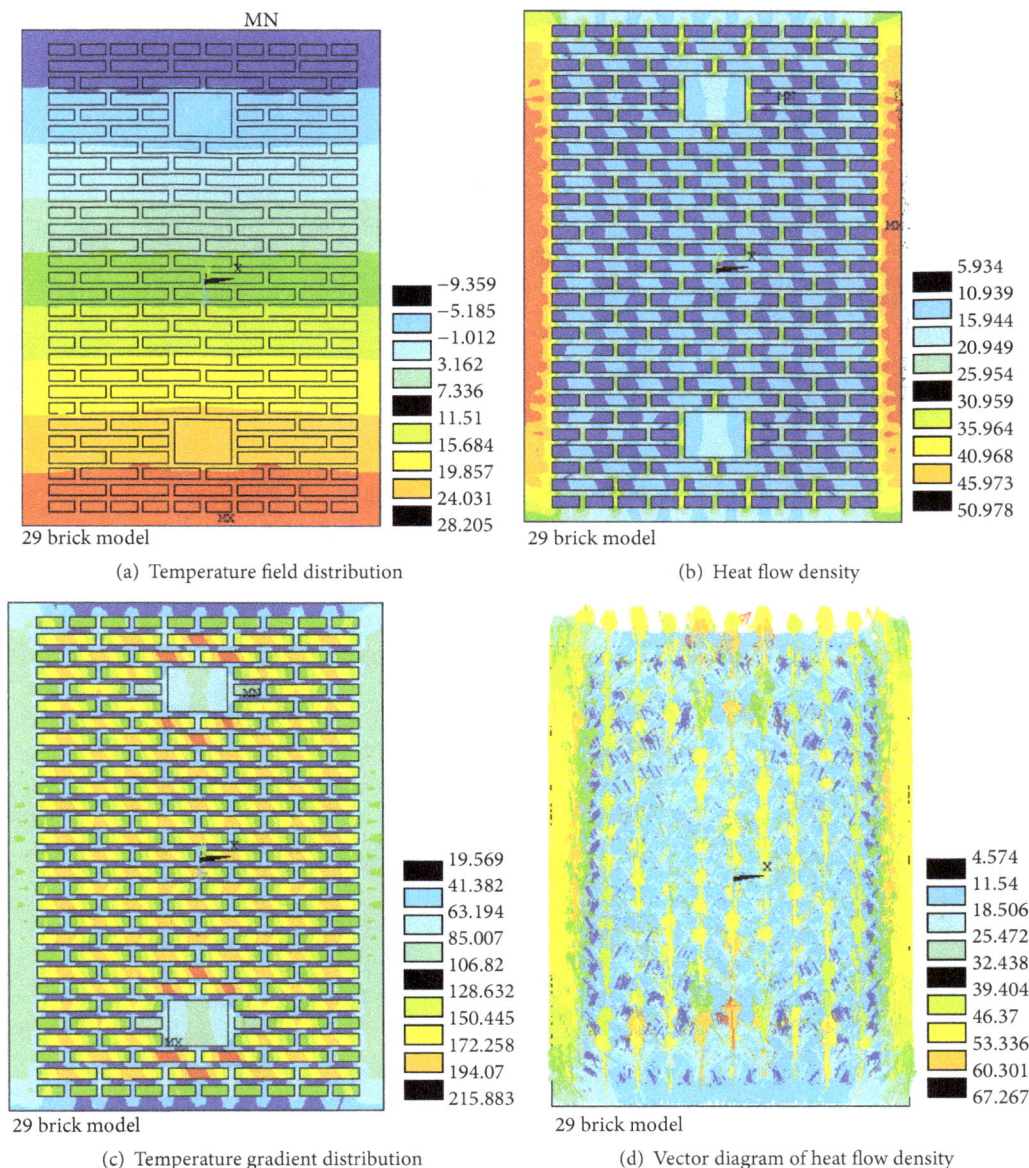

(a) Temperature field distribution

(b) Heat flow density

(c) Temperature gradient distribution

(d) Vector diagram of heat flow density

FIGURE 11: Simulation results for the hollow shale block.

properties of the masonry structure [24–26]. Few researches have been conducted on the thermal properties by this method. In the following studies, it is expected that homogenization strategy could be successively employed to predict the thermal properties of masonry walls staring from thermal properties and compositional structures of block and mortar.

6. Conclusion

This study investigates the thermal properties of hollow shale blocks using experimental testing, theoretical calculation, and FEM simulations. The following conclusions can be drawn from this research:

(i) The experimental heat transfer coefficient of hollow shale block walls is $0.726\,\text{W/m}^2\cdot\text{K}$, which meets

the design codes and shows their remarkable self-insulation characteristics compared with other wall materials.

(ii) Using the theoretical formula, the heat transfer coefficient of a single hollow shale block is $0.544\,\text{W/m}^2\cdot\text{K}$, and the heat transfer coefficient of a hollow shale block wall is $0.546\,\text{W/m}^2\cdot\text{K}$. Using the FEM simulation, the heat transfer coefficient of a hollow shale block wall is $0.671\,\text{W/m}^2\cdot\text{K}$. The simplification on both sides of the hollow shale blocks may contribute to the higher experimental heat transfer coefficient.

(iii) The strong heat flow and large temperature gradient mainly appear in the vertical junctions of two blocks because there is no air interlayer along the heat flow

(a) Temperature field distribution

(b) Heat flow density

(c) Temperature gradient distribution

(d) Vector diagram of heat flow density

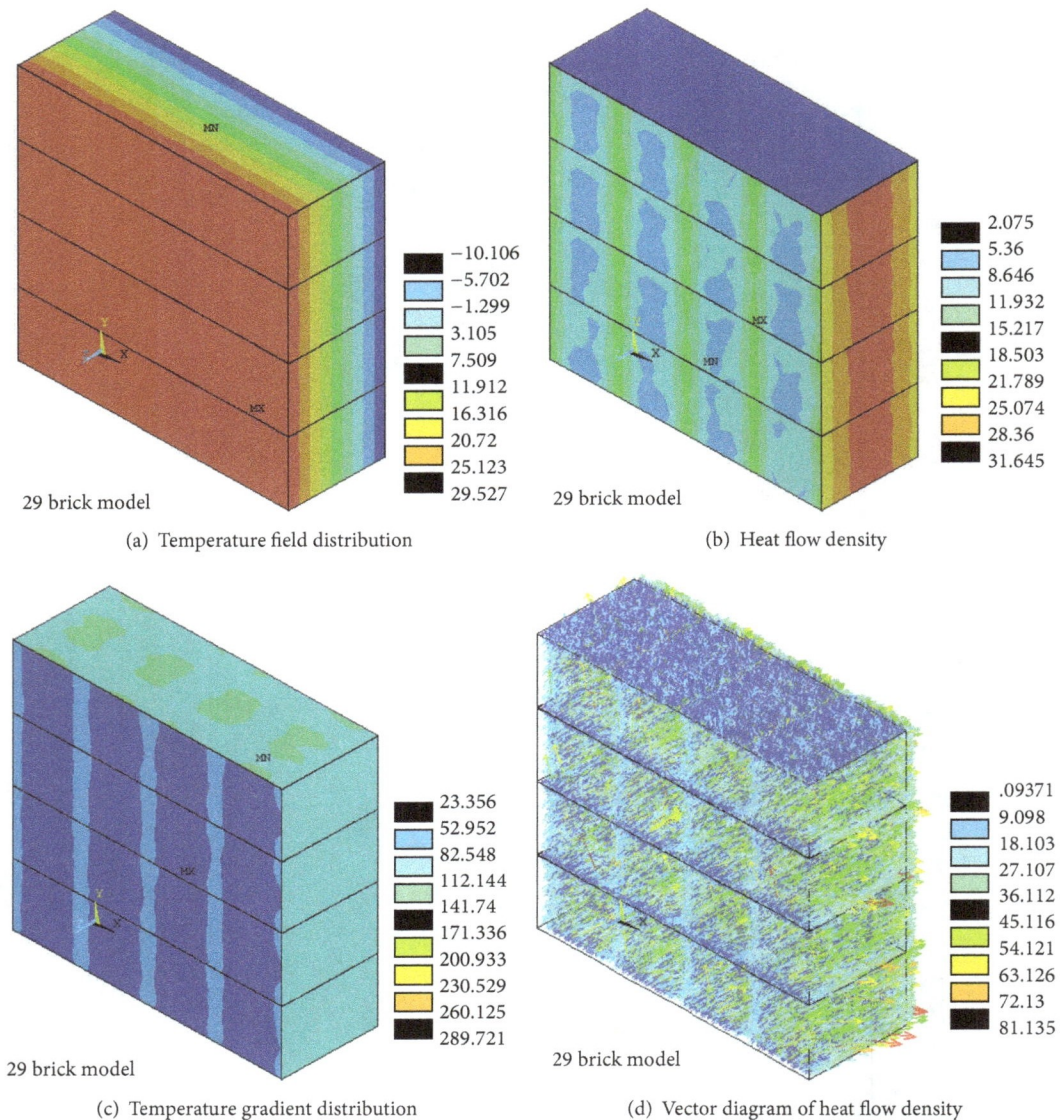

FIGURE 12: Simulation results for the hollow shale block wall.

direction. The thin mortar joints with a thickness of 2 mm are beneficial to the strong self-insulation performance of hollow shale block walls.

Conflicts of Interest

The authors declare that there are no conflicts of interest regarding the publication of this paper.

Acknowledgments

This research was supported by both the Innovation Team of Xi'an University of Architecture and Technology and the National Science-Technology Support Plan Projects "Energy-Saving Wall Materials Construction Technology Research" and "Cultivating Fund of Excerpts of Dissertation." The support of the Natural Science Foundation of China (Grant nos. 51478381, 51578444) and Key Laboratory Project of Shaanxi Province Education Department (15JS050) is also acknowledged.

References

[1] W. Lu and Y. Ma, "Image of energy consumption of well off society in China," *Energy Conversion and Management*, vol. 45, no. 9-10, pp. 1357–1367, 2004.

[2] Ministry of Housing and Urban-rural Development of People's Republic of China, *Notification on Further Strengthening of Banning the Use of Solid Clay Brick Work*, 2005.

[3] Xi'an Research & Design Institute of Wall & Roof Materials, The feasibility report of the application technologies for energy-saving sintering wall materials, 2010 (Chinese).

[4] L. Zhu, J. Dai, G. Bai, and F. Zhang, "Study on thermal properties of recycled aggregate concrete and recycled concrete blocks," *Construction and Building Materials*, vol. 94, article no. 6873, pp. 620–628, 2015.

[5] E. Sodupe-Ortega, E. Fraile-Garcia, J. Ferreiro-Cabello, and A. Sanz-Garcia, "Evaluation of crumb rubber as aggregate for automated manufacturing of rubberized long hollow blocks and bricks," *Construction and Building Materials*, vol. 106, pp. 305–316, 2016.

[6] Z. P. Zhang, S. W. Zhu, and G. P. Chen, "Study on thermal performance for straw fiber concrete hollow block," *Advanced Materials Research*, vol. 953-954, pp. 1596–1599, 2014.

[7] X. S. Fan, Y. L. Chen, X. L. Niu, X. C. Wang, and C. F. Liang, "Numerical analysis on hollow EPSRC block and its thermal insulation wall," *Advanced Materials Research*, vol. 724-725, pp. 1526–1530, 2013.

[8] J. J. Del Coz Díaz, F. P. Álvarez-Rabanal, O. Gencel et al., "Hygrothermal study of lightweight concrete hollow bricks: a new proposed experimental-numerical method," *Journal of Energy and Buildings*, vol. 70, pp. 194–206, 2014.

[9] J. J. Del Coz Díaz, P. J. García Nieto, A. M. Rodríguez, A. L. Martínez-Luengas, and C. B. Biempica, "Non-linear thermal analysis of light concrete hollow brick walls by the finite element method and experimental validation," *Applied Thermal Engineering*, vol. 26, no. 8-9, pp. 777–786, 2006.

[10] J. J. del Coz Díaz, P. J. García Nieto, C. Betegón Biempica, and M. B. Prendes Gero, "Analysis and optimization of the heat-insulating light concrete hollow brick walls design by the finite element method," *Applied Thermal Engineering*, vol. 27, no. 8-9, pp. 1445–1456, 2007.

[11] J. J. Del Coz Díaz, P. J. García Nieto, J. L. Suárez Sierra, and I. Peñuelas Sánchez, "Non-linear thermal optimization and design improvement of a new internal light concrete multi-holed brick walls by FEM," *Applied Thermal Engineering*, vol. 28, no. 8-9, pp. 1090–1100, 2008.

[12] A. Li, X. Xu, J. Xie, and Y. Sun, "Development of a simplified heat transfer model of hollow blocks by using finite element method in frequency domain," *Energy and Buildings*, vol. 111, pp. 76–86, 2016.

[13] J. Wu, G.-L. Bai, H.-Y. Zhao, and X. Li, "Mechanical and thermal tests of an innovative environment-friendly hollow block as self-insulation wall materials," *Construction and Building Materials*, vol. 93, pp. 342–349, 2015.

[14] G. Bai, G. Fu, Z. Quan, H. Wang, and X. Li, "Experimental study on basic mechanical properties of fired heat-insulation hollow block thin mortar joint masonry," *Journal of Building Structures*, vol. 34, pp. 151–158, 2013.

[15] G. Bai, G. Fu, Z. Quan, H. Wang, and X. Li, "Experimental study on seismic behavior of shale fired heat-insulation block walls with ultra-thin mortar joint," *Journal of Building Structures*, vol. 10, pp. 111–121, 2014.

[16] J. P. Liu, *Building Physics*, China Architecture and Building Press, Beijing, China, 2009 (Chinese).

[17] G. Yu, *Advantages And Development Direction of Fired Shale Blocks*, Brick and Tile World, 2006.

[18] "Thermal insulation-determination of steady-state thermal transmission properties-calibrated and guard hot box. Part 2: device," National Standard GB/T 13475-2008, 2008.

[19] "Energy efficiency of public buildings," National Standard GB50189-2005, 2005.

[20] Z. Yang, F. Zhang, and Z. Quan, "Thermal performance test on the masonry of recycled aggregate concrete block in three rows of holes," *Jorunal of Block-Brick-Tile*, vol. 6, pp. 37–39, 2012 (Chinese).

[21] "Technical specification for concrete small-sized hollow block masonry buildings of China," Construction Industry Standard JGJ/T2011, 2011.

[22] D. B. Crawley, L. K. Lawrie, F. C. Winkelmann et al., "EnergyPlus: creating a new-generation building energy simulation program," *Energy and Buildings*, vol. 33, no. 4, pp. 319–331, 2001.

[23] "Thermal design code for civil building of China," National Standard GB50176-93, 1993.

[24] A. Denisiewicz and M. S. Kuczma, "Two-scale numerical homogenization of the constitutive parameters of reactive powder concrete," *International Journal for Multiscale Computational Engineering*, vol. 12, no. 5, pp. 361–374, 2014.

[25] S. Rastkar, M. Zahedi, I. Korolev, and A. Agarwal, "A meshfree approach for homogenization of mechanical properties of heterogeneous materials," *Engineering Analysis with Boundary Elements*, vol. 75, pp. 79–88, 2017.

[26] E. Reccia, G. Milani, A. Cecchi et al., "Full 3D homogenization approach to investigate the behavior of masonry arch bridges: the venice trans-lagoon railway bridge," *Construction and Building Materials*, vol. 66, no. 36, pp. 567–586, 2014.

19

Additive Manufacturing Enabled by Electrospinning for Tougher Bio-Inspired Materials

Komal Agarwal,[1] Yinning Zhou,[2] Hashina Parveen Anwar Ali,[2] Ihor Radchenko,[2] Avinash Baji ⓘ,[1,3] and Arief S. Budiman ⓘ[2]

[1]Engineering Product Development Pillar, Singapore University of Technology and Design, Singapore 487372
[2]Xtreme Materials Laboratory (XML), Singapore University of Technology and Design, Singapore 487372
[3]Department of Engineering, School of Engineering and Mathematical Sciences (SEMS), La Trobe University, Bundoora 3086, Australia

Correspondence should be addressed to Avinash Baji; a.baji@latrobe.edu.au and Arief S. Budiman; suriadi@alumni.stanford.edu

Academic Editor: Jose M. Cabrera

Nature has taught us fascinating strategies to design materials such that they exhibit superior and novel properties. Shells of mantis club have protein fibres arranged in a 3D helicoidal architecture that give them remarkable strength and toughness, enabling them to absorb high-impact energy. This complex architecture is now possible to replicate with the recent advances in additive manufacturing. In this paper, we used melt electrospinning to fabricate 3D polycaprolactone (PCL) fibrous design to mimic the natural helicoidal structures found in the shells of the mantis shrimp's dactyl club. To improve the tensile deformation behavior of the structures, the surface of each layer of the samples were treated with carboxyl and amino groups. The toughness of the surface-treated helicoidal sample was found to be two times higher than the surface-treated unidirectional sample and five times higher than the helicoidal sample without surface treatment. Free amino groups (NH_2) were introduced on the surface of the fibres and membrane via surface treatment to increase the interaction and adhesion among the different layers of membranes. We believe that this represents a preliminary feasibility in our attempt to mimic the 3D helicoidal architectures at small scales, and we still have room to improve further using even smaller fibre sizes of the modeled architectures. These lightweight synthetic analogue materials enabled by electrospinning as an additive manufacturing methodology would potentially display superior structural properties and functionalities such as high strength and extreme toughness.

1. Introduction

Natural biological composites have captured tremendous scientific interest in recent years among researchers to develop high-performance synthetic composites [1–4]. Mimicking the structure and design of the natural composites that are found in connective tissues, bones, and exoskeletons can offer fascinating prospects for designing strong and novel synthetic composites [4, 5]. The natural composites are obtained by the self-assembly process and are made of macromolecular building blocks. At the structural levels, the molecules are self-assembled to obtain a higher order micro- or macrostructure [4].

In a recent study, Weaver et al. attributed the high toughness and strength of mantis shrimp's dactyl club to its microstructure and unique arrangement of protein fibres

within the structures. Investigations have also attributed the mechanical strength and toughness of the material to the complex interplay between structure, stiffness, strength, and impact mechanics of the dactyl club [6]. Their findings show that the club is divided into three distinct regions viz. outer crystalline apatite region, chitin/amorphous calcium carbonate (ACC) periodic region in the form of rotated plywood, and mineralized chitin fibrils striated region. The periodic mineralized chitin fibril region that shows a helicoidal architecture (Figure 1(b)) can dissipate energy from the impact region through quasi-plastic compressive responses. Therefore, it provides a fracture-toughening barrier that hinders the catastrophic propagation of microcracks when subjected to repeated impacts [6, 7]. Wilts et al. in their article demonstrated that such helicoidal architecture is

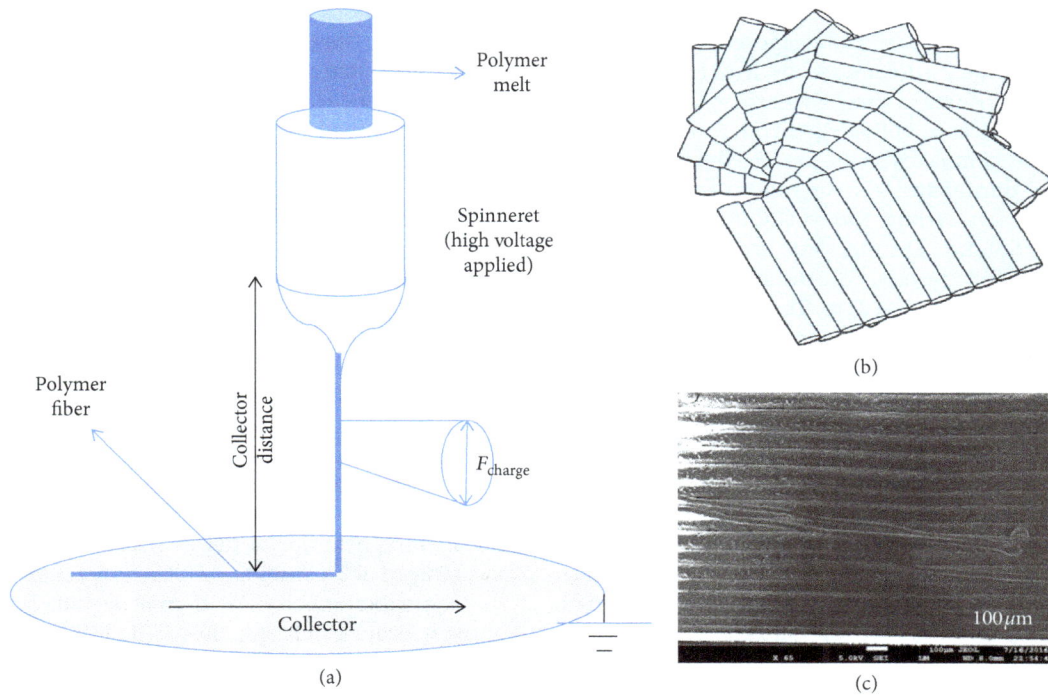

FIGURE 1: Methodology, design, and SEM images of one layer of 3D Helicoidal Structure. (a) Schematics of near-field melt electrospinning (NFES)—the process used to first lay down patterns of electrospun fibres on top a collector; (b) illustration of a top view of the hierarchical helicoidal design of the basic building blocks found in shells of mantis shrimp dactyl club, as reported in the literature; (c) scanning electron microscope (SEM) image of a layer of highly aligned, dense fibre-structured samples of melt electrospinning fibres with diameters around $10\,\mu m$ at 65x magnification.

commonly found in some plants and animal species [8]. They not only attribute the optimal properties and structural color to this helicoidal arrangement but also attribute the impact toughness to such structures. The studies indicate that the remarkable strength (rigidity) and toughness of the exoskeletons is attributed to the hierarchical design of the basic building blocks as illustrated in Figure 1(b). The fibres are assembled into bundles to form horizontal planes, which are then superimposed to obtain helicoidal stacking [9–11].

Mimicking this structured design could lead to the development of synthetic impact resistant materials. Although some studies have characterized the mechanical deformation behavior of the natural material, very few studies have concentrated their efforts on fabricating synthetic structures [6, 7, 12]. Gu et al. [13] created nacre-like polymer composite structures using 3D printing where layers were stacked with orientation angles of 0 and 90 degrees to generate a laminate construct. These nacre-like designs outperform the constituent materials in impact resistance. Yaraghi et al. [14] used 3D printing to mimic the natural 3D helicoidal architecture. However, the sizes of fibres in this study are in the millimeter size domain. In this study, we use near-field solution electrospinning as shown in Figure 1(a) to fabricate the synthetic 3D hierarchical structures by precisely depositing the fibres in a controlled fashion. By controlling the processing variable, the size of the fibres can be in smaller scales (hundreds to tens of microns). This approach of using electrospinning as an additive

manufacturing methodology offers the most versatile and promising technique that enables the fabrication of the hierarchical fibres even to the nanometer regimes. From the literature, both fibre structures [15, 16] as well as the smaller scaled microstructures [17–20] of material have been known to lead to unprecedented materials properties, including toughness and other damage-resistant behavior [21–26].

We believe that the helicoidal structure adapts to tensile stress by allowing the plane of fibres to rotate toward or away from the applied tensile load, prolonging the final catastrophic events of failures. The reason behind the unique strength and toughness of the structure is its ability to adapt to the loading stress. Thus, we establish a fundamental understanding of the fracture-delaying mechanisms in the 3D hierarchy microarchitectured advanced materials enabled by additive manufacturing to achieve remarkable structural properties and functionalities.

Strengthening the biomimetic 3D hierarchical nano- and microscale architectures, enabled by additive manufacturing, is the unique aspect of this article. The development of such low-cost, lightweight, high-performance materials will be highly useful for protective technologies such as in soldier body armors, sports/athletic gears, and aerospace/aircraft applications.

2. Experimental

2.1. Near-Field Melt Electrospinning (NFES). A custom-built near-field melt electrospinning setup was used to fabricate the samples. Briefly, polycaprolactone (PCL, molecular weight = 45 kDa, Sigma-Aldrich, Singapore) pellets were fed into a metallic syringe surrounded by a hot water jacket system (NanoNC, Korea). The temperature was raised to 80°C to melt the polymer. Following this step, a high voltage power supply was connected to the needle. The flow rate and the voltage used during the electrospinning were 20 μl/h and 9 kV, respectively. A syringe pump (EQ-500SP-H Syringe Pump, Premier Solution Pte Ltd.) was used to dispense the PCL from the metallic syringe. The distance between the tip of the needle and collector was 2 cm. The fibres were collected on an aluminum plate placed on a XY linear computer-integrated motorized stage (Zugo Photonics, Singapore). The speed of motorized stage was set at 100 mm/s at 30 μm microstep intervals. All the experiments were conducted at room temperature (17–22°C) within a closed chamber. The experimental conditions were kept constant for all the experiments. A vertical electrospinning setup was adopted (having only one axis to oscillate) to obtain highly uniform and dense fibres as shown in Figure 1(c). Seven layers of such aligned array of fibres were collected and manually stacked on top of each other to mimic the natural helicoidal structure. Each layer was placed such that the longitudinal direction of the fibres was rotated by 15° [24] with respect to the longitudinal direction of the fibre just below it. Following this, the sample was hot-pressed at 45°C and 0.2 MPa pressure using a manual hydraulic press with digital heating plates (Model 4386, Carver Inc. USA) to obtain the helicoidal structures as shown in Figure 2. A control sample was also prepared by stacking the seven layers of fibre arrays on top of each other followed by hot-pressing. In this set of samples, the fibres in each layer were unidirectionally oriented.

2.2. Surface Treatment. Another set of samples were prepared by functionalizing the surface of the fibres before they were hot-pressed. Briefly, once the aligned arrays of fibres were obtained, carboxyl groups were introduced on the surface of the fibres by treating the fibres with low-pressure air plasma for 5 minutes using the electrodeless, inductively coupled RFGD instrument (PDC-002; Harrick Plasma, Ithaca, NY, USA) [27]. Following this, amino groups were introduced by immersing the samples in 10% (w/v) solution of 1,6-hexanediamine prepared in isopropanol at 37°C for 1 h [28]. Distilled water was used to wash the samples. The samples were then dried under vacuum at room temperature for two hours before they were hot-pressed to obtain helicoidal samples. A control sample was prepared by stacking the surface-treated fibre array on top of each other followed by hot-pressing. The fibres in each layer were unidirectionally oriented.

2.3. Characterization. The cross section of the helicoidal samples is investigated by fracturing the samples under liquid nitrogen and imaging them using a scanning electron microscope (SEM). To view the microstructure of the samples, the samples were sputter-coated with gold (18 mA, 90 seconds) before their structures were examined using a scanning electron microscope (Tescan, accelerating voltage of 10 kV).

The tensile deformation behavior of the samples was investigated using a Zwick Roell (Zwick Roell, Ulm, Germany) Z0.5 static testing machine (Figure 3(a)). All four samples, viz. helicoidal PCL fibre samples with and without surface treatment and unidirectionally oriented PCL samples with and without surface treatment with identical dimensions ($4 \times 2.5 \times 0.6\ mm^3$ width, length, and thickness) were tested. For comparison purposes, a bulk PCL sample was also prepared by hot-pressing the PCL pellets and its deformation behavior was compared with the other samples. The sample ends were glued to acrylic slabs prior to the tests as shown in Figures 3(c) and 3(d). This setup was used to ensure that the samples were firmly gripped during the tension tests. The tensile test was performed at room temperature under displacement control at a strain rate of = 100 μm/min and a preload of 0.1 N.

A differential scanning calorimeter (DSC) from TA Instruments, DSC Q100, was used to determine the crystallinity and melting behavior of the PCL samples. The temperature was ramped at 3°C/min from 0°C temperature to 80°C under a nitrogen atmosphere. The thermal degradation temperature of the helicoidal PCL sample and bulk PCL was determined using thermogravimetric analyzer (TGA, TA Q50). The weight loss of the samples as a function of temperature was recorded at a heating rate of 10°C/min in the N_2 atmosphere.

3. Results and Discussion

Figure 2 clearly illustrates the cross section of partial 3D helicoidal structures with 15° rotations between each layer. The red dashed lines shown in the SEM image illustrate the different layers within the sample and demonstrate that the sample is composed of multiple layers of fibre arrays. It is evident that the cross section of the fibres in each region of the sample shows variation from circle to ellipse and then becomes a column.

Our objective is to investigate the mechanisms leading to the enhanced mechanical properties of the 3D hierarchical microarchitectured materials inspired by nature while maintaining other desirable properties and/or functionalities.

The tensile tester is used to investigate the tensile deformation behavior of the samples. Digital images of the samples are taken during the deformation process. Figure 4 shows various representative stages of the tensile test in (a) bulk PCL, (b) unidirectionally oriented PCL fibre sample, and (c) helicoidal PCL fibre sample. Figure 5 shows the representation of stress vs. strain curves recorded for the samples. Each of the tensile stress-strain curves shown in Figure 5 is the representative of a few samples (between 5 and 8 samples) in each of the groups specified (bulk, unidirectionally oriented, helicoidal, unidirectionally oriented with surface treatment, and helicoidal with surface treatment). Within each group, the stress-strain curves showed self-consistent characteristics as well as similar ranges of magnitudes of the strengths and ductility (with differences between the samples within the group not more than ±6% for the max stress values, as shown with the experimental error bars in Figure 5). For clarity and simplicity, Figure 5 shows only the representatives of each of the groups (but their variations are indicated by the

(a) (b)

FIGURE 2: SEM image of the 3D helicoidal structure with 15° rotations between each layer. (a) SEM image showing cross section of the samples with a few stacked layers of fibre aligned at different direction at 95x magnification (the red-dashed lines are to indicate the boundary between layers), demonstrating of the basic feasibility to synthesize the full 3D helicoidal fibre structures similar to the natural material; (b) SEM image of the sample viewed from the top.

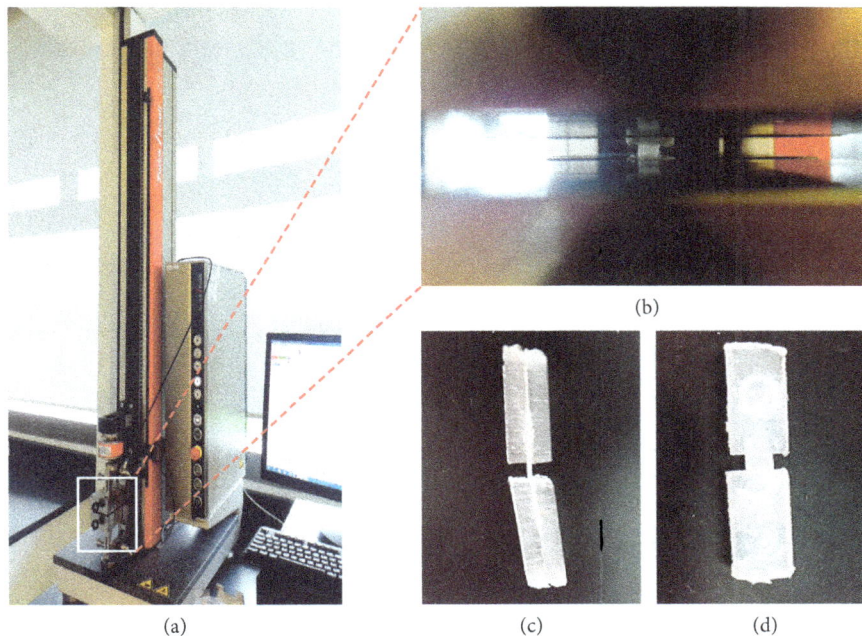

(a) (b) (c) (d)

FIGURE 3: Tensile test setup and sample preparation. (a) Zwick Roell static testing machine for tensile test. (b) Partial magnification of sample placing. (c) Side view of tensile test sample, four pieces of acrylic slabs with super glue were used to attach the samples. (d) Top view of tensile test sample.

experimental error bars shown at max stress values and subsequently at strain values of 200%, 400%, and 600%). They show unique characteristics, which are distinctive enough from group to group such as evident and represented in Figure 5. The toughness of the samples is determined by measuring the area under the stress-strain curves of the samples. It is clear that the bulk sample demonstrates elastic deformation as well as plastic deformation. For the bulk PCL, a crack initiated at the edge of the sample and propagated perpendicular to the applied tensile stress. For the sample with unidirectionally oriented fibre arrays, all the fibres arrays are aligned along with the tensile loading direction. Thus, this sample demonstrated higher yield strength as the load is shared equally by all the fibre arrays, as shown in Figure 5(a). The digital images show that not all the fibres fractured at the same time. This could be because of some misalignment.

The stress vs. strain curve of the helicoidal PCL fibre sample (without surface treatment) shows that the yield stress is lower than the bulk sample, as shown in Figure 5(a). We believe that this can be attributed to the poor adhesion between the fibres. Unlike the bulk PCL and the unidirectional PCL fibre samples, the helicoidal PCL fibre samples have significantly lower adhesion between the layers due to the different orientation of the fibres in each of the layers. To investigate this hypothesis, the surface of the fibres is functionalized and then their tensile deformation behavior is investigated as described in the subsequent paragraphs as well as shown in Figure 5(b).

Although the helicoidal PCL fibre samples showed lower yield strength, it is evident from Figure 5(a) that these samples exhibit extended ductility compared to the other two control samples. Furthermore, the digital images taken

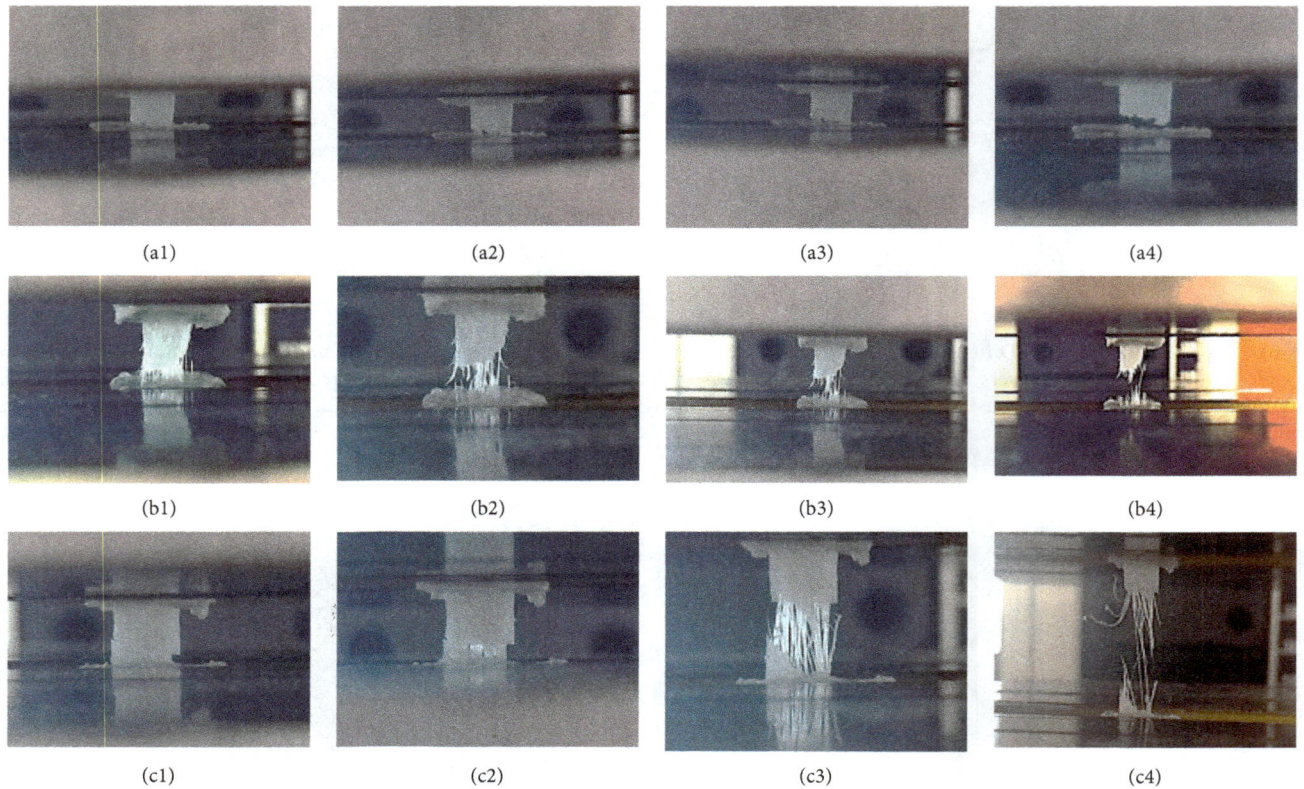

FIGURE 4: Various stages of tensile deformation of polycaprolactone (PCL) samples. (a1–a4) Bulk PCL. (b1–b4) Unidirectionally oriented PCL sample, composed of fibres aligned along the loading axis. (c1–c4) Helicoidal PCL sample, composed of fibres aligned at different angles forming helicoidal structure.

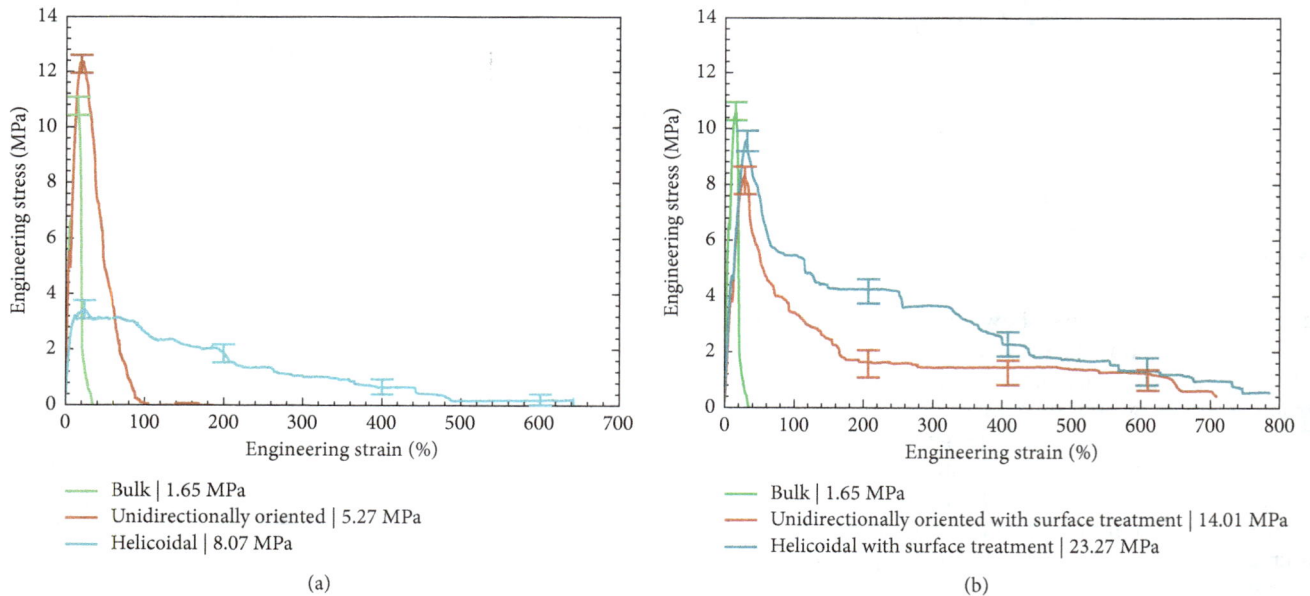

(a)

(b)

FIGURE 5: Stress-strain curves and toughness measurements. (a) Bulk PCL in comparison with unidirectionally oriented PCL fibre sample and helicoidal PCL fibre sample without surface treatment. (b) Bulk PCL in comparison with surface-treated unidirectionally oriented PCL fibre sample and surface-treated helicoidal PCL fibre sample. The experimental error bars are provided here to indicate variations within each group of the samples at max stress values and subsequently at strain values of 200%, 400%, and 600%. The variations of the max stress values in each of the groups are not more than ±6%. The toughness, as defined by the area under the stress-strain curve, for each of the groups of samples has been provided in the legend (the toughness is presented here with the unit of MPa × unitless strain).

during the deformation process show that the fibres within the different layers of the materials were reoriented along the loading direction in different stages during deformation. This explains why the fracture of the sample was delayed compared to the bulk sample. The fracture of the helicoidal PCL fibre samples took 5 times longer than the unidirectionally orientated samples and nearly 10 times longer than the bulk PCL samples. The final failure of the helicoidal PCL samples is preceded by twisting of the fibres and dramatical elongation of the fibres before fracture.

The areas under the stress vs. strain curve (i.e., the toughness) of the bulk PCL, unidirectionally oriented PCL, and the PCL helicoidal structures are measured to be 1.65, 5.27, and 8.07 MPa (the area under the stress-strain curve is presented here with the unit of MPa x unitless strain), respectively. Thus, the toughness of the PCL helicoidal structures is at least 5 times and 1.5 times higher than those of the bulk PCL and the unidirectionally oriented PCL samples, respectively. The helicoidal PCL fibre sample achieved nearly 650% strain deformation while unidirectionally oriented PCL fibre sample reached 100% strain deformation and bulk PCL only appeared to achieve about 30% strain deformation, as shown in Figure 5(a).

The effect of the surface treatment as alluded earlier in this manuscript is shown in Figure 5(b). The surface-treated unidirectionally oriented PCL samples showed increased tensile strength and extended ductility compared to their non-surface-treated counterparts, whereas the surface-treated PCL helicoidal samples showed comparable yield strength to the bulk PCL as well as surface-treated unidirectional PCL samples, as evident in Figure 5(b), while maintaining similar benefits in the tensile deformation behavior (i.e., delayed fracture events) with even some further improvements in extended ductility compared to their non-surface-treated counterparts. The surface-treated unidirectional PCL samples achieved extended ductility of nearly 800% as shown in Figure 5(b). The toughness value of the surface-treated unidirectionally oriented PCL sample and surface-treated PCL helicoidal sample is determined to be 14.01 and 23.27 MPa.

These results show that the surface treatment of the fibres improved the overall toughness of the sample during tensile testing. The toughness of surface-treated unidirectionally oriented PCL sample increased by 2.6 times compared to the unidirectionally oriented PCL sample. The surface-treated helicoidal PCL sample showed an increase in toughness by almost 3 times compared to the 3D helicoidal PCL sample without surface treatment and by almost 1.6 times higher than surface-treated unidirectionally oriented PCL samples, accompanied by a substantial increment of engineering stress.

Although the surface treatment seems to result in improvements in both the unidirectionally oriented PCL samples and helicoidal PCL samples (as represented by the red and blue curves in Figure 5(b), respectively), we observed a unique characteristic of the blue curves. This characteristic is signified by the broad "shoulder" that only gradually got lower and lower as each of the layers (with the different fibre orientations) was gradually extended to the maximum possible. In contrast, the red curves could be characterized by a much narrower "shoulder" that quickly gets down and

stays down until the last strain of fibres was broken. Although the total extended ductility of the two representative curves was similar as shown in Figure 5(b), the areas under the curves of the blue curves are always about 40–50% higher, which indicates the significant difference between these two groups of samples.

This unique characteristic is also evident from the tortuosity of the breaking paths as shown in images of Figure 4(c) of the helicoidal PCL samples especially. To substantially reduced extent, the unidirectional PCL samples exhibit similar behaviors, but as the evidence from the calculations of the areas under the curves, the toughness of the blue curves is always about 40–50% higher than those of the red curves. In addition, Figure 5(a) provides further evidence that the helicoidal architecture indeed uniquely leads to the higher performance. Even without surface treatment, the few samples in the "Helicoidal" group in Figure 5(a) have already consistently shown the characteristics of "broad shoulders" and gradual reduction of strength with extended strains, such as illustrated above with the blue curve in Figure 5(b). The surface treatment just elevates the overall stress-strain curves into a higher level of strengths of the materials. We, therefore, think all these evidences are strong indications that the helicoidal fibre orientation matters in addition to the surface treatment effects.

The stress-strain curves here thus showed promising results of the feasibility of the helicoidal PCL samples over both bulk and unidirectionally oriented PCL samples. As the PCL sample is being pulled in tension, a crack is initiated and propagated in the direction perpendicular to the loading axis. In theory, fibres in the transverse direction are least likely to break over longitudinal direction along the loading. For the unidirectionally oriented sample, since all the fibres were in the same direction as the loading axis, it was easy for the crack to be propagated through the fibres. The individual fibres were easy to rip due to the same orientation. Due to the rotated fibre alignment directions in the layers in the helicoidal samples, it is more difficult for the crack to propagate and further proceed to the final catastrophic events. As the layers of fibres keep changing directions, crack propagation from one layer to the next is effectively delayed as it keeps losing its primary driving force. Each layer of fibres is essentially propagating the crack only after further tensile straining effectively rotates its fibre alignment to become normal to the loading axis (i.e., the weakest configuration). But at the same time, as these layers rotate to weaker configurations, other layers rotate to stronger configurations (when the fibre alignment is parallel to the loading axis). These mechanisms are evident from the tortuosity of the breaking paths as shown in images of Figure 4(c) of the helicoidal PCL samples especially (although to some reduced extent, the unidirectional PCL samples also exhibit similar behaviors). These mechanisms basically represent an effective means of energy and thus damage dissipation, which could further lead to an effective mechanism for impact resistance.

Surface adhesion between the individual PCL fibres and layers plays a significant role in the mechanical performance of these biomimetic materials. This is evident from

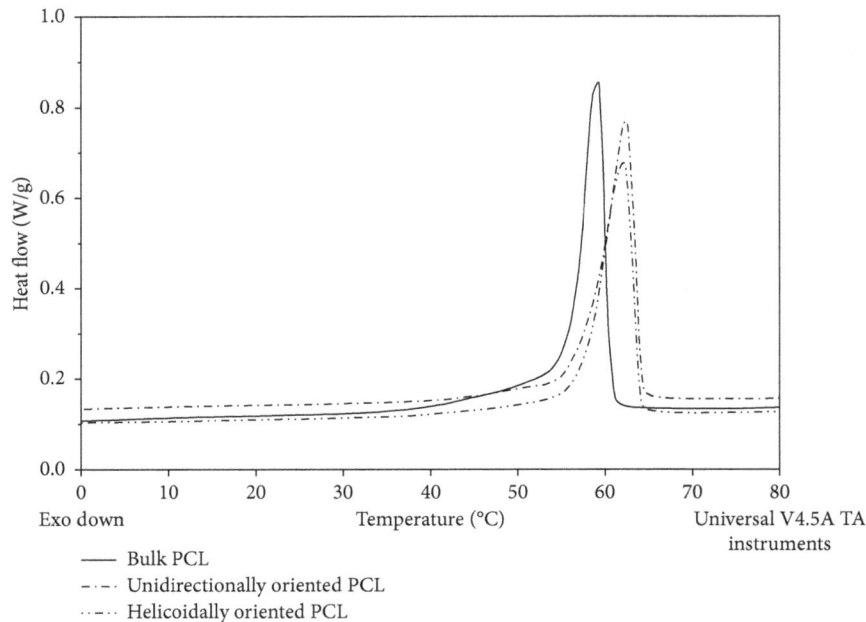

FIGURE 6: DSC scans of bulk, helicoidal, and unidirectionally oriented PCL samples.

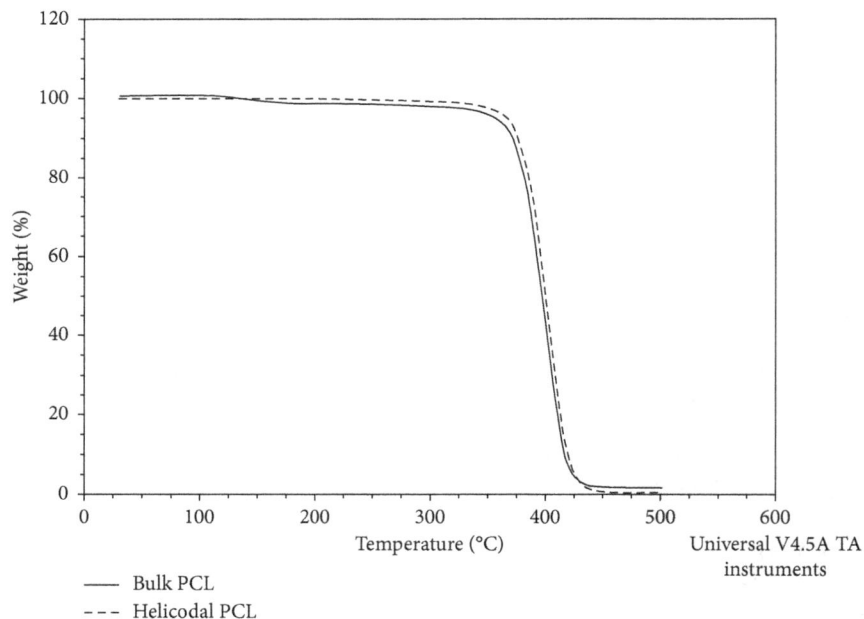

FIGURE 7: TGA curves of bulk and helicoidal PCL.

the fracture behavior of surface-treated versus the non-surface-treated samples, as illustrated in Figures 5(a) and 5(b). After the fibre surface treatment, free amino groups were introduced on the PCL membrane [28]. The interaction between the oppositely charged carboxyl group and amino groups helped to improve the adhesion between the fibres and the individual layers. This shows the importance of further studies in the interplay between all the micro-structural building blocks in mimicking the 3D helicoidal alignment of the exoskeleton structure of the mantis dactyl

club in the design of enhanced fracture toughness of the biomimetic synthetic materials.

The thermal properties of the samples are also in-vestigated and the effect of helicoidal architecture on the crystallinity, melting temperature, and thermal degradation is determined. Figure 6 shows the DSC thermal scans of the bulk PCL, unidirectionally oriented PCL, and helicoidal PCL samples. It is clear that the melting temperature of the helicoidal and unidirectionally oriented PCL is around 62°C while the melting temperature of bulk PCL is 59°C. The lower melting temperature recorded for bulk PCL indicates the presence of smaller crystallite structure within the sample. It is also obvious that the peak in the case of bulk PCL is sharper than the peaks recorded for the other two samples. This indicates that bulk PCL has smaller average crystallite size [29]. Melting enthalpy of the samples is measured from the DSC curves to determine the crystallinity. Melting enthalpy of pure crystalline PCL is obtained from literature to be 136 J/g [30]. The crystallinity of bulk PCL is determined to be 45.7%, while the crystallinity of helicoidal PCL and unidirectionally oriented PCL is determined to be 39.7% and 42%, respectively. This indicates that although the crystallite size in bulk PCL is smaller, the crystalline content is much higher than the other two samples. The lower crystallinity in helicoidal and unidirectionally oriented PCL can be attributed to the electrospinning process. During melt electrospinning, PCL is heated to a higher temperature and deposited on the collector at ambient temperature. This quenching of the material restrains the crystal formation and is responsible for the lower crystallinity displayed by the helicoidal and unidirectionally oriented PCL fibres. It is interesting to note that even though the helicoidal PCL has a lower percentage of crystallinity, it displays better tensile properties. This indicates that the helicoidal architecture plays a major in influencing the tensile properties of the helicoidal PCL. Figure 7 shows the TGA curves of helicoidal and bulk PCL. The curves show that the degradation behavior of both samples is identical and both samples degrade at identical temperatures.

4. Conclusion

In summary, we have proven the technical feasibility to create a synthetic analogue of 3D helicoidal fibre architecture mimicking the structures found in mantis dactyl club. Using electrospinning, we have produced uniform and highly dense microscale fibres in 3D helicoidal manner. The toughness of the helicoidal PCL sample is found to be nearly 5 times and 1.5 times higher than the bulk PCL and unidirectionally oriented PCL sample, respectively. Surface treatment of the helicoidal sample further enhances toughness by almost 3 times as compared to the normal 3D helicoidal sample. Mechanical testing has indeed shown some evidence of enhanced toughness in the case of the PCL helicoidal samples.

Conflicts of Interest

The authors declare that they have no conflicts of interest.

Acknowledgments

The authors would like to gratefully acknowledge the funding from the Ministry of Education (MOE) Academic Research Funds Tier 2 titled "Materials with Tunable Impact Resistance via Integrated Additive Manufacturing" (MOE2017-T2-2-175). KA acknowledges Prof. Roland Bouffanais, Assistant Professor, Engineering Product Development (EPD) Pillar, at Singapore University of Technology and Design, for his encouragement and mentorship during her PhD study. The authors gratefully acknowledge the critical support and infrastructure provided by Singapore University of Technology and Design (SUTD) especially through the Engineering Product Design (EPD) Pillar during research work as well as during the manuscript preparation. YZ, IR, HPAA, and ASB gratefully acknowledge the critical infrastructure and capabilities in nanomechanical characterization such as provided by the Xtreme Materials Laboratory (XML). The XML Nanomechanical Characterization Laboratory capability was built through the grant funding and support provided by SUTD-MIT International Design Centre (IDC), Singapore, through the project under IDC Grant "(IDG31400102)—Designing Nanomaterials Through Atomic Engineering of Interfaces." YZ, HPAA, IR, AB, and ASB also gratefully acknowledge receipt of funding and support from TEMASEK LAB@SUTD Singapore, through its SEED grant program for the project IGDS S15 01011: titled "Biomimetic, Strong yet Tough Composite through 3D Printing." KA, AB, and ASB also gratefully acknowledge receipt of funding and support from SMART (Singapore-MIT Alliance for Research and Technology) through its Ignition grant program for the project SMART ING-000067 ENG IGN: titled "Development of Novel Impact-Resistant Bio-Inspired Materials using Novel 3D Fabrication Technique." This work was supported by the SUTD-MIT International Design Centre (IDC), Singapore (IDG31400102); TEMASEK LAB@SUTD Singapore, Singapore (IGDS S15 01011); and the Singapore-MIT Alliance for Research and Technology, Singapore (SMART ING-000067), also by the Ministry of Education (MOE) Academic Research Funds Tier 2 titled "Materials with Tunable Impact Resistance via Integrated Additive Manufacturing" (MOE2017-T2-2-175).

References

[1] H. J. Gao, B. H. Ji, I. L. Jager, E. Arzt, and P. Fratzl, "Materials become insensitive to flaws at nanoscale: lessons from nature," *Proceedings of the National Academy of Sciences*, vol. 100, no. 10, pp. 5597–5600, 2003.

[2] D. Dimitrov, K. Schreve, and N. de Beer, "Advances in three dimensional printing - state of the art and future perspectives," *Rapid Prototyping Journal*, vol. 12, no. 3, pp. 136–147, 2006.

[3] E. E. de Obaldia, C. Jeong, L. K. Grunenfelder, D. Kisailus, and P. Zavattieri, "Analysis of the mechanical response of biomimetic materials with highly oriented microstructures through 3D printing, mechanical testing and modeling," *Journal of the Mechanical Behavior of Biomedical Materials*, vol. 48, pp. 70–85, 2015.

[4] A. Sellinger, P. M. Weiss, A. Nguyen et al., "Continuous self-assembly of organic-inorganic nanocomposite coatings that mimic nacre," *Nature*, vol. 394, no. 6690, pp. 256–260, 1998.

[5] R. Z. Wang, Z. Suo, A. G. Evans, N. Yao, and I. A. Aksay, "Deformation mechanisms in nacre," *Journal of Materials Research*, vol. 16, no. 9, pp. 2485–2493, 2001.

[6] J. C. Weaver, G. W. Milliron, A. Miserez et al., "The stomatopod dactyl club: a formidable damage-tolerant biological hammer," *Science*, vol. 336, no. 6086, pp. 1275–1280, 2012.

[7] S. Amini, M. Tadayon, S. Idapalapati, and A. Miserez, "The role of quasi-plasticity in the extreme contact damage tolerance of the stomatopod dactyl club," *Nature Materials*, vol. 14, no. 9, pp. 943–950, 2015.

[8] B. D. Wilts, H. M. Whitney, B. J. Glover, U. Steiner, and S. Vignolini, "Natural helicoidal structures: morphology, self-assembly and optical properties," *Materials Today: Proceedings*, vol. 1, pp. 177–185, 2014.

[9] J. Lian and J. Wang, "Microstructure and mechanical anisotropy of crab cancer magister exoskeletons," *Experimental Mechanics*, vol. 54, no. 2, pp. 229–239, 2014.

[10] P. Y. Chen, A. Y. M. Lin, J. McKittrick, and M. A. Meyers, "Structure and mechanical properties of crab exoskeletons," *Acta Biomaterialia*, vol. 4, no. 3, pp. 587–596, 2008.

[11] T. J. Zhang, Y. R. Ma, K. Chen et al., "Structure and mechanical properties of a pteropod shell consisting of interlocked helical aragonite nanofibres," *Angewandte Chemie International Edition*, vol. 50, no. 44, pp. 10361–10365, 2011.

[12] S. Amini, A. Masic, L. Bertinetti et al., "Textured fluorapatite bonded to calcium sulphate strengthen stomatopod raptorial appendages," *Nature Communications*, vol. 5, p. 12, 2014.

[13] G. X. Gu, M. Takaffoli, A. J. Hsieh, and M. J. Buehler, "Biomimetic additive manufactured polymer composites for improved impact resistance," *Extreme Mechanics Letters*, vol. 9, pp. 317–323, 2016.

[14] N. A. Yaraghi, N. Guarin-Zapata, L. K. Grunenfelder et al., "A sinusoidally architected helicoidal biocomposite," *Advanced Materials*, vol. 28, no. 32, pp. 6835–6844, 2016.

[15] A. Baji, Y. W. Mai, M. Abtahi, S. C. Wong, Y. Liu, and Q. Li, "Microstructure development in electrospun carbon nanotube reinforced polyvinylidene fluoride fibres and its influence on tensile strength and dielectric permittivity," *Composites Science and Technology*, vol. 88, pp. 1–8, 2013.

[16] A. Baji, Y. W. Mai, X. S. Du, and S. C. Wong, "Improved tensile strength and ferroelectric phase content of self-assembled polyvinylidene fluoride fibre yarns," *Macromolecular Materials and Engineering*, vol. 297, no. 3, pp. 209–213, 2012.

[17] A. Budiman, S. Han, J. Greer, N. Tamura, J. Patel, and W. Nix, "A search for evidence of strain gradient hardening in Au submicron pillars under uniaxial compression using synchrotron X-ray microdiffraction," *Acta Materialia*, vol. 56, no. 3, pp. 602–608, 2008.

[18] A. S. Budiman, G. Lee, M. J. Burek et al., "Plasticity of indium nanostructures as revealed by synchrotron X-ray microdiffraction," *Materials Science and Engineering: A*, vol. 538, pp. 89–97, 2012.

[19] I. Radchenko, S. Tippabhotla, N. Tamura, and A. Budiman, "Probing phase transformations and microstructural evolutions at the small scales: synchrotron X-ray microdiffraction for advanced applications in 3D IC (integrated circuits) and solar PV (photovoltaic) devices," *Journal of Electronic Materials*, vol. 45, no. 12, pp. 6222–6232, 2016.

[20] A. S. Budiman, S. M. Han, N. Li et al., "Plasticity in the nanoscale Cu/Nb single-crystal multilayers as revealed by synchrotron Laue x-ray microdiffraction," *Journal of Materials Research*, vol. 27, no. 3, pp. 599–611, 2012.

[21] U. Chakkingal, A. B. Suriadi, and P. F. Thomson, "The development of microstructure and the influence of processing route during equal channel angular drawing of pure aluminum," *Materials Science and Engineering: A*, vol. 266, no. 1-2, pp. 241–249, 1999.

[22] U. Chakkingal, A. B. Suriadi, and P. Thomson, "Microstructure development during equal channel angular drawing of Al at room temperature," *Scripta Materialia*, vol. 39, no. 6, pp. 677–684, 1998.

[23] V. Handara, I. Radchenko, S. Tippabhotla et al., "Probing stress and fracture mechanism in encapsulated thin silicon solar cells by synchrotron X-ray microdiffraction," *Solar Energy Materials and Solar Cells*, vol. 162, pp. 30–40, 2017.

[24] D. Ginzburg, F. Pinto, O. Lervolino, and M. Meo, "Damage tolerance of bio-inspired helicoidal composites under low velocity impact," *Composite Structures*, vol. 161, pp. 187–203, 2017.

[25] S. K. Tippabhotla, I. Radchenko, W. Song et al., "From cells to laminate: probing and modeling residual stress evolution in thin silicon photovoltaic modules using synchrotron X-ray micro-diffraction experiments and finite element simulations," *Progress in Photovoltaics: Research and Applications*, vol. 25, no. 9, pp. 791–809, 2017.

[26] H. P. Anwar Ali, I. Radchenko, J. Zhou, L. Qing, and A. Budiman, "Designing novel multilayered nanocomposites for high-performance coating materials with online strain monitoring capability," *Proceedings of the Institution of Mechanical Engineers, Part L: Journal of Materials: Design and Applications*, 2017.

[27] Z. W. Ma, W. He, T. Yong, and S. Ramakrishna, "Grafting of gelatin on electrospun poly(caprolactone) nanofibres to improve endothelial cell spreading and proliferation and to control cell orientation," *Tissue Engineering*, vol. 11, no. 7-8, pp. 1149–1158, 2005.

[28] Y. Zhu, C. Gao, X. Liu, and J. Shen, "Surface modification of polycaprolactone membrane via aminolysis and biomacromolecule immobilization for promoting cytocompatibility of human endothelial cells," *Biomacromolecules*, vol. 3, no. 6, pp. 1312–1319, 2002.

[29] S. E. Kim, A. M. Jordan, L. T. J. Korley, and J. K. Pokorski, "Drawing in poly(ε-caprolactone) fibres: tuning mechanics, fibre dimensions and surface-modification density," *Journal of Materials Chemistry B*, vol. 5, no. 23, pp. 4499–4506, 2017.

[30] F. B. Khambatta, F. Warner, T. Russell, and R. S. Stein, "Small-angle x-ray and light scattering studies of the morphology of blends of poly(ε-caprolactone) with poly(vinyl chloride)," *Journal of Polymer Science: Polymer Physics Edition*, vol. 14, no. 8, pp. 1391–1424, 1976.

Thermomechanical Properties of Jute/Bamboo Cellulose Composite and Its Hybrid Composites: The Effects of Treatment and Fiber Loading

Fui Kiew Liew,[1,2] Sinin Hamdan,[1] Md. Rezaur Rahman,[1] and Mohamad Rusop[3]

[1]*Faculty of Engineering, Universiti Malaysia Sarawak, 94300 Kota Samarahan, Sarawak, Malaysia*
[2]*Faculty of Applied Science, Universiti Teknologi MARA, 94300 Kota Samarahan, Sarawak, Malaysia*
[3]*NANO-SciTech Centre (NST), Institute of Science, Universiti Teknologi MARA, Shah Alam, Selangor, Malaysia*

Correspondence should be addressed to Fui Kiew Liew; liewsan2004@gmail.com

Academic Editor: Peter Chang

Jute cellulose composite (JCC), bamboo cellulose composite (BCC), untreated hybrid jute-bamboo fiber composite (UJBC), and jute-bamboo cellulose hybrid biocomposite (JBCC) were fabricated. All cellulose hybrid composites were fabricated with chemical treated jute-bamboo cellulose fiber at 1 : 1 weight ratio and low-density polyethylene (LDPE). The effect of chemical treatment and fiber loading on the thermal, mechanical, and morphological properties of composites was investigated. Treated jute and bamboo cellulose were characterized by Fourier transform infrared spectroscopy (FTIR) to confirm the effectiveness of treatment. All composites were characterized by tensile testing, thermogravimetric analysis (TGA), and differential scanning calorimetry (DSC). Additionally, surface morphology and water absorption test was reported. The FTIR results revealed that jute and bamboo cellulose prepared are identical to commercial cellulose. The tensile strength and Young's modulus of composites are optimum at 10 weight percentage (wt%) fibers loading. All cellulose composites showed high onset decomposition temperature. At 10 wt% fiber loading, JBCC shows highest activation energy followed by BCC and JCC. Significant reduction in crystallinity index was shown by BCC which reduced by 14%. JBCC shows the lowest water absorption up to 43 times lower compared to UJBC. The significant improved mechanical and morphological properties of treated cellulose hybrid composites are further supported by SEM images.

1. Introduction

Cellulose is the world's most abundant natural raw material with renewable sources for composite fabrication. Natural cellulose fibers have attracted global researchers owing to their unique properties such as biodegradability, low weight, easy availability, easy processing, being environmental friendly, flexibility, high strength, and stiffness [1–3].

Jute is commonly grown in India, China, and Bangladesh. Jute fibers contain mainly cellulose (58~63%), hemicellulose (20~24%), and lignin (12~15%) [4]. It has been used as packaging material, geotextiles, household textiles, and carpet backing. Bamboo is a fast-growing species and a high-yield renewable resource. Bamboo fiber consists of cellulose (73.8%), hemicellulose (12.5%), lignin (10.1%), pectin (0.4%), and aqueous extract (3.2%) [5]. Bamboo fibers have been widely used in the household, transport, and composite manufacturing industries [6]. However, all cellulose fibers have similar drawbacks which include being polar, hydrophilic in natural, low thermal stability and poor compatibility with polymer matrix. Those limitations contributed to weak fiber-matrix interfacial bonding, leading to decrease in mechanical properties [7].

To improve the interfacial bonding, researchers have attempted various surface treatments, such as alkali treatment [8–10], silane treatment [11, 12], acetylation [13], and different coupling agents. Among various treatments, alkali treatment has been found to be most effective method [14, 15]. The surface treatments of cellulose fibers provide advantages which include increased hydrophobicity, increased surface roughness, reduced water uptake, and increased reactivity towards polymeric matrices [16, 17]. Alkaline treatment or

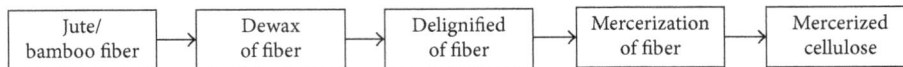

FIGURE 1: Flow diagram of cellulose preparation.

mercerization led to break down of fiber bundle into smaller fibers. Mercerization also modifies the lignin content by rupturing the ester bonds that form cross-links involving xylan and lignin, thus increasing the porosity of the lignocelluloses [18]. This treatment can effectively remove lignin and hemicellulose which indirectly increase cellulose contained in treated fiber. Mercerization increases the number of possible reactive sites and allows better fiber wetting [16]. As a result, mercerization had long-lasting effect on the mechanical properties of fibers, mainly on fiber strength and stiffness [19].

Hybrid cellulose composites are materials made from combining two or more different celluloses in a common matrix. Hybrid cellulose composites made up of two different cellulosic fibers are less common but they are potentially useful materials with respect to environmental concerns [20]. Hybrid cellulose composites fabrication by cellulosic fibers is economical and provides another dimension to the versatility of cellulosic fiber reinforced composites. The constituent fibers in a hybrid composite can be altered in many ways, leading to good mechanical properties [21]. Hybrid composites are one of the emerging fields that draw attention for application in various sectors ranging from automobile to the building industry. Previous research showed that cellulose composites have best mechanical performance at about 10–15 wt% fiber loading [22, 23].

Pöllänen et al. (2013) studied the influence of the viscose fibers and the microcrystalline cellulose on the morphological, mechanical, and thermal expansion properties of high density polyethylene (HDPE) composite. The filler content of 5, 10, 20, and 40 wt% was used. The interaction between the fillers and the matrix was improved with maleic anhydride grafted polyethylene. Both fillers increased Young's modulus and tensile strength of polyethylene with the highest values achieved by 40 wt% viscose fibers [24]. Both fillers show decreased coefficients of the linear thermal expansion of the HDPE matrix in the flow direction. Cellulose composites containing up to 10 wt% cellulose pulp fibers with low-density polyethylene (LDPE) were fabricated by Sdrobiş et al. (2012). Unbleached and bleached Kraft cellulose pulp fibers modified with oleic acid in cold plasma conditions have been used. It was found that thermal stability of cellulose pulp fibers composites is higher than that of pure LDPE [25]. Bleached cellulose pulp fibers are more efficient in enhancement of composites properties.

In the present work, the jute and bamboo cellulose had been extracted and treated with peracetic acid solution and then followed by mercerization. The effects of chemical treatment and cellulose fiber and hybrid cellulose fiber loading on thermal, mechanical, morphological, and water absorption properties of cellulose composites were investigated and reported. The outcome of investigation is expected to determine the suitability of composite for nonstructural and outdoor applications such as door trim panels, parts of automobile, and playground facilities.

2. Materials and Methods

Jute (*Corchorus olitorius*) and bamboo (*Dendrocalamus asper*) fibers were obtained from Bangladesh Jute Research Institute (BJRI), Dhaka, Bangladesh, and Forest Research Institute, Sarawak, Malaysia, respectively. Low-density polyethylene (LDPE) granula was obtained from the Siam Polyethylene Co., Ltd., Prakanong, Bangkok, Thailand. The density and the melting point were $0.935 \, \text{g·cm}^{-3}$ and between 105 and 125°C, respectively. The chemicals used were toluene (Sigma Aldrich, St. Louis, MO, USA), commercial cellulose (Sigma Aldrich, St. Louis, MO, USA), ethanol (Sigma Aldrich, St. Louis, MO, USA), hydrogen peroxide (Qeric), acetic acid glacial (JT Baker, Center Valley, PA, USA), titanium (IV) oxide (JT Baker, Center Valley, PA, USA), and sodium hydroxide (Merck KgaA, Darmstadt, Germany). All chemicals were of analytical grade.

2.1. Natural Fiber Preparation. The jute fibers were cleaned and air-dried for 48 h under direct sunlight. The middle portion of the jute fiber was removed and chopped into approximately 10-mm long and then oven-dried at 105°C for 24 h. Finally, the sections were ground and sieved using a 100-μm sieve. The length of the bamboo culm trim, excluding the bamboo internode, was 1 m in length. It was cut using a planner machine to produce chips and then ground to powdered form. The chips and powder mixture were dried in an oven at 70°C for 72 h. The oven-dried samples were grinded and sieved, using a 100-μm sieve, to obtain bamboo fibers with a mesh size of 100 μm.

2.2. Chemical Treatment of Jute and Bamboo Fiber. Jute and bamboo fiber were chemically treated to obtain treated cellulose. Initially, fibers were dewaxed. Then, cellulose fiber was obtained by oxidative delignification using a water solution containing acetic acid, hydrogen peroxide with titanium (IV) oxide, and TiO_2 catalysts adopted Kuznetsov method [26]. Cellulose fiber is alkaline treated to obtain mercerized cellulose fiber. The details of chemical treatments flow are shown in Figure 1 and steps are explained below.

Dewaxing of Fiber. The dewaxing was done by applying the Leavitt-Danzer method. In this process, two types of chemical were used, namely, toluene ($C_6H_5CH_3$) and ethanol (C_2H_6O) with ratios of 2:1 (v/v). The extraction process was done using the extraction column (Soxhlet extractor, round bottom flask, Liebig condenser, heater, membrane, and thermometer). Then, the jute/bamboo fibers were immersed

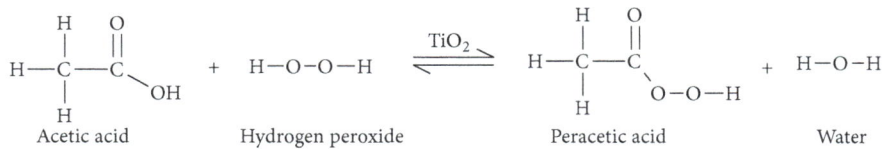

FIGURE 2: Peracetic acid solution reaction scheme.

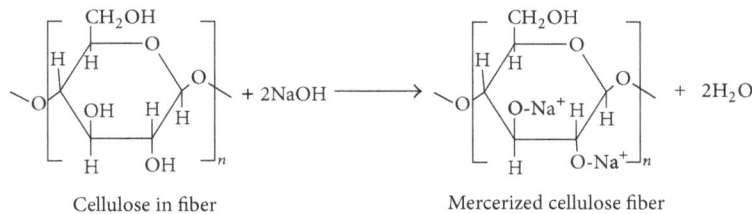

FIGURE 3: Mercerization reaction scheme.

Treated bamboo-jute cellulose fiber (1 : 1)   Hybrid jute-bamboo cellulose fiber + LDPE matrix   Hot pressing   Composite samples

FIGURE 4: Flow of hybrid composite preparation.

in the extraction column. This process was continued for 2 hours at 250°C. The collected fibers were later placed in the forced air convention oven for 24 hours at 75°C for drying overnight and were kept for delignification processes.

Delignification of Fiber. The delignification was done using peracetic acid solution which consists of 99.8 wt% acetic acid (CH_3COOH) and 35 wt% hydrogen peroxide (H_2O_2) in the present of titanium (IV) oxide (TiO_2) catalyst in a round bottom vector vessel. The reaction scheme is shown in Figure 2.

Then, the dewaxed jute/bamboo fibers were placed in the round bottom vessel. This process was continued for 2 hours at 130°C. After this, the collected fibers were carefully washed and placed in the forced air convention oven for 24 hours at 70°C.

Mercerization. Delignified fiber was immersed into alkaline solution to dissolve the pectin and hemicelluloses. 6 wt% sodium hydroxide was used to treat the cellulose fiber in a flask. The mixture was stirred using autoshaker at 150 rpm, heated to 80°C for 2 hours, and stopped after 8 hours of stirring. The mixture was rinsed with deionized (DI) water. The treated product was then filtered using Buchner flask and rinsed with DI water until the pH level reaches 7 and was freeze-dried at −85°C for 48 hours. The treated fiber was then ground using a shear grinder for 10 min to obtain fine mercerized cellulose fiber. The detail of reaction is shown in Figure 3.

2.3. Composite Preparation. Both mercerized jute and bamboo cellulose were kept in a convection oven at 100°C for 24 hours prior to composite fabrication. Desired weight fraction of hybrid cellulose fibers was equally distributed using sieve on correct weight fraction of LDPE laminate sheet to produce hybrid cellulose fibers composite. The LDPE laminate sheet was put in an aluminum mold using sandwich method and then placed in an electric hot press set at 180°C with 3.45 MPa (500 psi) for 45 min. Aluminum mold was cooled down under pressure for 1 hour and the prepared specimens were taken out to label. Specimens were cut into composite samples by tensile cutter. Table 1 shows the details of hybrid composites by weight percentages. Flow of hybrid composite preparation was shown in Figure 4.

3. Characterization Methods

3.1. Fourier Transform Infrared Spectroscopy (FTIR). The FTIR spectra were obtained using a spectrometer (spectrum 100, Perkin Elmer, Waltham, MA). The obtained spectra are described in the Results and Discussion. The transmittance range of the scan was 4000 to 650 cm^{-1}.

3.2. Tensile Test. Tensile testing was conducted according to the ASTM D 638-10 (2010) [27] testing standard using a universal testing machine (MSC-5/500, Shimadzu Company Ltd., Japan) at a crosshead speed of 5 mm·min^{-1}. The dimensions of the specimens were 115 mm (L) × 6.5 mm (W) ×

TABLE 1: Hybrid composite by weight percentage.

| Number | Hybrid composite | Composition by weight (wt.%) | | | Sample ID |
		Jute cellulose	Bamboocellulose	LDPE	
1	Untreated jute-bamboo/LDPE composites	2.5	2.5	95	F05UJBC
		5.0	5.0	90	F10UJBC
		7.5	7.5	85	F15UJBC
		10.0	10.0	80	F20UJBC
2	Jute cellulose/LDPE composites	5	—	95	F05JCC
		10	—	90	F10JCC
		15	—	85	F15JCC
		20	—	80	F20JCC
3	Bamboo cellulose/LDPE composites	—	5	95	F05BCC
		—	10	90	F10BCC
		—	15	85	F15BCC
		—	20	80	F20BCC
4	Jute-bamboo cellulose/LDPE composites	2.5	2.5	95	F05JBCC
		5.0	5.0	90	F10JBCC
		7.5	7.5	85	F15JBCC
		10.0	10.0	80	F20JBCC

3.1 mm (T). Five rectangular specimens were tested and the average value was reported for each series.

3.3. Thermogravimetric Analysis (TGA).
Thermogravimetric measurements were performed using a Perkin Elmer Pyris 1 TGA system. All measurements were obtained under a nitrogen flow rate of 20 mL·min^{-1} over a temperature range of 50 to 600°C. Then, an oxygen flow rate of 20 mL·min^{-1} and a temperature range of 600 to 700°C was applied, while maintaining a constant heating rate of 20°C·min^{-1}. Three specimens were tested and the average value was reported.

3.4. Differential Scanning Calorimetry (DSC).
The DSC tests were conducted using a differential scanning calorimeter (DSC; 8000, Perkin Elmer). Temperature programs for dynamic tests were run from 50 to 180°C at the heating rate of 20°C·min^{-1} under a 20 mL·min^{-1} nitrogen atmosphere. The degree of crystallinity was calculated according to the following formula.

$$\text{Degree of crystallinity, } \chi_c = \frac{\Delta H_m}{w \Delta H_m^o} \times 100, \quad (1)$$

where ΔH_m is enthalpy of fusion, w is the weight fraction of polymeric matrix in the composite, and $\Delta H_m^o = 290$ J/g (heat of fusion for 100% crystalline LDPE).

3.5. Scanning Electron Microscopy (SEM) Analysis.
The surface morphology was examined using a scanning electron microscope (TM 3030 pitch emission, Hitachi, Tokyo, Japan). The SEM specimens were sputter-coated with gold using auto-fine coater (JFC-1600, Joel Ltd.). The micrographs with a 500-time resolution were presented in the Results and Discussion.

3.6. Water Absorption Test.
In order to measure the water absorption characteristics of the composites, rectangular specimens were prepared. The water absorption test was conducted according to ASTM D570-99 [28]. The test specimens were immersed in a beaker containing 100 mL of deionized water at room temperature (27°C) for 22 days. The initial weights of the samples were determined. After 24-hour interval, samples were taken out from the beaker, wiped and dried, and weighed immediately. The percentages of water absorbed by the samples were calculated using the following formula. Three replicates were tested for each set and average data was recorded.

$$\text{Water absorption} = \frac{\text{Final weight} - \text{Original weight}}{\text{Original weight}} \quad (2)$$

4. Results and Discussions

4.1. Fourier Transform Infrared Spectroscopy (FTIR).
FTIR composition of bamboo fiber, jute fiber, commercial cellulose, jute cellulose, and bamboo cellulose was shown in Figure 5.

Figure 5 showed bamboo fiber and jute fiber were having similar peak intensity. Both bamboo and jute fiber consist of cellulose, hemicellulose, and lignin; therefore peak intensities are identical. However, there were still differences among bamboo fiber and jute fiber peak. Only jute fiber showed peak at 1324 cm^{-1}, which is attributed to the CH$_2$ bending. Both jute and bamboo fibers show peak at 1738 cm^{-1} represented either the acetyl and uronic ester groups or the ester linkage of carboxylic group of the ferulic and p-coumeric acids of hemicelluloses [29, 30]. The absence of this peak is observed in jute and bamboo cellulose, indicating the removal of most of the hemicellulose. The peak at 1242.4 cm^{-1} region is due to the stretching of phenolic hydroxyl groups in lignin [31]. Total disappear of this peak in jute and bamboo cellulose

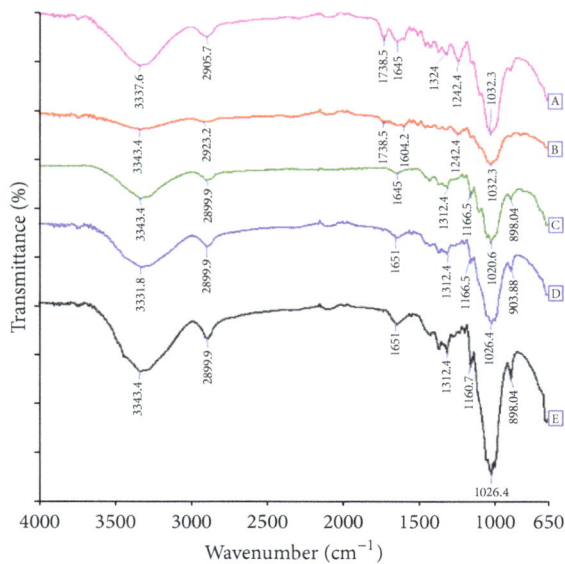

FIGURE 5: Wavelengths in FTIR of (A) jute fiber, (B) bamboo fiber, (C) commercial cellulose, (D) jute cellulose, and (E) bamboo cellulose.

samples revealed that lignin is removed. On the other hand, all cellulose samples show peak at 1166–1160 cm^{-1} region assigned as C–O–C stretching attributed to the β-(1→4)-glycosidic linkage in cellulose [32].

Figure 5 also shows all cellulose with broad peak intensity band at 3343–3331 cm^{-1} attributed to O–H stretching vibration. The peaks at 2923–2900 cm^{-1} and 1324–1312 cm^{-1} regions are characteristics of C–H stretching and –CH$_2$ bending, respectively. The peak at 1651–1646 cm^{-1} is attributed to the H–O–H stretching vibration of absorbed water in carbohydrate. The peak at 903–898 cm^{-1} region is related to glycosidic –C$_1$–H deformation, a ring vibration, and –O–H bending. These characters imply the β-glycosidic linkages between the anhydroglucose units in cellulose. The rise of intensity peak at 1026–1020 cm^{-1} region confirms that the cellulose content increases due to chemical treatment [33]. Both jute and bamboo cellulose show identical peaks compared to commercial cellulose. This further confirms that hemicelluloses and lignin in both jute and bamboo fiber were well removed during the chemical treatment.

4.2. Tensile Properties.

The tensile strength and Young's modulus of UJBC, JCC, BCC, and JBCC hybrid composite by weight percentages are shown in Figures 6(a) and 6(b), respectively. The results indicated that all treated cellulose composites have significant high tensile strength and Young's modulus compared to untreated hybrid composite (UJBC). Hybrid cellulose composite with 10 wt% fiber loading shows highest tensile strength and Young's modulus among the sample series tested. The tensile strength gradually decreases when fiber loading is greater than 10 wt%.

Among the cellulose hybrid composites, it is obviously shown that JBCC at 10 wt% was the optimized hybrid composite which increases tensile strength and Young's modulus

by 157% and 195%, respectively. The significant increase of tensile properties in JBCC is contributed by great improvements of interfacial adhesion between the treated hybrid cellulose fibers and LDPE matrix. The removal of hemicelluloses in hybrid cellulose fibers further enhances fiber rearrangement in composite which led to effective loads transfer [34]. JCC and BCC show similar tensile strengths which increase by 130% and 126.5%, respectively. On the other hand, Young's modulus of BCC increased by 144% is much higher than JCC which only increases by 64.3%. High resistance of elastic deformation on BCC attributed to high stiffness of bamboo cellulose in composite. Smaller cross-section of bamboo cellulose enhanced the stiffness of the composites [35]. The strong interface bonding between bamboo cellulose and LDPE enables effective stresses distribution on composite and contributed to high Young's modulus.

It is worth noting that all UJBC series show decreasing trend of mechanical properties with increased hybrid fibers loading. The incompatibility of hydrophilic fibers with hydrophobic polymer matrix causes relative weak fibers-matrix interfacial adhesion. Untreated hybrid fibers also exhibit poor resistance to moisture and led to high water absorption and subsequently contributed to poor mechanical properties of UJBC composites.

The stress-strain graph of UJBC, JCC, BCC, and JBCC composite at 10 wt% fiber loading is shown in Figure 7. The results revealed that all treated cellulose composites have high stress with greater strain compared with UJBC. The removal of hemicellulose and lignin in fiber provides strong adhesion between fiber and matrix. It contributed to enhancing strain by greater elongation of composites.

4.3. Thermogravimetric Analysis (TGA).

The thermogravimetric analysis of UJBC, JCC, BCC, and JBCC hybrid composites at 10 wt% fiber loading was summarized in Table 2. Previous studies have found that there is no degradation of jute-bamboo hybrid composite up to 300°C [36]. Above this temperature, the decomposition takes place and thermal stability decreased.

UJBC sample showed that low onset decomposition started at 304.9°C. Low onset decomposition of UJBC was mainly contributed by decomposition of hemicellulose [37]. On the other hand, all treated cellulose hybrid composites showed significant high onset decomposition temperature ranging from 346 to 415°C. Relative high onset decomposition was due to high thermal resistance of treated cellulose fibers in hybrid composite. The high onset decomposition also confirms that hemicellulose was well removed in cellulose composites. There is no significant difference between untreated hybrid composite and treated cellulose hybrid composite related to major decomposition. The major decomposition of hybrid composite occurs between 511°C and 529°C involving α-cellulose and LDPE resin [38]. At above 545°C, final decomposition takes place. The final decomposition of F10JCC, F10BCC, and F10JBCC showed remaining weight of 5.1%, 3.2%, and 3.5%, respectively. The final remaining weight was mainly due to char formation.

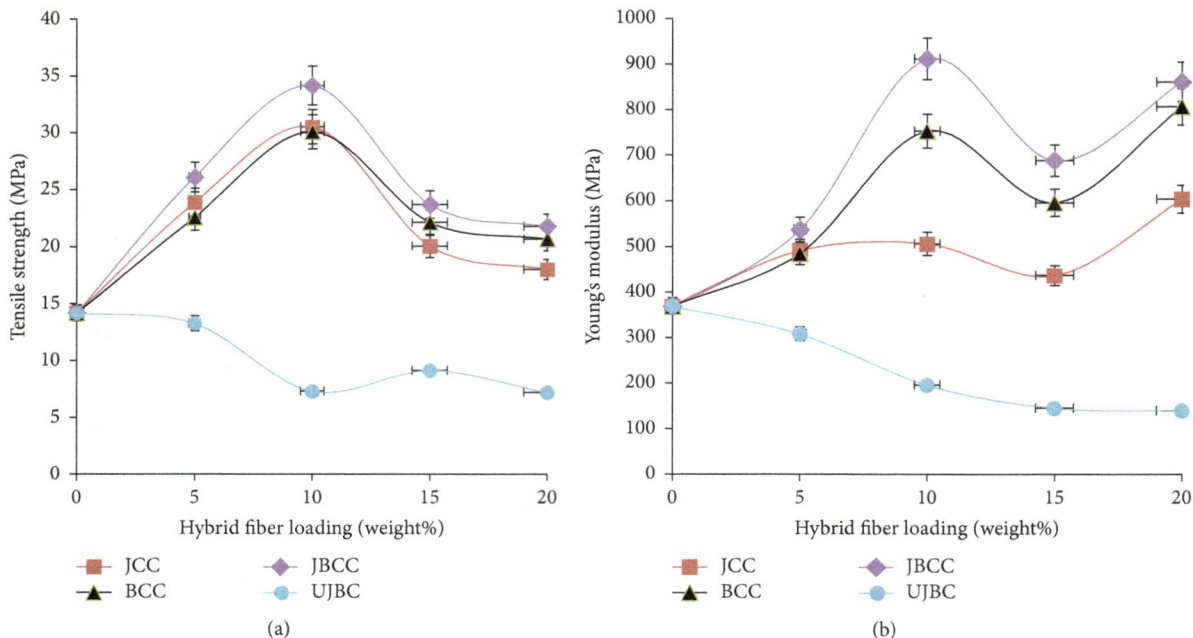

FIGURE 6: (a) Tensile strength of composite at different fiber loading. (b) Young's modulus of composite at different fiber loading.

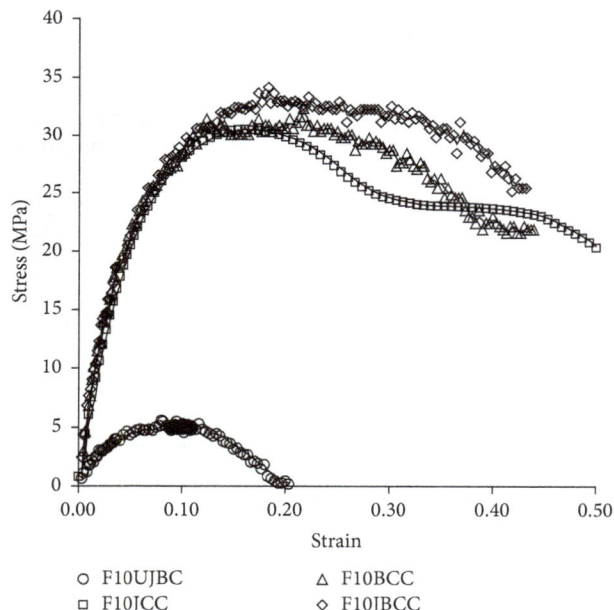

FIGURE 7: Stress-strain graph of hybrid composite at 10 wt% fiber loading.

The activation energy (E_a) was calculated from TGA graphs using Broido equation [39] as given below:

$$\ln\left(\ln\frac{1}{y}\right) = \frac{E_a}{RT} + \text{constant},\tag{3}$$

where y is the fraction of undecomposed nonvolatilized material, T is the absolute temperature (in Kelvin), and R

is the gas constant ($8.314\,\text{J}\,\text{mol}^{-1}\,\text{K}^{-1}$). The values of y have been taken from the TGA data. The values of $\ln(\ln(1/y))$ were plotted on y-axis, while the temperatures as $1/T$ (in Kelvin scale) were plotted on x-axis. Finally, from the slope of the trend line, the activation energy of the composite was calculated.

Table 2 shows that all treated cellulose composites obtain higher activation energy compared to F10UJBC. Among those composites, F10JBCC shows 2.5 times higher activation energy followed by F10BCC and F10JCC which are 2.0 and 1.1 times, respectively. The higher value of activation energy indicates the higher thermal stability of composite [40]. Higher activation energy in treated cellulose hybrid composite was attributed to rough surface and pores on the cellulose fibers become more prominent, which give better adhesion with the matrix [41].

4.4. Differential Scanning Calorimetry (DSC). The melting characteristic of 10 wt% cellulose composite is showed in Table 3. From Table 3, all cellulose and its hybrid composites show higher initial melting temperature compared to F10UJBC.

The crystallinity index was reduced from 55.9% for neat LDPE to 41.9%–50.3% for treated cellulose hybrid composites. Shortening of composite solidification time due to presence of cellulose fiber and the surface energy of fibers becoming closer to the surface energy of the matrix are reasons of reduced crystallinity index [42]. Significant reduction in crystallinity index was shown by F10BCC which reduced by 14%. The result revealed that F10BCC has better fiber-matrix interfacial bonding which further supports high Young's modulus results. On the other hand, F10UJBC, F10JBCC, and F10JCC show a marginal reduction in crystallinity index of

TABLE 2: Comparison of activation energy among hybrid composites at 10 wt% fiber loading.

Samples	T_i (°C)[a]	T_m (°C)[b]	T_f (°C)[c]	W_{Ti} (%)[d]	W_{Tm} (%)[e]	W_{Tf} (%)[f]	Activation energy, E_a (KJ/mol)
F10UJBC	304.9	525.2	550.5	99.1	19.3	2.8	64.55
F10JCC	355.7	511.0	553.1	97.9	28.6	5.1	72.96
F10BCC	346.4	524.5	545.1	99.0	22.0	3.2	129.86
F10JBCC	415.0	528.9	564.6	98.3	32.6	3.5	166.91

[a]Temperature corresponding to the onset of decomposition. [b]Temperature corresponding to maximum rate of mass loss. [c]Temperature corresponding to the end of decomposition. [d]Mass loss at temperature corresponding to the onset of decomposition. [e]Mass loss at temperature corresponding to the maximum rate of mass loss. [f]Mass loss at temperature corresponding to the end of decomposition.

TABLE 3: Melting characteristics of neat LDPE, UJBC, JCC, BCC, and JBCC at 10 wt% cellulose fiber loading.

Specimen	Initial melting temperature (°C)	Peak melting temperature (°C)	Final melting temperature (°C)	Heat flow ΔH_m (Jg^{-1})	Crystallinity χ_C
Neat LDPE	119.44	131.30	136.08	162.10	55.9%
F10UJBC	117.56	129.43	134.64	118.41	45.3%
F10JCC	119.15	128.46	132.64	131.38	50.3%
F10BCC	120.81	131.90	136.61	109.39	41.9%
F10JBCC	119.54	130.10	134.82	120.91	46.3%

ΔH_m: enthalpy of fusion and χ_C: degree of crystallinity.

9.9%, 9.6%, and 5.6%, respectively. It is worth noting that all DSC results for F10JBCC show a value between F10JCC and F10BCC. Those results indicated that F10JBCC hybrid composite shows average melting characteristic combination of F10JCC and F10BCC which is following rule of mixtures for hybrid composite [43].

4.5. Scanning Electron Microscopy (SEM) Analysis. Figure 8 shows SEM images of the cross-section surface of UJBC, JCC, BCC, and JBCC at 10 wt% fiber loading after tensile test. Figure 8(a) shows the delamination of F10UJBC. There were some microcracks because of the relatively weak interfacial bonding among the fibers and the matrix. Poor adhesion between the untreated hybrid fibers and the matrix creates further voids spaces around fibers in the hybrid composite [44].

Figure 8(b) shows the well-bonded jute cellulose composite. The good interfacial bonding between jute cellulose and LDPE matrix contributed to high tensile strength and modulus. Strong interfacial bonding was observed in F10JCC. Micrographs revealed that the jute cellulose fibers smoothly distributed in of LDPE resin.

Figure 8(c) shows a well-bonded interface between the bamboo cellulose fiber and LDPE matrix with some minor voids observed. This strong interfacial bond among cellulose fiber-LDPE matrix contributed to a higher tensile strength and modulus. This was mainly attributed to an elimination of the hydroxyl groups from chemical surface modified cellulose fibers.

Figure 8(d) shows strong bonding between the jute-bamboo cellulose fiber and LDPE matrix. The hydrophobic characteristics of treated hybrid cellulose fibers and well

bonding of LDPE both strengthen the fibers-matrix interfacial adhesion. Relative minor microvoids appear between fiber and matrix interfacial.

4.6. Water Absorption Test. Figure 9 showed that water absorption of UJBC, BCC, JBCC, and JCC of 10 wt% cellulose fibers loading for 22 days. The untreated hybrid fiber composite (UJBC) showed highest water absorption rate up to 30.2% in this duration. Water absorption rate of UJBC starts to saturate after 11 days. High water absorption was results of poor wettability and adhesion between hydrophilic fibers-LDPE [45]. Untreated fibers are hydrophilic with many hydroxyl group (–OH). The high percentage of –OH group in fibers tend to show low moisture resistance. This leads to dimensional variation of UJBC and poor interfacial bonding between the fiber and matrix. This also caused microcracks and voids formation in UJBC which trap and absorb greater percentage of water. As the microcracks propagated, capillarity and transport of water molecules via microcracks become active [46].

Among treated hybrid composites, F10JBCC shows the lowest water absorption rate of 0.7%. F10JBCC water absorption saturated after 10 days. The mercerization treatment reduced the polar groups in fiber by replacing the hydroxyl groups on the surfaces of fibers [47]. The low water absorption rate in F10JBCC is contributed by good adhesion among treated hybrid fibers-matrix composite. Good interfacial adhesion bonding among treated cellulose fiber-matrix reacts to retard water entering cellulose fiber. F10JCC attained a saturation level after 6 days with water absorption of 1.9%. It is worth noting that F10BCC shows highest water absorption rate among treated cellulose hybrid composites. The water

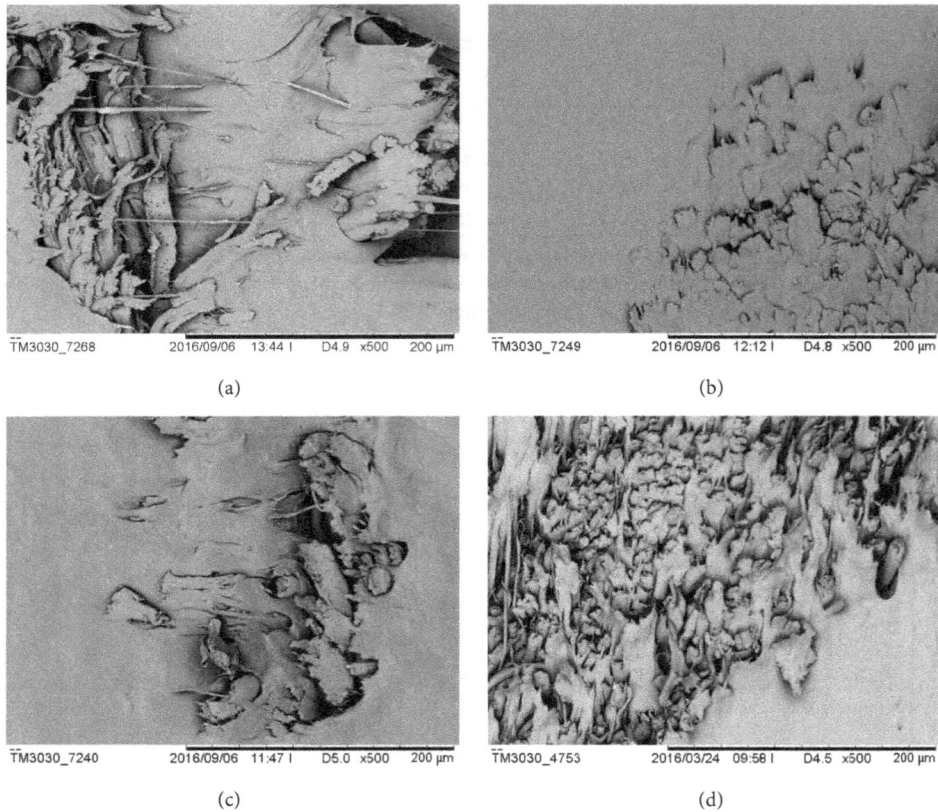

FIGURE 8: SEM images of the cross-section surface of (a) UJBC, (b) JCC, (c) BCC, and (d) JBCC at 10 wt% fiber loading.

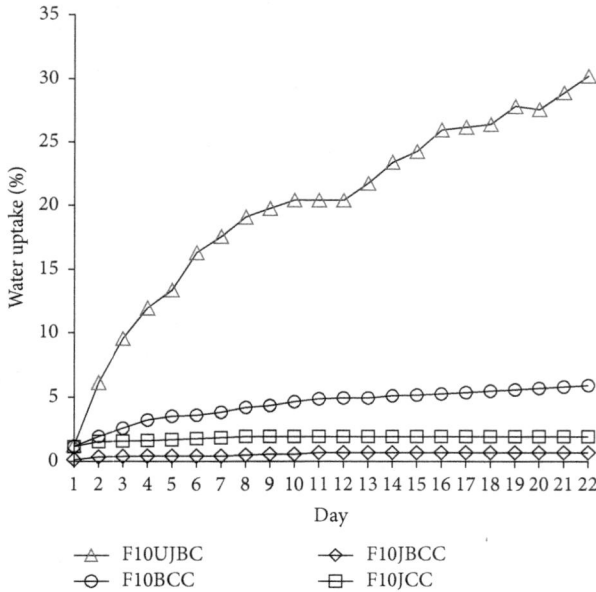

FIGURE 9: Water absorption percentage of UJBC, BCC, JBCC, and JCC for 22 days for 10 wt% fiber loading.

absorption is 5.9% and absorption rate reduced only after 10 days. This is mostly due to high surface porosity of bamboo cellulose combined with microvoid among treated cellulose and matrix interfaces. Overall, water absorption was significantly low for all treated cellulose hybrid composites compared with untreated hybrid fiber composite.

5. Conclusions

The present study reveals that both treated jute and bamboo cellulose were successfully used as reinforcement fiber in hybrid composites. The FTIR results show that jute and bamboo cellulose obtained are identical to commercial cellulose. The tensile strength and Young's modulus of hybrid cellulose composite are optimized at 10 wt% hybrid fibers loading. JBCC at 10 wt% reveals high tensile strength and Young's modulus increased by 157% and 195%, respectively. All treated cellulose hybrid composites showed high onset decomposition temperature ranging from 346 to 415°C due to elimination of hemicellulose. F10JBCC shows 2.5 times higher activation energy followed by F10BCC and F10JCC at 2.0 times and 1.1 times, respectively. Water absorption was significantly low for all treated cellulose hybrid composites compared with untreated hybrid fiber composite. F10JBCC shows lowest water absorption up to 43 times lower compared to F10UJBC. The significant improved mechanical and morphological properties of treated cellulose hybrid composites are further supported by SEM images that show better matrix-fiber interaction compared to untreated hybrid fiber composite. The outcome of the composite shows cellulose hybrid composite suitable for outdoor application such as street light enclosure, automobile signal light cover, and playground slides.

Conflicts of Interest

The authors declare that there are no conflicts of interest regarding the publication of this paper.

Acknowledgments

The authors would like to acknowledge the Center of Excellence and Renewable Energy (CoERE), UNIMAS for their financial support, Grant no. CoERE/Grant/2013/06 and Ph.D. seed fund F02(DPP30)/1245/2015(05). The authors would like to thank Forestry Department Sarawak for their technical support on tensile testing.

References

[1] G. Bogoeva-Gaceva, M. Avella, M. Malinconico et al., "Natural fiber eco-composites," *Polymer Composites*, vol. 28, no. 1, pp. 98–107, 2007.

[2] H. Ku, H. Wang, N. Pattarachaiyakoop, and M. Trada, "A review on the tensile properties of natural fiber reinforced polymer composites," *Composites Part B: Engineering*, vol. 42, no. 4, pp. 856–873, 2011.

[3] V. K. Thakur and M. K. Thakur, "Review: processing and characterization of natural cellulose fibers/thermoset polymer composites," *Carbohydrate Polymers*, vol. 109, pp. 102–117, 2014.

[4] H. Wang, L. Huang, and Y. Lu, "Preparation and characterization of micro- and nano-fibrils from jute," *Fibers and Polymers*, vol. 10, no. 4, pp. 442–445, 2009.

[5] P. Chaowana, "Bamboo: an alternative raw material for wood and wood-based composites," *Journal of Materials Science Research*, vol. 2, no. 2, pp. 1–13, 2013.

[6] H. P. S. Abdul Khalil, I. U. H. Bhat, M. Jawaid, and etal., "Bamboo fibre reinforced biocomposites: a review," *Materials & Design*, vol. 42, pp. 353–368, 2012.

[7] C. Fuentes, G. Brughmans, L. Q. N. Tran et al., "Mechanical behaviour and practical adhesion at a bamboo composite interface: physical adhesion and mechanical interlocking," *Composites Science and Technology*, vol. 109, pp. 40–47, 2015.

[8] R. C. Sun, X. F. Sun, P. Fowler, and J. Tomkinson, "Structural and physico-chemical characterization of lignins solubilized during alkaline peroxide treatment of barley straw," *European Polymer Journal*, vol. 38, no. 7, pp. 1399–1407, 2002.

[9] Y. Liu and H. Hu, "X-ray diffraction study of bamboo fibers treated with NaOH," *Fibers and Polymers*, vol. 9, no. 6, pp. 735–739, 2008.

[10] T. Lu, S. Liu, M. Jiang et al., "Effects of modifications of bamboo cellulose fibers on the improved mechanical properties of cellulose reinforced poly(lactic acid) composites," *Composites Part B: Engineering*, vol. 62, pp. 191–197, 2014.

[11] Y. Xie, C. A. S. Hill, Z. Xiao, H. Militz, and C. Mai, "Silane coupling agents used for natural fiber/polymer composites: a review," *Composites Part A: Applied Science and Manufacturing*, vol. 41, no. 7, pp. 806–819, 2010.

[12] M. R. Ismail, A. A. M. Yassene, and H. M. H. Abd El Bary, "Effect of silane coupling agents on rice straw fiber/polymer composites," *Applied Composite Materials*, vol. 19, no. 3-4, pp. 409–425, 2012.

[13] J. Cai, P. Fei, Z. Xiong, Y. Shi, K. Yan, and H. Xiong, "Surface acetylation of bamboo cellulose: Preparation and rheological properties," *Carbohydrate Polymers*, vol. 92, no. 1, pp. 11–18, 2013.

[14] D. Ray, B. K. Sarkar, R. K. Basak, and A. K. Rana, "Thermal behavior of vinyl ester resin matrix composites reinforced with alkali-treated jute fibers," *Journal of Applied Polymer Science*, vol. 94, no. 1, pp. 123–129, 2004.

[15] X. Li, L. G. Tabil, and S. Panigrahi, "Chemical treatments of natural fiber for use in natural fiber-reinforced composites: a review," *Journal of Polymers and the Environment*, vol. 15, no. 1, pp. 25–33, 2007.

[16] S. Kalia, A. Dufresne, B. M. Cherian et al., "Cellulose-based bio- and nanocomposites: a review," *International Journal of Polymer Science*, vol. 2011, Article ID 837875, 35 pages, 2011.

[17] L. F. Zemljic, O. Sauperl, T. Kreze, and S. Strnad, "Characterization of regenerated cellulose fibers antimicrobial functionalized by chitosan," *Textile Research Journal*, vol. 83, no. 2, pp. 185–196, 2013.

[18] P. B. Subhedar and P. R. Gogate, "Alkaline and ultrasound assisted alkaline pretreatment for intensification of delignification process from sustainable raw-material," *Ultrasonics Sonochemistry*, vol. 21, no. 1, pp. 216–225, 2014.

[19] J. Gassan and A. K. Bledzki, "Possibilities for improving the mechanical properties of jute/epoxy composites by alkali treatment of fibres," *Composites Science and Technology*, vol. 59, no. 9, pp. 1303–1309, 1999.

[20] M. Jawaid and H. P. S. Abdul Khalil, "Cellulosic/synthetic fibre reinforced polymer hybrid composites: a review," *Carbohydrate Polymers*, vol. 86, no. 1, pp. 1–18, 2011.

[21] S. Nunna, P. R. Chandra, S. Shrivastava, and A. K. Jalan, "A review on mechanical behavior of natural fiber based hybrid composites," *Journal of Reinforced Plastics and Composites*, vol. 31, no. 11, pp. 759–769, 2012.

[22] M. M. Rahman, M. R. Rahman, S. Hamdan, M. F. Hossen, J. C. H. Lai, and F. K. Liew, "Effect of silicon dioxide/nanoclay on the properties of jute fiber/polyethylene biocomposites," *Journal of Vinyl and Additive Technology*, 2015.

[23] K. P. Rajan, N. R. Veena, H. J. Maria, R. Rajan, M. Skrifvars, and K. Joseph, "Extraction of bamboo microfibrils and development of biocomposites based on polyhydroxybutyrate and bamboo microfibrils," *Journal of Composite Materials*, vol. 45, no. 12, pp. 1325–1329, 2011.

[24] M. Pöllänen, M. Suvanto, and T. T. Pakkanen, "Cellulose reinforced high density polyethylene composites—Morphology, Mechanical and thermal expansion properties," *Composites Science and Technology*, vol. 76, pp. 21–28, 2013.

[25] A. Sdrobiş, R. N. Darie, M. Totolin, G. Cazacu, and C. Vasile, "Low density polyethylene composites containing cellulose pulp fibers," *Composites Part B: Engineering*, vol. 43, no. 4, pp. 1873–1880, 2012.

[26] B. N. Kuznetsov, V. G. Danilov, S. A. Kuznetsova, O. V. Yatsenkova, and N. B. Aleksandrova, "Optimization of fir wood delignification by acetic acid in the presence of hydrogen peroxide and a TiO2 catalyst," *Theoretical Foundations of Chemical Engineering*, vol. 43, no. 4, pp. 499–503, 2009.

[27] "International A, ASTM D 638 Standard Test Method for Tensile Properties of Plastics. ASTM International, West Conshohocken, PA, USA, 2010".

[28] International A, ASTM D 570 Standard Test Method for Water Absorption of Plastics. ASTM International, West Conshohocken, PA, USA, 1998.

[29] A. Alemdar and M. Sain, "Isolation and characterization of nanofibers from agricultural residues—wheat straw and soy hulls," *Bioresource Technology*, vol. 99, no. 6, pp. 1664–1671, 2008.

[30] J. I. Morán, V. A. Alvarez, V. P. Cyras, and A. Vázquez, "Extraction of cellulose and preparation of nanocellulose from sisal fibers," *Cellulose*, vol. 15, no. 1, pp. 149–159, 2008.

[31] R. Sukmawan, H. Takagi, and A. N. Nakagaito, "Strength evaluation of cross-ply green composite laminates reinforced by bamboo fiber," *Composites Part B: Engineering*, vol. 84, pp. 9–16, 2016.

[32] S. Y. Oh, I. Y. Dong, Y. Shin et al., "Crystalline structure analysis of cellulose treated with sodium hydroxide and carbon dioxide by means of X-ray diffraction and FTIR spectroscopy," *Carbohydrate Research*, vol. 340, no. 15, pp. 2376–2391, 2005.

[33] H. D. Nguyen, T. T. Thuy Mai, N. B. Nguyen, T. D. Dang, M. L. Phung Le, and T. T. Dang, "A novel method for preparing microfibrillated cellulose from bamboo fibers," *Advances in Natural Sciences: Nanoscience and Nanotechnology*, vol. 4, no. 1, Article ID 015016, 2013.

[34] M. C. Symington, W. M. Banks, O. D. West, and R. A. Pethrick, "Tensile testing of cellulose based natural fibers for structural composite applications," *Journal of Composite Materials*, vol. 43, no. 9, pp. 1083–1108, 2009.

[35] R. B. Yusoff, H. Takagi, and A. N. Nakagaito, "Tensile and flexural properties of polylactic acid-based hybrid green composites reinforced by kenaf, bamboo and coir fibers," *Industrial Crops and Products*, vol. 94, pp. 562–573, 2016.

[36] F. K. Liew, S. Hamdan, M. R. Rahman et al., "4-methylcatechol-treated jute-bamboo hybrid composites: effects of ph on thermo-mechanical and morphological properties," *BioResources*, vol. 11, no. 3, pp. 6880–6895, 2016.

[37] H. Yang, R. Yan, H. Chen, D. H. Lee, and C. Zheng, "Characteristics of hemicellulose, cellulose and lignin pyrolysis," *Fuel*, vol. 86, no. 12-13, pp. 1781–1788, 2007.

[38] S. Mohanty, S. K. Verma, and S. K. Nayak, "Dynamic mechanical and thermal properties of MAPE treated jute/HDPE composites," *Composites Science and Technology*, vol. 66, no. 3-4, pp. 538–547, 2006.

[39] A. Broido, "A simple, sensitive graphical method of treating thermogravimetric analysis data," *Journal of Polymer Science Part A-2 Polymer Physics*, vol. 7, no. 10, pp. 1761–1773, 1969.

[40] M. R. Rahman, S. Hamdan, A. S. Ahmed et al., "Thermogravimetric analysis and dynamic Young's modulus measurement of N,N-dimethylacetamide-impregnated wood polymer composites," *Journal of Vinyl and Additive Technology*, vol. 17, no. 3, pp. 177–183, 2011.

[41] R. Agrawal, N. S. Saxena, K. B. Sharma, S. Thomas, and M. S. Sreekala, "Activation energy and crystallization kinetics of untreated and treated oil palm fibre reinforced phenol formaldehyde composites," *Materials Science and Engineering A*, vol. 277, no. 1-2, pp. 77–82, 2000.

[42] S. Kumar, V. Choudhary, and R. Kumar, "Study on the compatibility of unbleached and bleached bamboo-fiber with LLDPE matrix," *Journal of Thermal Analysis and Calorimetry*, vol. 102, no. 2, pp. 751–761, 2010.

[43] M. F. Ashby and Y. J. M. Bréchet, "Designing hybrid materials," *Acta Materialia*, vol. 51, no. 19, pp. 5801–5821, 2003.

[44] Z. N. Azwa, B. F. Yousif, A. C. Manalo, and W. Karunasena, "A review on the degradability of polymeric composites based on natural fibres," *Materials and Design*, vol. 47, pp. 424–442, 2013.

[45] P. K. Kushwaha and R. Kumar, "Bamboo fiber reinforced thermosetting resin composites: effect of graft copolymerization of fiber with methacrylamide," *Journal of Applied Polymer Science*, vol. 118, no. 2, pp. 1006–1013, 2010.

[46] H. N. Dhakal, Z. Y. Zhang, and M. O. W. Richardson, "Effect of water absorption on the mechanical properties of hemp fibre reinforced unsaturated polyester composites," *Composites Science and Technology*, vol. 67, no. 7, pp. 1674–1683, 2007.

[47] S. K. Saw, G. Sarkhel, and A. Choudhury, "Surface modification of coir fibre involving oxidation of lignins followed by reaction with furfuryl alcohol: Characterization and stability," *Applied Surface Science*, vol. 257, no. 8, pp. 3763–3769, 2011.

Experiment on Behavior of a New Connector Used in Bamboo (Timber) Frame Structure under Cyclic Loading

Junwen Zhou [1,2] **Dongsheng Huang,**[1] **Chun Ni,**[3] **Yurong Shen,**[1] **and Longlong Zhao**[1]

[1]*School of Civil Engineering, Nanjing Forestry University, Nanjing 210037, China*
[2]*School of Civil Engineering and Architecture, Changzhou Institute of Technology, Changzhou 213033, China*
[3]*FPInnovations, Vancouver, BC, Canada V6T 1Z4,*

Correspondence should be addressed to Dongsheng Huang; dshuang@njfu.edu.cn

Academic Editor: Ana S. Guimarães

Connection is an important part of the bamboo and timber structure, and it directly influences the overall structural performance and safety. Based on a comprehensive analysis of the mechanical performance of several wood connections, a new connector for the bamboo (timber) frame joint was proposed in this paper. Three full-scale T-type joint specimens were designed to study the mechanical performance under cyclic loading. The thickness of the hollow steel column was different among three specimens. The specimens were loaded under displacement control with a rate of 10 mm per minute until the specimens reach failure. It was observed that the failures of three specimens were caused by the buckling of flanges in the compression and that the steel of connections does not yield. The load-displacement hysteretic curve for three specimens is relatively plump, and the stiffness of connection degenerates with the increasing of cyclic load. The maximum rotation is 0.049 rad, and the energy dissipation coefficient is 1.77. The thickness of the hollow steel column of the connector has significant impact on the energy dissipation capacity and the strength of the connection. A simplified moment-rotation hysteresis model for the joint was proposed.

1. Introduction

Timber is a natural organic material, and people easily get it from nature and use it without much processing; therefore, timber was employed as a construction material a long time ago. Timber has higher strength in tension and compression parallel to grain, light mass, and good durability, some 1000-year timber buildings still stand well [1]. Because of friendly environment, graceful timber texture, and simple nature, timber building will still be enormously appealing to people. For the timber frame structure, the beam-to-column connection is usually the most unsubstantial part on account of fabricated construction. The mechanical behavior of timber joints directly influences the overall timber structural performance and safety. As a result, the design of timber joints is extremely important. In some Asian countries such as China, South Korea, and Japan, mortise and tenon joints [1–5] are traditionally used in timber buildings from dwelling houses to palaces. In these joints, steel fasteners are not used, therefore keeping the original beauty of the timber. However, due to the slippage between the mortise and tenon, the energy dissipation capacity of the whole structure is affected under earthquake load. The whole building can even collapse due to the separation between the mortise and tenon. In addition, the mortise-to-tenon connection wakens the column cross section at the connection, which reduces the vertical load-carrying capacity and also negatively affects the energy dissipation capacity of the column. To prevent the mortise-to-tenon connection from separation and improve the strength of connection, Bulleit et al. [6] used wooden pegs to fasten mortise-to-tenon connection. But the study showed that the split failure owes to prying force of the wood peg occurred along with a peg hole when there is a distance between the tenon and the sill of mortise, and either shear failure of the tenon appeared. Moreover, Hong et al. [7] employed a T-type steel plate to strengthen the mortise-to-tenon connection and not to enhance the energy dissipation capacity of the joint.

Steel plate-bolted connection is another joint commonly used to connect the beam and column in a timber structure, which often is applied in the heavy timber structure and has architecture beauty because the connecting steel plate is covered with wood. However, for these types of joints, the brittle failure mode was obvious [8–10]. Split failure along the bolt hole in the beam is the main failure mode in this kind of joint, and it reduces the load-carrying capacity of the joint. As the beam and column are slotted to accommodate the steel plate, the vertical load-carrying capacity is reduced. Under cyclic loading, the hysteretic curve shows a clear "pinching" effect because of the gap between the bolt and wall of the hole [8–10].

Glued-in-rod connection is also commonly used to connect the beam and column in a timber building. The rod is embedded in timber to connect the beam and column and to transfer the load from the beam to the column. With appropriate materials and good construction quality, the glued-in-rod connection has demonstrated good structural performance [11, 12]. The glued-in-rod connection has been widely studied and used in various projects [13]. The advantage of glued-in-rod connection is that the steel rod is embedded in timber to protect the steel rod from corrosion, and the durability is better. In order to get good construction quality, generally, the steel rod should be glued well with one timber first part in advance and then being connected with another part. Thus, the rod hole position on the beam and column must be precise. As timber and glue are brittle materials, the principle of energy dissipation is not obvious. Vašek [14, 15] used two U-type steel connectors located at upper and bottom edges of the beam to fasten the steel rods embedded in the timber beam and column. This new method could reduce stress concentration to the beam and column and increase the energy dissipation capacity of the joint by the deformation of U-type steel connectors. In addition, the U-type steel connectors are applied to connect the timber beam and column, and the position of the hole on the steel connector can easily be changed, which is convenient for connection construction on site.

Bolted timber-timber connection is a simple and practical connection for the timber beam-to-column joint. Only bolts are used to fasten the timber beam and column and to carry load from the girder to the post [16, 17]. Steel nails sometimes are used to substitute for bolts. The different mechanical performance of connection can be obtained by changing the quantity and arrangement style of bolts. Some special materials, such as steel plate or hard wood, are inserted into the contacted surface of the beam and column to improve the performance of connection as well.

For the timber beam-to-column joint, load-carrying capacity and energy dissipation are the two primary factors to judge the performance of the joint. Huang [18] gave a better energy dissipation connector for the timber frame joint, but the connection between the beam and column is not robust, slippage happens under earthquake, and shear stiffness of the joint is less. Based on the comprehensive analysis of the abovementioned joints, a new beam-to-column connector was developed and presented in this paper. The joint can connect top and bottom columns and also link beam and column parts without weakening the column cross section. As the mechanical behavior of this joint under the earthquake load is unknown, an experimental study was conducted to evaluate the stiffness, strength, energy

dissipation, and resilience of the joint with different thicknesses of the hollow steel column.

2. Joint Details and Fabrication

The joint consists of two parts. One part is beam and column members made of parallel strand bamboo (PSB) [19–21]. The beam is 55 mm in width, 200 mm in depth, and 1050 mm in length. Two beams were assembled in parallel. At the end of the beams where a bolt hole was located, carbon fiber-reinforced plastic, 8 mm in width, was used to wrap the beams to prevent it from cracking along the bolt hole. The top and bottom columns are 200 mm in width, 240 mm in depth, and 700 mm in length. The column depth is paralleled to the beam axis. The other part is a steel connector, which is the most important part of the joint. The details of the connector are shown in Figure 1. Four steel plates are welded together to form a hollow column, and then, a horizontal steel plate is welded in the middle of the hollow column to reinforce the hollow column and to transfer load from the top column to the bottom column. The top and bottom plates (flanges) of the I-shaped steel beam were bent into L shape and welded to the hollow steel column. A steel plate web was welded between the top and bottom flanges to form an I-shaped steel beam. Two shear connection plates were welded to the hollow steel column and fastened to the web of the I-shaped steel beam with a 16 mm diameter bolt. This is to ensure the connection between the hollow steel column and the I-shaped steel beam in case the welding between the hollow steel column and I-shaped steel beam flanges is broken.

The PSB columns were placed into the hollow steel column and connected to the hollow steel column with four 14 mm diameter bolts. Two PSB beams were connected to the steel web with four 16 mm diameter bolts.

3. Description of the Experiment

3.1. Specimen Designing. Three joint specimens were fabricated in site. Except for the thickness of the hollow steel column, the specimens are identical. Details of the specimens are shown in Figure 2. Holes in bamboo specimen (beam and column) and steel plate are 1.5 mm greater than those of bolts, which is easy to assemble. A detailed description of thickness of each tested steel connection plate is reported in Table 1.

3.2. Mechanical Properties of Materials. All the steel plates are of grade Q235B in accordance with the Chinese standard (GB/T700-2006) [22]. The material properties of the steel plates are determined according to EN10002-1 [23]. Table 2 lists the mechanical properties of the steel plates. The 14 mm diameter bolts, which were used to fasten the hollow steel column and PSB columns, had the average yield strength of 804 MPa under tension. The 16 mm diameter bolts had the average yield strength of 2241 MPa under bending.

According to the ASTM standard D143-09 [24], the ultimate compressive and tensile strength values of PSB parallel to grain are 65 MPa and 100 MPa, respectively, and the ultimate tensile strength value of PSB perpendicular to grain is 4.4 MPa.

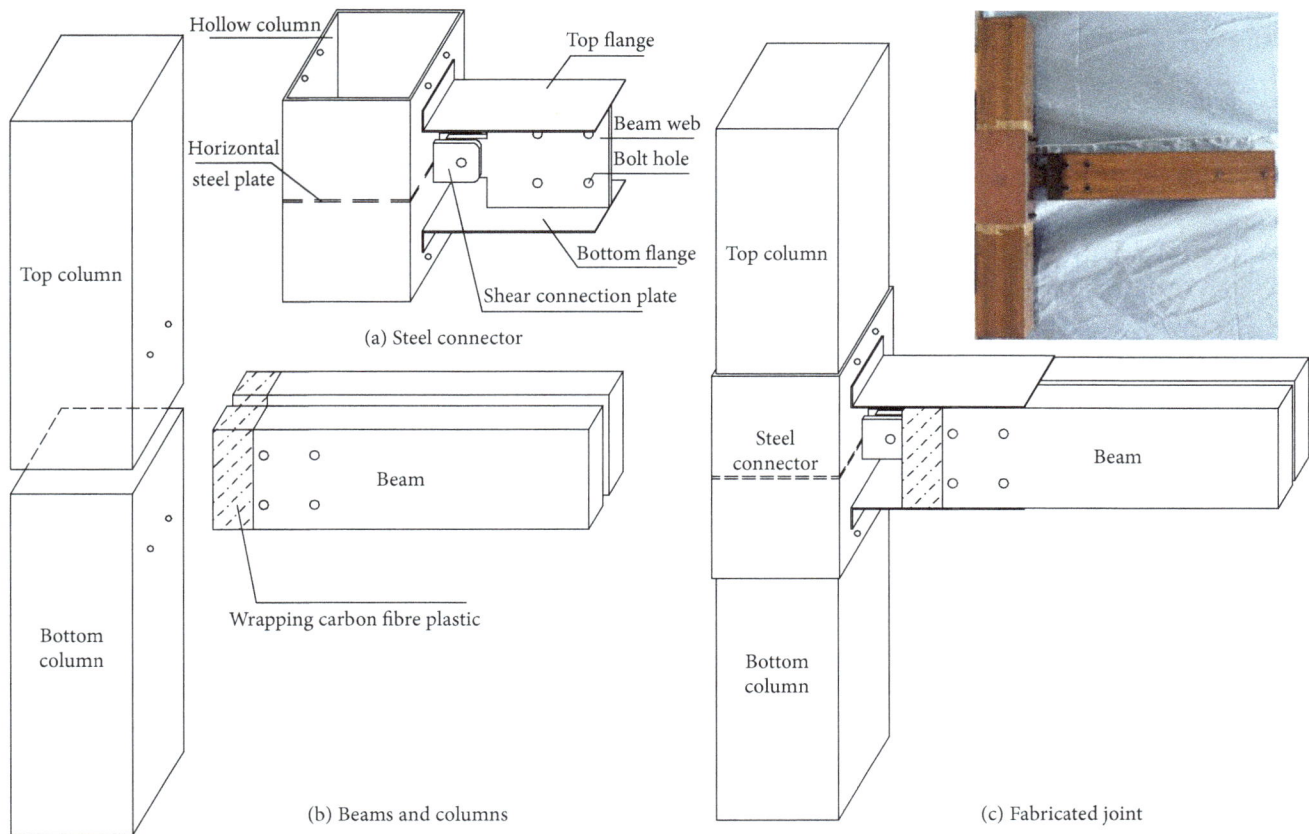

FIGURE 1: Constructional detail of the joint. (a) Steel connector, (b) beams and columns, and (c) fabricated joint.

3.3. Test Setup. A schematic illustration of the test setup is shown in Figure 3. The main purpose of this research is to study the mechanical performance of the connection specimens under cyclic loading. In order to load the specimen with the existing loading frame, the column was rotated an angle of 90° and was fixed to the floor channel with four 30 mm diameter bolts. A two-way 250 mm hydraulic actuator (100 kN capacity) was used to apply the cyclic load. The center of the actuator was 1000 mm above the top surface of the PSB column, and the head of the actuator was about 150 mm away from the PSB beam.

The beam displacements were monitored by a laser displacement sensor, and the corresponding load value was measured by a load cell mounted on the actuator rod. Two laser displacement sensors #3 and #4 were located at the beam flange to measure the rotation of the connection. The other two sensors #1 and #2 were located at the hollow steel column to measure the defection of the hollow column.

For the three specimens, the sensors were located at the same locations of the specimen.

3.4. Loading Schedule. In this study, a controlled cyclic displacement scheme was used. The maximum displacement was 10 mm in the first two cyclic loading, and the loading rate was 10 mm per minute. The maximum displacement was then increased by 10 mm after each step until specimen failure which showed specimen lost bearing loading capacity. The loading scheme is shown in Figure 4.

It was assumed that the actuator force in compression was positive, and the beam displacement moving away from the reaction wall was positive.

4. Test Results and Analysis

The test results of the three specimens are shown in Table 3. For each experiment, the failure mode, the energy dissipation, and the resilience model of each specimen will be discussed in the following sections.

4.1. Failure Mode. From Figure 3, it can be observed that one lateral beam flange was pulled and the opposite beam flange was compressed under loading. For the three specimens, the failure occurred when the beam flange buckled under compressive force. Figures 5–7 show the failure mode of the three specimens. It was noticed that the beam flange buckling occurred suddenly without yielding of the steel plate, which is a brittle failure and results in the failure of test specimens. Besides the beam flange buckling, for the specimen CJ2 in which the hollow steel column is 4 mm in thickness, the hollow steel column was also deformed around the beam flange because of tensile fore of the beam flange. The bolts and PSB members with larger cross-sectional dimensions were still in good condition after the tests.

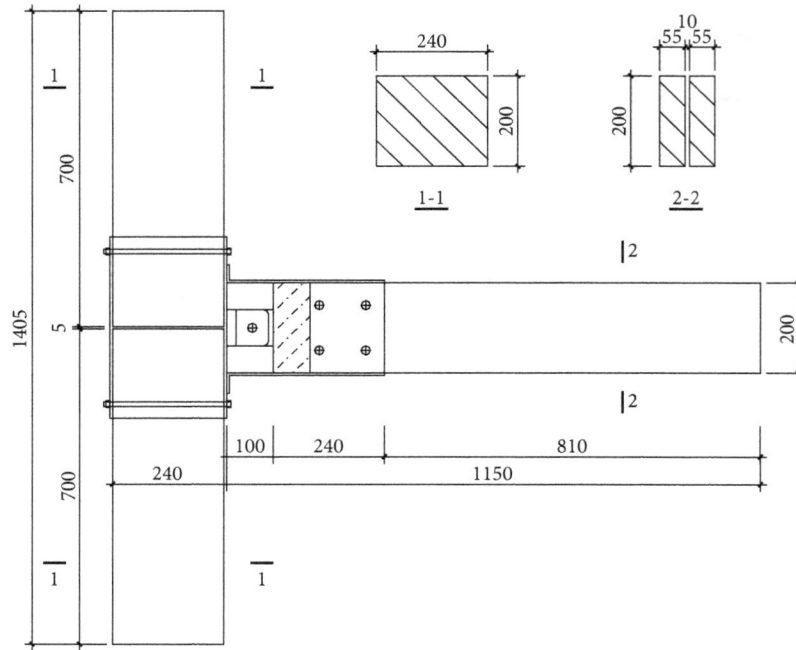

(a) Detailed drawing for joint.

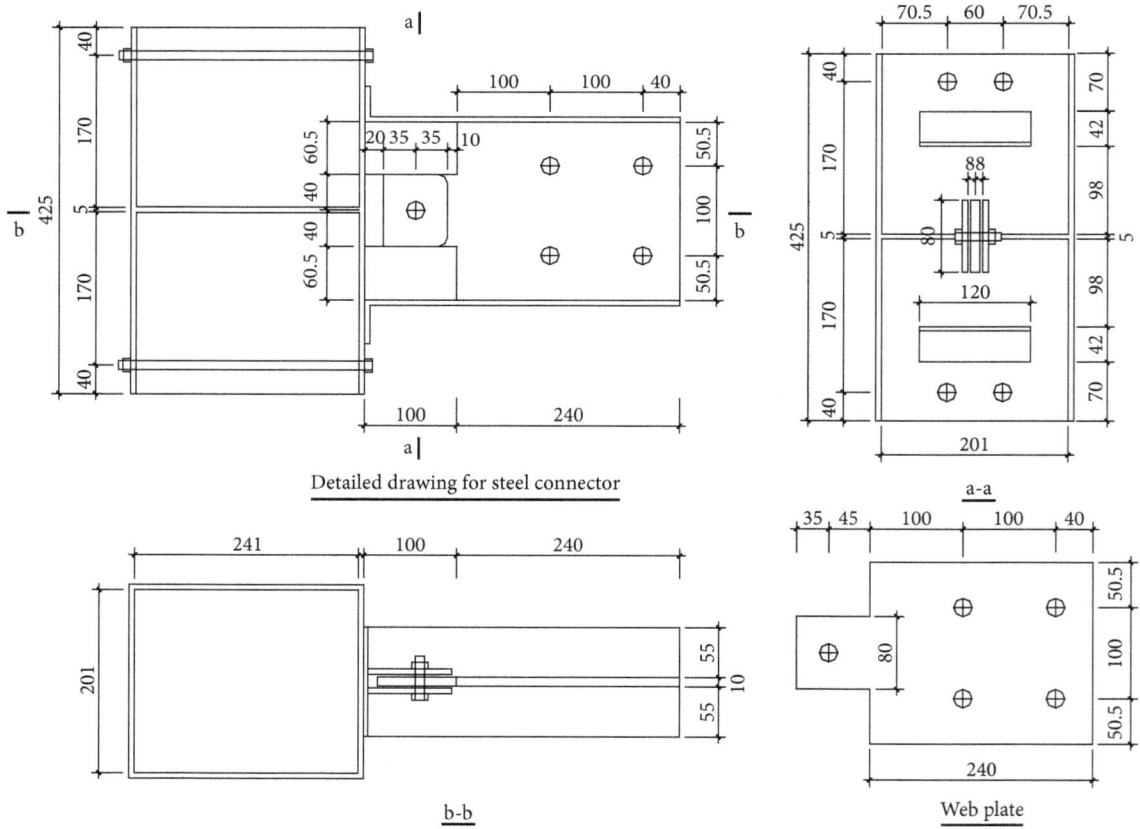

Detailed drawing for steel connector

a-a

b-b

Web plate

(b) Detailed diagramtic drawing for steel connector.

FIGURE 2: Detailed parameters for the specimen (mm).

TABLE 1: Steel plate thickness of the connector (mm).

Specimen	Hollow steel column	Beam flange	Beam web	Shear connection plate	Hollow column stiffener
CJ1	5	3	10	8	5
CJ2	4	3	10	8	5
CJ3	6	3	10	8	5

Grade Q235B is available for all steel plates.

TABLE 2: Mechanical property of the steel plate.

Thickness of the steel plate (mm)	σ_y (MPa)	σ_u (MPa)	ε_y (%)	ε_u (%)	E_s (MPa)
3	263	372	1.983	18.32	19523
4	256	381	1.831	20.23	19871
5	260	358	1.920	19.56	20522
6	249	363	1.853	18.92	20246
8	251	368	1.915	19.014	20042
10	238	375	2.052	18.76	19197

σ_y is the yield stress; σ_u is the limit stress; ε_y is the yield strain; ε_u is the limit strain; E_s is the elasticity modulus.

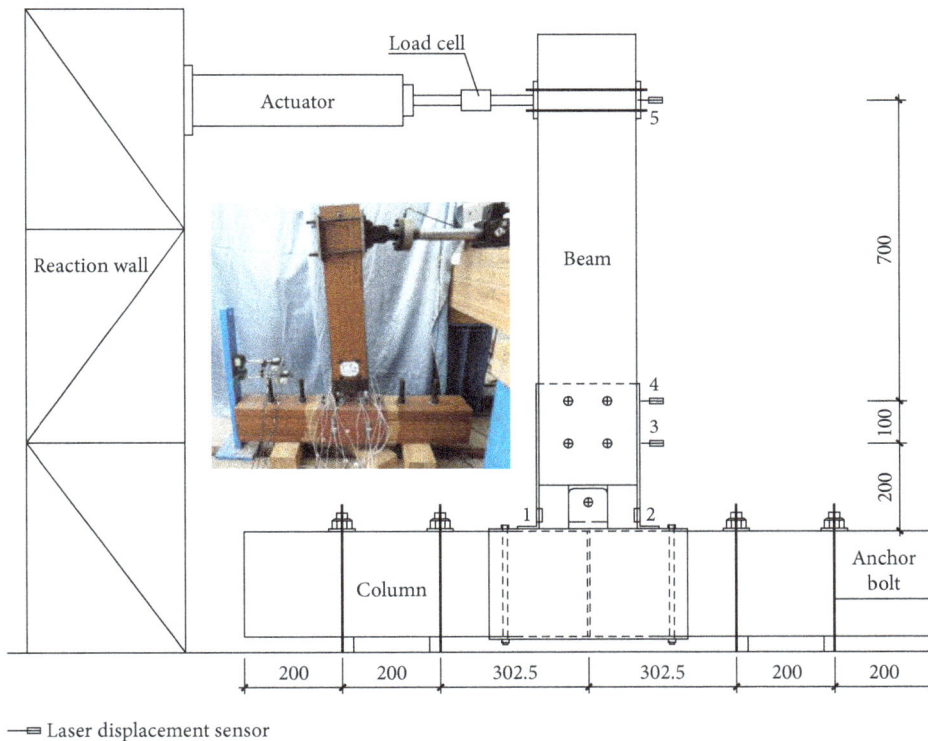

FIGURE 3: Experiment setup (mm).

According to tested results, it was also observed that the three specimens with the same thickness in the beam flange lost capacity in the same location and for the same reason, but they had obvious difference in ultimate bearing capacity; the main reason was that the boundary conditions of the compressed beam flange were not ideal situation under axial loading.

4.2. Load-Displacement Hysteretic Loop at the End of the Beam. The load-displacement hysteretic curve of the joint reflects the overall performance of connection. The load-displacement hysteretic curves at the end of the beam are shown in Figures 8–10. As can be seen from the figures, all three specimens exhibit good performance. The strength deterioration with the progress of the cyclic load was not observed. The rotational stiffness, however, deteriorated with the increase of cyclic load.

For the specimen CJ2 (4 mm thick hollow steel column), the beam flange in compression buckled suddenly after 9 displacement cycles. For the specimen CJ1 (5 mm thick hollow steel column) and the specimen CJ3 (6 mm thick hollow steel column), the beam flange in compression buckled suddenly after 8 displacement cycles.

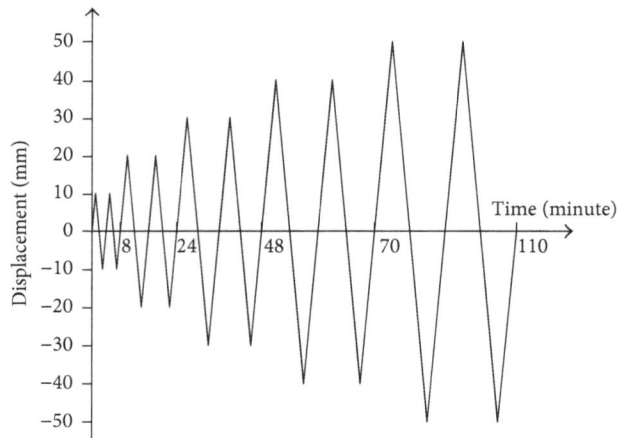

FIGURE 4: Loading scheme for the experiment.

TABLE 3: Test results of the three specimens.

Specimen	P_{max} (kN)	U_{max} (mm)	Failure mode
CJ1	−8.56	−39.19	Compressed buckling of the beam flange
CJ2	10.20	−50.00	Compressed buckling of the beam flange
CJ3	11.92	48.57	Compressed buckling of the beam flange

P_{max} is the maximum load value documented in a test. U_{max} is the maximum displacement value documented at summit from sensor #5 in a test.

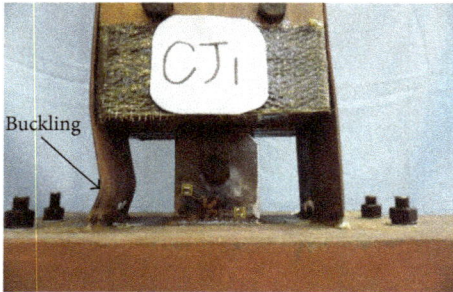

FIGURE 5: Beam flange buckled under compression for the specimen CJ1.

FIGURE 7: Beam flange buckled under compression for the specimen CJ3.

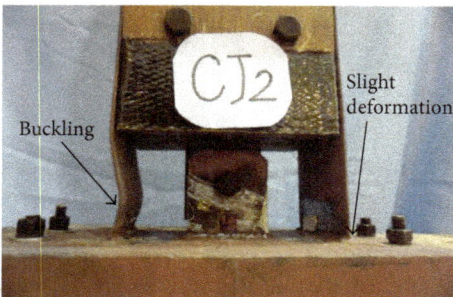

FIGURE 6: Beam flange buckled under compression for the specimen CJ2.

Figure 9 shows that the load in the specimen CJ2 is asymmetric. This indicates that the stiffness in positive direction is larger than that in negative direction.

Slight pinching of load-deflection hysteretic curves was observed in all three specimens. Such a phenomenon may be contributed to the following reasons: on one hand, because of the gap of the bolt connection between the shear connection plate and the bolt, slippage is inevitable under the cyclic loading; on the other hand, shear deformation is not ignored due to less effective antishear section dimension in the bear-to-column connection; of course, the deformation of the hollow column frame plate also contributes to the slippage.

4.3. Moment-Rotation Relationship Curve. Because of the large cross section of the PSB column, the stiffness of the column is quite large. As a result, it is assumed that there is no rotation in the column. Furthermore, as the PSB columns were fixed to the floor by four 30 mm diameter bolts, it was assumed that the PSB column did not slide under the cyclic loading. Based on the above assumptions, the rotation of

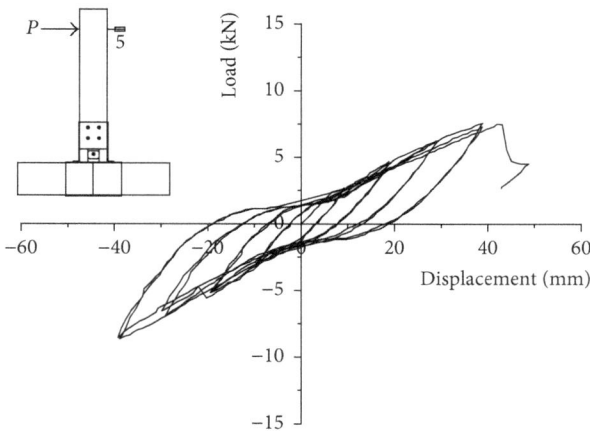

FIGURE 8: Load-displacement hysteretic loop of the specimen CJ1.

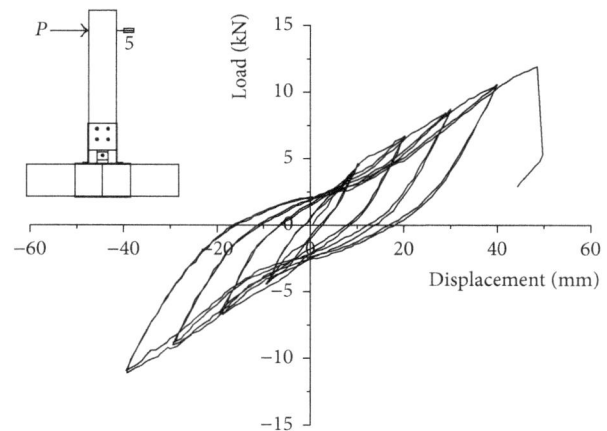

FIGURE 10: Load-displacement hysteretic loop of the specimen CJ3.

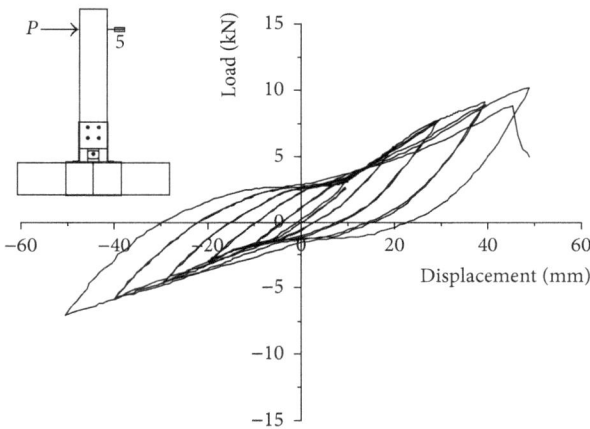

FIGURE 9: Load-displacement hysteretic loop of the specimen CJ2.

connection is the sum of the rotation due to the deflection of the hollow steel column, deflection of the beam flange, and the gap between the bolt and shear connection plate. In this test, the rotation of connection can be calculated as follows:

$$\varphi = \arctan\left(\frac{\Delta_4}{300}\right), \qquad (1)$$

where Δ_4 is the displacement of laser sensor #4 under cyclic loading and 300 is the distance from sensor #4 to the surface of the column.

When calculating the rotation of the joint, the displacements from sensors #3, #4, and #5 all can be used. However, the displacement of sensor #5 includes the elastic deformation of the PSB beam, so the result from sensor #5 cannot be used to calculate the joint rotation. As for sensor #3, which was placed at 100 mm away from the beam flange where beam web ends, the velocity of reciprocating deformation of the beam flange which is lagging behind the loading velocity in the beam tip affects the result of sensor #3; therefore, the rotation of the joint based on sensor #3 is not used.

The maximum rotation for each specimen is shown in Table 4. It is noted that the rotation for the specimen CJ2 with a 4 mm thick hollow steel column is the biggest among

the three specimens. The specimen CJ3 with a 6 mm thick hollow steel column has the least rotation.

Because the deformation of the column is not taken into account, the moment of the joint is obtained by multiplying the load and the length from the loading point to the surface of the column. The equation for joint moment is given as follows:

$$M = P \times L_{\text{load}}, \qquad (2)$$

where P is the load applied to the beam and L_{load} is the distance from the load point to the upper surface of the column.

Figures 11–13 show the moment-rotation hysteretic behavior of the three specimens. It shows that, for all three specimens, the hysteretic behavior is stable until the buckling of the beam flange.

The initial stiffness K_{ini} of each specimen, which is the ratio of the maximum moment to the corresponding rotation in the first cycle, is shown in Table 4. The initial stiffness increases with the increasing thickness of the hollow steel column. The relative parameters for the moment-rotation curve of the three specimens are showed in Table 4.

4.4. Envelope Curve. Envelope curve is defined as the line which connects the maximum rotation and corresponding moment in the first cycle of the displacement step. Figure 14 shows moment-rotation envelope curves of the three specimens. It is noted that the envelope curve of each specimen is similar.

Yielding phenomenon was not observed in the whole envelope curve, but the load increase appears to slow down with the increase of cyclic displacement. Such a phenomenon is mainly due to substantial losses in stiffness of the beam flange and hollow steel column under cyclic loading. It is evident that the thickness of the hollow steel column has a significant influence on the envelope curve.

From the envelope curves in the third quadrant shown in Figure 14, the specimen CJ3 has the largest load and the specimen CJ1 has the smallest load at the same displacement. This corresponds to the thickness of the hollow steel column of

TABLE 4: Test results.

Specimen	Loading direction	M_{max} (kN·m)	Φ_{max} (rad)	K_{ini} (kN·m/rad)
CJ1	Push	7.58	0.0375	333
	Pull	−8.65	−0.0375	
CJ2	Push	10.2	0.049	221
	Pull	−7.00	−0.0465	
CJ3	Push	11.92	0.044	354
	Pull	−11.14	−0.039	

Note. Φ_{max} corresponds to M_{max}.

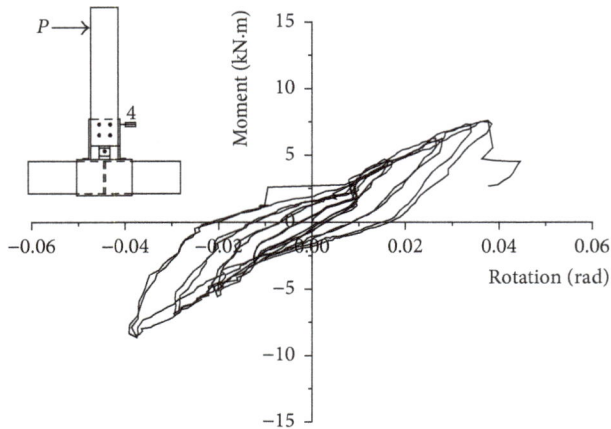

FIGURE 11: Moment-rotation hysteretic curve of the specimen CJ1.

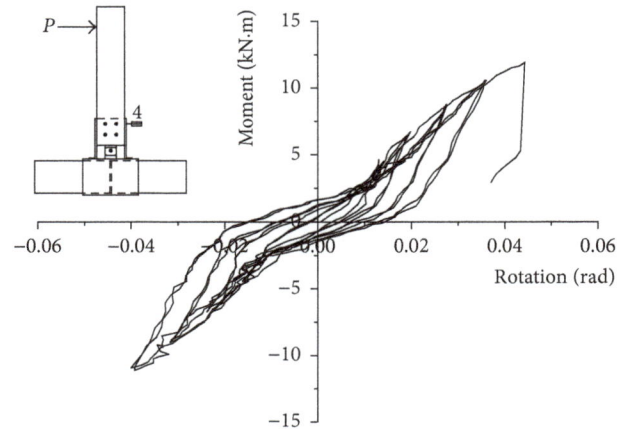

FIGURE 13: Moment-rotation hysteretic curve of the specimen CJ3.

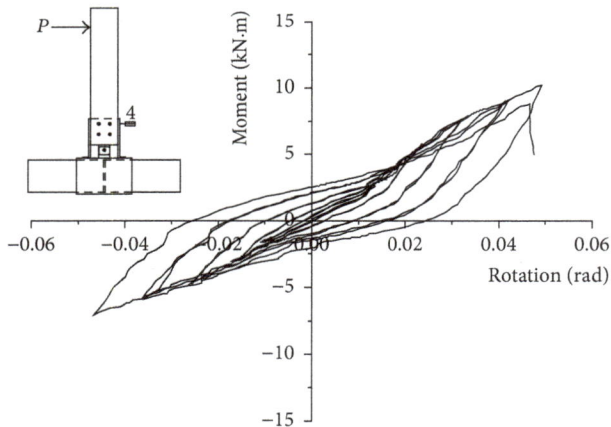

FIGURE 12: Moment-rotation hysteretic curve of the specimen CJ2.

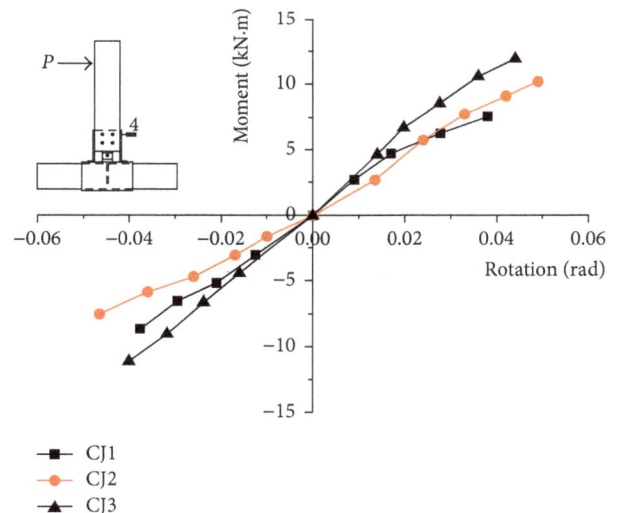

CJ1
CJ2
CJ3

FIGURE 14: Envelope curve of the specimen under cyclic loading.

each specimen. This indicates that the thickness of the hollow steel column has an influence on the strength of connection.

It is also observed that the envelope curve under compressive load is different from the envelope curve under tensile load. This is due to the stiffness degradation of the steel plate.

4.5. Energy Dissipation.
The energy dissipation of the connection is the enclosed area under cyclic loading, as shown in Figure 15, and the displacement in Figure 15 is the maximum beam tip displacement. It is an important index for evaluating seismic performance of the joint.

According to literature [25], the energy dissipation capacity coefficient E can be expressed in the following equation:

$$E = \frac{S_{ABD} + S_{CBD}}{S_{OAE} + S_{OCF}}, \tag{3}$$

Coefficient E for the three specimens CJ1, CJ2, and CJ3 is 1.70, 1.77, and 1.67, respectively. It can be seen that the specimen CJ2 has the biggest coefficient E, and the specimen

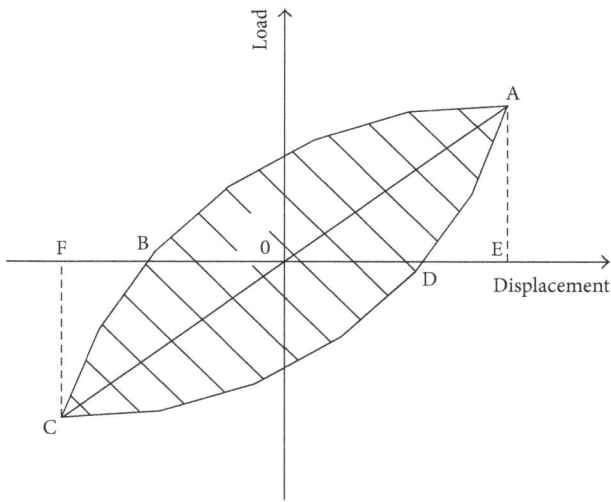

FIGURE 15: Enclosed area under cyclic loading.

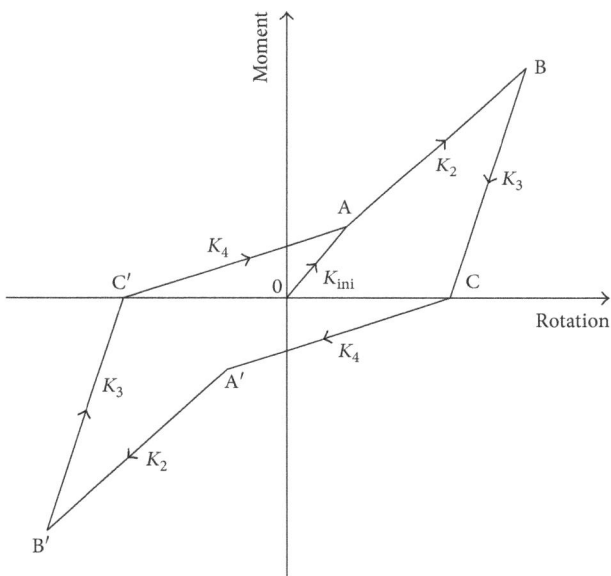

FIGURE 16: Hysteresis model of connection.

CJ3, which has the thickest hollow steel column, has the lowest coefficient E.

4.6. Hysteresis Model of Connections. Based on the moment-rotation curves showed in Figures 11–13, the simplified hysteresis model of the joint under cyclic loading is showed in Figure 16. Point A is the summit point of the first step cycle, and from point 0 to point A, the loading stiffness is expressed with secant stiffness K_{ini}. After passing point A, loading stiffness K_2 becomes smaller. This is due to the shear deformation of the beam and bending deformation of the hollow steel column. K_3 is the unloaded stiffness from point B, which is obviously larger than the loading stiffness K_2. At the start of unloading, only beam flanges act in tension and compression, and shear deformation of the beam and

TABLE 5: Model parameters of the specimens.

Specimen	Stiffness (kN·m/rad)			
	K_{ini}	K_2	K_3	K_4
CJ1	333	166	498	170
CJ2	221	205	309	116
CJ3	354	263	501	137

bending deformation of the column plate were not happening. When the unloading reaches zero, the loading in the opposite direction starts.

When the load starts to load in the opposite direction, the shear deformation of the beam and deflection of the hollow steel column recover first along with the close of the gap between the shear connection plate and the bolt, followed by the load increase of the joint. Therefore, the hysteretic curve of connection has slight pinching, and the stiffness of the slope is called as slippage stiffness K_4. Based on the tested results, the average stiffness values K_{ini}, K_2, K_3, and K_4 of each specimen are given in Table 5, in which K_2, K_3, and K_4 are the slopes of regression lines.

5. Conclusions

In this study, the mechanical performance of a new type connector in timber frame building was investigated. Three full-scale specimens with different hollow steel column thicknesses were subjected to cyclic loading up to failure, and the primary conclusions can be summarized as follows.

For the three specimens, failures were due to the compression buckling of the beam flange, and yielding in the beam flange was not observed. In addition, deformation perpendicular to the hollow steel column was observed in the specimen CJ2; however, no obvious deformation was noted in the specimens CJ1 and CJ3. Buckling of the beam flange governs the moment resistance and rotation capacity of connection.

It is also noticed that the hysteresis loops for the three specimens appear full, with slight pinching in every specimen. The energy dissipation coefficient E for the three specimens CJ1, CJ2, and CJ3 is 1.70, 1.77, and 1.67, respectively, which indicates that the three specimens have good energy dissipation. The hollow steel column thickness has significant influence on the energy dissipation of the connection. Although the thinner hollow steel column has better energy dissipation, it has less connection stiffness.

The maximum rotations of the specimens CJ1, CJ2, and CJ3 under cyclic loading are 0.0375, 0.049, and 0.044, respectively. The thinner the hollow steel column, the larger the rotation of the connection.

Under the cyclic loading, the connection undergoes substantial losses in stiffness.

According to the envelope curve and moment-rotation curve, a hysteresis model for this type of connection under cyclic loading is presented. Employing a thick hollow steel column will have bigger first secant stiffness K_{ini} under cyclic loading.

Conflicts of Interest

The authors declare that they have no conflicts of interest.

Acknowledgments

This research was supported by the National Natural Science Foundation of China (nos. 51778299 and 51578291), the Priority Academic Program Development of Jiangsu Higher Education Institutions, the Project of Housing and Urban-Rural Development Ministry of China (no. 2014-K2-014), and the Science Foundation of Changzhou Institute of Technology (no. YN1615).

References

[1] D. P. Fang, S. Iwasaki, M. H. Yu, Q. P. Shen, Y. Miyamoto, and H. Hikosaka, "Ancient Chinese timber architecture. I: experimental study," *Journal of Structural Engineering*, vol. 127, no. 11, pp. 1348–1357, 2001.

[2] W. S. King, J. Y. Richard Yen, and Y. N. Alex Yen, "Joint characteristics of traditional Chinese wooden frames," *Engineering Structures*, vol. 18, no. 8, pp. 635–644, 1996.

[3] Q. Chun, Z. Yue, and J. W. Pan, "Experimental study on seismic characteristics of typical mortise-tenon joints of Chinese southern traditional timber frame buildings," *Science China Technological Sciences*, vol. 54, no. 9, pp. 2404–2411, 2011.

[4] W. S. Chang, J. Shanks, A. Kitamori, and Kohei Komatsu, "The structural behavior of timber joints subjected to bi-axial bending," *Earthquake Engineering & Structural Dynamics*, vol. 38, no. 6, pp. 739–757, 2009.

[5] J. M. Seo, I. K. Choi, and J. R. Lee, "Static and cyclic behavior of wooden frames with tenon joints under lateral load," *Journal of Structural Engineering*, vol. 125, no. 3, pp. 344–349, 1999.

[6] W. M. Bulleit, L. Bogue Sandberg, M. W. Drewek, and T. L. O'Bryant, "Behavior and modeling of wood-pegged timber frames," *Journal of Structural Engineering*, vol. 125, no. 1, pp. 3–9, 1999.

[7] J. P. Hong, J. David Barrett, and F. Lam, "Three-dimensional finite element analysis of the Japanese traditional post-and-beam connection," *Journal of Wood Science*, vol. 57, no. 2, pp. 119–125, 2011.

[8] F. Lam, M. Schulte-Wrede, C. C. Yao et al., "Moment resistance of bolted timber connections with perpendicular to grain reinforcements," in *Proceedings of the 10th World Conference on Timber Engineering (WCTE)*, Miyazaki, Japan, June 2008.

[9] F. Lam, M. Gehloff, and M. Closen, "Moment-resisting bolted timber connections," *Proceedings of the Institution of Civil Engineers-Structures and Buildings*, vol. 163, no. 4, pp. 267–274, 2010.

[10] M. Q. Wang, X. B. Song, X. L. Gu, Y. Zhang, and L. Luo, "Rotational behavior of bolted beam-to-column connections with locally cross-laminated glulam," *Journal of Structural Engineering*, vol. 141, no. 4, pp. 1–7, 2015.

[11] M. Fragiacomo and M. Batchelar, "Timber frame moment joints with glued-in steel rods. I: design," *Journal of Structural Engineering*, vol. 138, no. 6, pp. 789–801, 2012.

[12] M. Fragiacomo and M. Batchelar, "Timber frame moment joints with glued-in steel rods. II: experimental investigation of long-term performance," *Journal of Structural Engineering*, vol. 138, no. 6, pp. 802–811, 2012.

[13] G. Tlustochowicz, E. Serrano, and R. Steiger, "State-of-the-art review on timber connections with glued-in steel rods," *Materials and Structures*, vol. 44, no. 5, pp. 997–1020, 2011.

[14] M. Vašek and R. Vyhnálek, "Timber semi rigid frame with glued-in-rods joints," in *Proceedings of the 9th World Conference on Timber Engineering (WCTE)*, August 2006.

[15] M. Vašek, "Semi rigid timber frame and space structure connections by glued-in rods," in *Proceedings of the 10th World Conference on Timber Engineering (WCTE)*, Miyazaki, Japan, June 2008.

[16] A. Bouchaïr, P. Racher, and J. F. Bocquet, "Analysis of dowelled timber to timber moment-resisting joints," *Materials and Structures*, vol. 40, no. 10, pp. 1127–1141, 2007.

[17] T. Tsuchimoto, N. Ando, T. Arima et al., "Effect of clearance on the mechanical properties of timber joint," in *Proceeding of Pacific timber engineering conference*, vol. 2, pp. 204–210pp. 204–210, Rotorua, New Zealand, March 1999.

[18] D. S. Huang, A. P. Zhou, Q. S. Zhang et al., "Quasi-static experimental research on energy dissipating joints for assembled timber frame structures," *Journal of Building Structures*, vol. 32, no. 7, pp. 87–92, 2011, in Chinese.

[19] D. S. Huang, A. P. Zhou, and Y. L. Bian, "Experimental and analytical study on the nonlinear bending of parallel strand bamboo beams," *Construction and Building Materials*, vol. 44, pp. 191–196, 2013.

[20] D. S. Huang, Y. L. Bian, A. P. Zhou, and B. Sheng, "Experimental study on stress-strain relationships and failure mechanisms of parallel strand bamboo made from phyllostachys," *Construction and Building Materials*, vol. 77, pp. 130–138, 2015.

[21] D. S. Huang, Y. L. Bian, D. M. Huang, A. Zhou, and B. Sheng, "An ultimate-state-based-model for inelastic analysis of intermediate slenderness PSB columns under eccentrically compressive load," *Construction and Building Materials*, vol. 94, pp. 306–314, 2015.

[22] GB/T 700-2006, "Carbon structural steels" (in Chinese)," 2007.

[23] European Standard EN 10002-1, *Metallic Materials–Tensile Testing–Part 1: Method of Testing at Ambient Temperature*, CEN, Brussels, Belgium, 2001.

[24] The American Society for Testing and Materials, ASTM D143–09, *Standard Test Methods for Small Clear Specimens of Timber*, ASTM, West Conshohocken, PA, USA, 2009.

[25] JGJ/T 101-2015, *Specification for Seismic Test of Building*, China Construction Industry Press, Beijing, 2015, in Chinese.

Permissions

All chapters in this book were first published in AMSE, by Hindawi Publishing Corporation; hereby published with permission under the Creative Commons Attribution License or equivalent. Every chapter published in this book has been scrutinized by our experts. Their significance has been extensively debated. The topics covered herein carry significant findings which will fuel the growth of the discipline. They may even be implemented as practical applications or may be referred to as a beginning point for another development.

The contributors of this book come from diverse backgrounds, making this book a truly international effort. This book will bring forth new frontiers with its revolutionizing research information and detailed analysis of the nascent developments around the world.

We would like to thank all the contributing authors for lending their expertise to make the book truly unique. They have played a crucial role in the development of this book. Without their invaluable contributions this book wouldn't have been possible. They have made vital efforts to compile up to date information on the varied aspects of this subject to make this book a valuable addition to the collection of many professionals and students.

This book was conceptualized with the vision of imparting up-to-date information and advanced data in this field. To ensure the same, a matchless editorial board was set up. Every individual on the board went through rigorous rounds of assessment to prove their worth. After which they invested a large part of their time researching and compiling the most relevant data for our readers.

The editorial board has been involved in producing this book since its inception. They have spent rigorous hours researching and exploring the diverse topics which have resulted in the successful publishing of this book. They have passed on their knowledge of decades through this book. To expedite this challenging task, the publisher supported the team at every step. A small team of assistant editors was also appointed to further simplify the editing procedure and attain best results for the readers.

Apart from the editorial board, the designing team has also invested a significant amount of their time in understanding the subject and creating the most relevant covers. They scrutinized every image to scout for the most suitable representation of the subject and create an appropriate cover for the book.

The publishing team has been an ardent support to the editorial, designing and production team. Their endless efforts to recruit the best for this project, has resulted in the accomplishment of this book. They are a veteran in the field of academics and their pool of knowledge is as vast as their experience in printing. Their expertise and guidance has proved useful at every step. Their uncompromising quality standards have made this book an exceptional effort. Their encouragement from time to time has been an inspiration for everyone.

The publisher and the editorial board hope that this book will prove to be a valuable piece of knowledge for researchers, students, practitioners and scholars across the globe.

List of Contributors

Ning Tang, Kaikai Yang, Limei Wu, Qing Wang and Yanwen Chen
School of Materials Science and Engineering, Shenyang Jianzhu University, Shenyang, China

Wenhao Pan
School of Materials Science and Engineering, Shenyang Jianzhu University, Shenyang, China
School of Materials Science and Engineering, Northeast University, Shenyang, China

Jie Chen and Fan-Long Jin
Department of Polymer Materials, Jilin Institute of Chemical Technology, Jilin City 132022, China

Tian-Yi Zhang
Key Laboratory of Ministry of Education for Enhancing Oil and Gas Recovery Ratio, Northeast Petroleum University, Daqing 163318, China

Soo-Jin Park
Department of Chemistry, Inha University, Nam-gu, Incheon 402-751, Republic of Korea

Pengfei Li, Tao Zhang and Chengzhi Wang
Chongqing Jiaotong University, Chongqing 400074, China

Meng Guo and Yubo Jiao
The Key Laboratory of Urban Security and Disaster Engineering of Ministry of Education, Beijing University of Technology, Beijing 100124, China

Yiqiu Tan
School of Transportation Science and Engineering, Harbin Institute of Technology, Harbin 150090, China

Daisong Luo
School of Transportation Science and Engineering, Harbin Institute of Technology, Harbin 150090, China
Research & Consulting Department of Road Structure & Materials Research Center, China Academy of Transportation Sciences, Beijing 100029, China

Yafei Li
Research & Consulting Department of Road Structure & Materials Research Center, China Academy of Transportation Sciences, Beijing 100029, China

Asim Farooq
University of Science and Technology Beijing, Beijing 100083, China

Liantong Mo
State Key Lab of Silicate Materials for Architectures, Wuhan University of Technology, Wuhan 430070, China

Eman S. Al-Hwaitat, Sami H. Mahmood and Mahmoud Al–Hussein
Department of Physics, e University of Jordan, Amman 11942, Jordan

Ibrahim Bsoul
Department of Physics, Al al-Bayt University, Mafraq 13040, Jordan

Wallace Matizamhuka
Vaal University of Technology, Department of Metallurgical Engineering andries Potgieter Blvd, Vanderbijlpark, South Africa

Wonsuk Jung
School of Mechanical Engineering, Chungnam National University, Daejeon, Republic of Korea

Se-Jin Choi
Department of Architectural Engineering, Wonkwang University, 460 Iksan-daero, Iksan 54538, Republic of Korea

Wassim Dridi and Mohamed Faouzi Zid
Laboratory of Materials, Crystal Chemistry and Applied Thermodynamics, Faculty of Sciences of Tunis, University of Tunis El Manar, El Manar II, 2092 Tunis, Tunisia

Miroslaw Maczka
Institute of Low Temperature and Structure Research, Polish Academy of Sciences, 50-950Wrocław 2, Poland

David Saucedo-Jimenez, Isaac Medina-Sanchez and Carlos Couder Castañeda
Centro de Desarrollo Aeroespacial, Instituto Politécnico Nacional, Belisario Dominguez 22, Centro, 06610 Ciudad de México, Mexico

Fabiane Salles Ferro and Milena Maria Van Der Neut de Almeida
Department of Forest Engineering, State University of the Midwest of Paraná, PR 153 Km, 84500-000 Irati, PR, Brazil

Amós Magalhães Souza, Isabella Imakawa de Araujo and Francisco Antonio Rocco Lahr
Department of Science and Engineering Materials, University of São Paulo, Av. Trabalhador São Carlense, 13566-590 São Carlos, SP, Brazil

André Luis Christoforo
Department of Civil Engineering, Federal University of São Carlos, Rodovia Washington Luís, 13565-905 São Carlos, SP, Brazil

Yoshitaka Fujimoto
Department of Physics, Tokyo Institute of Technology, Tokyo, Japan

Ming-liang Chen and Jia-wen Zhou
State Key Laboratory of Hydraulics and Mountain River Engineering, Sichuan University, Chengdu 610065, China

Gao-jian Wu and Wan-hong Jiang
Sinohydro Bureau 5 Co., Ltd., Power Construction Corporation of China, Chengdu 610066, China

Bin-rui Gan
College of Water Resource and Hydropower, Sichuan University, Chengdu 610065, China

Dunya Mahammad Babanly, Dilgam Babir Tagiyev and Mahammad Baba Babanly
Institute of Catalysis and Inorganic Chemistry, ANAS, 1143 Baku, Azerbaijan

Qorkhmaz Mansur Huseynov
ANAS, Nakhchivan Branch, 7000 Nakhchivan, Azerbaijan

Ziya Saxaveddin Aliev
Azerbaijani State Oil and Industrial University, Azadlıg Av. 16/21, 1010 Baku, Azerbaijan

Junwen Zhou
School of Civil and Architecture Engineering, Changzhou Institute of Technology, Changzhou 213032, China
School of Civil Engineering, Nanjing Forestry University, Nanjing 210037, China

Dongsheng Huang
School of Civil Engineering, Nanjing Forestry University, Nanjing 210037, China

Yang Song
School of Civil and Architecture Engineering, Changzhou Institute of Technology, Changzhou 213032, China

Chun Ni
FPInnovations, Vancouver, BC, Canada V6T 1Z4

Chun IL Kim
Department of Mechanical Engineering, University of Alberta, Edmonton, Alberta T6G 1H9, Canada

Jingjing Zhang, Junyang Tu, Ruonan Shi, Chuanxi Xiong and Ming Jiang
School of Materials Science and Engineering, State Key Lab for New Textile Materials and Advanced Processing Technology, Wuhan Textile University, Wuhan 430200, China

Hairong Li
Mechanical Metrology Division, Hubei Institute of Measurement and Testing Technology, Wuhan 430223, China

Zhiping Luo
School of Materials Science and Engineering, State Key Lab for New Textile Materials and Advanced Processing Technology, Wuhan Textile University, Wuhan 430200, China
Department of Chemistry and Physics, Fayetteville State University, Fayetteville, NC 28301, USA

Chao Chang and Zhu Zhang
School of Applied Science, Taiyuan University of Science and Technology, Taiyuan 030024, China

M. A. Garrido
Departamento de Tecnología Química y Ambiental, Tecnología Química y Energética y Tecnología Mecànica, Escuela Superior de Ciencias Experimentales y Tecnología, Universidad Rey Juan Carlos, c/Tulipán s/n Móstoles, Madrid, Spain

J. Ruiz-Hervias
Departamento de Ciencia de Materiales, UPM, E.T.S.I. Caminos, Canales y Puertos, c/Professor Aranguren s/n, 28040 Madrid, Spain

Le-le Zhang
School of Mechanical, Electronic and Control Engineering, Beijing Jiaotong University, Beijing 100044, China

Guo-liang Bai, Ning-jun Du, Ya-zhou Xu and Chao-gang Qin
School of Civil Engineering, Xi'an University of Architecture and Technology, Shaanxi 710055, China

Komal Agarwal
Engineering Product Development Pillar, Singapore University of Technology and Design, Singapore 487372

Yinning Zhou, Hashina Parveen Anwar Ali, Ihor Radchenko and Arief S. Budiman
Xtreme Materials Laboratory (XML), Singapore University of Technology and Design, Singapore 487372

Avinash Baji
Engineering Product Development Pillar, Singapore University of Technology and Design, Singapore 487372
Department of Engineering, School of Engineering and Mathematical Sciences (SEMS), La Trobe University, Bundoora 3086, Australia

Fui Kiew Liew
Faculty of Engineering, Universiti Malaysia Sarawak, 94300 Kota Samarahan, Sarawak, Malaysia
Faculty of Applied Science, Universiti Teknologi MARA, 94300 Kota Samarahan, Sarawak, Malaysia

Sinin Hamdan and Md. Rezaur Rahman
Faculty of Engineering, Universiti Malaysia Sarawak, 94300 Kota Samarahan, Sarawak, Malaysia

Mohamad Rusop
NANO-SciTech Centre (NST), Institute of Science, Universiti Teknologi MARA, Shah Alam, Selangor, Malaysia

Junwen Zhou
School of Civil Engineering, Nanjing Forestry University, Nanjing 210037, China
School of Civil Engineering and Architecture, Changzhou Institute of Technology, Changzhou 213033, China

Dongsheng Huang, Yurong Shen and Longlong Zhao
School of Civil Engineering, Nanjing Forestry University, Nanjing 210037, China

Index

A

Aliphatic Glycidyl Ether Resin, 33, 43
Artificial Neural Network, 159, 166
Asphalt Concrete, 1-7
Atomic Vacancies, 96-97, 99
Axial Compression, 18-19, 24, 28, 30-31, 137

B

Bolt Diameter, 131-137
Bulk Density, 46-47, 90-91, 93, 121

C

Carbon Nanofoam Synthesis, 79, 86
Catalytic Mixture, 79-80, 86-87
Complex Impedance Spectroscopy, 72
Concrete-filled Steel Tube, 18-20
Crystalline Nanocelluloses, 140
Curie Temperature, 45, 51-53, 57-59

D

Deformation Gradient Tensor, 140, 146, 148
Dielectric Behavior, 46, 54
Dowel-bearing Strength, 132, 134-136, 138

E

Elastic Modulus, 9-12, 15, 20, 143, 160, 165-167
Electromotive Force, 122, 129-130
Electron Beam Irradiation, 97, 99
Energy Spectrum Analysis, 33, 43
Epoxy Resin, 16, 35, 39-41
Euler Equilibrium Equations, 139-140, 148

F

Fiber-reinforced Composites, 139-140, 197
Flexural Strength, 8-12, 15
Fluorescence Microscope, 33, 41-43
Fly Ash, 65-66, 70, 168, 170
Focused Ion Beam, 82
Fracture Toughness, 8-10, 12-13, 15, 17, 166, 186
Free Amino Groups, 180, 186

G

Gibbs Free Energies, 126-127
Ground Granulated Blast-furnace Slag, 65, 70

H

Halloysite Nanotubes, 9, 16
Helicoidal Structures, 180, 182, 185, 188
Hexaferrites, 45-46, 50-51, 53-55
Hexagonal Boron Nitride, 96, 100
Hollow Shale Block, 168-178
Hydraulic Jack, 20

I

Intermolecular Friction, 150, 154, 156
Ionic Conductivity, 71, 73, 77-78

L

Lead-free Piezoceramics, 1, 7
Linear Model, 139, 146-147

M

Macromolecular Entanglement, 150, 154, 156
Magnetocrystalline Anisotropy, 45, 50, 53, 60
Methylene Diphenyl Diisocyanate, 90, 92

N

Neodymium-iron-boron, 56-57

O

Optoelectronics, 96, 99
Oriented Strand Board, 89, 94-95

P

Parallel Strand Bamboo, 131-132, 137, 200, 208
Particle Size Distribution, 47, 102, 104-105, 111-112, 117, 121
Piezo-asphalt Concrete, 1, 3
Piezoelectric Material, 1
Piola Stresses, 140, 146-148
Plastic Behavior, 18
Polarized Optical Microscopy, 151
Polycaprolactone, 8, 180, 182, 184, 188
Polycrystalline Powder, 72, 77
Pressure Aging Vessel, 34
Pulsed Electric Arc, 79-80, 88
Pulsed Laser Deposition, 79-80, 88

R

Raman Spectra, 71, 75-78, 82-84
Rock-socketed Columns, 18, 29, 31

S

Saturation Magnetization, 45-46, 50-51, 53
Scaling Distribution, 102, 120
Scanning Electron Microscopy, 8-10, 43, 45, 81-83, 87, 151, 192, 195
Silane Coupling Agent, 8-10, 15-16
Silicon Carbide, 8-9, 16-17
Spinner Anode, 79-80
Standard Entropy, 124-126
Static Load Capacity, 19
Strain Gauge, 19-20
Stress Constraint Factor, 159-161, 163, 165
Stress Intensity Factor, 10
Surface Tension Forces, 150, 155-156

T

Tabor's Equation, 159, 163-165
Tensile Deformation Behavior, 180, 182-183, 185
Thallium Chalcohalides, 122-123, 125, 127-129
Thermodynamic Functions, 122-129

V

Valence-band Maximum, 97
Vertical Load, 18-19, 31, 200
Vibrating Sample Magnetometer, 45-46

W

Weibull Function, 102, 109, 120
Wind Turbine Generators, 58-59

X

X-ray Diffraction, 45-46, 71-72, 77, 197-198

Y

Young's Modulus, 159-161, 163-165, 189-190, 193-194, 196, 198

Z

Z-type Hexaferrite Phase, 45-46, 50